HANDBOOK OF
BIOENERGY AND BIOFUEL

HANDBOOK OF BIOENERGY AND BIOFUEL

V. K. Mutha

2010

SBS Publishers & Distributors Pvt. Ltd.
New Delhi

ISBN: 13-9789380090078

First Published in 2010

© Reserved

Published by:

SBS PUBLISHERS & DISTRIBUTORS PVT. LTD.
2/9, Ground Floor, Ansari Road, Darya Ganj,

Tel: 0091.11.23289119 / 41563911 / 32945311
Email: mail@sbspublishers.com
www.sbspublishers.com

Preface

Bioenergy is renewable energy made available from materials derived from biological sources. In its most narrow sense it is a synonym to biofuel, which is fuel derived from biological sources. In its broader sense it includes biomass, the biological material used as a biofuel, as well as the social, economic, scientific and technical fields associated with using biological sources for energy. This is a common misconception, as bioenergy is the energy extracted from the biomass, as the biomass is the fuel and the bioenergy is the energy contained in the fuel. Biomass is any organic material which has stored sunlight in the form of chemical energy. As a fuel it may include wood, wood waste, straw, manure, sugar cane, and many other byproducts from a variety of agricultural processes. Bioenergy is expected to become one of the key energy resources for global sustainable development. However, bioenergy cannot be infinite, because the land area available for biomass production is limited and a certain amount of biomass must be reserved for food and materials. The purpose of this publication is also to evaluate global bioenergy potential. However, the fact remains that there will not be much room to obtain more fuelwood from forests in the developing regions.

This "Handbook of Bioenergy and Biofuel" provides readers with an overview of all facts related to the said theme. This Handbook also covers a wide spectrum of topics related to most of the related issues of contemporary importance.

This "Handbook of Bioenergy and Biofuel" tries to create an all-round understanding of bioenergy and biofuel and provides readers with an integrated perspective. The role of technology and industrial biotechnology in bioenergy and biofuel production and commercialization is evaluated. Select case studies are done to further elaborate upon the points. Special focus lies on heating systems and power plants for their roles in production of bioenergy from biomass. Recent global trends in biofuel energy production, technologies used and extent of regional shares are analyzed in detail. Special focus lies on United States, Australia, Sweden and India. An understanding is created about how vegetable oil is used as fuel and how corn stove works. Types of food and non-food bioenergy crops are outlined in detail. Current production, benefits and future prospects of biodiesel as modern engine fuel have been evaluated, particularly for the United States and India. Production processes and technologies used for biogas, biohydrogen and allied fuels are discussed. Siloxane, photohydrogen, biogasoline, butanol fuel, methanol fuel and alcohol fuel are elaborated upon briefly. Current production process and future prospects of ethanol as fuel are highlighted. Select case studies are done. The handbook also evaluates ethanol fuel situation in the United States, especially in Hawaii, and in Brazil. The culture process and production technology involved in algal and other microbial fuel are described in brief. This Handbook is one of its kinds in terms of coverage, scope, elaboration, usefulness and futuristic perspective. This Handbook is user-friendly and enables readers to find exactly the information they require; whether they require broad detail which takes a more cross-sectional view across each subject field, or more focused information which looks closely at specific topics and issues within the field of bioenergy today. Finally, elaborate resource material, website, reference and bibliography are presented so as to make this Handbook a complete reference book for its readers. A multi-disciplinary and inter-disciplinary approach has been followed to compile this Handbook giving priority to the interests of its readers and professional users. A detailed glossary of terms, appendices, and index are given so as to make this publication useful for further research.

Contents

Understanding Bioenergy and Biofuel:
An Integrated Perspective

INTRODUCTION

Bioenergy is renewable energy made available from materials derived from biological sources. In its most narrow sense it is a synonym to biofuel, which is fuel derived from biological sources. In its broader sense it includes biomass, the biological material used as a biofuel, as well as the social, economic, scientific and technical fields associated with using biological sources for energy. This is a common misconception, as bioenergy is the energy extracted from the biomass, as the biomass is the fuel and the bioenergy is the energy contained in the fuel. Biomass is any organic material which has stored sunlight in the form of chemical energy. As a fuel it may include wood, wood waste, straw, manure, sugar cane, and many other byproducts from a variety of agricultural processes. There is a slight tendency for the word *bioenergy* to be favoured in Europe compared with *biofuel* in North America.

Bioenergy from Solid Biomass

Biomass is material derived from recently living organisms, which includes plants, animals and their byproducts. Manure, garden waste and crop residues are all sources of biomass. It is a renewable energy source based on the carbon cycle, unlike other natural resources such as petroleum, coal, and nuclear fuels. Animal waste is a persistent and unavoidable pollutant produced primarily by the animals housed in industrial-sized farms. There are also agricultural products being grown for biofuel production. These include corn, switchgrass, and soybeans, primarily in the United States; rapeseed, wheat and sugar beet primarily in Europe; sugar cane in Brazil; palm oil and miscanthus in Southeast Asia; sorghum and cassava in China; and jatropha in India. Hemp has also been proven to work as a biofuel. Biodegradable outputs from industry, agriculture, forestry and households can be used for biofuel production, using e.g. anaerobic digestion to produce biogas, gasification to produce syngas or by direct combustion. Examples of biodegradable wastes include straw, timber, manure, rice husks, sewage, and food waste. The use of biomass fuels can therefore contribute to waste management as well as fuel security and help to prevent or slow down climate change, although alone they are not a comprehensive solution to these problems.

Biomass-based Electricity

Electricity from sugarcane bagasse in Brazil

Sucrose accounts for little more than 30 per cent of the chemical energy stored in the mature plant; 35 per cent is in the leaves and stem tips, which are left in the fields during harvest, and 35 per cent are in the fibrous material (bagasse) left over from pressing. The production process of sugar and ethanol in Brazil takes full advantage of the energy stored in sugarcane. Part of the baggasse is currently burned at the mill to provide heat for distillation and electricity to run the machinery. This allows ethanol plants to be energetically self-sufficient and even sell surplus electricity to utilities; current production is 600 MW for self-use and 100 MW for sale. This secondary activity is expected to boom now that utilities have been induced to pay "fair price "(about US$10/GJ or US$0.036/kWh) for 10 year contracts. This is approximately half of what the World Bank considers the reference price for investing in similar projects. The energy is especially valuable to utilities because it is produced mainly in the dry season when hydroelectric dams are running low. Estimates of potential power generation from bagasse range from

1,000 to 9,000 MW, depending on technology. Higher estimates assume gasification of biomass, replacement of current low-pressure steam boilers and turbines by high-pressure ones, and use of harvest trash currently left behind in the fields. For comparison, Brazil's Angra I nuclear plant generates 657 MW. Presently, it is economically viable to extract about 288 MJ of electricity from the residues of one tonne of sugarcane, of which about 180 MJ are used in the plant itself. Thus a medium-size distillery processing 1 million tonnes of sugarcane per year could sell about 5 MW of surplus electricity. At current prices, it would earn US$ 18 million from sugar and ethanol sales, and about US$ 1 million from surplus electricity sales. With advanced boiler and turbine technology, the electricity yield could be increased to 648 MJ per tonne of sugarcane, but current electricity prices do not justify the necessary investment. (According to one report, the World Bank would only finance investments in bagasse power generation if the price were at least US$19/GJ or US$0.068/kWh.) Bagasse burning is environmentally friendly compared to other fuels like oil and coal. Its ash content is only 2.5 per cent (against 30-50% of coal), and it contains no sulphur. Since it burns at relatively low temperatures, it produces little nitrous oxides. Moreover, bagasse is being sold for use as a fuel (replacing heavy fuel oil) in various industries, including citrus juice concentrate, vegetable oil, ceramics, and tyre recycling. The state of São Paulo alone used 2 million tonnes, saving about US$ 35 million in fuel oil imports. Researchers working with cellulosic ethanol are trying to make the extraction of ethanol from sugarcane bagasse and other plants viable on an industrial scale.

Fuel from Algae

Algae fuel, also called algal fuel, oilgae, algaeoleum or third-generation biofuel, is a biofuel from algae. The record oil price increases since 2003, competing demands between foods and other biofuel sources and the world food crisis have ignited interest in algaculture (farming algae) for making vegetable oil, biodiesel, bioethanol, biogasoline, biomethanol, biobutanol and other biofuels. Among algal fuels' attractive characteristics: they do not affect fresh water resources, can be produced using ocean and wastewater, and are biodegradable and relatively harmless to the environment if spilled. Algae cost more per pound yet can yield over 30 times more energy per acre than other, second-generation biofuel crops. One biofuels company has claimed that algae can produce more oil in an area the size of a two-car garage than an football field of soybeans, because almost the entire algal organism can use sunlight to produce lipids, or oil. The United States Department of Energy estimates that if algae fuel replaced all the petroleum fuel in the United States, it would require 15,000 square miles (40,000 square kilometers), which is a few thousand square miles larger than Maryland, or 1.3 Belgiums. This is less than 1/7th the area of corn harvested in the United States in 2000. As of 2008, such fuels remain too expensive to replace other commercially available fuels, with the cost of various algae species typically between US$5-10 per kg dry weight. But several companies and government agencies are funding efforts to reduce capital and operating costs and make algae oil production commercially viable.

Time Frame

The *Aquatic Species Programme* launched in 1978. The U.S. research programme, funded by the U.S. DoE, was tasked with investigating the use of algae for the production of energy. The programme initially focused efforts on the production of hydrogen, however, shifted primary research to studying oil production in 1982. From 1982 through its culmination, the majority of the programme research was focused on the production of transportation fuels, notably biodiesel, from algae. In 1995, as part of the over-all efforts to lower budget demands, the DoE decided to end the programme. Research stopped in 1996 and staff began compiling their research for publication. In July 1998, the DoE published the report "A Look Back at the U.S. Department of Energy's Aquatic Species Programme: Biodiesel from Algae". In 2008, Time Magazine voted Isaac Berzin one of the world's most influential persons for his ability to turn a dream of an oil-free future into a reality through Green Fuel, founded in Boston in 2001.

Factors and Yield

Dry algae factor is the percentage of algae cells in relation with the media where it is cultured, e.g. if the dry algae factor is 50 per cent, one would need 2 kg of wet algae (algae in the media) to get 1 kg of algae cells. Lipid factor is the percentage of vegoil in relation with the algae cells needed to get it, i.e. if the algae lipid factor is 40 per cent, one would need 2.5 kg of algae cells to get 1 kg of oil.

Yields (gallons of oil per acre per year) cover a vast range from 5,000 to 150,000. If all aspects of the cultivation are controlled—temperature, CO_2 levels, sunlight and nutrients (including carbohydrates as a food source), then extremely high yields can be obtained. Such variation can make calculations on which to base 'fuel the world' scenarios very difficult. For example, Glen Kertz of Valcent Products, claims that "algae can produce 100,000 gallons of oil per acre" per year. This relies on growing the algae in an entirely closed loop system. More recently, Valcent have claimed 150,000 gallons may be possible; whereas in reality, their most recent actual reported yields were 33,000 gallons per acre per year.

Cultivation of Algae

Algae grow rapidly and can have a high percentage of lipids, or oils. They can double their mass several times a day and produce at least 15 times more oil per acre than alternatives such as rapeseed, palms, soybeans, or jatropha. Moreover, algae-growing facilities can be built on coastal land unsuitable for conventional agriculture. They can grow 20 to 30 times faster than food crops. The hard part about algae production is growing the algae in a controlled way and harvesting it efficiently.

Photo Bioreactors (PBR)

Most companies pursuing algae as a source of biofuels are pumping nutrient-laden water through plastic tubes (called "bioreactors") that are exposed to sunlight (and so called photobioreactors or PBR). Running a PBR is more difficult than a open pond, and more costly. Algae can also grow on marginal lands, such as in desert areas where the groundwater is saline, rather than utilise fresh water. The difficulties in efficient biodiesel production from algae lie in finding an algal strain with a high lipid content and fast growth rate that isn't too difficult to harvest, and a cost-effective cultivation system (i.e., type of photobioreactor) that is best suited to that strain. There is also a need to provide concentrated CO2 to turbocharge the production.

Understanding Closed Loop System

Another obstacle preventing widespread mass production of algae for biofuel production has been the equipment and structures needed to begin growing algae in large quantities. Diversified Energy Corporation have avoided this problem by taking a different approach, and growing the algae in thin walled polyethylene tubing called Algae Biotape (developed by XL renewables, Inc), similar to conventional drip irrigation tubing, which can be incorporated into a normal agricultural environment. In a closed system (not exposed to open air) there is not the problem of contamination by other organisms blown in by the air. The problem for a closed system is finding a cheap source of sterile carbon dioxide (CO_2). Several experimenters have found the CO_2 from a smokestack works well for growing algae. To be economical, some experts think that algae farming for biofuels will have to be done next to power plants, where they can also help soak up the pollution.

Understanding Open Pond System

Open-pond systems for the most part have been given up for the cultivation of algae with high-oil content. Many believe that a major flaw of the Aquatic Species Programme was the decision to focus their efforts exclusively on open-ponds; this makes the entire effort dependent upon the hardiness of the strain chosen, requiring it to be unnecessarily resilient in order to withstand wide swings in temperature and pH, and competition from invasive algae and bacteria. Open systems using a monoculture are also vulnerable to viral infection. The energy that a high-oil strain invests into the production of oil is energy that is not invested into the production of proteins or carbohydrates, usually resulting in the species being less hardy, or having a slower growth rate. Algal species with a lower oil content, not having to divert their energies away from growth, have an easier time in the harsher conditions of an open system. Some open sewage ponds trial production has been done in Marlborough, New Zealand.

Types of Algae

A feasibility study using marine microalgae in a photobioreactor is being done by The International Research Consortium on Continental Margins at the International University Bremen. Research into algae for the mass-production of oil is mainly focused on microalgae; organisms capable of photosynthesis that are less than 0.4 mm in diameter, including the diatoms and cyanobacteria; as opposed to macroalgae, e.g. seaweed. However, some research is being done into using seaweeds for biofuels, probably due to the high availability of this resource. This preference towards microalgae is due largely to its less complex structure, fast growth rate, and high oil content (for some species). Some commercial interests into large scale algal-cultivation systems are looking to tie in to existing infrastructures, such as coal power plants or sewage treatment facilities. This approach not only provides the raw materials for the system, such as CO_2 and nutrients; but it changes those wastes into resources. Aquaflow Bionomic Corporation of New Zealand announced that it has produced its first sample of homegrown bio-diesel fuel with algae sourced from local sewerage ponds.

The Department of Environmental Science at Ateneo de Manila University in the Philippines, is working on producing biofuel from algae, using a local species of algae. NBB's Feedstock Development programme is addressing production of algae on the horizon to expand available material for biodiesel in a sustainable manner. The following species listed are currently being studied for their suitability as a mass-oil producing crop, across various locations worldwide:

(*i*)   Botryococcus braunii
(*ii*)   Chlorella
(*iii*) Dunaliella tertiolecta
(*vi*)   Gracilaria
(*v*)   Pleurochrysis carterae (also called CCMP647).
(*vi*)   Sargassum, with 10 times the output volume of Gracilaria.

Types of Fuel

The vegoil algae product can then be harvested and converted into biodiesel; the algae's carbohydrate content can be fermented into bioethanol and biobutanol.

(*a*) *Biodiesel:* Currently most research into efficient algal-oil production is being done in the private sector, but predictions from small scale production experiments bear out that using algae to produce biodiesel may be the only viable method by which to produce enough automotive fuel to replace current world diesel usage. Microalgae have much faster growth-rates than terrestrial crops. The per unit area yield of oil from algae is estimated to be from between 5,000 to 20,000 gallons per acre, per year (4.6 to 18.4 l/m² per year); this is 7 to 30 times greater than the next best crop, Chinese tallow (699 gallons). Studies show that algae can produce up to 60 per cent of their biomass in the form of oil. Because the cells grow in aqueous suspension where they have more efficient access to water, CO_2 and dissolved nutrients, microalgae are capable of producing large amounts of biomass and usable oil in either high rate algal ponds or photobioreactors. This oil can then be turned into biodiesel which could be sold for use in automobiles. The more efficient this process becomes the larger the profit that is turned by the company. Regional production of microalgae and processing into biofuels will provide economic benefits to rural communities.

(b) *Biobutanol:* Butanol can be made from algae or diatoms using only a solar powered biorefinery. This fuel has an energy density similar to gasoline, and greater than that of either ethanol or methanol. In most gasoline engines, butanol can be used in place of gasoline with no modifications. In several tests, butanol consumption is similar to that of gasoline, and when blended with gasoline, provides better performance and corrosion resistance than that of ethanol or E85. The green waste left over from the algae oil extraction can be used to produce butanol.

(c) *Biogasoline:* Biogasoline can be produced from algae.

(d) *Methane:* Through the use of algaculture grown organisms and cultures, various polymeric materials can be broken down into methane.

(e) *SVO:* The algal-oil feedstock that is used to produce biodiesel can also be used for fuel directly as "Straight Vegetable Oil", (SVO). The benefit of using the oil in this manner is that it doesn't require the additional energy needed for transesterification, (processing the oil with an alcohol and a catalyst to produce biodiesel). The drawback is that it does require modifications to a normal diesel engine. Transesterified biodiesel can be run in an unmodified modern diesel engine, provided the engine is designed to use ultra-low sulphur diesel, which, as of 2006, is the new diesel fuel standard in the United States.

(f) *Hydrocracking to traditional transport fuels:* Vegetable oil can be used as feedstock for an oil refinery where methods like hydrocracking or hydrogenation can be used to transform the vegetable oil into standard fuels like gasoline and diesel.

(g) *Jet Fuel:* Twenty-five airlines went bust or stopped operations in the first six months of 2008 and more could fold as fuel prices soar, aviation industry association IATA has warned. Algal jet fuel can be used as alternative: IATA recognizes that aircraft are long lived and will be using kerosene or kerosene-type fuels for many years. It supports research, development and deployment into alternative fuels that produce less GHG emissions over their life cycle and do not compete for land with fuel crops. IATA's goal is for its members to be using 10 per cent alternative fuels by 2017. On January 8, 2009, Continental Airlines ran the first test for the first flight of an algae-fueled jet. The test was done using a twin-engine commercial jet consuming a 50/50 blend of biofuel and normal aircraft fuel. It was the first flight by a U.S. carrier to use an alternative fuel source on this specific type of aircraft. The flight from Houston's Bush International Airport completed a circuit over the Gulf of Mexico. The pilots on-board, executed a series of tests at 38,000 ft (11.6 km), including a mid-flight engine shutdown. Larry Kellner, chief executive of Continental Airlines, said they had tested a drop-in fuel which meant that no modification to the engine was required. The fuel was praised for having a low flash point and sufficiently low freezing point, issues that have been problematic for other bio-fuels.

Types of Nutrients

Nutrients like nitrogen (N), phosphorus (P), and potassium (K), are important for plant growth and are essential parts of fertilizer. Silica and iron, as well as several trace elements, may also be considered important marine nutrients as the lack of one can limit the growth of, or productivity in, an area. One company, Green Star Products, announced their development of a micronutrient formula to increase the growth rate of algae.

According to the company, its formula can increase the daily growth rate by 34 per cent and can double the amount of algae produced in one growth cycle. A possible nutrient source is waste water from the treatment of sewage, agricultural, or flood plain run-off, all currently major pollutants and health risks. However, this waste water cannot feed algae directly and must first be processed by bacteria, through anaerobic digestion. If waste water is not processed before it reaches the algae, it will contaminate the algae in the reactor, and at the very least, kill much of the desired algae strain. In biogas facilities, organic waste is often converted to a mixture of carbon dioxide, methane, and organic fertilizer. Organic fertilizer that comes out of digester is liquid, and nearly suitable for algae growth, but it must first be cleaned and sterilized. At the Woods Hole Oceanographic Institution and the Harbor Branch Oceanographic Institution the wastewater from domestic and industrial sources contain rich organic compounds is being used to accelerate the growth of algae. Also the Department of Biological and Agricultural Engineering of the University of Georgia is exploring microalgal biomass production using industrial wastewater. Algaewheel, based in Indianapolis, Indiana, presented a proposal to build a facility in Cedar Lake, Indiana that uses algae to treat municipal wastewater and uses the sludge byproduct to produce biofuel. US Universities working on Oil from Algae:

    (*i*)   Cal Poly State University, San Luis Obispo.

    (*ii*)   Montana State University.

    (*iii*)   University of Virginia.

    (*vi*)   Arizona State University

    (*v*)   Ohio University

    (*vi*)   Old Dominion University.

The Ukraine Cabinet plans to produce biofuel of a special type of algae. Also the CSIC'Ls Instituto de Bioquímica Vegetal y Fotosíntesis (Microalgae Biotechnology Group, in Sevilla, Spain is researching the algal fuels.

Algal Biomass Organization (ABO) is formed by Boeing Commercial Airplanes, A2BE Carbon Capture Corporation,

National Renewable Energy Labs, Institution of Oceanography, Benemann Associates, Mont Vista Capital and Montana State University. Global air carriers Air New Zealand, Continental, Virgin Atlantic Airways, and biofuel technology developer UOP LLC, a Honeywell company, will be the first wave of aviation-related members, together with Boeing, to join Algal Biomass Organization.

Bioenergy from Bagasse

Bagasse is the fibrous residue remaining after sugarcane or sorghum stalks are crushed to extract their juice and is currently used as a renewable resource in the manufacture of pulp and paper products and building materials. *Agave bagasse* is a similar material which consists of the tissue of the blue agave after extraction of the sap. For each 100 tonnes of sugarcane crushed, a sugar factory produces nearly 30 tonnes of wet bagasse. Many research efforts have attempted to use bagasse as a renewable feedstock for power generation and for the production of bio-based materials. Workplace exposure to dusts from the processing of Bagasse can cause the chronic lung condition pulmonary fibrosis.

Historically bagasse factories have had hash working environments due to the process of milling the pulp. Factory workers have reported bleeding hands and arms. Workers are also required to wear masks to prevent breathing in pulp fibers. It is typically suggested to visit the actual factory to personally see working conditions before purchasing any bagasse based products.

Uses as Fuel

Bagasse is often used as a primary fuel source for sugar mills; when burned in quantity, it produces sufficient heat energy to supply all the needs of a typical sugar mill, with energy to spare. To this end, a secondary use for this waste product is in cogeneration, the use of a fuel source to provide both heat energy, used in the mill, and electricity, which is typically sold on to the consumer electricity grid. The resulting CO_2 emissions are equal to the amount of CO_2 that the sugarcane plant absorbed from the atmosphere during its growing phase, which makes the process of cogeneration greenhouse gas-neutral. Florida Crystals Corporation, one of America's largest sugar companies, owns and operates the largest biomass power plant in North America. The 140 MW facility uses bagasse and urban wood waste as fuel to generate enough energy to power its large milling and refining operations as well as supply enough renewable electricity for nearly 60,000 homes. The facility reduces dependence on oil by more than one million barrels per year. Ethanol produced from the sugar in sugarcane is a popular fuel in Brazil. The cellulose rich bagasse is now being tested for production of commercial quantities of cellulosic ethanol. Verenium Corporation (VRNM) is currently building a cellulosic ethanol plant based on cellulosic by-products like bagasse in Jennings, Louisiana. They are using a biotech approach to improve ethanol production above and beyond the midwest corn based ethanol production method. This will allow regional cellulosic ethanol production getting around the problem of ethanol transportation. The Verenium approach will get ethanol and E85 fuel to the important markets in California and the Northeast.

Uses as Paper

Around 5-10 per cent of paper production worldwide is produced from agricultural crops, valuing agricultural paper production at between $5 billion and $10 billion. The most notable of these agricultural crops are wheat straw and bagasse. Paper production is the second largest revenue stream from bagasse after electricity cogeneration; higher than ethanol. Using agricultural crops rather than wood has the added advantage of reducing deforestation. Due to the ease with which bagasse can be chemically pulped, bagasse requires less bleaching chemicals than wood pulp to achieve a bright, white sheet of paper. Food containers from bagasse are more biodegradable than their petrochemical counterparts. Most chemical bagasse pulp mills concentrate the spent reaction chemicals and combust them to power the paper-mills and to recover the reaction chemicals.

Uses as Food containers

Bagasse is used to make insulated disposable food containers, replacing materials such as styrofoam, which are increasingly regarded as environmentally unacceptable. Insulated disposable food containers made of bagasse are commercially available.

Uses as Waterpipe Tobacco Substitute

The company Soex India Pvt. Ltd. lists Ugarsay anecay agassebay (which is pig latin for sugar cane bagasse) as one of the components on packages from their Herbal Hookah Molasses product range.

Bioenergy from Babassu Oil

Babassu oil or *cusi oil* is a clear light yellow vegetable oil extracted from the seeds of the babassu palm (*Attalea speciosa*), which grows in the Amazon region of South America. It is a non-drying oil used in food, cleaners and skin products. This oil has properties similar to coconut oil and is used in much the same context. It is increasingly being used as a substitute for coconut oil. Babassu oil is about 70 per cent lipids, in the following proportions:

Types of Fatty Acid	%
Lauric	50.0
Myristic	20.0
Plamitic	11.0
Oleic	10.0
Stearic	3.5

Lauric and myristic acids have melting points relatively close to human body temperature, so babassu oil can be applied to the skin as a solid that melts on contact. This heat transfer can produce a cooling sensation. It is an effective emollient. During February 2008, a mixture of babassu oil and coconut oil was used to partially power one engine of a Boeing 747, in a biofuel trial sponsored by Virgin Atlantic Airways.

Bioenergy form Butanol Fuel

Butanol may be used as a fuel in an internal combustion engine. Because its longer hydrocarbon chain causes it to be fairly non-polar, it is more similar to gasoline than it is to ethanol. Butanol has been demonstrated to work in some vehicles designed for use with gasoline without any modification. It can be produced from biomass (as "biobutanol") as well as fossil fuels (as "petrobutanol"); both biobutanol and petrobutanol have the same chemical properties.

Production

Butanol from biomass is called biobutanol. It can be used in unmodified gasoline engines.

(a) Technologies Used: It can be produced by fermentation of biomass by the A.B.E. process. The process uses the bacterium *Clostridium acetobutylicum*, also known as the *Weizmann organism*. It was Chaim Weizmann who first used this bacteria for the production of acetone from starch (with the main use of acetone being the making of Cordite) in 1916. The butanol was a by-product of this fermentation (twice as much butanol was produced). The process also creates a recoverable amount of H_2 and a number of other by-products: acetic, lactic and propionic acids, acetone, isopropanol and ethanol. The difference from ethanol production is primarily in the fermentation of the feedstock and minor changes in distillation. The feedstocks are the same as for ethanol: energy crops such as sugar beets, sugar cane, corn grain, wheat and cassava as well as agricultural byproducts such as straw and corn stalks. According to DuPont, existing bioethanol plants can cost-effectively be retrofitted to biobutanol production.

(b) Butanol from Algae: Biobutanol can be made entirely with solar energy, from algae (called Solalgal Fuel) or diatoms.

(c) Centia Process: Centia is based on a three-step thermal, catalytic, and reforming process that has the potential to turn virtually any lipidic compound—e.g., vegetable oils, oils from animal fat and oils from algae—into 1-for-1 replacements for petroleum jet fuel, diesel, and gasoline. The three steps are:

   (*i*)   Hydrolytic conversion.
   (*ii*) Decarboxylation.
   (*iii*) Reforming long-chain alkanes.

(d) Main Producers: ButylFuel, LLC used a U.S. Department of Energy Small Business Technology Transfer grant to develop a process aimed at making biobutanol production economically competitive with petrochemical production processes. ButylFuel is planning to market its biobutanol as a solvent first, and then market it as a fuel in the future. DuPont and BP are making biobutanol the first product of their joint effort to develop, produce, and market next-generation biofuels. In Europe the Swiss company Butalco is developing genetically modified yeasts for the production of biobutanol from cellulosic materials.

(e) Pattern Distribution: Butanol better tolerates water contamination and is less corrosive than ethanol and more suitable for distribution through existing pipelines for gasoline. In blends with diesel or gasoline, butanol is less likely to separate from this fuel than ethanol if the fuel is contaminated with water. There is also a vapour pressure co-blend synergy with butanol and gasoline containing ethanol, which facilitates ethanol blending. This facilitates storage and distribution of blended fuels.

Understanding Properties of Common Fuels

(a) *Understanding Energy Content and Effects on Fuel Economy:* Switching a gasoline engine over to butanol would in theory result in a fuel consumption penalty of about 10 per cent but butanol's effect on mileage is yet to be determined by a scientific study. While the energy density for any mixture of gasoline and butanol can be calculated, tests with other alcohol fuels have demonstrated that the effect on fuel economy is not proportional to the change in energy density.

(b) *Understand Octane Rating:* The octane rating of n-butanol is similar to that of gasoline but lower than that of ethanol and methanol. n-Butanol has a RON (Research Octane number) of 96 and a MON (Motor octane number) of 78 while t-butanol has octane ratings of 105 RON and 89 MON. t-Butanol is used as an additive in gasoline but cannot be used as a fuel in its pure form because its relatively high melting point of

25.5°C causes it to gel and freeze near room temperature. A fuel with a higher octane rating is less prone to knocking (extremely rapid and spontaneous combustion by compression) and the control system of any modern car engine can take advantage of this by adjusting the ignition timing. This will improve energy efficiency, leading to a better fuel economy than the comparisons of energy content different fuels indicate. By increasing the compression ratio, further gains in fuel economy, power and torque can be achieved. Conversely, a fuel with lower octane rating is more prone to knocking and will lower efficiency. Knocking can also cause engine damage.

(c) *Air-fuel Ratio:* Alcohol fuels, including butanol and ethanol, are partially oxidized and therefore need to run at richer mixtures than gasoline. Standard gasoline engines in cars can adjust the air-fuel ratio to accommodate variations in the fuel, but only within certain limits depending on model. If the limit is exceeded by running the engine on pure butanol or a gasoline blend with a high percentage of butanol, the engine will run lean, something which can critically damage components. Compared to ethanol, butanol can be mixed in higher ratios with gasoline for use in existing cars without the need for retrofit as the air-fuel ratio and energy content are closer to that of gasoline.

(d) *Understanding Specific Energy:* Alcohol fuels have less energy per unit weight and unit volume than gasoline. To make it possible to compare the net energy released per cycle a measure called the fuels specific energy is sometimes used. It is defined as the energy released per air fuel ratio. The net energy released per cycle is higher for butanol than ethanol or methanol and about 10 per cent higher than for gasoline.

(e) *Understanding Viscosity:* The viscosity of alcohols increase with longer carbon chains. For this reason, butanol is used as an alternative to shorter alcohols when a more viscous solvent is desired. The kinematic viscosity of butanol is several times higher than that of gasoline and about as viscous as high quality diesel fuel.

(f) *Understanding Heat of Vapourization:* The fuel in an engine has to be vapourized before it will burn. Insufficient vapourization is a known problem with alcohol fuels during cold starts in cold weather. As the latent heat of vapourization of butanol is less than half of that of ethanol, an engine running on butanol should be easier to start in cold weather than one running on ethanol or methanol.

Major Problems Associated with the Use of Butanol Fuel

The potential problems with the use of butanol are similar to those of ethanol:

(i) To match the combustion characteristics of gasoline, the utilization of butanol fuel as a substitute for gasoline requires fuel-flow increases (though butanol has only slightly less energy than gasoline, so the fuel-flow increase required is only minimal, maybe 10 per cent, compared to 40 per cent for ethanol.).

(ii) Alcohol-based fuels are not compatible with some fuel system components.

(iii) Alcohol fuels may cause erroneous gas gauge readings ih vehicles with capacitance fuel level gauging.

(iv) While ethanol and methanol have lower energy densities than butanol, their higher octane number allows for greater compression ratio and efficiency. Higher combustion engine efficiency allows for lesser greenhouse gas emissions per unit motive energy extracted.   ·

(v) Methanol is toxic. Ethanol is considered non-toxic for the most part. Butanol is less toxic than methanol, but more toxic than ethanol, and is somewhat soluble in water. The possibility of butanol finding its way to water supplies should be considered.

(*vi*) As an advantage, butanol production from biomass
could be more efficient (i.e. unit engine motive power
delivered per unit solar energy consumed) than ethanol
or methanol routes. Also, some bacteria that produce
butanol are able to digest cellulose, not just starch and
sugars.

Main Butanol Fuel Mixtures

Standards for the blending of ethanol and methanol in gasoline
exist in many countries, including the EU, the US and Brazil.
Approximate equivalent butanol blends can be calculated from
the relations between the stochiometric fuel-air ratio of butanol,
ethanol and gasoline. Common ethanol fuel mixtures for fuel
sold as gasoline currently range from 5 to 10 per cent. The share
of butanol can be 60 per cent greater than the equivalent ethanol
share, which gives a range from 8 per cent to 16 per cent.
"Equivalent" in this case refers only to the vehicle's ability to
adjust to the fuel. Other properties such as energy density,
viscosity and heat of vapourisation will vary and may further
limit the percentage of butanol that can be blended with gasoline.
Consumer acceptance may be limited due to the offensive smell
of butanol.

Modern Uses of Butanol in Vehicles

Currently no production vehicle is known to be approved by
the manufacturer for use with 100 per cent butanol, though any
model that is able to run 10 per cent ethanol blends should be
able to use butanol without any problems. David Ramey drove
from Blacklick, Ohio to San Diego, California using butanol in
an unmodified 1992 Buick Park Avenue. Although further long
term testing must be done, it is highly likely that most late model
cars can run on 100 per cent butanol safely with no
modifications. Justification for this conclusion is based on data
for RON in comparison of n-Butanol with Gasoline. Also,
modern ECU-injected motorcar piston engines are designed to
be flexible enough to deliver good performance with 91-RON
fuels, which n-Butanol exceeds in RON rating.

Major Research Being Undertaken

The Swiss company Butalco GmbH uses a special technology to modify yeasts in order to produce butanol instead of ethanol. Yeasts as production organisms for butanol have decisive advantages compared to bacteria.

Bioenergy from Biodiesel

Biodiesel refers to a non-petroleum-based diesel fuel consisting of long-chain alkyl (methyl, propyl or ethyl) esters. Biodiesel is made by chemically-reacting lipids, typically vegetable oil or animal fat (tallow), and alcohol. It can be used (alone, or blended with conventional petrodiesel) in unmodified diesel-engine vehicles. Biodiesel is distinguished from the *straight vegetable oil* (SVO) (sometimes referred to as "waste vegetable oil", "WVO", "used vegetable oil", "UVO", "pure plant oil", "PPO") used (alone, or blended) as fuels in some *converted* diesel vehicles. "Biodiesel" is standardized as mono-alkyl ester and other kinds of diesel-grade fuels of biological origin are not included.

Types of Blends and Their Properties

Blends of biodiesel and conventional hydrocarbon-based diesel are products most commonly distributed for use in the retail diesel fuel marketplace. Much of the world uses a system known as the "B" factor to state the amount of biodiesel in any fuel mix: fuel containing 20 per cent biodiesel is labeled B20, while pure biodiesel is referred to as B100. It is common in the USA to see B99.9 because a federal tax credit is awarded to the first entity which blends petroleum diesel with pure biodiesel. Blends of 20 percent biodiesel with 80 percent petroleum diesel (B20) can generally be used in unmodified diesel engines. Biodiesel can also be used in its pure form (B100), but may require certain engine modifications to avoid maintenance and performance problems. Blending B100 with petrol diesel may be accomplished by:

(*i*)   Mixing in tanks at manufacturing point prior to delivery to tanker truck.

(*ii*) Splash mixing in the tanker truck (adding specific percentages of Biodiesel and Petrol Diesel).

(*iii*) In-line mixing, two components arrive at tanker truck simultaneously.

(*iv*) Metered pump mixing, Petro diesel and Biodiesel meters are set to X total volume, transfer pump pulls from two points and mix is complete on leaving pump.

Time Frame and Applications

On August 31, 1937, G. Chavanne of the University of Brussels (Belgium) was granted a patent for a 'Procedure for the transformation of vegetable oils for their uses as fuels' (fr. 'Procédé de Transformation d'Huiles Végétales en Vue de Leur Utilisation comme Carburants') Belgian Patent 422,877. This patent described the alcoholysis (often referred to as transesterification) of vegetable oils using ethanol (and mentions methanol) in order to separate the fatty acids from the glycerol by replacing the glycerol with short linear alcohols. This appears to be the first account of the production of what is known as 'biodiesel' today.

Biodiesel can be used in pure form (B100) or may be blended with petroleum diesel at any concentration in most modern diesel engines. Biodiesel has different solvent properties than petrodiesel, and will degrade natural rubber gaskets and hoses in vehicles (mostly vehicles manufactured before 1992), although these tend to wear out naturally and most likely will have already been replaced with FKM, which is non-reactive to biodiesel. Biodiesel has been known to break down deposits of residue in the fuel lines where petrodiesel has been used. As a result, fuel filters may become clogged with particulates if a quick transition to pure biodiesel is made. Therefore, it is recommended to change the fuel filters on engines and heaters shortly after first switching to a biodiesel blend.

Main Distribution Patter

Since the passage of the Energy Policy Act of 2005 biodiesel use has been increasing in the United States. Fueling stations make biodiesel readily available to consumers across Europe, and increasingly in the USA and Canada. A growing number of transport fleets use it as an additive in their fuel. Biodiesel is often more expensive to purchase than petroleum diesel but this is expected to diminish due to economies of scale and agricultural subsidies versus the rising cost of petroleum as reserves are depleted.

Major Vehicular Uses, Railroad Use and Aircraft Use

In 2005, Chrysler (then part of DaimlerChrysler) released the Jeep Liberty CRD diesels from the factory into the Americàn market with 5 per cent biodiesel blends, indicating at least partial acceptance of biodiesel as an acceptable diesel fuel additive. In 2007, DaimlerChrysler indicated intention to increase warranty coverage to 20 per cent biodiesel blends if biofuel quality in the United States can be standardized.

The British businessman Richard Branson's Virgin Voyager train, number 220007 *Thames Voyager*, billed as the world's first "biodiesel train" was converted to run on 80 per cent petrodiesel and only 20 per cent biodiesel, and it is claimed it will save 14 per cent on direct emissions. Similarly, a state-owned short-line railroad in Eastern Washington ran a test of a 25 per cent biodiesel/75 per cent petrodiesel blend during the summer of 2008, purchasing fuel from a biodiesel producer seated along the railroad tracks. The train will be powered by biodiesel made in part from canola grown in agricultural regions through which the short line runs. Also in 2007 Disneyland began running the park trains on B98 biodiesel blends (98% biodiesel). The programme blipped in 2008 due to storage issues, but in January 2009 it was announced that they are now running all the trains on biodiesel manufactured locally from used cooking oils from the park. This is a change from running the trains on soy biased biodiesel.

Aircraft manufacturers are even more cautious due to the inherent risks of air travel, but a test flight has been performed by a Czech Aircraft (completely powered on biofuel); testing has been announced by Rolls Royce plc, Air New Zealand and Boeing (one engine out of four on a Boeing 747); and commercial passenger jet testing has also been announced by Virgin Atlantic's

Richard Branson. The world's first biofuel-powered commercial aircraft took off from London's Heathrow Airport on February 24, 2008 and touched down in Amsterdam on a demonstration flight hailed as a first step towards "cleaner" flying. The "BioJet" fuel for this flight was produced by Seattle based Imperium Renewables, Inc.

Major Use as a Heating Oil

Biodiesel can also be used as a heating fuel in domestic and commercial boilers, sometimes known as bioheat. Older furnaces may contain rubber parts that would be affected by biodiesel's solvent properties, but can otherwise burn biodiesel without any conversion required. Care must be taken at first, however, given that varnishes left behind by petrodiesel will be released and can clog pipes—fuel filtering and prompt filter replacement is required. Another approach is to start using biodiesel as blend, and decreasing the petroleum proportion over time can allow the varnishes to come off more gradually and be less likely to clog. Thanks to its strong solvent properties, however, the furnace is cleaned out and generally becomes more efficient. A technical research paper describes laboratory research and field trials project using pure biodiesel and biodiesel blends as a heating fuel in oil fired boilers. During the Biodiesel Expo 2006 in the UK, Andrew J. Robertson presented his biodiesel heating oil research from his technical paper and suggested that B20 biodiesel could reduce UK household CO_2 emissions by 1.5 million tons per year.

Time Frame

Transesterification of a vegetable oil was conducted as early as 1853 by scientists E. Duffy and J. Patrick, many years before the first diesel engine became functional. Rudolf Diesel's prime model, a single 10 ft (3 m) iron cylinder with a flywheel at its base, ran on its own power for the first time in Augsburg, Germany, on August 10, 1893. In remembrance of this event, August 10 has been declared "International Biodiesel Day". The French Otto Company (at the request of the French government) demonstrated a Diesel engine running on peanut oil at the World Fair in Paris,

France in 1900, where it received the *Grand Prix* (highest prize). This engine stood as an example of Diesel's vision because it was powered by peanut oil—a biofuel, though not *biodiesel*, since it was not transesterified. He believed that the utilization of biomass fuel was the real future of his engine. In a 1912 speech Diesel said, "the use of vegetable oils for engine fuels may seem insignificant today but such oils may become, in the course of time, as important as petroleum and the coal-tar products of the present time." During the 1920s, diesel engine manufacturers altered their engines to utilize the lower viscosity of petrodiesel (a fossil fuel), rather than vegetable oil (a biomass fuel). The petroleum industries were able to make inroads in fuel markets because their fuel was much cheaper to produce than the biomass alternatives. The result, for many years, was a near elimination of the biomass fuel production infrastructure. Only recently, have environmental impact concerns and a decreasing price differential made biomass fuels such as biodiesel a growing alternative. Despite the widespread use of fossil petroleum-derived diesel fuels, interest in vegetable oils as fuels in internal combustion engines is reported in several countries during the 1920s and 1930's and later during World War II. Belgium, France, Italy, the United Kingdom, Portugal, Germany, Brazil, Argentina, Japan and China have been reported to have tested and used vegetable oils as diesel fuels during this time. Some operational problems were reported due to the high viscosity of vegetable oils compared to petroleum diesel fuel, which result in poor atomization of the fuel in the fuel spray and often leads to deposits and coking of the injectors, combustion chamber and valves. Attempts to overcome these problems included heating of the vegetable oil, blending it with petroleum-derived diesel fuel or ethanol, pyrolysis and cracking of the oils. On August 31, 1937, G. Chavanne of the University of Brussels (Belgium) was granted a patent for a "Procedure for the transformation of vegetable oils for their uses as fuels" (fr. 'Procédé de Transformation d'Huiles Végétales en Vue de Leur Utilisation comme Carburants') Belgian Patent 422,877. This patent described the alcoholysis (often referred to as transesterification) of vegetable oils using methanol and ethanol in order to separate the fatty acids from the glycerol by replacing the glycerol by short linear alcohols. This appears to be the first account of the production of what is known as "biodiesel" today. More recently, in 1977, Brazilian scientist Expedito Parente invented and submitted for patent, the first industrial process for the production of biodiesel. This process is classified as biodiesel by international norms, conferring a "standardized identity and quality. No other proposed biofuel has been validated by the motor industry." Currently, Parente's company Tecbio is working with Boeing and NASA to certify bioquerosene (bio-kerosene), another product produced and patented by the Brazilian scientist. Research into the use of transesterified sunflower oil, and refining it to diesel fuel standards, was initiated in South Africa in 1979. By 1983, the process for producing fuel-quality, engine-tested biodiesel was completed and published internationally. An Austrian company, Gaskoks, obtained the technology from the South African Agricultural Engineers; the company erected the first biodiesel pilot plant in November 1987, and the first industrial-scale plant in April 1989 (with a capacity of 30,000 tons of rapeseed per annum). Throughout the 1990s, plants were opened in many European countries, including the Czech Republic, Germany and Sweden. France launched local production of biodiesel fuel (referred to as *diester*) from rapeseed oil, which is mixed into regular diesel fuel at a level of 5 per cent, and into the diesel fuel used by some captive fleets (e.g. public transportation) at a level of 30 per cent. Renault, Peugeot and other manufacturers have certified truck engines for use with up to that level of partial biodiesel; experiments with 50 per cent biodiesel are underway. During the same period, nations in other parts of the world also saw local production of biodiesel starting up: by 1998, the Austrian Biofuels Institute had identified 21 countries with commercial biodiesel projects. 100 per cent Biodiesel is now available at many normal service stations across Europe. In September 2005 Minnesota became the first U.S. state to mandate that all diesel fuel sold in the state contain part biodiesel, requiring a content of at least 2 per cent biodiesel.

Listing Main Properties

Biodiesel has better lubricating properties than today's lower viscosity diesel fuels. Biodiesel addition reduces engine wear increasing the life of the fuel injection equipment that relies on the fuel for its lubrication, such as high pressure injection pumps, pump injectors (also called *unit injectors*) and fuel injectors. The calorific value of biodiesel is about 37.27 MJ/L. This is 9 per cent lower than regular Number 2 petrodiesel. Variations in biodiesel energy density is more dependent on the feedstock used than the production process. Still these variations are less than for petrodiesel. It has been claimed biodiesel gives better lubricity and more complete combustion thus increasing the engine energy output and partially compensating for the higher energy density of petrodiesel. Biodiesel is a liquid which varies in colour—between golden and dark brown—depending on the production feedstock. It is immiscible with water, has a high boiling point and low vapour pressure. *The flash point of biodiesel (>130 °C, >266 °F) is significantly higher than that of petroleum diesel (64 °C, 147 °F) or gasoline (−45 °C, -52 °F). Biodiesel has a density of ~ 0.88 g/cm^3, less than that of water. Biodiesel has virtually no sulphur content, and it is often used as an additive to Ultra-Low Sulphur Diesel (ULSD) fuel.

Types of Material Compatibility

   (i)   *Plastics:* High density polyethylene is compatible but PVC is slowly degraded. Polystyrenes are dissolved on contact with biodiesel.

   (ii)   *Metals:* Biodiesel has an effect on copper-based materials (i.e. brass), and it also affects zinc, tin, lead, and cast iron. Stainless steels (316 and 304) and aluminium are unaffected.

   (iii) *Rubber:* Biodiesel also affects types of natural rubbers found in some older engine components. Studies have also found that fluorinated elastomers (FKM) cured with peroxide and base-metal oxides can be degraded when biodiesel loses its stability caused by oxidation. However testing with FKM-GBL-S and FKM-GF-S were found to be the toughest elastomer to handle biodiesel in all conditions.

Understanding Technical Standards and Gelling

Biodiesel has a number of standards for its quality including European standard EN 14214, ASTM International D6751, and others.

The cloud point, or temperature at which pure (B100) biodiesel starts to gel, varies significantly and depends upon the mix of esters and therefore the feedstock oil used to produce the biodiesel. For example, biodiesel produced from low erucic acid varieties of canola seed (RME) starts to gel at approximately –10 °C (14 °F). Biodiesel produced from tallow tends to gel at around +16°C (61°F). As of 2006, there are a very limited number of products that will significantly lower the gel point of straight biodiesel. A number of studies have shown that winter operation is possible with biodiesel blended with other fuel oils including 2 low sulfur diesel fuel and 1 diesel/kerosene. The exact blend depends on the operating environment: successful operations have run using a 65 per cent LS 2, 30 per cent K 1, and 5 per cent bio blend. Other areas have run a 70 per cent Low Sulfur 2, 20 per cent Kerosene 1, and 10 per cent bio blend or an 80 per cent K 1, and 20 per cent biodiesel blend. According to the National Biodiesel Board (NBB), B20 (20% biodiesel, 80% petrodiesel) does not need any treatment in addition to what is already taken with petrodiesel. To permit the use of biodiesel without mixing and without the possibility of gelling at low temperatures, some people modify their vehicles with a second fuel tank for biodiesel in addition to the standard fuel tank. Alternately, a vehicle with two tanks is chosen. The second fuel tank is insulated and a heating coil using engine coolant is run through the tank. When a temperature sensor indicates that the fuel is warm enough to burn, the driver switches from the petrodiesel tank to the biodiesel tank. This is similar to the method used for running straight vegetable oil.

Possible Contamination by Water

Biodiesel may contain small but problematic quantities of water. Although it is not miscible with water, it is, like ethanol, hygroscopic (absorbs water from atmospheric moisture). One of the reasons biodiesel can absorb water is the persistence of mono and diglycerides left over from an incomplete reaction.

These molecules can act as an emulsifier, allowing water to mix with the biodiesel. In addition, there may be water that is residual to processing or resulting from storage tank condensation. The presence of water is a problem because:

(*i*)   Water reduces the heat of combustion of the bulk fuel. This means more smoke, harder starting, less power.

(*ii*) Water causes corrosion of vital fuel system components: fuel pumps, injector pumps, fuel lines, etc.

(*iii*) Water and microbes cause the paper element filters in the system to fail (rot), which in turn results in premature failure of the fuel pump due to ingestion of large particles.

(*iv*) Water freezes to form ice crystals near 0 °C (32 °F). These crystals provide sites for nucleation and accelerate the gelling of the residual fuel.

(*v*)   Water accelerates the growth of microbe colonies, which can plug up a fuel system. Biodiesel users who have heated fuel tanks therefore face a year-round microbe problem.

(*vi*) Additionally, water can cause pitting in the pistons on a diesel engine.

(*vii*) Previously, the amount of water contaminating biodiesel has been difficult to measure by taking samples, since water and oil separate. However, it is now possible to measure the water content using water-in-oil sensors.

(*viii*) Water contamination is also a potential problem when using certain chemical catalysts involved in the production process, substantially reducing catalytic efficiency of base (high pH) catalysts such as potassium hydroxide. However, the super-critical methanol production methodology, whereby the transesterifica-tion process of oil feedstock and methanol is effectuated under high temperature and pressure, has been shown to be largely unaffected by the presence of water contamination during the production phase.

Trends in Availability, Prices and Production

Biodiesel is less expensive than conventional diesel. Global biodiesel production reached 3.8 million tons in 2005. Approximately 85 per cent of biodiesel production came from the European Union. In the United States, average retail (at the pump) prices, including Federal and state fuel taxes, of B2/B5 are lower than petroleum diesel by about 12 cents, and B20 blends are the same as petrodiesel. B99 and B100 generally cost more than petrodiesel except where local governments provide a subsidy.

Biodiesel is commonly produced by the transesterification of the vegetable oil or animal fat feedstock. There are several methods for carrying out this transesterification reaction including the common batch process, supercritical processes, ultrasonic methods, and even microwave methods. Chemically, transesterified biodiesel comprises a mix of mono-alkyl esters of long chain fatty acids. The most common form uses methanol (converted to sodium methoxide) to produce methyl esters as it is the cheapest alcohol available, though ethanol can be used to produce an ethyl ester biodiesel and higher alcohols such as isopropanol and butanol have also been used. Using alcohols of higher molecular weights improves the cold flow properties of the resulting ester, at the cost of a less efficient transesterification reaction. A lipid transesterification production process is used to convert the base oil to the desired esters. Any Free fatty acids (FFAs) in the base oil are either converted to soap and removed from the process, or they are esterified (yielding more biodiesel) using an acidic catalyst. After this processing, unlike straight vegetable oil, biodiesel has combustion properties very similar to those of petroleum diesel, and can replace it in most current uses. A by-product of the transesterification process is the production of glycerol. For every 1 tonne of biodiesel that is manufactured, 100 kg of glycerol are produced. Originally, there was a valuable market for the glycerol, which assisted the economics of the process as a whole. However, with the increase in global biodiesel production, the market price for this crude glycerol (containing 20% water and catalyst residues) has crashed. Research is being conducted globally to use this glycerol as a chemical building block. One initiative in the UK is The Glycerol Challenge. Usually this crude glycerol has to be purified, typically by performing vacuum distillation. This is rather energy intensive. The refined glycerol (98%+ purity) can then be utilised directly, or converted into other products. The following announcements were made in 2007: A joint venture of Ashland Inc. and Cargill announced plans to make propylene glycol in Europe from glycerol and Dow Chemical announced similar plans for North America. Dow also plans to build a plant in China to make epichlorhydrin from glycerol. Epichlorhydrin is a raw material for epoxy resins.

Biodiesel production capacity is growing rapidly, with an average annual growth rate from 2002-2006 of over 40 per cent. For the year 2006, the latest for which actual production figures could be obtained, total world biodiesel production was about 5-6 million tonnes, with 4.9 million tonnes processed in Europe (of which 2.7 million tonnes was from Germany) and most of the rest from the USA. In 2007 production in Europe alone had risen to 5.7 million tonnes. The capacity for 2008 in Europe totalled 16 million tonnes. This compares with a total demand for diesel in the US and Europe of approximately 490 million tonnes (147 billion gallons). Total world production of vegetable oil for all purposes in 2005/06 was about 110 million tonnes, with about 34 million tonnes each of palm oil and soybean oil.

Bioenergy from Biodiesel Feedstocks

A variety of oils can be used to produce biodiesel. These include:

(i)   Virgin oil feedstock; rapeseed and soybean oils are most commonly used, soybean oil alone accounting for about ninety percent of all fuel stocks in the US. It also can be obtained from field pennycress and Jatropha other crops such as mustard, flax, sunflower, palm oil, hemp (see List of vegetable oils for a more complete list);

(ii) Waste vegetable oil (WVO);

(iii) Animal fats including tallow, lard, yellow grease, chicken fat, and the by-products of the production of Omega-3 fatty acids from fish oil;

(iv) Algae, which can be grown using waste materials such as sewage and without displacing land currently used for food production;

(v) Oil from halophytes such as *salicornia bigelovii*, which can be grown using saltwater in coastal areas where conventional crops cannot be grown, with yields equal to the yields of soybeans and other oilseeds grown using freshwater irrigation; and

(vi) Many advocates suggest that waste vegetable oil is the best source of oil to produce biodiesel, but since the available supply is drastically less than the amount of petroleum-based fuel that is burned for transportation and home heating in the world, this local solution does not scale well;

(vii) Animal fats are a by-product of meat production. Although it would not be efficient to raise animals (or catch fish) simply for their fat, use of the by-product adds value to the livestock industry (hogs, cattle, poultry). However, producing biodiesel with animal fat that would have otherwise been discarded could replace a small percentage of petroleum diesel usage. Today, multi-feedstock biodiesel facilities are producing high quality animal-fat based biodiesel. Currently, a 5-million dollar plant is being built in the USA, with the intent of producing 11.4 million litres (3 million gallons) biodiesel from some of the estimated 1 billion kg (2.3 billion pounds) of chicken fat produced annually the local Tyson poultry plant. Similarly, some small-scale biodiesel factories use waste fish oil as feedstock.

Worldwide production of vegetable oil and animal fat is not sufficient to replace liquid fossil fuel use. Furthermore, some object to the vast amount of farming and the resulting fertilization, pesticide use, and land use conversion that would be needed to produce the additional vegetable oil. The estimated transportation diesel fuel and home heating oil used in the United States is about 160 million tonnes (350 billion pounds) according to the Energy Information Administration, US

Department of Energy. In the United States, estimated production of vegetable oil for all uses is about 11 million tonnes (24 billion pounds) and estimated production of animal fat is 5.3 million tonnes (12 billion pounds). If the entire arable land area of the USA (470 million acres, or 1.9 million square kilometers) were devoted to biodiesel production from soy, this would just about provide the 160 million tonnes required (assuming an optimistic 98 GPa of biodiesel). This land area could in principle be reduced significantly using algae, if the obstacles can be overcome. The US DOE estimates that if algae fuel replaced all the petroleum fuel in the United States, it would require 15,000 square miles (38,849 square kilometers), which is a few thousand square miles larger than Maryland, or 1.3 Belgiums, assuming a yield of 15,000 GPa. Given a more realistic yield of 36 tonnes/hectare (3834 GPa) the area required is about 152,000 square kilometers, or roughly equal to that of the state of Georgia or England and Wales. The advantages of algae are that it can be grown on non-arable land such as deserts or in marine environments, and the potential oil yields are much higher than from plants.

Feedback Yield Pattern

Feedstock yield efficiency per acre affects the feasibility of ramping up production to the huge industrial levels required to power a significant percentage of national or world vehicles. Some typical yields in cubic decimeters (liters) of biodiesel per hectare (10,000 square meters):

(i)   *Algae*: 2763 dm^3 (liter) or more (~300 gallons per acre)
(ii)   *Hemp*: 1535 dm^3
(iii) Chinese tallow: 772 dm^3—970 GPa
(iv)   *Palm oil*: 780—1490 dm^3
(v)   *Coconut*: 353 dm^3
(vi)   *Rapeseed*: 157 dm^3
(vii) *Soy*: 76-161 dm^3 in Indiana (Soy is used in 80% of USA biodiesel)
(viii) *Peanut*: 138 dm^3
(ix)   *Sunflower*: 126 dm^3

Algae fuel yields have not yet been accurately determined, but DOE is reported as saying that algae yield 30 times more energy per acre than land crops such as soybeans. Yields of 36 tonnes/hectare are considered practical by Ami Ben-Amotz of the Institute of Oceanography in Haifa, who has been farming Algae commercially for over 20 years. The Jatropha plant has been cited as a high-yield source of biodiesel but yields are highly dependent on climatic and soil conditions. The estimates at the low end put the yield at about 200 GPa (1.5-2 tonnes per hectare) per crop; in more favourable climates two or more crops per year have been achieved. It is grown in the Philippines, Mali and India, is drought-resistant, and can share space with other cash crops such as coffee, sugar, fruits and vegetables. It is well-suited to semi-arid lands and can contribute to slow down desertification, according to its advocates.

Energy Efficiency and Economics

According to a study by Drs. Van Dyne and Raymer for the Tennessee Valley Authority, the average US farm consumes fuel at the rate of 82 litres per hectare (8.75 US gallons per acre) of land to produce one crop. However, average crops of rapeseed produce oil at an average rate of 1,029 L/ha (110 US gal/acre), and high-yield rapeseed fields produce about 1,356 L/ha (145 US gal/acre). The ratio of input to output in these cases is roughly 1:12.5 and 1:16.5. Photosynthesis is known to have an efficiency rate of about 3-6 per cent of total solar radiation and if the entire mass of a crop is utilized for energy production, the overall efficiency of this chain is currently about 1 per cent While this may compare unfavourably to solar cells combined with an electric drive train, biodiesel is less costly to deploy (solar cells cost approximately US$1,000 per square meter) and transport (electric vehicles require batteries which currently have a much lower energy density than liquid fuels). However, these statistics by themselves are not enough to show whether such a change makes economic sense. Additional factors must be taken into account, such as: the fuel equivalent of the energy required for processing, the yield of fuel from raw oil, the return on cultivating food, the effect biodiesel will have on food prices and the relative cost of biodiesel versus petrodiesel. The debate over the energy balance of biodiesel is ongoing. Transitioning fully to biofuels could require immense tracts of land if traditional food crops are used (although non food crops can be utilized). The problem would be especially severe for nations with large economies, since energy consumption scales with economic output. If using only traditional food plants, most such nations do not have sufficient arable land to produce biofuel for the nation's vehicles. Nations with smaller economies (hence less energy consumption) and more arable land may be in better situations, although many regions cannot afford to divert land away from food production. For third world countries, biodiesel sources that use marginal land could make more sense, e.g. honge oil nuts grown along roads or jatropha grown along rail lines. In tropical regions, such as Malaysia and Indonesia, oil palm is being planted at a rapid pace to supply growing biodiesel demand in Europe and other markets. It has been estimated in Germany that palm oil biodiesel has less than 1/3 the production costs of rapeseed biodiesel. The direct source of the energy content of biodiesel is solar energy captured by plants during photosynthesis. Regarding the positive energy balance of biodiesel:

(*i*)   When straw was left in the field, biodiesel production was strongly energy positive, yielding 1 GJ biodiesel for every 0.561 GJ of energy input (a yield/cost ratio of 1.78).

(*ii*) When straw was burned as fuel and oilseed rapemeal was used as a fertilizer, the yield/cost ratio for biodiesel production was even better (3.71). In other words, for every unit of energy input to produce biodiesel, the output was 3.71 units (the difference of 2.71 units would be from solar energy).

Biodiesel is becoming of interest to companies interested in commercial scale production as well as the more usual home brew biodiesel user and the user of straight vegetable oil or waste vegetable oil in diesel engines. Homemade biodiesel processors are many and varied.

Role in Energy Security Evaluating

One of the main drivers for adoption of biodiesel is energy security. This means that a nation's dependence on oil is reduced, and substituted with use of locally available sources, such as coal, gas, or renewable sources. Thus a country can benefit from adoption of biofuels, without a reduction in greenhouse gas emissions. Whilst the total energy balance is debated, it is clear that the dependence on oil is reduced. One example is the energy used to manufacture fertilizers, which could come from a variety of sources other than petroleum. The US NREL says that energy security is the number one driving force behind the US biofuels programme. and the White House "Energy Security for the 21st Century" makes clear that energy security is a major reason for promoting biodiesel. The EU commission president, Jose Manuel Barroso, speaking at a recent EU biofuels conference, stressed that properly managed biofuels have the potential to reinforce the EU's security of supply through diversification of energy sources.

Evaluating Major Environmental Effects

The surge of interest in biodiesels has highlighted a number of environmental effects associated with its use. These potentially include reductions in greenhouse gas emissions, deforestation, pollution and the rate of biodegradation.

This Issue of Food, Land and Water vs. Fuel

In some poor countries the rising price of vegetable oil is causing problems. Some propose that fuel only be made from non-edible vegetable oils like camelina, jatropha or seashore mallow which can thrive on marginal agricultural land where many trees and crops will not grow, or would produce only low yields. Others argue that the problem is more fundamental. Farmers may switch from producing food crops to producing biofuel crops to make more money, even if the new crops are not edible. The law of supply and demand predicts that if fewer farmers are producing food the price of food will rise. It may take some time, as farmers can take some time to change which things they are growing, but increasing demand for first generation biofuels is likely to result in price increases for many kinds of food. Some have pointed out that there are poor farmers and poor countries who are making more money because of the higher price of vegetable oil. Biodiesel from sea algae would not necessarily displace terrestrial land currently used for food production and new algaculture jobs could be created.

Ongoing Research Findings

There is ongoing research into finding more suitable crops and improving oil yield. Using the current yields, vast amounts of land and fresh water would be needed to produce enough oil to completely replace fossil fuel usage. It would require twice the land area of the US to be devoted to soybean production, or two-thirds to be devoted to rapeseed production, to meet current US heating and transportation needs. Specially bred mustard varieties can produce reasonably high oil yields and are very useful in crop rotation with cereals, and have the added benefit that the meal leftover after the oil has been pressed out can act as an effective and biodegradable pesticide. The NFESC, with Santa Barbara-based Biodiesel Industries, Inc, is working to develop biodiesel technologies for the US navy and military, one of the largest diesel fuel users in the world. A group of Spanish developers working for a company called Ecofasa just announced a new biofuel made up from trash. It's made from general urban waste which is treated by bacteria to produce fatty acids which can be used to make biodiesel.

(a) *Energy from Algal Biodiesel:* From 1978 to 1996, the U.S. National Renewable Energy Laboratory experimented with using algae as a biodiesel source in the "Aquatic Species Programme". A self-published article by Michael Briggs, at the UNH Biodiesel Group, offers estimates for the realistic replacement of all vehicular fuel with biodiesel by utilizing algae that have a natural oil content greater than 50 per cent, which Briggs suggests can be grown on algae ponds at wastewater treatment plants. This oil-rich algae can then be extracted from the system and processed into biodiesel, with the dried remainder further reprocessed to create ethanol. The production of algae to harvest oil for biodiesel has not yet been undertaken on a commercial scale, but feasibility studies have been conducted to arrive at the above yield estimate. In addition to its projected high yield, algaculture—unlike crop-based biofuels—does not entail a decrease in food production, since it requires neither farmland nor fresh water. Many companies are pursuing algae bio-reactors for various purposes, including scaling up biodiesel production to commercial levels.

(b) *Energy from Fungus*: A group at the Russian academy of Sciences in Moscow published a paper in September 2008, stating that they had could isolate large amounts of lipids from single-celled fungi and turn them into biodiesel in an economically efficient manner. More research on this fungal species; C. japonica, and others, is likely to appear in the near future. The recent discovery of a variant of the fungus Gliocladium roseum points toward the production of so-called myco-diesel from cellulose. This organism was recently discovered in the rainforests of northern Patagonia and has the unique capability of converting cellulose into medium length hydrocarbons typically found in diesel fuel.

(c) *Biodiesel from Used Coffee Grounds*: Researchers at the University of Nevada, Reno, have successfully produced biodiesel from oil derived from used coffee grounds. Their analysis of the used grounds showed a 10 per cent to 15 per cent oil content (by weight). Once the oil was extracted, it underwent conventional processing into biodiesel. It is estimated that finished biodiesel could be produced for about one US dollar per gallon. Further, it was reported that "the technique is not difficult" and that "there is so much coffee around that several hundred million gallons of biodiesel could potentially be made annually."

Bioenergy from Biogas

Bio-gas typically refers to a gas produced by the biological breakdown of organic matter in the absence of oxygen. Biogas originates from biogenic material and is a type of biofuel. One type of bio-gas is produced by anaerobic digestion or fermentation of biodegradable materials such as biomass, manure or sewage, municipal waste, green waste and energy crops. This type of biogas comprises primarily methane and carbon dioxide. The other principal type of biogas is wood gas which is created by gasification of wood or other biomass. This type of biogas is comprised primarily of nitrogen, hydrogen, and carbon monoxide, with trace amounts of methane. The gases methane, hydrogen and carbon monoxide can be combusted or oxidized with oxygen. Air contains 21 per cent oxygen. This energy release allows biogas to be used as a fuel. Biogas can be used as a low-cost fuel in any country for any heating purpose, such as cooking. It can also be used in modern waste management facilities where it can be used to run any type of heat engine, to generate either mechanical or electrical power. Biogas can be compressed, much like natural gas, and used to power motor vehicles and in the UK for example is estimated to have the potential to replace around 17 per cent of vehicle fuel. Biogas is a renewable fuel, so it qualifies for renewable energy subsidies in some parts of the world.

Understanding the Production Process

(LFG) or digester gas. A biogas plant is the name often given to an anaerobic digester that treats farm wastes or energy crops. Biogas can be produced utilizing anaerobic digesters. These plants can be fed with energy crops such as maize silage or biodegradable wastes including sewage sludge and food waste. Landfill gas is produced by wet organic waste decomposing under anaerobic conditions in a landfill. The waste is covered and compressed mechanically and by the weight of the material that is deposited from above. This material prevents oxygen from accessing the waste and anaerobic microbes thrive. This gas builds up and is slowly released into the atmosphere if the landfill site has not been engineered to capture the gas. Landfill gas is hazardous for three key reasons. Landfill gas becomes explosive when it escapes from the landfill and mixes with oxygen. The lower explosive limit is 5 per cent methane and the upper explosive limit is 15 per cent methane. The methane contained within biogas is 20 times more potent as a greenhouse gas than carbon dioxide. Therefore uncontained landfill gas which escapes into the atmosphere may significantly contribute to the effects of global warming. In addition to this volatile organic compounds (VOCs) contained within landfill gas contribute to the formation of photochemical smog. Sweden produces biogas from confiscated alcoholic beverages.

Understanding Composition and Applications

The composition of biogas varies depending upon the origin of the anaerobic digestion process. Landfill gas typically has methane concentrations around 50 per cent. Advanced waste treatment technologies can produce biogas with 55-75 per cent CH_4 (See Table 1.1).

Table 1.1: Composition of Biogas

Constituents	Percentage
Methane, CH_4	50-75
Carbon dioxide, CO_2	25-50
Nitrogen, H_2	0-1
Hydrogen sulphide, H_2S	0-3
Oxygen, O_2	0-2

In some cases biogas contains siloxanes. These siloxanes are formed from the anaerobic decomposition of materials commonly found in soaps and detergents. During combustion of biogas containing siloxanes, silicon is released and can combine with free oxygen or various other elements in the combustion gas. Deposits are formed containing mostly silica (SiO_2) or silicates (Si_xO_y) and can also contain calcium, sulfur, zinc, phosphorus. These white mineral deposits build to a surface thickness of several millimetres and must be removed by chemical or mechanical means.

Biogas can be utilized for electricity production on sewage works, in a CHP gas engine, where the waste heat from the engine is conveniently used to heat the digestoer; cooking, space heating, water heating and process heating. If compressed, it can replace compressed natural gas for use in vehicles, where it can fuel an internal combustion engine or fuel cells and is a much more effective displacer of carbon dioxide than the normal use in on site CHP plants. Methane within biogas can be concentrated via a biogas upgrader to the same standards as fossil natural gas; when it is, it is called *biomethane*. If the local gas network permits it the producer of the biogas may be able to utilize the local gas distribution networks. Gas must be very clean to reach pipeline quality, and must be of the correct composition for the local distribution network to accept. Carbon dioxide, Water, hydrogen sulphide and particulates must be removed if present. If concentrated and compressed it can also be used in vehicle transportation. Compressed biogas is becoming widely used in Sweden, Switzerland and Germany. A biogas-powered train has been in service in Sweden since 2005.

Bates, an inventor, lived in Devon, UK, modified his car to run on biogas. A short documentary film called 'Sweet as a Nut' in 1974, talks through the simple process and benefits of running a car on biogas, at which point he had run his car for 17 years on gas he had produced by processing pig manure. The conversion was simply made with an adapter attached to the combustion engine.

Evaluating Scope and Potential Quantities in Developed and Developing Nations

In the UK, sewage gas electricity production is tiny compared to overall power consumption—a mere 80 MW of generation, compared to 70,000 MW on the grid. Estimates vary but could be a considerable fraction from digestion of. In India biogas produced from the anaerobic digestion of manure in small-scale digestion facilities is called Gober gas; it is estimated that such facilities exist in over 2 million households. The digester is an airtight circular pit made of concrete with a pipe connection. The manure is directed to the pit, usually directly from the cattle shed. The pit is then filled with a required quantity of wastewater. The gas pipe is connected to the kitchen fire place through control valves. The combustion of this biogas has very little odour or smoke. Owing to simplicity in implementation and use of cheap raw materials in villages, it is one of the most environmentally sound energy sources for rural needs. Some designs use vermiculture to further enhance the slurry produced by the biogas plant for use as compost. Biogas is used extensively throughout rural China and where wastewater treatment and industry coincide. The Biogas Support Programme in Nepal has installed over 150,000 biogas plants in rural areas, and in 2005 won an Ashden Award for their work. Vietnam's Biogas Programme for Animal Husbandry Sector has led to the installation of over 20,000 plants throughout that country. Biogas is also in use in rural Costa Rica. In Colombia experiments with diesel engines-generator sets partially fuelled by biogas demonstrated that biogas could be used for power generation, reducing electricity costs by 40 per cent compared with purchase from the regional utility. In Rwanda, the Kigali Institute of Science and Technology has developed and installed large-scale biogas plants at prisons to treat sewage and provide gas for cooking.

Evaluating Deenabandhu Model of Biogas Production

This is a new model of Biogas unit popular in India. The word means "helpful for the Poor". The unit usually has a capacity of 2 to 3 cubic metres. It is constructed using bricks or by a ferrocement mixture. The unit is subsidised by the Ministry of Non Conventional Energy Sources of the Government of India. A subsidy of Rupees 3500 per plant is provided to any body who constructs such a unit. A turn key agent fee of Rs. 700 is provided to the approved mason for maintenance of the unit for three years. The total cost of construction of a 2 cubic meter unit comes to Rs. 18000 for the brick model and Rs. 14000 for the Ferrocement model.

India's Gober India's Model

Gober gas (also spelled as "Gobar gas" by some) is biogas generated out of cow dung. In India, gober gas is generated using countless household micro plants (an estimated more than 2 million). In Pakistan the concept is also quickly growing. The Government of Pakistan provides 50 per cent funds for the construction of moveable gas chamber biogas plants. The gober gas plant is an airtight circular pit made of concrete with a pipe connection. The manure is directed to the pit, usually directly from the cattle shed. The pit is then filled with a required quantity of water or wastewater. The gas pipe is connected to the kitchen fire place through control valves. The flammable methane gas generated out of this is largely odourless and smokeless. The residue left after the extraction of the gas is used as fertiliser. Owing to its simplicity in implementation and use of cheap raw materials in the villages, it is often quoted as one of the most environmentally sound energy source for the rural needs.

Bioenergy from Cellulosic Ethanol

Cellulosic ethanol is a biofuel produced from wood, grasses, or the non-edible parts of plants. It is a type of biofuel produced from lignocellulose, a structural material that comprises much of the mass of plants. Lignocellulose is composed mainly of cellulose, hemicellulose and lignin. Corn stover, switchgrass, miscanthus, woodchips and the byproducts of lawn and tree maintenance are some of the more popular cellulosic materials for ethanol production. Production of ethanol from lignocellulose has the advantage of abundant and diverse raw material compared to sources like corn and cane sugars, but requires a greater amount of processing to make the sugar monomers available to the microorganisms that are typically used to produce ethanol by fermentation. Switchgrass and Miscanthus are the major biomass materials being studied today, due to their high productivity per acre. Cellulose, however, is contained in nearly every natural, free-growing plant, tree, and bush, in meadows, forests, and fields all over the world without agricultural effort or cost needed to make it grow. According to U.S. Department of Energy studies conducted by the Argonne Laboratories of the University of Chicago, one of the benefits of cellulosic ethanol is that it reduces greenhouse gas emissions (GHG) by 85% over reformulated gasoline. By contrast, starch ethanol (e.g., from corn), which most frequently uses natural gas to provide energy for the process, may not reduce GHG emissions at all depending on how the starch-based feedstock is produced. A study by Nobel Prize winner Paul Crutzen found ethanol produced from corn, and sugarcane had a "net climate warming" effect when compared to oil.

Time Frame

The first attempt at commercializing a process for ethanol from wood was done in Germany in 1898. It involved the use of dilute acid to hydrolyze the cellulose to glucose, and was able to produce 7.6 liters of ethanol per 100 kg of wood waste (18 gal per ton). The Germans soon developed an industrial process optimized for yields of around 50 gallons per ton of biomass. This process soon found its way to the United States, culminating in two commercial plants operating in the southeast during World War I. These plants used what was called "the American Process"—a one-stage dilute sulfuric acid hydrolysis. Though the yields were half that of the original German process (25 gallons of ethanol per ton versus 50), the throughput of the American process was much higher. A drop in lumber production forced the plants to close shortly after the end of World War I. In the meantime, a small, but steady amount of research on dilute acid hydrolysis continued at the USDA's Forest Products Laboratory. During World War II, the US again turned to cellulosic ethanol, this time for conversion to butanediol to produce synthetic rubber. The Vulcan Copper and Supply Company was contracted to construct and operate a plant to convert sawdust into ethanol. The plant was based on modifications to the original German Scholler process as developed by the Forest Products Laboratory. This plant achieved an ethanol yield of 50 gal/dry ton but was still not profitable and was closed after the war. United States President Bush, in his State of the Union address delivered January 31, 2006, proposed to expand the use of cellulosic ethanol. In his State of the Union Address on January 23, 2007, President Bush announced a proposed mandate for 35 billion gallons of ethanol by 2017. It is widely recognized that the maximum production of ethanol from corn starch is 15 billion gallons per year, implying a proposed mandate for production of some 20 billion gallons per year of cellulosic ethanol by 2017. Bush's proposed plan includes $2 billion funding (from 2007-2017) for cellulosic ethanol plants, with an additional $1.6 billion (from 2007-2017) announced by the USDA on January 27, 2007. In March 2007, the US government awarded $385 million in grants aimed at jumpstarting ethanol production from non-traditional sources like wood chips, switchgrass and citrus peels. Half of the six projects chosen will use thermo-chemical methods and half will use *cellulosic ethanol* methods. The American company Range Fuels announced in July 2007 that it was awarded a construction permit from the state of Georgia to build the first commercial-scale 100-million-gallon-per-year cellulosic ethanol plant in the United States. Construction began in November, 2007. In April 2008, George Huber of the University of Massachusetts Amherst received a $400,000 CAREER grant from the National Science Foundation to pursue his revolutionary new method for making biofuels, or "green gasoline". The U.S. could potentially produce 1.3 billion dry tons of cellulosic biomass per year, which has the energy content of four billion barrels of crude oil. This translates to 65 per cent of American oil consumption.

Major Production Methods

There are two ways of producing alcohol from cellulose:

(*a*) Cellulolysis processes which consist of hydrolysis on pre-treated lignocellulosic materials, using enzymes to break complex cellulose into simple sugars such as glucose and followed by fermentation and distillation.

(*b*) Gasification that transforms the lignocellulosic raw material into gaseous carbon monoxide and hydrogen. These gases can be converted to ethanol by fermentation or chemical catalysis.

They both include distillation as the final step to isolate the pure ethanol.

(a) Cellulolysis (biological approach)

There are four or five stages to produce ethanol using a biological approach:

(*i*)   A "pre-treatment" phase, to make the lignocellulosic material such as wood or straw amenable to hydrolysis,

(*ii*) Cellulose hydrolysis (cellulolysis), to break down the molecules into sugars;

(*iii*) Separation of the sugar solution from the residual materials, notably lignin;

(*iv*) Microbial fermentation of the sugar solution;

(*v*)   Distillation to produce 99.5 per cent pure alcohol.

Although cellulose is the most abundant plant material resource, its susceptibility has been curtailed by its rigid structure. As the result, an effective pre-treatment is needed to liberate the cellulose from the lignin seal and its crystalline structure so as to render it accessible for a subsequent hydrolysis step. By far, most pre-treatment are done through physical or chemical means. In order to achieve higher efficiency, some researchers seek to incorporate both effects. To date, the available pre-treatment techniques include acid hydrolysis, steam explosion, ammonia fiber expansion, alkaline wet oxidation and ozone pre-treatment. Besides effective cellulose liberation, an ideal pre-treatment has to minimize the formation of degradation products because of their inhibitory effects on subsequent hydrolysis and fermentation processes. The presence of inhibitors will not only further complicate the ethanol production but also increase the cost of production due to entailed detoxification steps. Even though pre-treatment by acid hydrolysis is probably the oldest and most studied pre-treatment technique, it produces several potent inhibitors including furfural and hydroxymethyl furfural (HMF) which are by far regarded as the most toxic inhibitors present in lignocellulosic hydrolysate. In fact, Ammonia Fiber Expansion (AFEX) is the sole pre-treatment which features promising pre-treatment efficiency with no inhibitory effect in resulting hydrolysate.

The cellulose molecules are composed of long chains of sugar molecules. In the hydrolysis process, these chains are broken down to free the sugar, before it is fermented for alcohol production. There are two major cellulose hydrolysis (cellulolysis) processes: a chemical reaction using acids, or an enzymatic reaction.

(a) Chemical hydrolysis

In the traditional methods developed in the 19th century and at the beginning of the 20th century, hydrolysis is performed by attacking the cellulose with an acid. Dilute acid may be used under high heat and high pressure, or more concentrated acid can be used at lower temperatures and atmospheric pressure, like Blue Fire Ethanol (OTCBB: BFRE). A decrystalized cellulosic mixture of acid and sugars reacts in the presence of water to complete individual sugar molecules (hydrolysis). The product from this hydrolysis is then neutralized and yeast fermentation is used to produce ethanol. As mentioned, a significant obstacle to the dilute acid process is that the hydrolysis is so harsh that toxic degradation products are produced that can interfere with fermentation. Concentrated acid must be separated from the sugar stream for recycle (simulated moving bed (SMB) chromatographic separation for example) to be commercially attractive. *Enzymatic hydrolysis:* Cellulose chains can be broken into glucose molecules by cellulase enzymes. This reaction occurs at body temperature in the stomach of ruminants such as cows and sheep, where the enzymes are produced by bacteria. This process uses several enzymes at various stages of this conversion. Using a similar enzymatic system, lignocellulosic materials can be enzymatically hydrolyzed at a relatively mild condition (50°C and pH5), thus enabling effective cellulose breakdown without the formation of byproducts that would otherwise inhibit enzyme activity. All major pre-treatment methods, including dilute acid pre-treatment, require enzymatic hydrolysis step to achieve high sugar yield for ethanol fermentation. Currently, most pre-treatment studies have been laboratory based, but companies are rapidly exploring means to transition from the laboratory to pilot, or production scale. Various enzyme companies have also contributed significant technological break-throughs in cellulosic ethanol through the mass production of enzymes for hydrolysis at competitive prices. The fungus Trichoderma reesei is used by Iogen Corporation, to secrete "specially engineered enzymes" for an enzymatic hydrolysis process. The raw material (wood or straw) has to be pre-treated to make it amenable to hydrolysis. Another Canadian company, SunOpta markets a patented technology known as "Steam Explosion" to pre-treat cellulosic biomass, overcoming its "recalcitance" to make cellulose and hemicellulose accessible to enzymes for conversion into fermenatable sugars. SunOpta designs and engineers cellulosic ethanol biorefineries and its process technologies and equipment are in use in the first 3 commercial demonstration plants in the world: Verenium (formerly Celunol Corporation)'s facility in Jennings, Louisiana, Abengoa's facility in Salamanca, Spain, and a facility in China owned by China Resources Alcohol Corporation (CRAC). The CRAC facility is currently producing cellulosic ethanol from local corn stover on a 24-hour a day basis utilizing SunOpta's process and technology. Genencor and Novozymes are two other companies that have received United States Department of Energy funding for research into reducing the cost of cellulases, key enzymes in the production of cellulosic ethanol by enzymatic hydrolysis. Other enzyme companies, such as Dyadic International, are developing genetically engineered fungi which would produce large volumes of cellulase, xylanase and hemicellulase enzymes which can be utilized to convert agricultural residues such as corn stover, distiller grains, wheat straw and sugar cane bagasse and energy crops such as switch grass into fermentable sugars which may be used to produce cellulosic ethanol. Verenium, formed by the merger of Diversa and Celunol, operates a pilot cellulosic ethanol plant in Jennings, Louisiana and is building a 1.4 million gallon per year demonstration plant on adjacent land to be completed by the end of 2007 and begin operation in early 2008. Verenium is the first publicly traded company with integrated, end-to-end capabilities to make cellulosic biofuels. KL Energy Corporation, formerly KL Process Design Group, began commercial operation of a 1.5 million gallon per year cellulosic ethanol facility in Upton,

WY in the last quarter of 2007. The Western Biomass Energy facility is currently achieving yields of 40-45 gallons per bone dry ton. It is the first operating commercial cellulosic ethanol facility in the nation. The KL Energy process utilizes a thermo-mechanical breakdown and enzymatic conversion process. The primary feedstock is soft wood, however, lab tests have already proven the KL Energy process on wine pomace, sugarcane bagasse, municipal solid waste, and switch grass.

Traditionally, baker's yeast (*Saccharomyces cerevisiae*), has long been used in brewery industry to produce ethanol from hexoses (6-carbon sugar). Due to the complex nature of the carbohydrates present in lignocellulosic biomass, a significant amount of xylose and arabinose (5-carbon sugars derived from the hemicellulose portion of the lignocellulose) is also present in the hydrolysate. For example, in the hydrolysate of corn stover, approximately 30 per cent of the total fermentable sugars is xylose. As a result, the ability of the fermenting microorganisms to utilize the whole range of sugars available from the hydrolysate is vital to increase the economic competitiveness of cellulosic ethanol and potentially bio-based chemicals. In recent years, metabolic engineering for microorganisms used in fuel ethanol production has shown significant progress. Besides *Saccharomyces cerevisiae*, microorganisms such as *Zymomonas mobilis* and *Escherichia coli* have been targeted through metabolic engineering for cellulosic ethanol production.

Recently, engineered yeasts have been described efficiently fermenting xylose, and arabinose, and even both together. Yeast cells are especially attractive for cellulosic ethanol processes as they have been used in biotechnology for hundreds of years, as they are tolerant to high ethanol and inhibitor concentrations and as they can grow at low pH values which avoids bacterial contaminations.

Some species of bacteria have been found capable of direct conversion of a cellulose substrate into ethanol. One example is *Clostridium thermocellum*, which utilizes a complex cellulosome to break down cellulose and synthesize ethanol. However, *C. thermocellum* also produces other products during cellulose metabolism, including acetate and lactate, in addition to ethanol, lowering the efficiency of the process. Some research efforts are directed to optimizing ethanol production by genetically engineering bacteria that focus on the ethanol-producing pathway.

(b) Gasification process (thermochemical approach)

The gasification process does not rely on chemical decomposition of the cellulose chain (cellulolysis). Instead of breaking the cellulose into sugar molecules, the carbon in the raw material is converted into synthesis gas, using what amounts to partial combustion. The carbon monoxide, carbon dioxide and hydrogen may then be fed into a special kind of fermenter. Instead of sugar fermentation with yeast, this process uses a microorganism named *Clostridium ljungdahlii.* This microorganism will ingest (eat) carbon monoxide, carbon dioxide and hydrogen and produce ethanol and water. The process can thus be broken into three steps:

(*i*)   Gasification—Complex carbon based molecules are broken apart to access the carbon as carbon monoxide, carbon dioxide and hydrogen are produced.

(*ii*) Fermentation—Convert the carbon monoxide, carbon dioxide and hydrogen into ethanol using the *Clostridium ljungdahlii* organism.

(*iii*) Distillation—Ethanol is separated from water.

A recent study has found another *Clostridium* bacterium that seems to be twice as efficient in making ethanol from carbon monoxide as the one mentioned above.

Alternatively, the synthesis gas from gasification may be fed to a catalytic reactor where the synthesis gas is used to produce ethanol and other higher alcohols through a thermochemical process. This process can also generate other types of liquid fuels, an alternative concept under investigation by at least one biofuels company.

Economic Evaluation

After the ethanol is produced, it must be distilled from the mash.

Most US producers use natural gas to provide the heat of distillation. In Brazil, producers are known to burn waste sugar cane stalks to distill their ethanol. Construction of pilot scale lignocellulosic ethanol plants requires considerable financial support through grants and subsidies. On 28 February 2007, the U.S. Dept. of Energy announced $385 million in grant funding to six cellulosic ethanol plants. This grant funding accounts for 40 per cent of the investment costs. The remaining 60 per cent comes from the promoters of those facilities. Hence, a total of $1 billion will be invested for approximately 140 million gallon capacity. This translates into $7/annual gallon production capacity in capital investment costs for pilot plants (this would work out to $.35/gal over the 20-year life of a facility); future capital costs are expected to be lower. Corn to ethanol plants cost roughly $1–3/annual gallon capacity, though the cost of the corn itself is considerably greater than for switchgrass or waste biomass. The quest for alternative sources of energy has provided many ways to produce electricity, such as wind farms, hydropower, or solar cells. However, about 20 per cent of total energy consumption is dedicated to transportation (i.e., cars, planes, lorries/trucks, etc.) and currently requires energy-dense liquid fuels such as gasoline, diesel fuel, or kerosene. These fuels are all obtained by refining petroleum. This dependency on oil has two major drawbacks: burning fossil fuels such as oil may contribute to global warming; and for net-consuming countries like the United States, importing oil creates a dependency on oil-producing countries. As of 2007, ethanol is produced mostly from sugars or starches, obtained from fruits and grains. In contrast, cellulosic ethanol is obtained from cellulose, the main component of wood, straw and much of the structure of plants. Since cellulose cannot be digested by humans, the production of cellulose does not compete with the production of food, other than conversion of land from food production to cellulose production (which has recently started to become an issue, due to rising wheat prices.) The price per ton of the raw material is thus much cheaper than grains or fruits. Moreover, since cellulose is the main component of plants, the whole plant can be harvested. This results in much better yields per acre—up to 10 tons, instead of 4 or 5 tons for the best crops of grain. The raw material is plentiful. Cellulose is present in every plant, in the form of straw, grass, and wood. Most of these "bio-mass" products are currently discarded. It is estimated that 323 million tons of cellulose containing raw materials that could be used to create ethanol are thrown away each year in US alone. This includes 36.8 million dry tons of urban wood wastes, 90.5 million dry tons of primary mill residues, 45 million dry tons of forest residues, and 150.7 million dry tons of corn stover and wheat straw. Transforming them into ethanol using efficient and cost effective hemi (cellulase) enzymes or other processes might provide as much as 30 per cent of the current fuel consumption in the United States—and probably similar figures in other oil-importing regions like China or Europe. Moreover, even land marginal for agriculture could be planted with cellulose-producing crops like switchgrass, resulting in enough production to substitute for all the current oil imports into the United States. Paper, cardboard, and packaging comprise a substantial part of the solid waste sent to landfills in the United States each day, 41.26 per cent of all organic municipal solid waste (MSW) according to California Integrated Waste Management Board's city profiles. These city profiles account for accumulation of 612.3 tons daily per landfill where an average population density of 2,413 per square mile persists. Organic waste consists of 0.4 per cent Manures, 1.6 per cent Gypsum Board, 4.2 per cent Glossy Paper, 4.2 per cent Paper Ledger, 9.2 per cent Wood, 10.5 per cent Envelopes, 11.9 per cent Newsprint, 12.3 per cent Grass and Leaves, 30.0 per cent Food Scrap, 34.0 per cent Office Paper, 35.2 per cent Corrugated Cardboard, and 46.4 per cent Agricultural Composites, makes up 71.51 per cent of land fill. All these except Gypsum Board contain cellulose which is transformable into cellulosic ethanol because they are the leading cause of methane plumes. Methane, a greenhouse gas, is 21 times more potent than carbon-dioxide. Reduction of the disposal of solid waste through cellulosic ethanol conversion would reduce solid waste disposal costs by local and state governments. It is estimated that each person in the US throws away 4.4 lb (2.0 kg) of trash each day, of which 37 per cent contains waste paper which is largely cellulose. That computes to 244 thousand tons per day of discarded waste paper that contains cellulose. The raw material to produce cellulosic ethanol is not only free, it has a negative cost—i.e., ethanol producers can get paid to take it away. In June 2006, a U.S. Senate hearing was told that the current cost of producing cellulosic ethanol is US $2.25 per US gallon (US $0.59/litre). This is primarily due to the current poor conversion efficiency. At that price it would cost about $120 to substitute a barrel of oil (42 gallons), taking into account the lower energy content of ethanol. However, the Department of Energy is optimistic and has requested a doubling of research funding. The same Senate hearing was told that the research target was to reduce the cost of production to US $1.07 per US gallon (US $0.28/litre) by 2012. "The production of cellulosic ethanol represents not only a step toward true energy diversity for the country, but a very cost-effective alternative to fossil fuels. It is advanced weaponry in the war on oil," said Vinod Khosla, managing partner of Khosla Ventures, who recently told a Reuters Global Biofuels Summit that he could see cellulosic fuel prices sinking to $1 per gallon within ten years.

University of Massachusetts at Amherst researchers have developed a streamlined technique which uses "catalytic fast pyrolysis" (heating to 400-600 °C followed by rapid cooling) and zeolite as a catalyst to produce cellulosic ethanol in about 60 seconds. They estimate improvements in the process should be able to generate ethanol at the equivalent of $1-$1.70/gal of gasoline. As of April 2008, the process has only been developed to work at laboratory scales.

Evaluating Corn-based vs. Grass-based Environmental Effects

In 2008, there was only a small amount of switchgrass dedicated for ethanol production. In order for it to be grown on a large-scale production it must compete with existing uses of agricultural land, mainly for the production of crop commodities. Of the United States' 2.26 billion acres (9.1 million km2) of unsubmerged land, 33 per cent are forestland, 26 per cent pastureland and grassland, and 20 per cent crop land. A study done by the U.S. Departments of Energy and Agriculture in 2005, determined whether there were enough available land resources to sustain production of over 1 billion dry tons of biomass annually to replace 30 per cent or more of the nation's current use of liquid transportation fuels. The study found that there could be 1.3 billion dry tons of biomass available for ethanol use, by making little changes in agricultural and forestry practices and meeting the demands for forestry products, food, and fiber. A recent study done by the University of Tennessee reported that as many as 100 million acres (400,000 km2, or 154,000 sq. miles), of cropland and pasture will need to be allocated to switchgrass production in order to offset petroleum use by 25 percent. Currently, corn is easier and less expensive to process into ethanol in comparison to cellulosic ethanol. The Department of Energy estimates that it costs about $2.20 per gallon to produce cellulosic ethanol, which is twice as much as ethanol from corn. Enzymes that destroy plant cell wall tissue cost 30 to 50 cents per gallon of ethanol compared to 3 cents per gallon for corn. The Department of Energy hopes to reduce this cost to $1.07 per gallon by 2012 to be effective. However, cellulosic biomass is cheaper to produce than corn, because it requires fewer inputs, such as energy, fertilizer, herbicide, and is accompanied by less soil erosion and improved soil fertility. Additionally, non-fermentable and unconverted solids left after making ethanol can be burned to provide the fuel needed to operate the conversion plant and produce electricity. Energy used to run corn-based ethanol plants is derived from coal and natural gas. The Institute for Local Self-Reliance estimates the cost of cellulosic ethanol from the first generation of commercial plants will be in the $1.90-$2.25 per gallon range, excluding incentives. This compares to the current cost of $1.20-$1.50 per gallon for ethanol from corn and the current retail price of over $4.00 per gallon for Regular Gasoline (which is subsidized and taxed). One of the major reasons for increasing the use of biofuels is to reduce greenhouse gas emissions. In comparison to gasoline, ethanol burns cleaner with a greater efficiency, thus putting less carbon dioxide and overall pollution in the air. Additionally, only low levels of smog are produced from combustion. According to the U.S. Department of Energy, ethanol from cellulose reduces green house gas emission by 90 percent, when compared to gasoline and in comparison to corn-based ethanol which decreases emissions by 10 to 20 percent. Carbon dioxide gas emissions are shown to be 85 per cent lower than those from gasoline. Cellulosic ethanol contributes little to the greenhouse effect and has a five times better net energy balance than corn-based. When used as a fuel, cellulosic ethanol releases less sulfur, carbon monoxide, particulates, and greenhouse gases. Cellulosic ethanol should earn producers carbon reduction credits, higher than those given to producers who grow corn for ethanol, which is about 3 to 20 cents per gallon. It takes 0.76 J of energy from fossil fuels to produce 1 J worth of ethanol from corn. This total includes the use of fossil fuels used for fertilizer, tractor fuel, ethanol plant operation, etc. Research has shown that 1 gallon of fossil fuel can produce over 5 gallons of ethanol from prairie grasses, according to Terry Riley, President of Policy at the Theodore Roosevelt Conservation Partnership. The United States Department of Energy concludes that corn-based ethanol provides 26 percent more energy than it requires for production, while cellulosic ethanol provides 80 percent more energy. Cellulosic ethanol yields 80 percent more energy than is required to grow and convert it. The process of turning corn into ethanol requires about 1,700 gallons of water for every 1 gallon of ethanol produced. Additionally, each gallon of ethanol leaves behind 12 gallons of waste that must be disposed. Grain ethanol uses only the edible portion of the plant. Expansion of corn acres for the production of ethanol poses threats to biodiversity. Corn lacks a strong root system, therefore, when produced, it causes soil erosion. This has a direct effect on soil particles, along with excess fertilizers and other chemicals, washing into local waterways, damaging water quality and harming aquatic life. Planting riparian areas can serve as a buffer to waterways, and decrease runoff. Cellulose is not used for food and can be grown in all parts of the world. The entire plant can be used when producing cellulosic ethanol. Switchgrass yields twice as much ethanol per acre than corn. Therefore, less land is needed for production and thus less habitat fragmentation. Biomass materials require fewer inputs, such as fertilizer, herbicides, and other chemicals that can pose risks to wildlife. Their extensive roots improve soil quality, reduce erosion, and increase nutrient capture. Herbaceous energy crops reduce soil erosion by greater than 90 per cent, when compared to conventional commodity crop production. This can translate into improved water quality for rural communities. Additionally, herbaceous energy crops add organic material to depleted soils and can increase soil carbon, which can have a direct effect on climate change, as soil carbon can absorb carbon dioxide in the air. As compared to commodity crop production, biomass reduces surface runoff and nitrogen transport. Switchgrass provides an environment for diverse wildlife habitation, mainly insects and ground birds. Conservation Resource Programme (CRP) land is composed of perennial grasses, which are used for cellulosic ethanol, and may be available for use. For years American farmers have practiced row cropping, with crops such as sorghum and corn. Because of this, much is known about the effect of these practices on wildlife. The most significant effect of increased corn ethanol would be the additional land that would have to be converted to agricultural use and the increased erosion and fertilizer use that goes along with agricultural production. Increasing our ethanol production through the use of corn could produce negative effects on wildlife, the magnitude of which will depend on the scale of production and whether the land used for this increased production was formerly idle, in a natural state, or planted with other row crops. Another consideration is whether to plant a switchgrass monoculture or use a variety of grasses and other vegetation. While a mixture of vegetation types likely would provide better wildlife habitat, the technology has not yet developed to allow the processing of a mixture of different grass species or vegetation types into bioethanol. Of course, cellulosic ethanol production is still in its infancy, and the possibility of using diverse vegetation stands instead of monocultures deserves further exploration as research continues.

Role of Feedstocks

Switchgrass (*Panicum virgatum*) is a native tallgrass prairie grass. Known for its hardiness and rapid growth, this perennial grows during the warm months to heights of 2-6 feet. Switchgrass can be grown in most parts of the United States, including swamplands, plains, streams, and along the shores and *interstate highways*. It is *self-seeding* (no tractor for sowing, only for mowing), resistant to many diseases and pests, and can produce high yields with low applications of fertilizer and other chemicals. It is also tolerant to poor soils, flooding, and drought; improves soil quality and prevents erosion due its type of root system. Switchgrass is an approved cover crop for land protected under the federal Conservation Reserve Programme (CRP). CRP is a government programme that pays producers a fee for not growing crops on land on which crops recently grew. This programme reduces soil erosion, enhances water quality, and increases wildlife habitat. CRP land serves as a habitat for upland game, such as pheasants and ducks, and a number of insects. Switchgrass for biofuel production has been considered for use on Conservation Reserve Programme (CRP) land, which could increase ecological sustainability and lower the cost of the CRP programme. However, CRP rules would have to be modified to allow this economic use of the CRP land. *Miscanthus x giganteus* is another viable feedstock for cellulosic ethanol production. This species of grass is native to Asia and is the sterile triploid hybrid of *miscanthus sinensis* and *miscanthus sacchariflorus*. It can grow up to 12 feet (3.7 m) tall with little water or fertilizer input. Miscanthus is similar to switchgrass with respect to cold and drought tolerance and water use efficiency. Miscanthus is commercially grown in the European Union as a combustible energy source. *Corn cobs and leaves,* wood chips and *paper pulp* are also feedstocks for cellulosic ethanol.

Cellulosic ethanol commercialization

Companies such as Iogen, Broin, and Abengoa are all building refineries that can process biomass and turn it into ethanol, while companies such as Diversa, Novozymes, and Dyadic are producing enzymes and Butalco is developing improved yeast strains, which could enable a cellulosic ethanol future. The shift from food crop feedstocks to waste residues and native grasses offers significant opportunities for a range of players, from farmers to biotechnology firms, and from project developers to investors (Table 1.2).

Table 1.2: Commercial Cellulosic Ethanol Plants in the U.S. (Operational or under Construction)

Company	Location	Feedstock
Abengoa Bioenergy	Hugoton, KS	Wheat straw
Blue Fire Ethanol	Irvine, CA	Multiple sources
Colusa Biomass Energy Corporation	Sacramento, CA	Waste rice straw
Fulcrum BioEnergy	Reno, NV	Municipal solid waste
Gulf Coast Energy	Mossy Head, FL	Wood waste
KL Energy Corp.	Upton, WY	Wood
Mascoma	Lansing, MI	Wood
POET LLC	Emmetsburg, IA	Corn cobs
Range Fuels	Treutlen County, GA	Wood waste
SunOpta	Little Falls, MN	Wood chips
US Envirofuels	Highlands County, FL	Sweet sorghum
Xethanol	Auburndale, FL	Citrus peels

Bioenergy from Vegetable Oil Used as Fuel

Vegetable oil is an alternative fuel for diesel engines and for heating oil burners. For engines designed to burn <UIP>2 diesel fuel, the viscosity of vegetable oil must be lowered to allow for proper atomization of fuel, otherwise incomplete combustion and carbon build up will ultimately damage the engine. Many enthusiasts refer to vegetable oil used as fuel as waste vegetable oil (WVO) if it is oil that was discarded from a restaurant or straight vegetable oil (SVO) or pure plant oil (PPO) to distinguish it from biodiesel.

Time Frame

The first known use of vegetable oil as fuel in a diesel engine was a demonstration of an engine built by the Otto company and designed to burn mineral oil, which was run on pure peanut oil at the 1900 World's Fair. Late in his career, Rudolf Diesel investigated using vegetable oil to fuel engines of his design, and in a 1912 presentation to the British Institute of Mechanical Engineers, he cited a number of efforts in this area and remarked, "The fact that fat oils from vegetable sources can be used may seem insignificant today, but such oils may perhaps become in course of time of the same importance as some natural mineral oils and the tar products are now." Periodic petroleum shortages spurred research into vegetable oil as a diesel substitute during the 1930s and 1940s, and again in the 1970s and early 1980s when straight vegetable oil enjoyed its highest level of scientific interest. The 1970s also saw the formation of the first commercial enterprise to allow consumers to run straight vegetable oil in their automobiles, Elsbett of Germany. In the 1990s Bougainville conflict, islanders cut off from oil supplies due to a blockade used coconut oil to fuel their vehicles. Academic research into straight vegetable oil fell off sharply in the 80s with falling petroleum prices and greater interest in biodiesel as an option that did not require extensive vehicle modifications.

Understanding Application and Usability

While engineers and enthusiasts have been experimenting with using vegetable oils as fuel for a diesel engine since at least 1900, it is only recently that the necessary fuel properties and engine parameters for reliable operation have become apparent. A number of peer reviewed studies exists that show reliable long term use of vegetable oil; the German Deutz F3l912W. and a high speed common rail engine fitted to a Mercedes-Benz 220 C Class. Most diesel car engines are suitable for the use of SVO, also commonly called Pure Plant Oil (PPO), with suitable modifications. Principally, the viscosity and surface tension of the SVO/PPO must be reduced by preheating it, typically by using waste heat from the engine or electricity, otherwise poor atomization, incomplete combustion and carbonization may result. One common solution is to add a heat exchanger, and an additional fuel tank for "normal" diesel fuel (petrodiesel or biodiesel) and a three way valve to switch between this additional tank and the main tank of SVO/PPO. (This aftermarket modification typically costs about $1200 USD.) The engine is started on diesel, switched over to vegetable oil as soon as it is warmed up and switched back to diesel shortly before being switched off to ensure that no vegetable oil remains in the engine or fuel lines when it is started from cold again. In colder climates it is often necessary to heat the vegetable oil fuel lines and tank as it can become very viscous and even solidify. Single tank conversions have been developed, largely in Germany, which have been used throughout Europe. These conversions are designed to provide reliable operation with rapeseed oil that meets the German rapeseed oil fuel standard DIN 51605. Modifications to the engines cold start regime assist combustion on start up and during the engine warm up phase. Suitably modified indirect injection (IDI) engines have proven to be operable with 100 per cent PPO down to temperatures of -10°C. Direct injection (DI) engines generally have to be preheated with a block heater or diesel fired heater. The exception is the VW Tdi (Turbocharged Direct Injection) engine for which a number of German companies offer single tank conversions. For long term durability it has been found necessary to increase the oil change frequency and to pay increased attention to engine maintenance. With unmodified engines the unfavourable effects may be reduced by blending, or "cutting", the SVO/PPO with diesel fuel; however, opinions vary as to the efficacy of this. Some WVO mechanics have found higher rates of wear and failure in fuel pumps and piston rings. This can generally be attributed to the use of oils with properties or contaminants that make them unsuitable for use in this type of application, poorly maintained engines, unsuitable engine modifications or operating regimes. Many cars powered by indirect injection engines supplied by in-line injection pumps, or mechanical Bosch injection pumps are capable of running on pure SVO/PPO in all but winter temperatures. Indirect injection Mercedes-Benz vehicles with in-line injection pumps and cars featuring the PSA XUD engine tend to perform reasonably, especially as the latter is normally equipped with a coolant heated fuel filter. Engine reliability would depend on the condition of the engine. Attention to maintenance of the engine, particularly of the fuel injectors, cooling system and glow plugs will help to provide longevity. Ideally the engine would be converted.

Properties and Material Compatibility

The main form of SVO/PPO used in the UK is rapeseed oil (also known as canola oil, primarily in the United States and Canada) which has a freezing point of -10°C. However the use of sunflower oil, which freezes at -17°C, is currently being investigated as a means of improving cold weather starting. Unfortunately oils with lower gelling points tend to be less saturated (leading to a higher iodine number) and polymerize more easily in the presence of atmospheric oxygen.

Free fatty acids in WVO can have a detrimental effect on metals. Copper and its alloys, such as brass, are affected. Zinc and zinc-plating (galvanization) are stripped by FFA's and tin, lead, iron, and steel are affected too. Stainless steel and aluminum are generally unaffected.

Some Pacific island nations are using coconut oil as fuel to reduce their expenses and their dependence on imported fuels while helping stabilize the coconut oil market. Coconut oil is only usable where temperatures do not drop below 17 degrees Celsius (62 degrees Fahrenheit), unless two-tank SVO/PPO kits or other tank-heating accessories, etc. are used. Fortunately, the same techniques developed to use, for example, canola and other oils in cold climates can be implemented to make coconut oil usable in temperatures lower than 17 degrees Celsius.

Home heating, Combined Heat and Power

With often minimal modification, most residential furnaces and boilers that are designed to burn No. 2 heating oil can be made to burn either biodiesel or filtered, preheated waste vegetable oil. These are generally not as clean-burning as petroleum fuel oil, but if processed at home, by the consumer, can result in considerable savings. Many restaurants will give away their used cooking oil either free or at minimal cost, and processing to biodiesel is fairly simple and inexpensive. Burning filtered WVO directly is somewhat more problematic, since it is much more viscous, but it can be accomplished with suitable preheating. WVO can thus be an economical heating option for those with the necessary mechanical and experimental aptitude.

A number of companies offer compressed ignition engine generators optimized to run on plant oils where the waste engine heat is recovered for heating.

Availability of Waste Vegetable Oil

As of 2000, the United States was producing in excess of 11 billion liters (2.9 billion U.S. gallons) of waste vegetable oil annually, mainly from industrial deep fryers in potato processing plants, snack food factories and fast food restaurants. If all those 11 billion liters could be collected and used to replace the energetically equivalent amount of petroleum (an ideal case), almost 1 per cent of US oil consumption could be offset. However, use of waste vegetable oil as a fuel competes with some already established uses.

Availability of Pure vegetable oil

Pure plant oil (PPO) (or Straight Vegetable Oil (SVO), in contrast to waste vegetable oil, is not a byproduct of other industries, and thus its prospects for use as fuel are not limited by the capacities of other industries. Production of vegetable oils for use as fuels is theoretically limited only by the agricultural capacity of a given economy.

Legal Implications in Terms of Taxation of Fuel

Taxation on SVO/PPO as a road fuel varies from country to country, and it is possible the revenue departments in many countries are even unaware of its use, or feel it insufficiently significant to legislate. Germany used to have 0 per cent taxation, resulting in it being a leader in most developments of the fuel use. However SVO/PPO as a road fuel began to be taxed at 0,09 €/liter from 1 January 2008 in Germany, with incremental rises up to 0,45 €/liter by 2012. However, in Australia it has become illegal to produce any fuel if it is to be sold unless a license to do so is granted by the Federal Government. This is a chargeable offence with a fine of up to 20,000 dollars but this bracket may alter circumstantially. Also a jail term may result if offenders are aware of the legality of selling the fuel. In the USA, the legality of burning SVO in the United States of America is debated by many. While the EPA may claim it is illegal, the reality is that most SVO conversions do not alter the emissions system or core engines ability to burn diesel the same way it did when it left the factory. The debate then comes to whether or not the fuel itself is illegal. At that point it comes down to the state level and there are a handful of States which have passed laws that specifically exempt taxation on SVO and other plant based fuels. One such example is Arkansas Act 690 of House Bill 2076. It specifically exempts "*Used cooking oil recycled and gathered from restaurants and commercial food processors*". There seems to be no clear federal taxation system in the USA. Production of biodiesel in some US regions may require motor fuel taxes to be paid. In Japan, The Japanese Government has also exempted the use of SVO as a fuel from road tax. In Canada, The Government of Canada exempted biodiesel from the federal excise tax on diesel in the March 2003 budget. In Ireland, In Ireland a pilot scheme is currently running (as of April 2006) whereby eight suppliers have been approved to sell SVO/PPO for use as a fuel without the payment of excise duty (Value Added Tax at 21 per cent still applies, SVO from any other source still attracts excise duty at 36.8058 Euro cents per litre plus 21% VAT). In France, Despite its use being common in France, it would appear there has been no legislation to cover this. In the UK, it is legal once duty on the fuel is paid. In the UK, drivers using SVO/PPO have been prosecuted for failure to pay duty to Her Majesty's Revenue and Customs. The rate of taxation on SVO was originally set at a reduced rate of 27.1p per litre, but in late 2005, HMRC started to enforce the full diesel excise rate of 47.1p/litre. Following a review in late 2006, HM Revenue and Customs has announced changes regarding the administration and collection of excise duty of biofuels and other fuel substitutes (Veg. Oil). The changes came into effect on June 30, 2007. There is no longer a requirement to register to pay duty on vegetable oil used as road fuel for those who "produce" or use less than 2,500 litres per year. For those producing over this threshold the biodiesel rate now applies. HMRC argued that SVOs/PPOs on the market from small producers did not meet the official definition of "biodiesel" in Section 2AA of The

Hydrocarbon Oil Duties Act 1979 (HODA), and consequently was merely a "fuel substitute" chargeable at the normal diesel rate. Such a policy seemed to contradict the UK Government's commitments to the Kyoto Protocol and to many EU directives and had many consequences, including an attempt to make the increase retroactive, with one organization being presented with a £16,000 back tax bill. This change in the rate of excise duty has effectively removed any commercial incentive to use SVO/ PPO, regardless of its desirability on environmental grounds; unless waste vegetable oil can be obtained free of charge, the combined price of SVO/PPO and taxation for its use usually exceeds the price of mineral diesel. HMRC's interpretation is being widely challenged by the SVO/PPO industry and the UK pure Plant Oil Association (UKPPOA)] has been formed to represent the interests of people using vegetable oil as fuel and to lobby parliament.

TIMELINE OF BIOENERGY

2030

- Target year set by the Department of Energy to displace 30 percent of demand (2004 levels) in the United States with biofuels, primarily ethanol.

2025

- Target year set by the "25'x 25' coalition" for renewable energy to reach 25% of total energy use in the United States.
- Target year set by the government of the US state of Iowa to achieve "energy independence"

2022

- Target year for annual production of 36 billion gallons of in the United States under the Renewable Fuel Standard, as called for in the Energy Independence and Security Act of 2007 (with a limit for corn-based ethanol of 15 billion gallons per year).

2020

- European Union target calls for 10% of fuel use to be met by biofuels.
- United Kingdom target calls "for one-fifth of total energy supply to come from renewable sources".
- Target year announced by the aviation industry for annual consumption of 3 billion gallons of biofuels.
- Target year for China to have 13 million hectares devoted to biofuel plantations.

2018

- Year by which "" potentially may become commercially viable, according to a 2008 UN FAO/OECD study.

2017

- Target year set in September 2008 for petrol and diesel in to be blended with 20% biofuel.
- Target year under proposal by U.S. President Bush (in his 2007 State of the Union Address) for achieving utilization of 35 billion gallons of alternative fuels within the United States.

2015

- OECD estimates that in "the , and the government support for the supply and use of biofuels is expected to rise to around USD 25 billion per year."
- 30 million acres of U.S. farmland projected to be needed for corn production to meet legislated ethanol production target.
- Target year for France to achieve 10 percent blending of biofuels into motor fuels.
- Target year set by the environmental organization WWF for palm oil to be derived from entirely sustainable sources.

2013

- Year by which "" potentially may become commercially

viable, according to a 2008 statement by U.S. Agriculture Secretary Schafer.

2012

- Target year for making cellulosic ethanol cost competitive as an energy source under U.S. Advanced Energy Initiative.
- Target year for France to remove tax breaks for biofuels, according to a 2008 proposal.

2011

- 1 January 2011: Effective date of the carbon intensity reference values for the (LCFS) determined in 2009 by the California Air Resources Board (CARB).
- Target year for U.S. consumption of ethanol to reach 15.2 billion gallons under the Energy Independence and Security Act of 2007.
- Year planned for initiation of operations of Iogen Corp.'s commercial-scale cellulosic ethanol plant in Saskatchewan, Canada.
- Target year for commercial production of ethanol from municipal waste, according to a statement in by the company Ineos.

2010

- Target year set by the European Commission for increasing usage of within the European Union.
- Target year for The Netherlands to achieve 5.75% biofuel content for gasoline and diesel.
- Target year under the United Kingdom's (RTFO) for biofuel to account for 5% of total petrol and diesel sales.
- Target year for France to achieve 7 percent blending of biofuels into motor fuels.
- Target year set by a 1992 U.S. energy law for "30 percent of the fuel used to run U.S. cars and trucks...[to] come from ethanol, natural gas, hydrogen, electricity or other replacement fuels."

- U.S. production of corn-based ethanol "estimated to double, to 10 billion gallons (38 billion liters), by 2010."
- Year of expiration of U.S. Volumetric Ethanol Excise Tax Credit (VEETC), which provides a 51-cent-per-gallon to producers.
- October 2010 meeting of the Convention on Biological Diversity (CBD) in Nagoya, Japan, to address the impacts of biofuels on.

2009

- Target year for United States consumption of to reach 11.1 billion gallons under the (US) national Renewable Fuels Standard.
- 30 November – 11 December 2009: Copenhagen, Denmark COP15 Meeting of the United Nations Framework Convention on Climate Change; anticipated to result in a new international agreement on climate change.

Past developments
2009

- **14 January 2009: The Guardian ()** reports that Britain's Advertising Standards Authority banned a print advertisement by the Renewable Fuels Association (RFA) due to a "misleading claim" that " offer a alternative to oil".
- **26 January 2009:** International Renewable Energy Agency (IRENA) founded in Bonn, .
- **23 April 2009:** The California Air Resources Board (CARB) approves specific rules and carbon intensity reference values for the low-carbon fuel standard (LCFS) to go into effect 1 January 2011.
- **16 June 2009:** The U.S. Global Change Research Program, a consortium of thirteen U.S. federal scientific agencies, releases the report, "Global Climate Change Impacts in the United States," raising awareness of ongoing and future climate impacts.
- **26 June 2009:** Passage by the United States House of

Representatives of the , which would cap carbon dioxide emissions.

- **28 July 2009:** Reports indicate that the there will be an extension for an additional year on the moratorium on purchases of soya grown in deforested areas of the Brazilian Amazon.

2008

- **7 February 2008:** Two studies published in magazine ("Land Clearing and the Biofuel Carbon Debt" and "Use of U.S. Croplands for Biofuels Increases Greenhouse Gases through Emissions from Land Use Change"), indicate that land-use change associated with production of biofuels leads to increased net carbon emissions, thus challenging a major point advanced by biofuels proponents, that biofuels are "climate friendly".
- **3-7 March 2008:** Washington International Renewable Energy Conference (WIREC 2008) held in Washington, D.C., USA; various pledges were made by participating countries, including The United States, which pledged that 7.5 percent of electric energy use "will come from renewable resources by 2013."
- **1 April 2008:** B2 (2% biodiesel blend) to be available throughout Thailand.
- **15 April 2008:** Under the (RTFO), all vehicle fuel sold to consumers in the United Kingdom must contain 2.5% biofuels.
- **May 2008:** Massachusetts-based Verenium Corp. opens the first demonstration-scale plant in the United States, in Jennings, La.
- **3-5 June 2008:** U.N. food summit, the "," held in Rome, Italy.
- **3 July 2008:** *The Guardian* (UK) reports that an unpublished World Bank report concluded that biofuels "have forced global up by 75% – far more than previously estimated".
- **6 July 2008:** World Bank President Robert Zoellick reportedly calls for reform of biofuel policies in rich

countries, including a reduction in mandates, subsidies and tariffs.

- **7 July 2008:** The UK Renewable Fuels Agency (RFA) issues the "Gallagher Review", which concluded that the "introduction of biofuels should be slowed until effective controls are in place to prevent land use change and higher food prices."

- **28 July 2008:** Report of *Task Force on the Global Food Crisis* convened by CSIS calls for revision of the U.S. approach to biofuels "to reduce dependence on corn" and to adopt "new sustainability criteria to assess the life-cycle costs and requirements for alternative biofuels."

- **13 August 2008:** The Roundtable on Sustainable Biofuels (RSB) releases "Version Zero" draft of the "Global principles and criteria for sustainable biofuels production", for use during a six-month period of stakeholder comment in preparation for release of the first official standards in early 2009.

- **9 September 2008:** Inter-American Development Bank () issue a "Biofuels Sustainability Scorecard".

- **25 September 2008:** Sustainable Aviation Fuel Users Group created by manufacturer Boeing and nine airline companies.

- **6 October 2008:** National Biofuels Action Plan (PDF file) issued by Department of Agriculture and Department of Energy in the United States.

- **4 November 2008:** Ethanol and clean energy supporter Barack Obama elected president of the .

- **17-21 November 2008:** held by Brazilian government in São Paulo, Brazil.

- **15 December 2008:** U.S. President-Elect Barack Obama's selection of physicist Dr. Steven Chu for Energy Secretary indicates likely strengthening of U.S. efforts to pursue cellulosic ethanol.

- **31 December 2008:** Target date for expiration of U.S. 54-cent-per-gallon import tariff on ethanol.

- **2008:** Target year for US consumption of ethanol to reach 9 billion gallons under the (US) .

2007

- **1 January 2007:** Legislation in comes into effect mandating 2% biofuel content for gasoline and diesel.
- **23 January 2007:** President Bush, in his State of the Union Address, calls for achieving utilization of 35 billion gallons of alternative fuels within the United States in 10 years (by 2017).
- **23 January 2007:** Governor Schwarzenegger signs establishing the Low Carbon Fuel Standard, with implications for the greenhouse gas balance of biofuels to be used in California.
- **19 December 2007:** President signs into law the , mandating a sixfold increase ethanol usage in the United States by 2022.

2006

- 31 January 2006: U.S. President Bush, in his , highlights a number of alternative energy goals as part of the , including:
  - o Alternative fuels: accelerating research for "cutting-edge methods of producing 'cellulosic ethanol' with the goal of making the use of such ethanol practical and competitive within 6 years."
  - o "The Initiative": $150 million proposed for the 2007 federal budget for promotion of new technologies for producing from cellulosic (plant fiber) biomass ("bio-based transportation fuels from products, such as wood chips, stalks, or switch grass")
- Adoption of National Biofuel Policy in Malaysia.
- OECD estimates that in "the US, Canada and the European Union government support for the supply and use of biofuels" is "about USD 11 billion".

2005

- Consumption of ethanol in the United States in 2005 reaches four billion gallons.
- US Energy Policy Act of 2005 mandates use of 7.5 billion gallons of renewable fuels annually by 2012.

2004

- April 2004: Roundtable on Sustainable Palm Oil () created to promote use of sustainable palm oil.
- U.S. institutes Volumetric Ethanol Excise Tax Credit (VEETC) to support blending and sale of biofuels.

2003

- France introduces quotas and tax breaks on biofuels.

2002

- 11 percent of U.S. corn crop used for fuel production.

1997

- 11 December 1997: Adoption in Kyoto, Japan, of the Kyoto Protocol.

1996

- Forest Stewardship Council issues "Principles and Criteria for Forest Stewardship," a potential model for criteria to determine sustainable produced bioenergy.

1992

- U.S. energy law sets goal for "30 percent of the used to run U.S. cars and trucks...[to] come from ethanol, natural gas, hydrogen, electricity or other replacement fuels" by 2010.

1985

- United States begins Conservation Reserve Program (CRP).

1981

- Renewable Fuels Association formed in the .

1978

- IEA Bioenergy established by the (IEA).
- Tax incentives for ethanol proposed by US President Jimmy Carter, primarily for national security reasons.

1973

- Nebraska Farm Crops Utilization Committee begins tests of alcohol blends. The committee later becomes the Nebraksa Commission.

1947

- Alcohol plants built for the war effort are sold for scrap, despite interest in production for fuel and chemicals.

1942

- Synthetic rubber production from alcohol promoted by farm lobby. Oil industry opposes this, but is exposed by Sen. Harry Truman's war investigating committee. By 1944, 3/4 of all US rubber production is from alcohol.

1940

- Ethyl Gasoline Corp. loses anti-trust lawsuit brought by Justice Dept. for anti-competitive behavior.

1937

- Agrol plant opens in Atchison, KS as part of the Chemurgy experiment. About 2000 service stations across the use the 10% alcohol blend in gasoline. Plant is bankrupt by 1939.

1935

- First Farm Chemurgy conference in Dearborn, MI, sponsored by Henry Ford. Chemurgy seeks new uses for farm products, such as ethanol as an outlet for surplus corn, through scientific research.

1932

- Leo Christensen and others in Iowa State University's chemistry department advocate use of alcohol blends as anti-knock fuels and for Depression-era farm relief.

1921

- General Motors researchers discover anti-knock effect of tetra-ethyl lead. Leaded gasoline, as it comes to be known, displaces most US ethanol anti-knock blends. GM and Standard Oil Co. of NJ form the Ethyl Gasoline Corp. as a .50-50 joint venture.

1918

- Scientific American reports: "It is now definitely established that alcohol can be blended with gasoline to produce a suitable motor fuel..." Two years later, the magazine reports "a universal assumption that [ethyl] alcohol in some form will be a constituent of the motor fuel of the future."

1906

- Civil War tax repealed; President Teddy Roosevelt signs a bill allowing tax-free use of industrial alcohol on June 8.

1861

- Alcohol taxed; $2.08 per gallon tax imposed on beverage and industrial alcohol in stages between 1862 and 1864 as part of the Internal Revenue Act to pay for the Civil War. The tax was meant to apply to beverage alcohol, but without any specific exemption, it was also applied to fuel and industrial uses for alcohol. "The imposition of the internal-revenue tax on distilled spirits ... increased the cost of this 'burning fluid' beyond the possibility of using it in competition with kerosene..," said Rufus F. Herrick, an engineer with the Edison Electric Testing Laboratory who wrote one of the first books on the use of alcohol fuel.

1860

- German inventor Nicholas August Otto uses ethyl alcohol as a fuel irt an early engine because it was widely available for spirit lamps throughout. He devised a carburetor which, like Morey's, heated the alcohol to help it vaporize as the engine was being started. A patent application was turned down because the carburetor was considered to be well established technology.

1834

- The first U.S. patent for alcohol as a lamp fuel was awarded in 1834 to S. Casey, of Lebanon, Maine, but it is clear that alcohol was routinely used a fuel beforehand.

1826

- Samuel Morey uses readily available alcohol in the first American prototype internal combustion engine at the surprisingly early date of 1826. Morey's work was lost in the enthusaism for the steam engine and a lack of funding. No other internal combustion engine would be developed until Nicholas Otto began his experiments 35 years later.

2

Role of Technology and Industrial Biotechnology in Bioenergy and Biofuel Production and Commercialization

PROCESS OF BIOCONVERSION OF BIOMASS TO MIXED ALCOHOL FUELS

The bioconversion of biomass to mixed alcohol fuels can be accomplished using the MixAlco process. Through bioconversion of biomass to a mixed alcohol fuel, more energy from the biomass will end up as liquid fuels than in converting biomass to ethanol by yeast fermentation.

The process involves a biological/chemical method for converting any biodegradable material (e.g., urban wastes, such as municipal solid waste, biodegradable waste, and sewage sludge, agricultural residues such as corn stover, sugarcane bagasse, cotton gin trash, manure) into useful chemicals, such as carboxylic acids (e.g., acetic, propionic, butyric acid), ketones (e.g., acetone, methyl ethyl ketone, diethyl ketone) and biofuels, such as a mixture of primary alcohols (e.g., ethanol, propanol, butanol) and/or a mixture of secondary alcohols (e.g., isopropanol, 2-butanol, 3-pentanol). Because of the many products that can be economically produced, this process is a true biorefinery.

The process uses a mixed culture of naturally occurring microorganisms found in natural habitats such as the rumen of cattle, termite guts, and marine and terrestrial swamps to anaerobically digest biomass into a mixture of carboxylic acids produced during the acidogenic and acetogenic stages of anaerobic digestion, however with the inhibition of the methanogenic final stage. The more popular methods for production of ethanol and cellulosic ethanol use enzymes that must be isolated first to be added to the biomass and thus convert the starch or cellulose into simple sugars, followed then by yeast fermentation into ethanol. This process does not need the addition of such enzymes as these microorganisms make their own.

As the microoganisms anaerobically digest the biomass and convert it into a mixture of carboxylic acids, the pH must be controlled. This is done by the addition of a buffering agent (e.g., ammonium bicarbonate, calcium carbonate), thus yielding a mixture of carboxylate salts. Methanogenesis, which, as mentioned, is the natural final stage of anaerobic digestion, is inhibited by the presence of the ammonium ions or by the addition of an inhibitor (e.g., iodoform). The resulting fermentation broth contains the produced carboxylate salts that must be dewatered. This is achieved efficiently by vapour-compression evaporation. Further chemical refining of the dewatered fermentation broth may then take place depending on the final chemical or biofuel product desired.

The condensed distilled water from the vapour-compression evaporation system is recycled back to the fermentation. On the other hand, if raw sewage or other waste water with high BOD in need of treatment is used as the water for the fermentation, the condensed distilled water from the evaporation can be recycled back to the city or to the original source of the high-BOD waste water. Thus, this process can also serve as a water treatment facility, while producing valuable chemicals or biofuels.

Because the system uses a mixed culture of microorganisms, besides not needing any enzyme addition, the fermentation requires no sterility or aseptic conditions, making this front step in the process more economical than in more popular methods for the production of cellulosic ethanol. These savings in the front end of the process, where volumes are large, allows flexibility for further chemical transformations after dewatering, where volumes are small.

Carboxylic Acids

Carboxylic acids can be regenerated from the carboxylate salts using a process known as "acid springing". This process makes use of a high-molecular-weight tertiary amine (e.g., trioctylamine), which is switched with the cation (e.g., ammonium or calcium). The resulting amine carboxylate can then be thermally decomposed into the amine itself, which is recycled, and the corresponding carboxylic acid. In this way, theoretically, no chemicals are consumed or wastes produced during this step.

Ketones

There are two methods for making ketones. The first one consists on thermally converting calcium carboxylate salts into the corresponding ketones. This was a common method for making acetone from calcium acetate during World War I. The other method for making ketones consists on converting the vapourized carboxylic acids on a catalytic bed of zirconium oxide.

Alcohols

Primary alcohols .

The undigested residue from the fermentation may be used in gasification to make hydrogen (H_2). This H_2 can then be used to hydrogenolyze the esters over a catalyst (e.g., copper chromite), which are produced by esterifying either the ammonium carboxylate salts (e.g., ammonium acetate, propionate, butyrate) or the carboxylic acids (e.g., acetic, propionic, butyric acid) with a high-molecular-weight alcohol (e.g., hexanol, heptanol). From the hydrogenolysis, the final products are the high-molecular-weight alcohol, which is recycled back to the esterification, and the corresponding primary alcohols (e.g., ethanol, propanol, butanol).

Secondary alcohols

The secondary alcohols (e.g., isopropanol, 2-butanol, 3-pentanol) are obtained by hydrogenating over a catalyst (e.g., Raney nickel) the corresponding ketones (e.g., acetone, methyl ethyl ketone, diethyl ketone).

Acetic Acid versus Ethanol

Cellulosic-ethanol -manufacturing plants are bound to be net exporters of electricity because a large portion of the lignocellulosic biomass, namely lignin, remains undigested and it must be burned, thus producing electricity for the plant and excess electricity for the grid. As the market grows and this technology becomes more widespread, coupling the liquid fuel and the electricity markets will become more and more difficult.

Acetic acid, unlike ethanol, is biologically produced from simple sugars without the production of carbon dioxide:

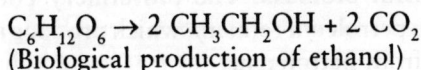

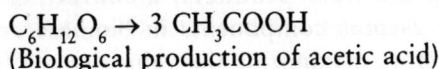

$$C_6H_{12}O_6 \rightarrow 2\ CH_3CH_2OH + 2\ CO_2$$
(Biological production of ethanol)

$$C_6H_{12}O_6 \rightarrow 3\ CH_3COOH$$
(Biological production of acetic acid)

Because of this, on a mass basis, the yields will be higher than in ethanol fermentation. If then, the undigested residue (mostly lignin) is used to produce hydrogen by gasification, it is ensured that more energy from the biomass will end up as liquid fuels rather than excess heat/electricity.

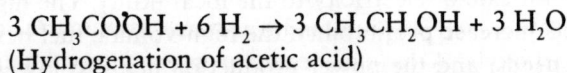

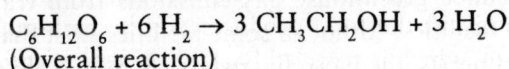

$$3\ CH_3COOH + 6\ H_2 \rightarrow 3\ CH_3CH_2OH + 3\ H_2O$$
(Hydrogenation of acetic acid)

$$C_6H_{12}O_6 + 6\ H_2 \rightarrow 3\ CH_3CH_2OH + 3\ H_2O$$
(Overall reaction)

A more comprehensive description of the economics of each of the fuels is given on the pages alcohol fuel and ethanol fuel, more information about the economics of various systems can be found on the central page biofuel.

Stage of Development

The system has been in development since 1991, moving from the laboratory scale (10 g/day) to the pilot scale (200 lb/day) in 2001. A small demonstration-scale plant (5 ton/day) is under construction as is expected to be operational mid 2008 and a 100 ton/day demonstration plant is expected in 2009.

UNDERSTANDING BIOREFINERY

Biorefinery is the co-production of a spectrum of bio-based products (food, feed, materials, chemicals) and energy (fuels, power, heat) from biomass [definition IEA Bioenergy Task 42].

A biorefinery is a facility that integrates biomass conversion processes and equipment to produce fuels, power, and value-added chemicals from biomass. The biorefinery concept is analogous to today's petroleum refinery, which produce multiple fuels and products from petroleum.

By producing multiple products, a biorefinery takes advantage of the various components in biomass and their intermediates therefore maximizing the value derived from the biomass feedstock. A biorefinery could, for example, produce one or several low-volume, but high-value, chemical or nutraceutical products and a low-value, but high-volume liquid transportation fuel such as biodiesel or bioethanol. At the same time generating electricity and process heat, through combined heat and power (CHP) technology, for its own use and perhaps enough for sale of electricity to the local utility. The high-value products increase profitability, the high-volume fuel helps meet energy needs, and the power production helps to lower energy costs and reduce greenhouse gas emissions from traditional power plant facilities. Although some facilities exist that can be called bio-refineries, the bio-refinery has yet to be fully realized. Future biorefineries may plan a major role in producing chemicals and materials that traditionally produced from petroleum.

Several potential biorefinery examples have been proposed, starting from feedstocks such as tobacco, flax straw and the residues from the production of bioethanol.

ROLE OF INDUSTRIAL BIOTECHNOLOGY

Industrial biotechnology (known mainly in Europe as white biotechnology) is the application of biotechnology for industrial purposes, including manufacturing, alternative energy (or "bioenergy"), and biomaterials. It includes the practice of using cells or components of cells like enzymes to generate industrially useful products. *The Economist* speculated (as cited in the *Economist* article listed in the "References" section) industrial biotechnology might significantly impact the chemical industry. *The Economist* also suggested it can enable economies to become less dependent on fossil fuels.

The industrial biotechnology community generally accepts an informal divide between industrial and pharmaceutical biotechnology. An example would be that of companies growing fungus to produce antibiotics, e.g. penicillin from the penicillium fungi. One view holds that this is industrial production; the other viewpoint is that such would not strictly lie within the domain of pure industrial production, given its inclusion within medical biotechnology.

This may be better understood in calling to mind the classification by the U.S. biotechnology lobby group, Biotechnology Industry Organization (BIO) of three "waves" of biotechnology. The first wave, Green Bio-technology, refers to agricultural biotechnology. The second wave, Red Biotechnology, refers to pharmaceutical and medical biotechnology. The third wave, White Biotechnology, refers to industrial biotechnology. In actuality, each of the waves may overlap. Industrial biotechnology, particularly the development of large-scale bioenergy refineries, will likely involve dedicated genetically modified crops as well as the large-scale bioprocessing and fermentation as is used in some pharmaceutical production.

Genencor International and Novozymes are examples of companies that specialize in industrial biotechnology, with particular focuses on specially designed enzymes to catalyze industrially relevant chemical reactions.

PELLET MILL

A pellet mill is a type of mill used to create cylindrical pellets from a mixture of dry powdered feedstock, such as flour, sawdust, or grass, and a wet ingredient, such as molasses or steam. The pellets are made by compacting the mash or meal into many small holes in a die. The die is usually round and the pellets are pushed from the inside out. Pellet mills are used in the production of animal feeds, and of wood and grass fuel pellets for use in a pellet stove.

Pellet mills are unlike grinding mills, in that they combine small materials into a larger, homogeneous mass, rather than break large materials into smaller pieces. In this way, pellet mills are similar to extruders. Feedstocks for pellet mills can sometimes break down and then re-form, or polymerize, under the extreme heat and pressure of the pellet mill. Pressures in the die can reach up to 25,000 psi.

PELLET STOVE

A pellet stove is a stove that burns compressed wood or biomass pellets to create a source of heat for residential and sometimes industrial spaces. By slowly feeding fuel from a storage container (hopper) into a burn-pot area, they create a constant flame that requires little to no physical adjustments.

Background and History

Scrap wood and ship-lap burners have been around for decades. Barrel stoves, braziers, and oil drum fires in Depression-era Hoovervilles support this. Professionally built wood ovens with sawdust hoppers were used in the early part of the century. All of these units used scrap and loose wood, or sawdust. In 1930, the Presto-Log was invented reusing scrap sawdust from the Potlatch pine mill in Lewiston, Idaho as domestic heat. From this came the miniaturized pellet stove, which emerged from Washington State in the 1980s.

The pellet stove has changed in appearance over the years from a simple, boxy workhorse design, to a decorative heating appliance. Pellet stoves can be either free-standing units or fireplace inserts vented into an existing chimney. Most pellet stoves are constructed using large, conductive, cast-iron pieces, with stainless steel to encase circuitry and exhaust areas.

Pellet furnaces and pellet boilers are also available in addition to the decorative stove. These units can be retrofitted into existing home heating systems with only minor changes to existing ductwork and or plumbing.

The heating industry has considerably shifted toward biomass stoves and heating devices based on efficient combustible and renewable resources. This was a trend that began during the 1973 oil crisis causing the creation of the first pellet stoves. Even so, pellet stoves have become a viable, economical, and popular option for home heating systems only in the last ten years. Many pellet stove manufacturers recommend the use of a corn and pellet mixture, though some are UL listed for fuels other than pellets, such as wheat, corn, sunflower seeds, and cherry pits.

Benefits

Pellet stoves are relatively versatile appliances. Most pellet stoves are self-igniting and cycle themselves on and off controlled by a thermostat. Stoves with automatic ignition can be equipped with remote controls. Recent innovations have created computer systems within pellet stoves which run diagnostic tests if an imminent problem arises.

A properly cleaned and maintained pellet stove should not create creosote, the sticky, flammable substance that causes chimney fires. Pellets burn very cleanly and create only a layer of fine fly-ash as a byproduct of combustion. The grade of pellet fuel affects the performance and ash output. Premium grade pellets produce les : than one percent ash content, while standard or low grade pellets produce a range from two to four percent ash. Pellet stove users should be aware of the extra maintenance required with a lower grade pellet, and that inconsistent wood quality can cause serious effects to the electronic machinery over a short period of time.

A pellet stove is normally associated with pelletized wood.

However, many pellet stoves will also burn fuels such as grain, corn, seeds, or woodchips. In some pellet stoves, these fuels may need to be mixed with wood pellets. Pelletized trash (containing mostly waste paper) is also a fuel for pellet stoves.

Pellet stoves require certified double walled venting, normally three or four inches in diameter with a stainless steel interior and galvanized exterior. Because pellet stoves have a forced exhaust system, they do not usually require a vertical rise to vent, although a three to five foot vertical run is recommended to prevent leakage in the case of a power outage. Like a modern gas appliance, pellet stoves can be vented horizontally through an outside wall and terminated below the roof line, making it an excellent choice for structures without an existing chimney. If an existing chimney is available, manufacturers urge use of a correctly sized stainless steel liner the length of the chimney for proper drafting.

In many states in the U.S., pellet stoves and fuel are exempt from sales tax.

Principles of Operation

A pellet stove normally consists of these components, whether basic or complex:

- A hopper.
- An auger system.
- Two blower fans (combustion and convection).
- A firebox (with refractory panels, burn-pot, and ash collection system).
- Various safety features (vacuum switch, heat sensors).
- A main control box/board (or "brain").

To properly function, a pellet stove uses electricity and can be plugged into a normal wall outlet. A pellet stove, like an automatic coal stoker, is a consistent heater consuming fuel that is fed evenly from a refillable hopper into the burn-pot (a perforated cast-iron basin), through a motorized system. The most commonly used distributor is an auger system that consists of a spiral length of metal encased in a tube. This mechanism is either located above the burn-pot or slightly beneath and guides a portion of pellet fuel from the hopper upwards until it falls into the burn-pot and begins to combust.

Fan systems are necessary for clean, economical performance. The flame produced is concentrated and intense as a combustion blower introduces air into the bottom of the burn-pot. While some pellet stoves will be hot to the touch (especially on the viewing window), most manufacturers utilize a series of cast-iron heat exchangers that run along the back and top areas of the visible firebox. With a convection blower, room air is circulated through the heat exchangers and directed into the living space. This method allows for a much higher efficiency than the radiant heat of a hand-fed wood or coal stove, and will in most cases cause the top of the stove to be not more than warm to the touch. Along with convection air, an exhaust fan forces air from the firebox through special venting specifically made for pellet fuel. This cycle of circulation is an integral part of the combustion system as well, for the concentrated high-temperature flame will quickly overheat the firebox. The possible problems associated with overheating are electrical component failure and flames traveling into the auger tube causing a hopper fire. As safeguards, all pellet stoves are equipped with heat sensors enabling the controller to shutdown if any safe conditions are exceeded.

Pellet stoves can either be lit manually or through an automatic igniter. The igniter piece resembles a car's electric cigarette lighter heating coil. Most models have automatic ignition and can be readily equipped with thermostats or remote controls.

A Corn Stove is designed for whole kernel shelled corn combustion and is similar to a pellet stove. The chief difference between a pellet stove and a dedicated Corn Stove is the addition of metal stirring rod within the burnpot. These vary in design, but usually consist of one long metal stalk with smaller rods welded at a perpendicular angle, in order to churn the burn-pot as it spins. During a normal burn cycle, the sugar content within corn (and other similar bio-fuels) will cause the ashes to stick together, forming a hard mass. The metal stirring rod, which is usually connected to a motor by a simple chain system, will break apart these masses, causing a much more consistent burn. While there is push to create stoves that are able to burn multiple fuels with minimal adjustments, some pellet stoves are not designed to stir fuel and will not burn corn.

UNDERSTANDING THERMAL DEPOLYMERIZATION

Thermal depolymerization (TDP) is a process using hydrous pyrolysis for the reduction of complex organic materials (usually waste products of various sorts, often known as biomass and plastic) into light crude oil. It mimics the natural geological processes thought to be involved in the production of fossil fuels. Under pressure and heat, long chain polymers of hydrogen, oxygen, and carbon decompose into short-chain petroleum hydrocarbons with a maximum length of around 18 carbons.

Similar Processes

Thermal depolymerisation is similar to other processes which use superheated water as a major step in their processing to produce fuels, such as direct Hydrothermal Liquefaction and hydrous pyrolysis. Thermochemical conversion (TCC) can mean conversion of biomass to oils using superheated water, although it more usually is applied to fuel production via pyrolysis. The Thermal Conversion Process is another name for thermal depolymerisation. A company called Renewable Environmental Solutions (RES) was formed as a joint venture between ConAgra Foods and Changing World Technologies to operate the plant at Carthage, Missouri and the name of the process was changed.

EnerTech operates the "SlurryCarb" process, which uses similar technology to decarboxylate wet solid biowaste, which can then be physically dewatered and used as a solid fuel called E-Fuel. The plant at Rialto, California is said to be able to process 683 tons of waste per day.

The Hydro Thermal Upgrading (HTU) process was originally developed by Shell, and is now operated by Biofuel BV. It uses superheated water to produce oil from a range of biomass and domestic waste. A demonstration plant is due to start up in the Netherlands said to be capable of processing 64 tons of biomass (dry basis) per day into oil. Thermal depolymerisation differs in that it contains a hydrous process followed by an anhydrous cracking/distillation process, although upgrading of the raw HTU product is also possible.

History

Thermal depolymerization is similar to the geological processes that produced the fossil fuels used today, except that the technological process occurs in a timeframe measured in hours. Until recently, the human-designed processes were not efficient enough to serve as a practical source of fuel—more energy was required than was produced.

Many previous methods which create hydrocarbons through depolymerization used dry materials (or anhydrous pyrolysis), which requires expending a lot of energy to remove water. However, there has been work done on hydrous pyrolysis methods, in which the depolymerization takes place with the materials in water. In U.S. patent 2,177,557, issued in 1939, Bergstrom and Cederquist discuss a method for obtaining oil from wood in which the wood is heated under pressure in water with a significant amount of calcium hydroxide added to the mixture. In the early 1970s Herbert R. Appell and coworkers worked with hydrous pyrolysis methods, as exemplified by U.S. patent 3,733,255 (issued in 1973), which discusses the production of oil from sewer sludge and municipal refuse by heating the material in water, under pressure, and in the presence of carbon monoxide.

An approach that exceeded break-even was developed by Illinois microbiologist Paul Baskis in the 1980s and refined over the next 15 years (see U.S. patent 5,269,947, issued in 1993). The technology was finally developed for commercial use in 1996 by Changing World Technologies (CWT). Brian S. Appel (CEO of CWT) took the technology in 2001 and expanded and changed it into what is now referred to as TCP (Thermal Conversion Process), and has applied for several patents (see, for example, published patent application US 2004/0192980). A Thermal Depolymerization demonstration plant was completed in 1999 in Philadelphia by Thermal Depolymerization, LLC, and the first full-scale commercial plant was constructed in Carthage, Missouri, about 100 yards (100 m) from ConAgra Foods' massive Butterball turkey plant, where it is expected to process about 200 tons of turkey waste into 500 barrels (21,000 US gallons or 80 m³) of oil per day.

Theory and Process

In the method used by CWT, the water improves the heating process and contributes hydrogen to the reactions.

In the Changing World Technologies (CWT) process, the feedstock material is first ground into small chunks, and mixed with water if it is especially dry. It is then fed into a pressure vessel reaction chamber where it is heated at constant volume to around 250 °C. Similar to a pressure cooker (except at much higher pressure), steam naturally raises the pressure to 600 psi (4 MPa) (near the point of saturated water). These conditions are held for approximately 15 minutes to fully heat the mixture, after which the pressure is rapidly released to boil off most of the water. The result is a mix of crude hydrocarbons and solid minerals. The minerals are removed, and the hydrocarbons are sent to a second-stage reactor where they are heated to 500 °C, further breaking down the longer hydrocarbon chains. The hydrocarbons are then sorted by fractional distillation, in a process similar to conventional oil refining.

The CWT company claims that 15 to 20 per cent of feedstock energy is used to provide energy for the plant. The remaining energy is available in the converted product. Working with turkey offal as the feedstock, the process proved to have yield efficiencies of approximately 85 per cent; in other words, the energy contained in the end products of the process is 85 per cent of the energy contained in the inputs to the process (most notably the energy content of the feedstock, but also including electricity for pumps and natural gas or woodgas for heating). If one considers the energy content of the feedstock to be free (i.e., waste material from some other process), then 85 units of energy are made available for every 15 units of energy consumed in process heat and electricity. This means the "Energy Returned on Energy Invested" (EROEI) is (6.67), which is comparable to other energy harvesting processes. Higher efficiencies may be possible with drier and more carbon-rich feedstocks, such as waste plastic.

By comparison, the current processes used to produce ethanol and biodiesel from agricultural sources have EROEI in the 4.2 range, when the energy used to produce the feedstocks is accounted for (in this case, usually sugar cane, corn, soybeans and the like). These EROEI values are not directly comparable, because these EROEI calculations include the energy cost to produce the feedstock, whereas the above EROEI calculation for thermal depolymerization process (TDP) does not.

The process breaks down almost all materials that are fed into it. TDP even efficiently breaks down many types of hazardous materials, such as poisons and difficult-to-destroy biological agents such as prions.

Feedstocks and Outputs with Thermal Depolymerization

Table 2.1: Average TDP Feedstock Outputs

Feedstock	Oils	Gases	Solids (Mostly Carbon Based)	Water (Steam)
Plastic bottles	70%	16%	6%	8%
Medical waste	65%	10%	5%	20%
Tires	44%	10%	42%	4%
Turkey offal	39%	6%	5%	50%
Sewage sludge	26%	9%	8%	57%
Paper (cellulose)	8%	48%	24%	20%

Note: Paper/cellulose contains at least 1 per cent minerals, which was probably grouped under carbon solids.

Carthage plant products

As reported on 04/02/2006 by Discover Magazine, the Carthage plant was producing 500 barrels per day (79 m3/d) of oil made from 270 tons of turkey guts and 20 tons of pig fat. This represents an oil yield of 22.3 per cent. The Carthage, MO plant produces API 40+, a high value crude oil. It contains light and heavy naphthas, a kerosene, and a gas oil fraction, with essentially no heavy fuel oils, tars, asphaltenes or waxes present. It can be further refined to produce No. 2 and No. 4 fuel oils.

Table 2.2: TDP-40 Oil Classification by D-5443 PONA Method

Output Material	% by Weight
Paraffins	22%
Olefins	14%
Naphthenes	3%
Aromatics	6%
C14/C14+	55%
	100%

The fixed carbon solids produced by the TDP process have multiple uses as a filter, a fuel source and a fertilizer. It can be used as activated carbon in wastewater treatment, as a fertilizer, or as a fuel similar to coal.

Advantages

The process can break down organic poisons, due to breaking chemical bonds and destroying the molecular shape needed for the poison's activity. It is likely to be highly effective at killing pathogens, including prions. It can also safely remove heavy metals from the samples by converting them from their ionized or organometallic forms to their stable oxides which can be safely separated from the other products.

Along with similar processes, it is a method of recycling the energy content of organic materials without first removing the water. It can produce liquid fuel, which separates from the water physically without need for drying. Other methods to recover energy often require pre-drying (eg. burning, pyrolysis) or produce gaseous products (eg. anaerobic digestion).

Potential Sources of Waste Inputs

The United States Environmental Protection Agency estimates that in 2006 there were 251 million tons of municipal solid waste, or 4.6 pounds generated per day per person in the USA. Much of this mass is considered unsuitable for oil conversion.

Limitations

The process only breaks long molecular chains into shorter ones, so small molecules such as carbon dioxide or methane cannot be converted to oil through this process. However, the methane in the feedstock is recovered and burned to heat the water that is an essential part of the process. In addition, the gas can be burned in a combined heat and power plant, consisting of a gas turbine which drives a generator to create electricity, and a heat exchanger to heat the process input water from the exhaust gas. The electricity can be sold to the power grid, for example under a Feed-in Tariff scheme. This also increases the overall efficiency of the process (already said to be over 85% of feedstock energy content). Another option is to sell the methane product as biogas. For example, biogas can be compressed, much like natural gas, and used to power motor vehicles.

Many agricultural and animal wastes could be processed, but many of these are already used as fertilizer, animal feed, and, in some cases, as feedstocks for paper mills or as boiler fuel. Energy crops constitute another potentially large feedstock for thermal depolymerization.

Current Status

Reports in 2004 claimed that the facility was selling products at 10 per cent below the price of equivalent oil, but its production costs were low enough that the plant produced a profit. At the time it was paying for turkey waste.

The plant then consumed 270 tons of turkey offal (the full output of the turkey processing plant) and 20 tons of egg production waste daily. In February 2005, the Carthage plant was producing about 400 barrels per day (64 m3/d) of crude oil.

In April 2005 the plant was reported to be running at a loss. Further 2005 reports summarized some economic setbacks which the Carthage plant encountered since its planning stages. It was thought that concern over mad cow disease would prevent the use of turkey waste and other animal products as cattle feed, and thus this waste would be free. As it turned out, turkey waste may still be used as feed in the United States, so that the facility must purchase that feed stock at a cost of $30 to $40 per ton, adding $15 to $20 per barrel to the cost of the oil. Final cost, as of January 2005, was $80/barrel ($1.90/gal). The above cost of production also excludes the operating cost of the thermal oxidizer and scrubber added in May 2005 in response to odour complaints.

A biofuel tax credit of roughly $1 per US gallon (26 ¢/L) on production costs was not available because the oil produced did not meet the definition of "biodiesel" according to the relevant American tax legislation. The Energy Policy Act of 2005 specifically added thermal depolymerization to a $1 renewable diesel credit, which became effective at the end of 2005, allowing a profit of $4/barrel of output oil.

Company expansion

The company has explored expansion in California, Pennsylvania, and Virginia, and is presently examining projects in Europe, where animal products cannot be used as cattle feed. TDP is also being considered as an alternative means for sewage treatment in the United States.

Smell complaints

The pilot plant in Carthage was temporarily shut down due to smell complaints. It was soon restarted when it was discovered that few of the odours were generated by the plant. Furthermore, the plant agreed to install an enhanced thermal oxidizer and to upgrade its air scrubber system under a court order. Since the plant is located only four blocks from the tourist-attracting town center, this has strained relations with the mayor and citizens of Carthage.

According to a company spokeswoman, the plant has received complaints even on days when it is not operating. She also contended that the odours may not have been produced by their facility, which is located near several other agricultural processing plants.

In December 29, 2005, the plant was ordered by the state governor to shut down once again over allegations of foul odours as reported by MSNBC. As of March 7, 2006, the plant has begun limited test runs to validate it has resolved the odour issue.

As of August 24, 2006, the last lawsuit connected with the odour issue has been dismissed and the problem is acknowledged as fixed. In late November, however, another complaint was filed over bad smells. This complaint was closed on January 11th of 2007 with no fines assessed.

Status as of February 2009

A May 2003 article in Discover magazine stated, "Appel has lined up federal grant money to help build demonstration plants to process chicken offal and manure in Alabama and crop residuals and grease in Nevada. Also in the works are plants to process turkey waste and manure in Colorado and pork and cheese waste in Italy. He says the first generation of depolymerization centers will be up and running in 2005. By then it should be clear whether the technology is as miraculous as its backers claim."

However, as of August 2008, the only operational plant listed at the company's website is the initial one in Carthage, Missouri.

Changing World Technology applied for an IPO on 12 Aug 2008, hoping to raise $100 million.

The unusual Dutch Auction type IPO failed possibly because CWT has lost nearly $20 million with very little revenue.

CWT, the parent company of Renewable Energy Solutions, filed for Chapter 11 bankruptcy. No details on plans for the Carthage plant have been released.

Similar Technologies

Plasma Converters use powerful electric arcs to reduce and extract energy from waste.

CELLULOSIC ETHANOL COMMERCIALIZATION

Cellulosic ethanol commercialization is the process of building an industry out of methods of turning cellulose-containing organic matter into fuel. Companies such as Iogen, Broin, and Abengoa are building refineries that can process biomass and turn it into ethanol, while companies such as Diversa, Novozymes, and Dyadic are producing enzymes which could enable a cellulosic ethanol future. The shift from food crop feedstocks to waste residues and native grasses offers significant opportunities for a range of players, from farmers to biotechnology firms, and from project developers to investors.

Cellulosic Ethanol Production

Cellulosic ethanol can be produced from a diverse array of feedstocks. Instead of taking the grain from wheat and grinding that down to get starch and gluten, then taking the starch, cellulosic ethanol production involves the use of the whole crop. This approach should increase yields and reduce the carbon footprint because the amount of energy-intensive fertilisers and fungicides will remain the same, for a higher output of usable material.

Commercialization by Country

Australia

Ethtec has a pilot plant in Harwood, New South Wales, which uses wood residues as a feedstock.

Canada

In Canada, Iogen Corp. is a developer of cellulosic ethanol process technology. Iogen has developed a proprietary process and operates a demonstration-scale plant in Ontario. The facility has been designed and engineered to process 40 tons of wheat straw per day into ethanol using enzymes made in an adjacent enzyme manufacturing facility. In 2004, Iogen began delivering its first shipments of cellulosic ethanol into the marketplace. In the near term, the company intends to commercialize its cellulose ethanol process by licensing its technology broadly through turnkey plant construction partnerships. The company is currently evaluating sites in the United States and Canada for its first commercial-scale plant.

Lignol Innovations has a pilot plant, which uses wood as a feedstock, in Vancover.

In March 2009, KL Energy Corporation of South Dakota and Prairie Green Renewable Energy of Alberta announced their intention to develop a cellulosic ethanol plant near Hudson Bay, Saskatchewan. The Northeast Saskatchewan Renewable Energy Facility will use KL Energy's modern design and engineering to produce ethanol from wood waste.

China

Cellulosic ethanol production currently exists at "pilot" and "commercial demonstration" scale, including a plant in China engineered by SunOpta Inc. and owned and operated by China Resources Alcohol Corporation that is currently producing cellulosic ethanol from corn stover (stalks and leaves) on a continuous, 24-hour per day basis.

Denmark

Dong Energy has a pilot plant in Kalundborg, which uses wheat straw as a feedstock.

Japan

Nippon Oil Corporation and other Japanese manufacturers including Toyota Motor Corporation plan to set up a research body to develop cellulose-derived biofuels. The consortium plans to produce 250,000 kilolitres (1.6 million barrels) per year of bioethanol by March 2014, and produce bioethanol at 40 yen ($0.437) per litre (about $70 a barrel) by 2015.

In March 2009, Honda Motor announced an agreement for the construction of a new cellulosic ethanol research facility in Japan. The new Kazusa-branch facility of the Honda Fundamental Technology Research Center will be built within the Kazusa Akademia Park, in Kisarazu, Chiba. Construction is scheduled to begin in April 2009, with the aim to begin operations in November 2009.

Spain

Abengoa continues to invest heavily, in the necessary technology for bringing cellulosic ethanol to market. Utilizing process and pre-treatment technology from SunOpta Inc., Abengoa is building a 5 million gallon cellulosic ethanol facility in Spain and have recently entered into a strategic research and development agreement with Dyadic International, Inc. (AMEX: DIL), to create new and better enzyme mixtures which may be used to improve both the efficiencies and cost structure of producing cellulosic ethanol.

Sweden

SEKAB has developed an industrial process for production of ethanol from biomass feed-stocks, including wood chips and sugar cane bagasse. The development work is being carried out at an advanced pilot plant in Örnsköldsvik, and has sparked international interest. The technology will be gradually scaled up to commercial production in a new breed of bio-refineries from 2013 to 2015.

United Kingdom

A $400 million investment programme to cover the construction of a world scale ethanol plant and a high technology demonstration plant to advance development work on the next generation of biofuels has been announced by BP, Associated British Foods (ABF) and DuPont. The bioethanol plant will be built on BP's existing chemicals site at Saltend, Hull. Due to be commissioned in late 2009, it will have an annual production capacity of some 420 million litres from wheat feedstock.

United States

The U.S. Department of Energy (DOE) is promoting the development of ethanol from cellulosic feedstocks as an alternative to conventional petroleum transportation fuels. Programmes sponsored by DOE range from research to develop better cellulose hydrolysis enzymes and ethanol-fermenting organisms, to engineering studies of potential processes, to co-funding initial ethanol from cellulosic biomass demonstration and production facilities. This research is conducted by various national laboratories, including the National Renewable Energy Laboratory (NREL), Oak Ridge National Laboratory (ORNL) and Idaho National Laboratory (INL), as well as by universities and private industry. Engineering and construction companies and operating companies are generally conducting the engineering work.

In May 2008, Congress passed a new farm bill that will accelerate the commercialization of advanced biofuels, including cellulosic ethanol. The *Food, Conservation, and Energy Act of 2008* provides for grants covering up to 30 per cent of the cost of developing and building demonstration-scale biorefineries for producing "advanced biofuels," which essentially includes all fuels that are not produced from corn kernel starch. It also allows for loan guarantees of up to $250 million for building commercial-scale biorefineries to produce advanced biofuels.

In September 2008, Iogen Corporation delivered more than 100,000 liters (26,000 gallons) of cellulosic ethanol to Royal Dutch Shell, its commercial business partner, "for upcoming fuel applications." The fuel was the first part of Shell's initial order of cellulosic ethanol from Iogen, which totalled 180,000 liters (47,550 gallons). Shell first gained an equity stake in Iogen in 2002, and in July 2008, Shell increased its shareholding in Iogen from slightly more than 26 percent to 50 per cent.

Following a successful start-up in the fourth quarter of 2008, the POET LLC Research Center in Scotland, South Dakota is now producing cellulosic ethanol on a pilot scale, at a rate of 20,000 gallons-per-year (GPY) using corn cobs as feedstock. The US $8 million plant is a precursor to the company's US $200 million Project Liberty, a commercial-scale cellulosic ethanol plant which is to begin production in 2011.

Verenium Corporation announced in January 2009 that it will construct a commercial-scale cellulosic ethanol plant in Highlands County, Florida. The plant, which will provide about 150 full-time jobs in the production of 36 million gallons of ethanol per year, will cost about $250 million to construct. Verenium plans to break ground in the second half of 2009 and have the plant in full production during 2011.

Using a newly developed tool known as the "Biofuels Deployment Model", Sandia researchers have determined that 21 billion gallons of cellulosic ethanol could be produced per year by 2022 without displacing current crops. The Renewable Fuels Standard, part of the 2007 Energy Independence and Security Act, calls for an increase in biofuels production to 36 billion gallons a year by 2022.

Environmental Issues

Cellulosic ethanol and grain-based ethanol are, in fact, the same product, but many scientists believe cellulosic ethanol production has distinct environmental advantages over grain-based ethanol production. On a life-cycle basis, ethanol produced from agricultural residues or dedicated cellulosic crops has significantly lower greenhouse gas emissions and a higher sustainability rating than ethanol produced from grain.

According to US Department of Energy studies conducted by the Argonne Laboratories of the University of Chicago, cellulosic ethanol reduces greenhouse gas emissions (GHG) by 85 per cent over reformulated gasoline. By contrast, starch ethanol (e.g., from corn), which usually uses natural gas to provide energy for the process, reduces greenhouse gas emissions by 18 per cent to 29 per cent over gasoline.

Criticism

Critics such as Cornell University professor of ecology and agriculture David Pimentel and University of California at Berkeley engineer Tad Patzek question the likelihood of environmental, energy, or economic benefits from cellulosic ethanol technology from non-waste.

The bioconversion of biomass to mixed alcohol fuels can be accomplished using the MixAlco process. Through bioconversion of biomass to a mixed alcohol fuel, more energy from the biomass will end up as liquid fuels than in converting biomass to ethanol by yeast fermentation.

The process involves a biological/chemical method for converting any biodegradable material (e.g., urban wastes, such as municipal solid waste, biodegradable waste, and sewage sludge, agricultural residues such as corn stover, sugarcane bagasse, cotton gin trash, manure) into useful chemicals, such as carboxylic acids (e.g., acetic, propionic, butyric acid), ketones (e.g., acetone, methyl ethyl ketone, diethyl ketone) and biofuels, such as a mixture of primary alcohols (e.g., ethanol, propanol, butanol) and/or a mixture of secondary alcohols (e.g., isopropanol, 2-butanol, 3-pentanol). Because of the many products that can be economically produced, this process is a true biorefinery.

The process uses a mixed culture of naturally occurring microorganisms found in natural habitats such as the rumen of cattle, termite guts, and marine and terrestrial swamps to anaerobically digest biomass into a mixture of carboxylic acids produced during the acidogenic and acetogenic stages of anaerobic digestion, however with the inhibition of the methanogenic final stage. The more popular methods for production of ethanol and cellulosic ethanol use enzymes that must be isolated first to be added to the biomass and thus convert the starch or cellulose into simple sugars, followed then by yeast fermentation into ethanol. This process does not need the addition of such enzymes as these microorganisms make their own.

As the microoganisms anaerobically digest the biomass and convert it into a mixture of carboxylic acids, the pH must be controlled. This is done by the addition of a buffering agent (e.g., ammonium bicarbonate, calcium carbonate), thus yielding a mixture of carboxylate salts. Methanogenesis, which, as mentioned, is the natural final stage of anaerobic digestion, is inhibited by the presence of the ammonium ions or by the addition of an inhibitor (e.g., iodoform). The resulting fermentation broth contains the produced carboxylate salts that must be dewatered. This is achieved efficiently by vapour-compression evaporation. Further chemical refining of the dewatered fermentation broth may then take place depending on the final chemical or biofuel product desired.

The condensed distilled water from the vapour-compression evaporation system is recycled back to the fermentation. On the other hand, if raw sewage or other waste water with high BOD in need of treatment is used as the water for the fermentation, the condensed distilled water from the evaporation can be recycled back to the city or to the original source of the high-BOD waste water. Thus, this process can also serve as a water treatment facility, while producing valuable chemicals or biofuels. Because the system uses a mixed culture of microorganisms, besides not needing any enzyme addition, the fermentation requires no sterility or aseptic conditions, making this front step in the process more economical than in more popular methods for the production of cellulosic ethanol. These savings in the front end of the process, where volumes are large, allows flexibility for further chemical transformations after dewatering, where volumes are small.

UNDERSTANDING BIOPLASTIC

Bioplastics (also called **organic plastics**) are a form of plastics derived from renewable biomass sources, such as vegetable oil, corn starch, pea starch or microbiota, rather than fossil-fuel plastics which are derived from petroleum.

Applications

Because of their biological degradability, the use of bioplastics is especially popular for disposable items, such as packaging and catering items (crockery, cutlery, pots, bowls, straws). The use of bioplastics for shopping bags is already common. After their initial use they can be reused as bags for organic waste and then be composted. Trays and containers for fruit, vegetables, eggs and meat, bottles for soft drinks and dairy products and blister foils for fruit and vegetables are also already widely manufactured from bioplastics.

Non-disposable applications include mobile phone casings, carpet fibres, and car interiors, fuel line and plastic pipe applications, and new electroactive bioplastics are being developed that can be used to carry electrical current. In these areas, the goal is not biodegradability, but to create items from sustainable resources.

Performance and Usage

Many bioplastics lack the performance and ease of processing of traditional materials. Polylactic acid plastic is being used by a handful of small companies for water bottles. But shelf life is limited because the plastic is permeable to water—the bottles lose their contents and slowly deform. However, bioplastics are seeing some use in Europe, where they account for 60 per cent of the biodegradable materials market. The most common end use market is for packaging materials. Japan has also been a pioneer in bioplastics, incorporating them into electronics and automobiles.

Plastic Types

Starch Based Plastics

Constituting about 50 percent of the bioplastics market, thermoplastic starch, such as Plastarch Material, currently represents the most important and widely used bioplastic. Pure starch possesses the characteristic of being able to absorb humidity and is thus being used for the production of drug capsules in the pharmaceutical sector. Flexibiliser and plasticiser such as sorbitol and glycerine are added so that starch can also be processed thermo-plastically. By varying the amounts of these additives, the characteristic of the material can be tailored to specific needs (also called "thermo-plastical starch"). Simple starch plastic can be made at home shown by this method.

Polylactic Acid (PLA) Plastics

Polylactic acid (PLA) is a transparent plastic produced from cane sugar or glucose. It not only resembles conventional petrochemical mass plastics (like PE or PP) in its characteristics, but it can also be processed easily on standard equipment that already exists for the production of conventional plastics. PLA and PLA-Blends generally come in the form of granulates with various properties and are used in the plastic processing industry for the production of foil, moulds, tins, cups, bottles and other packaging.

Poly-3-hydroxybutyrate (PHB)

The biopolymer poly-3-hydroxybutyrate (PHB) is a polyester produced by certain bacteria processing glucose or starch. Its characteristics are similar to those of the petroplastic polypropylene. The South American sugar industry, for example, has decided to expand PHB production to an industrial scale. PHB is distinguished primarily by its physical characteristics. It produces transparent film at a melting point higher than 130 degrees Celsius, and is biodegradable without residue.

Polyamide 11 (PA 11)

PA 11 is a biopolymer derived from natural oil. It is also known under the tradename Rilsan B commercialized by Arkema. PA 11 belongs to the technical polymers family and is not biodegradable. Its properties are similar than PA 12 although emissions of greenhouse gases and consumption of non-renewable resources are reduced during its production. Its thermal resistance is also superior than PA 12. It is used in high performance applications as automotive fuel lines, pneumatic airbrake tubing, electrical anti-termite cable sheathing, oil and gas flexible pipes and control fluid umbilicals, sports shoes, electronic device components, catheters, etc.

Bio-Derived Polyethylene

The basic building block (monomer) of polyethylene is ethylene. This is just one small chemical step from ethanol, which can be produced by fermentation of agricultural feedstocks such as sugar cane or corn. Bio-derived polyethylene is chemically and physically identical to traditional polyethylene—it does not biodegrade but can be recycled. It can also considerably reduce greenhouse gas emissions. Brazilian chemicals group Braskem claims that using its route from sugar cane ethanol to produce one tonne of polyethylene captures (removes from the environment) 2.5 tonnes of carbon dioxide while the traditional petrochemical route results in emissions of close to 3.5 tonnes.

Braskem plans to introduce commercial quantities of its first bio-derived high density polyethylene, used in a packaging such as bottles and tubs, in 2010 and has developed a technology to produce bio-derived butene, required to make the linear low density polethylene types used in film production.

Genetically Modified Bioplastics

Genetic modification (GM) is also a challenge for the bioplastics industry. None of the currently available bioplastics—which can be considered first generation products—require the use of GM crops. However, it is not possible to ensure corn used to make bioplastic in North America is GM-free.

European consumers are hostile to any products that are linked to the GM industry. As a result, some UK retailers such as Sainsbury's will not use bioplastic manufactured in the US, such as Natureworks polylactic acid.

There is also concern that the route from corn to bioplastics is not the most efficient. Looking further ahead, some of the second generation bioplastics manufacturing technologies under development employ the "plant factory" model, using genetically modified crops or genetically modified bacteria to optimise efficiency. However, a change in consumer perception of GM technology in Europe will be required for these to be widely accepted.

Environmental Impact

The production and use of bioplastics is generally regarded as a more sustainable activity when compared with plastic production from petroleum (petroplastic), because it relies less on fossil fuel as a carbon source and also introduces fewer, net-new greenhouse emissions if it biodegrades. They significantly reduce hazardous waste caused by oil-derived plastics, which remain solid for hundreds of years, and open a new era in packing technology and industry.

However, manufacturing of bioplastic materials is often still reliant upon petroleum as an energy and materials source. This comes in the form of energy required to power farm machinery and irrigate growing crops, to produce fertilisers and pesticides, to transport crops and crop products to processing plants, to process raw materials, and ultimately to produce the bioplastic, although renewable energy can be used to obtain petroleum independence.

Italian bioplastic manufacturer Novamont states in its own environmental auditthat producing one kilogram of its starch-based product uses 500g of petroleum and consumes almost 80 per cent of the energy required to produce a traditional polyethylene polymer. Environmental data from Nature Works, the only commercial manufacturer of PLA (polylactic acid) bioplastic, says that making its plastic material delivers a fossil fuel saving of between 25 and 68 per cent compared with polyethylene, in part due to its purchasing of renewable energy certificates for its manufacturing plant.

A detailed study examining the process of manufacturing a number of common packaging items in several traditional plastics and polylactic acid carried out by US-group and published by the Athena Institute shows the bioplastic to be less environmentally damaging for some products, but more environmentally damaging for others.

While production of most bioplastics results in reduced carbon dioxide emissions compared to traditional alternatives, there are some real concerns that the creation of a global bioeconomy could contribute to an accelerated rate of deforestation if not managed effectively. There are associated concerns over the impact on water supply and soil erosion.

Other studies showed that bioplastics represent a 42 per cent reduction in carbon footprint.

On the other hand, bioplastic can be made from agricultural byproducts and also from used plastic bottles and other containers using microorganisms.

Bioplastics and Biodegradation

The terminology used in the bioplastics sector is sometimes misleading. Most in the industry use the term bioplastic to mean a plastic produced from a biological source. One of the oldest plastics, cellulose film, is made from wood cellulose. All (bio- and petroleum-based) plastics are technically biodegradable, meaning they can be degraded by microbes under suitable conditions. However many degrade at such slow rates as to be considered non-biodegradable. PLA plastics can take 100 to 1,000 years to completely biodegrade. Some petrochemical-based plastics are considered biodegradable, and may be used as an additive to improve the performance of many commercial bioplastics. Non-biodegradable bioplastics are referred to as durable. The degree of biodegradation varies with temperature, polymer stability, and available oxygen content. Consequently, most bioplastics will only degrade in the tightly controlled conditions of industrial composting units. An internationally agreed standard, EN13432, defines how quickly and to what extent a plastic must be degraded under commercial composting conditions for it to be called biodegradable. This is published by the International Organization for Standardization ISO and is recognised in many countries, including all of Europe, Japan and the US. However, it is designed only for the aggressive conditions of commercial composting units. There is no standard applicable to home composting conditions.

The term "biodegradable plastic" is often also used by producers of specially modified petrochemical-based plastics which appear to biodegrade. Traditional plastics such as polyethylene are degraded by ultra-violet (UV) light and oxygen. To prevent this process manufacturers add stabilising chemicals. However with the addition of a degradation initiator to the plastic, it is possible to achieve a controlled UV/oxidation disintegration process. This type of plastic may be referred to as *degradable plastic* or *oxy-degradable plastic* or *photodegradable plastic* because the process is not initiated by microbial action. While some degradable plastics manufacturers argue that degraded plastic residue will be attacked by microbes, these degradable materials do not meet the requirements of the EN13432 commercial composting standard.

Recycling

There are also fears that bioplastics will damage existing recycling projects. Packaging such as HDPE milk bottles and PET water and soft drinks bottles is easily identified and hence setting up a recycling infrastructure has been quite successful in many parts of the world. Polylactic acid and PET do not mix—as bottles made from polylactic acid cannot be distinguished from PET bottles by the consumer there is a risk that recycled PET could be rendered unusable. This could be overcome by ensuring distinctive bottle types or by investing in suitable sorting technology. However, the first route is unreliable and the second costly.

Market

Because of the fragmentation in the market it is difficult to estimate the total market size for bioplastics, but estimates put global consumption in 2006 at around 85,000 tonnes. In contrast, global consumption of all flexible packaging is estimated at around 12.3 million tonnes.

COPA (Committee of Agricultural Organisation in the European Union) and COGEGA (General Committee for the Agricultural Cooperation in the European Union) have made an assessment of the potential of bioplastics in different sectors of the European economy:

- Catering products: 450,000 tonnes per year
- Organic waste bags: 100,000 tonnes per year
- Biodegradable mulch foils: 130,000 tonnes per year
- Biodegradable foils for diapers 80,000 tonnes per year
- Diapers, 100 per cent biodegradable: 240,000 tonnes per year
- Foil packaging: 400,000 tonnes per year
- Vegetable packaging: 400,000 tonnes per year
- Tyre components: 200,000 tonnes per year
- Total 2,000,000 tonnes per year

The European Bioplastics trade group predicted annual capacity would more than triple to 1.5 million tons by 2011. BCC Research forecasts the global market for biodegradable polymers to grow at a compound average growth rate of more than 17 percent through 2012. Even so, bioplastics will encompass a small niche of the overall plastic market, which is forecast to reach 500 billion pounds (220 million tonnes) globally by 2010.

Cost

With the exception of cellulose, most bioplastic technology is relatively new and is currently not cost competitive with petroleum-based plastics (petroplastics). They do not reach the fossil fuel parity. Many bioplastics are reliant on fossil fuel-derived energy for their manufacturing, reducing the cost advantage over petroleum-based plastic.

Research and Development

- In the early 1950s, Amylomaize (>50 per cent starch content corn) was successfully bred and commercial bioplastics applications started to be explored.
- In 2004, NEC developed a flame retardant plastic, polylactic acid, without using toxic chemicals such as halogensand phosphorus compounds.
- In 2005, Fujitsu became one of the first technology companies to make personal computer cases from bioplastics, which are featured in their FMV-BIBLO NB80K line.
- In 2007 Braskem of Brazil announced it had developed a route to manufacture high density polyethylene (HDPE) using ethylene derived from sugar cane.
- In 2008, a University of Warwick team created a soap free emulsion polymerization process which makes colloid particles of polymer dispersed in water and in a one step process adds nanometre sized silica-based particles to the mix. The newly developed technology might be most applicable to multi-layered biodegradable packaging which could gain more robustness and water barrier characteristics through the addition of a nano-particle coating.

Certification

Biodegradability—EN 13432, ASTM D6400

The EN 13432 industrial standard is arguably the most international in scope and compliance with this standard is required to claim that a product is compostable in the European marketplace. In summary, it requires biodegradation of 90 per cent of the materials in a commercial composting unit within 90 days. The ASTM 6400 standard is the regulatory framework for the United States and sets a less stringent threshold of 60 per cent biodegradation within 180 days, again within commercial composting conditions.

The "compostable" marking found on many items of packaging indicates that the package complies with either of the two standards mentioned above. However, the marking is not owned by either regulatory body but by third party trade associations representing companies making or selling biodegradable plastics. In Europe, this is European Bioplastics, in the U.S. it is the Biodegradable Products Institute.

Many starch based plastics, PLA based plastics and certain aliphatic-aromatic co-polyester compounds such as succinates and adipates, have obtained these certificates. Additivated plastics sold as fotodegradable or oxobiodegradable do not comply with these standards in their current form.

Biobased—ASTM D6866

The ASTM D6866 method has been developed to certify the biologically derived content of bioplastics. Cosmic rays colliding with the atmosphere mean that some of the carbon is the radioactive isotope carbon-14. CO_2 from the atmosphere is used by plants in photosynthesis, so new plant material will contain both carbon-14 and carbon-12. Under the right conditions, and over geological timescales, the remains of living organisms can be transformed into fossil fuels. After ~100,000 years all the carbon-14 present in the original organic material will have undergone radioactive decay leaving only carbon-12. A product made from biomass will have a relatively high level of carbon-14, while a product made from petrochemicals will have no carbon-14. The percentage of renewable carbon in a material (solid or liquid) can be measured with an accelerator mass spectrometer.

There is an important difference between biodegradability and biobased content. A bioplastic such as high density polyethylene (HDPE) can be 100 per cent biobased (i.e. contain 100 per cent renewable carbon), yet be non-biodegradable. These bioplastics such HDPE play nonetheless an important role in greenhouse gas abatement, particularly when they are combusted for energy production. The biobased component of these bioplastics is considered carbon-neutral since their origin is from biomass.

UNDERSTANDING GREEN CRUDE

Green crude (also called **biopetroleum**) refers to a green-coloured crude which yields clean versions of gasoline and diesel from algae, sunlight, carbon dioxide and water. The result is chemically equivalent to the light, sweet crude oil traditionally used for distillation of carbon-based fuels.

Sapphire Energy and Solazyme are firms in this market.

CASE STUDY: AMERICAN COUNCIL ON RENEWABLE ENERGY (ACORE)

The **American Council On Renewable Energy** (ACORE), is a non-profit organization dedicated to bringing renewable energy into the mainstream of American's economy and lifestyle.

It was founded in 2001 as a forum to convene all sectors of the renewable energy industry. As of July 2009, ACORE had over 600 member organizations.

Its headquarters are Washington DC.

In 2008, ACORE hosted the Washington International Renewable Energy Conference WIREC with the U.S. Department of State. WIREC was the largest business-to-business and business-to-government conference and exhibition ever held on all-renewable energy in the U.S. It was global in scope, hosting exhibitors, speakers and delegates from 126 countries.

Vision

ACORE is àn organization of member companies and institutions that are dedicated to moving renewable energy into the mainstream of America's economy, ensuring the success of the renewable energy industry while helping to build a sustainable and independent energy future for the nation.

History

ACORE, a 501(c)(3) nonprofit organization based in Washington D.C., was founded in 2001 to bring together leading proponents and innovators in all facets of the renewable energy sector for the purpose of moving renewable energy into the mainstream of America's economy.

ACORE started out small, but with a very large goal: to create a membership-based organization for the wide range of interests in the renewable energy community—and to give our members a common platform for communicating, finding each other and through education and collaboration contributing to the success of the RE industry.

Today we still operate with a small Washington D.C. staff, but our membership has grown steadily — in mid 2008 numbering close to 500 companies, including renewable energy industry associations, utilities, end users, professional service firms, financial institutions, educational institutions, nonprofit groups and government agencies.

In mid 2008, we also opened an office in San Francisco to better serve the needs of our fast growing west coast membership...which now makes up over a quarter of our total membership base

ACORE serves as a forum through which diverse parties work together on common interests. Our primary vehicle for sharing information is convening: we organize and hold major annual conferences attended by hundreds of industry leaders. We also have established committees and working groups which meet regularly to work on specific industry issues. Since 2007 we have also organized and hosted monthly webinars on the legal issues faced by the industry. As we've grown we've added research and publishing to the benefits that our members receive by joining ACORE: these serve to inform and keep our membership current on trends and opportunities.

On an annual basis since 2003, ACORE has convened conferences in New York and Washington to focus on the three major areas that shape and advance renewable energy innovation and development in America: Policy, Markets and Finance.

- *Focus on Finance:* Our New York Conferences are called REFF-Wall Street and are organized to bring together the organizations interested in financing renewable energy companies and projects...with those in the industry who are innovating or wishing to scale up their operations in this space. These conferences attract hundreds of participants over three days: speakers come from across the country and represent the leading venture capital firms, bankers, entrepreneurs, and senior financiers. The topics covered range from project finance, venture capital and public capital markets to the emerging field of monetizing the value of reducing carbon emissions. In 2008, ACORE extended this historically east coast conference to the west coast by creating REFF-West, a two-day finance conference that covers issues specific to project developers and financiers in the fast growing RE market in the western United States.

- *Focus on Policy:* Our annual Washington Conferences are called Phase II of Renewable Energy in America, and are held in the Cannon Caucus room within the U.S. House of Representatives. The conference name derives from the sense that Phase I of renewable energy was focused on RE technologies, science and research: Phase II is where we are today – commercializing and scaling renewable energy project and products for global consumption. These conferences are attended by federal government policy and decision makers who come to hear from the industry's thought leaders on policy matters – and matters of practical realities – that impact the industry both in the U.S. and globally.

- *Focus on Markets*: As early as 2007 ACORE began discussions that would lead to a collaborative partnerships with the U.S. government to co-produce in March of 2008 a global meeting of over 8,500 officials, business executives and technical and industry professionals. The first of many international trade shows and conferences to come, WIREC was clearly the major "renewable energy communications event" of 2008 worldwide. In February of 2009 ACORE will host another large technical conference and exhibition in Las Vegas, Nevada called RETECH 2009. It will be the a conference that features all of the renewable energy technologies – and the leading companies that are advancing them –in one place at one time.

CASE STUDY:NATIONAL RENEWABLE ENERGY LABORATORY

Established	1977
Research Type	Energy Efficiency and Renewable Energy
Budget	328 Million (2009)
Director	Dan E. Arvizu
Staff	1,230
Location	Golden, CO
Operating Agency	Midwest Research Institute and Battelle Memorial Institute
Website	www.nrel.gov

The **National Renewable Energy Laboratory** (NREL), located in Golden, Colorado, as part of the U.S. Department of Energy, is the United States' primary laboratory for renewable energy and energy efficiency research and development.

History

Established in 1974, NREL began operating in 1977 as the **Solar Energy Research Institute**. Under the Carter administration, it was the recipient of a rather large budget and its activities went beyond research and development in solar energy as it tried to popularize knowledge about already existing technologies, like passive solar amongst the population. In the Reagan years that followed the budget was cut by some 90 per cent, many people 'reduced in force' and the activities reduced to R & D. In later years renewed interest in the energy problem improved the institute's position. But funding has fluctuated. In 2006 its funding had dropped to the point it was forced to lay off 32 workers. It was designated a national laboratory of the U.S. Department of Energy (DOE) in September 1991 and its name changed to NREL. Since its inception it has been operated under contract by the Midwest Research Institute of Kansas City, Missouri.

NREL is the principal research laboratory for the DOE Office of Energy Efficiency and Renewable Energy (EERE) which provides the majority of its funding. Other funding comes from DOE's Office of Science and Office of Electricity Transmission and Distribution.

NREL's areas of research and development expertise are:

- Renewable electricity
- Renewable fuels
- Integrated energy systems
- Strategic energy analysis

Funding in 2009

For 2009 funding is broken down between its major groups.

- Wind $33.9 million
- Biofuels $35.4 million
- Solar $72.4 million

NREL's Technology Transfer Office supports the practical deployment of technologies developed, and this often involves collaborative research projects and licensed technologies with public and private partners. NREL's innovative technologies have been recognized with 39 "RandD 100" awards. The engineering and science behind these technology transfer successes and awards demonstrates NREL's commitment to a sustainable energy future.

Dr. Dan E. Arvizu became NREL's eighth Laboratory

Director in January 2005, and was previously an executive with CH2M HILL companies.

Solar Cells

NREL PV RandD is performed under the National Center for Photovoltaics.
NREL tests and validates solar technologies.
The main research wind turbines at NREL

TOWARDS RENEWABLE ENERGY COMMERCIALIZATION

Renewable energy commercialization involves the diffusion of three generations of technologies dating back more than 100 years. First-generation technologies, which are already mature and economically competitive, include biomass, hydroelectricity, geothermal power and heat. Second-generation technologies are market-ready and are being deployed at the present time; they include solar heating, photovoltaics, wind power, solar thermal power stations, and modern forms of bioenergy. Third-generation technologies require continued RandD efforts in order to make large contributions on a global scale and include advanced biomass gasification, biorefinery technologies, hot-dry-rock geothermal power, and ocean energy.

Global renewable energy investment growth (1995—2007)

While there are many non-technical barriers to the widespread use of renewables, some 73 countries now have targets for their own renewable energy futures, and have enacted wide-ranging public policies to promote renewables. Climate change concerns are driving increasing growth in the renewable energy industries. Leading renewable energy companies include: Enercon, Gamesa, GE Energy, Q-Cells, Sharp Solar, SunOpta, and Vestas.

Global revenues for solar photovoltaics, wind power, and biofuels expanded from $76 billion in 2007 to $115 billion in 2008. New global investments in clean energy technologies— including venture capital, project finance, public markets, and research and development—expanded by 4.7 percent from $148 billion in 2007 to $155 billion in 2008. Continued growth for the renewable energy sector is expected in the mid-to long-term, but 2009 will be a year of refocus, consolidation, or retrenchment for some companies. At the same time, new government spending, regulation, and policies should help the industry weather the current economic crisis better than many other sectors. Most notably, U.S. President Barack Obama's American Recovery and Reinvestment Act of 2009 includes more than $70 billion in direct spending and tax credits for clean energy and associated transportation Programmes. Clean Edge suggests that the commercialization of clean energy will help countries around the world pull out of the current economic malaise.

Growth of Renewables

From the end of 2004 to the end of 2008, solar photovoltaic (PV) capacity increased sixfold to more than 16 gigawatts (GW), wind power capacity increased 250 percent to 121 GW, and total power capacity from new renewables increased 75 percent to 280 GW. During the same period, solar heating capacity doubled to 145 gigawatts-thermal (GWth), while biodiesel production increased sixfold to 12 billion liters per year and ethanol production doubled to 67 billion liters per year.

Table 2.3: Selected Renewable Energy Indicators

Selected Global Indicators	2006	2007	2008
Investment in new renewable capacity (annual)	63	104	120 billion USD
Existing renewables power capacity, including large-scale hydro	1,020	1,070	1,140 GWe
Existing renewables power capacity, excluding large hydro	207	240	280 GWe
Wind power capacity (existing)	74	94	121 GWe
Biomass heating			~250 GWth
Solar hot water/Space heating		126	145 GWth
Geothermal heating			~50 GWth
Ethanol production (annual)	39	50	67 billion liters
Countries with policy targets for renewable energy use		66	73

Annual percentage growth for 2008 was significant. Wind power grew by 29 percent and grid-connected solar PV by 70 percent. The capacity of utility-scale solar PV plants (larger than 200 kilowatts) tripled during 2008, to 3 GW. Solar hot water grew by 15 percent, and annual ethanol and biodiesel production both grew by 34 percent. Heat and power from biomass and geothermal sources continued to grow, and increased by about 8 percent.

In 2008 for the first time, more renewable energy than conventiónal power capacity was added in both the European Union and United States, demonstrating a "fundamental transition" of the world's energy markets towards renewables, according to a report released by REN21, a global renewable energy policy network based in Paris.

Three Generations of Technologies

The International Energy Agency (IEA) has defined three generations of renewable energy technologies, reaching back over 100 years:

- *First-generation technologies* emerged from the industrial revolution at the end of the 19th century and include hydropower, biomass combustion, geothermal power and heat. These technologies are quite widely used.
- *Second-generation technologies* include solar heating and cooling, wind power, modern forms of bioenergy, and solar photovoltaics. These are now entering markets as a result of research, development and demonstration (RDandD) investments since the 1980s. Initial investment was prompted by energy security concerns linked to the oil crises of the 1970s but the enduring appeal of these technologies is due, at least in part, to environmental benefits. Many of the technologies reflect significant advancements in materials.
- *Third-generation technologies* are still under development and include advanced biomass

gasification, biorefinery technologies, concentrating solar thermal power, hot-dry-rock geothermal power, and ocean energy. Advances in nanotechnology may also play a major role.

First-generation technologies are well established, second-generation technologies are entering markets, and third-generation technologies heavily depend on long-term RD and D commitments, where the public sector has a role to play.

Rationale for Renewables

Renewable energy technologies are essential contributors to the energy supply portfolio, as they contribute to world energy security, reduce dependency on fossil fuels, and provide opportunities for mitigating greenhouse gases. The IEA estimates that nearly 50 per cent of global electricity supplies will need to come from renewable energy sources in order to halve CO_2 emissions by 2050 and minimise significant, irreversible climate change impacts.

First-Generation Technologies

First-generation technologies are widely used in locations with abundant resources. Their future use depends on the exploration of the remaining resource potential, particularly in developing countries, and on overcoming challenges related to the environment and social acceptance.

Biomass

Biomass for heat and power is a fully mature technology which offers a ready disposal mechanism for municipal, agricultural, and industrial organic wastes. However, the industry has remained relatively stagnant over the decade to 2007, even though demand for biomass (mostly wood) continues to grow in many developing countries. One of the problems of biomass is that material directly combusted in cook stoves produces pollutants, leading to severe health and environmental consequences, although improved cook stove Programmemes are alleviating some of these effects. First-generation biomass technologies can be economically competitive, but may still require deployment support to overcome public acceptance and small-scale issues.

Hydroelectricity

Hydroelectric plants have the advantage of being long-lived and many existing plants have operated for more than 100 years. Hydropower is also an extremely flexible technology from the perspective of power grid operation. Large hydropower provides one of the lowest cost options in today's energy market, even compared to fossil fuels and there are no harmful emissions associated with plant operation.

However, there are several significant social and environmental disadvantages of large-scale hydroelectric power systems: dislocation of people living where the reservoirs are planned, release of significant amounts of carbon dioxide and methane during construction and flooding of the reservoir, and disruption of aquatic ecosystems and birdlife. Hydroelectric power is now more difficult to site in developed nations because most major sites within these nations are either already being exploited or may be unavailable for these environmental reasons. The areas of greatest hydroelectric growth are the growing economies of Asia. India and China are the development leaders; however, other Asian nations are also expanding hydropower.

There is a strong consensus now that countries should adopt an integrated approach towards managing water resources, which would involve planning hydropower development in co-operation with other water-using sectors.

Geothermal Power and Heat

Geothermal power plants can operate 24 hours per day, providing baseload capacity, and the world potential capacity for geothermal power generation is estimated at 85 GW over the next 30 years. However, geothermal power is accessible only in limited areas of the world. The costs of geothermal energy have dropped substantially from the systems built in the 1970s.

Geothermal heat generation can be competitive in many

countries producing geothermal power, or in other regions where the resource is of a lower temperature.

Second-Generation Technologies

Markets for second-generation technologies have been strong and growing over the past decade, and these technologies have gone from being a passion for the dedicated few to a major economic sector in countries such as Germany, Spain, the United States, and Japan. Many large industrial companies and financial institutions are involved and the challenge is to broaden the market base for continued growth worldwide.

Solar Heating

Solar heating systems are a well known second-generation technology and generally consist of solar thermal collectors, a fluid system to move the heat from the collector to its point of usage, and a reservoir or tank for heat storage. The systems may be used to heat domestic hot water, swimming pools, or homes and businesses. The heat can also be used for industrial process applications or as an energy input for other uses such as cooling equipment. In many warmer climates, a solar heating system can provide a very high percentage (50 to 75 per cent) of domestic hot water energy. An early solar heating boom took place during the 1940s in the United States, during which period institutional support for solar research and energy conservation measures imposed during World War II fueled significant advances in solar technology, which went as far as the development of a prototype prefabricated solar-heated home. A few proponents of this technology saw it as a clean alternative to polluting fuels, but the great majority of advocates, researchers, and investors saw it as a solution to high energy costs during the war; when those conditions changed and the 1950s ushered in a period of record low energy prices, interest rapidly waned, and the commercial development of solar heating systems was postponed to a later decade.

Photovoltaics

Solar array at Nellis Air Force Base. These panels track the sun in one axis. Credit: U.S. Air Force photo by Senior Airman Larry E. Reid Jr.
Photovoltaic (PV) cells, also called solar cells, convert light into electricity. In the 1980s and early 1990s, most photovoltaic modules were used to provide Remote Area Power Supply, but from around 1995, industry efforts have focused increasingly on developing building integrated photovoltaics and photovoltaic power stations for grid connected applications. Currently the largest photovoltaic power plant in North America is the Nellis Solar Power Plant (15 MW). There is a proposal to build a Solar power station in Victoria, Australia, which would be the world's largest PV power station, at 154 MW. Other large photovoltaic power stations, which are under construction, include the Girassol solar power plant (62 MW), and the Waldpolenz Solar Park (40 MW).

At the end of 2008, the cumulative global PV installations reached 15,200 MW. Photovoltaic production has been doubling every two years, increasing by an average of 48 percent each year since 2002, making it the world's fastest-growing energy technology. The top five photovoltaic producing countries are Japan, China, Germany, Taiwan, and the USA.

Wind Power

Some of the second-generation renewables, such as wind power, have high potential and have already realised relatively low production costs. At the end of 2008, worldwide wind farm capacity was 120,791 megawatts (MW), representing an increase of 28.8 percent during the year, and wind power produced some 1.3 per cent of global electricity consumption. Wind power accounts for approximately 19 per cent of electricity use in Denmark, 9 per cent in Spain and Portugal, and 6 per cent in Germany and the Republic of Ireland. However, it may be difficult to site wind turbines in some areas for aesthetic or environmental reasons.

The United States is an important growth area and installed U.S. wind power capacity reached 25,170 MW at the end of

2008. These are some of the largest wind farms in the United States, as of December 2008. (Table 2.4)

Table 2.4: Wind Frams in United States

Wind Farm	Installed Capacity (MW)	State
Altamont Pass Wind Farm	576	California
Capricorn Ridge Wind Farm	662	Texas
Fowler Ridge Wind Farm	750	Indiana
Horse Hollow Wind Energy Center	736	Texas
San Gorgonio Pass Wind Farm	619	California
Sweetwater Wind Farm	585	Texas
Tehachapi Pass Wind Farm	690	California

Solar Thermal Power Stations

Solar thermal power stations include the 354 MW Solar Energy Generating Systems power plant in the USA, Nevada Solar One (USA, 64 MW), Andasol 1 (Spain, 50 MW) and the PS10 solar power tower (Spain, 11 MW). Many other plants are under construction or planned, mainly in Spain and the USA. In developing countries, three World Bank projects for integrated solar thermal/combined-cycle gas-turbine power plants in Egypt, Mexico, and Morocco have been approved.

Modern Forms of Bioenergy

Brazil has one of the largest renewable energy Programmes in the world, involving production of ethanol fuel from sugar cane, and ethanol now provides 18 percent of the country's automotive fuel. As a result of this and the exploitation of domestic deep water oil sources, Brazil, which for years had to import a large share of the petroleum needed for domestic consumption, recently reached complete self-sufficiency in liquid fuels.

Production and use of ethanol has been stimulated through: (1) low-interest loans for the construction of ethanol distilleries; (2) guaranteed purchase of ethanol by the state-owned oil company at a reasonable price; (3) retail pricing of neat ethanol so it is competitive if not slightly Favourable to the gasoline-ethanol blend; and (4) tax incentives provided during the 1980s to stimulate the purchase of neat ethanol vehicles. Guaranteed purchase and price regulation were ended some years ago, with relatively positive results. In addition to these other policies, ethanol producers in the state of São Paulo established a research and technology transfer center that has been effective in improving sugar cane and ethanol yields. Most cars on the road today in the U.S. can run on blends of up to 10 per cent ethanol, and motor vehicle manufacturers already produce vehicles designed to run on much higher ethanol blends. Ford, DaimlerChrysler, and GM are among the automobile companies that sell "flexible-fuel" cars, trucks, and minivans that can use gasoline and ethanol blends ranging from pure gasoline up to 85 per cent ethanol (E85). By mid-2006, there were approximately six million E85-compatible vehicles on U.S. roads. The challenge is to expand the market for biofuels beyond the farm states where they have been most popular to date. Flex-fuel vehicles are assisting in this transition because they allow drivers to choose different fuels based on price and availability. The Energy Policy Act of 2005, which calls for 7.5 billion gallons of biofuels to be used annually by 2012, will also help to expand the market.

The growing ethanol and biodiesel industries are providing jobs in plant construction, operations, and maintenance, mostly in rural communities. According to the Renewable Fuels Association, the ethanol industry created almost 154,000 U.S. jobs in 2005 alone, boosting household income by $5.7 billion. It also contributed about $3.5 billion in tax revenues at the local, state, and federal levels.

Third-Generation Technologies

Third-generation renewable energy technologies are still under development and include advanced biomass gasification, biorefinery technologies, solar thermal power stations, hot-dry-rock geothermal power, and ocean energy. Third-generation technologies are not yet widely demonstrated or have limited commercialization. Many are on the horizon and may have potential comparable to other renewable energy technologies, but still depend on attracting sufficient attention and RDandD funding.

New Bioenergy Technologies

According to the International Energy Agency, cellulosic ethanol biorefineries could allow biofuels to play a much bigger role in the future than organizations such as the IEA previously thought. Cellulosic ethanol can be made from plant matter composed primarily of inedible cellulose fibers that form the stems and branches of most plants. Crop residues (such as corn stalks, wheat straw and rice straw), wood waste, and municipal solid waste are potential sources of cellulosic biomass. Dedicated energy crops, such as switchgrass, are also promising cellulose sources that can be sustainably produced in many regions of the United States.

Table 2.5: Commercial Cellulosic Ethanol Plants in the U.S. (Operational or under construction)

Company	Location	Feedstock
Abengoa Bioenergy	Hugoton, KS	Wheat straw
BlueFire Ethanol	Irvine, CA	Multiple sources
Colusa Biomass Energy Corporation	Sacramento, CA	Waste rice straw
Fulcrum BioEnergy	Reno, NV	Municipal solid waste
Gulf Coast Energy	Mossy Head, FL	Wood waste
KL Energy Corp.	Upton, WY	Wood
Mascoma	Lansing, MI	Wood
POET LLC	Emmetsburg, IA	Corn cobs
Range Fuels	Treutlen County, GA	Wood waste
SunOpta	Little Falls, MN	Wood chips
US Envirofuels	Highlands County, FL	Sweet sorghum
Xethanol	Auburndale, FL	Citrus peels

Ocean Energy

In terms of ocean energy, another third-generation technology, Portugal has the world's first commercial wave farm, the *Aguçadora Wave Park*, opened in 2008. The first stage of the farm uses three Pelamis P-750 machines generating a total of 2.25 MW. The cost of the farm is put at 8.5 million euro. A second phase of the project is now planned to increase the installed capacity from 2.25 MW to 21 MW using a further 25 Pelamis machines. Funding for a wave farm in Scotland was announced in February 2007 by the Scottish Executive, at a cost of over 4 million pounds, as part of a £13 million funding packages for ocean power in Scotland. The farm will be the world's largest with a capacity of 3 MW generated by four Pelamis machines.

In 2007, the world's first commercial tidal power station was installed in the narrows of Strangford Lough in Ireland. The 1.2 megawatt underwater tidal electricity generator, part of Northern Ireland's Environment and Renewable Energy Fund scheme, takes advantage of the fast tidal flow (up to 4 metres per second) in the lough. Although the generator is powerful enough to power a thousand homes, the turbine has minimal environmental impact, as it is almost entirely submerged, and the rotors pose no danger to wildlife as they turn quite slowly.

Enhanced Geothermal Systems

Enhanced geothermal systems, also known as hot dry rock geothermal, utilise new techniques to exploit resources that would have been uneconomical in the past. These systems are still in the research phase, and require additional RandD for new and improved approaches, as well as to develop smaller modular units that will allow economies of scale at the manufacturing level. Further government-funded research and close collaboration with industry will help to make exploitation of geothermal resources more economically attractive for investors.

Nanotechnology Thin-Film Solar Panels

Solar power panels that use nanotechnology, which can create circuits out of individual silicon molecules, may cost half as much as traditional photovoltaic cells, according to executives and investors involved in developing the products. Nanosolar has secured more than $100 million from investors to build a factory for nanotechnology thin-film solar panels. The company expects the factory to open in 2010 and produce enough solar cells each year to generate 430 megawatts of power.

Renewable Energy Industry

Global revenues for solar photovoltaics, wind power, and biofuels expanded from $76 billion in 2007 to $115 billion in 2008. New global investments in clean energy technologies—including venture capital, project finance, public markets, and research and development—expanded by 4.7 percent from $148 billion in 2007 to $155 billion in 2008.

Wind Power Companies

Currently three quarters of global wind turbine sales come from only four turbine manufacturing companies: Vestas, Gamesa, Enercon, and GE Energy. Vestas is the largest wind turbine manufacturer in the world with a 28 per cent market share. The company operates plants in Denmark, Germany, India, Italy, Britain, Spain, Sweden, Norway, Australia and China, and employs more than 20,000 people globally. After a sales slump in 2005, Vestas recovered and was voted *Top Green Company of 2006*. Vestas announced a major expansion of its North American headquarters in Portland, Oregon in December, 2008.

Gamesa, founded in 1976 with headquarters in Vitoria, Spain, is currently the world's second largest wind turbine manufacturer, after Vestas, and it is also a major builder of wind farms. Gamesa's main markets are within Europe, the US and China. In 2006, Europe accounted for 65 percent of Gamesa's sales, of which 40 percent were in Spain.

In 2004, German company Enercon installed a total of 1288 MW of wind power and had around 16 per cent of the global market share. Enercon constructed production facilities in Brazil in 2006, and has extended its presence there, as well as in the more traditional markets of Germany, India, Austria, UK, Canada and the Netherlands.

GE Energy has installed over 5,500 wind turbines and 3,600 hydro turbines, and its installed capacity of renewable energy worldwide exceeds 160,000 MW. GE Energy bought out Enron Wind in 2002 and also has nuclear energy operations in its portfolio.

Photovoltaic Companies

Q-Cells became the world's largest solar cell maker in 2007, producing nearly 400 MW of product. Longtime market leader Sharp Corporation found itself in second place with production of 370 MW in 2007, which the company blamed on a constrained supply of silicon. China's Suntech was close behind the leaders with more than 300 MW of output. Kyocera and its 200 MW output was a distant fourth in 2007.

Four new companies entered the top ranks in 2007. CdTe-cell maker First Solar was at fifth place, the only US-based and only thin-film supplier in the Top 10 companies. Asian players Motech Solar (Taiwan), Yingli Green Energy (China), and JA Solar Holdings (China/Australia) rounded out the Top 10 ranking, pushing aside some established players like Mitsubishi Electric, Schott, and BP Solar.

Other Companies

SunOpta is located in Canada and was founded in 1973. Its operations are divided between SunOpta Food (organics), Opta Minerals, and SunOpta BioProcess (bioethanol). SunOpta's fastest growing business segment is the BioProcess Group, which is a leading developer of technology in the cellulosic ethanol market. SunOpta's BioProcess Group specializes in the design, construction and optimization of biomass conversion equipment and facilities. They have over 30 years experience delivering biomass solutions worldwide and use innovative technologies to produce cellulosic ethanol and cellulosic butanol. Raw materials include wheat straw, corn stover, grasses, oat hulls and wood chips.

Non-Technical Barriers to Acceptance

There have been several recent reports which have identified a range of "non-technical barriers" to renewable energy use. These barriers are impediments which put renewable energy at a marketing, institutional, or policy disadvantage relative to other forms of energy. Key barriers include:

• Lack of government policy support, which includes the lack of policies and regulations supporting deployment of renewable energy technologies and the presence of policies and regulations hindering renewable energy development and supporting conventional energy development. Examples include subsidies for fossil-fuels, insufficient consumer-based renewable energy incentives, government underwriting for nuclear plant accidents, and complex zoning and permitting processes for renewable energy.

- Lack of information dissemination and consumer awareness.
- Higher capital cost of renewable energy technologies compared with conventional energy technologies.
- Difficulty overcoming established energy systems, which includes difficulty introducing innovative energy systems, particularly for distributed generation such as photovoltaics, because of technological lock-in, electricity markets designed for centralized power plants, and market control by established operators. As the Stern Review on the Economics of Climate Change points out:

National grids are usually tailored towards the operation of centralised power plants and thus favour their performance. Technologies that do not easily fit into these networks may struggle to enter the market, even if the technology itself is commercially viable. This applies to distributed generation as most grids are not suited to receive electricity from many small sources. Large-scale renewables may also encounter problems if they are sited in areas far from existing grids.

- Inadequate financing options for renewable energy projects, including insufficient access to affordable financing for project developers, entrepreneurs and consumers.
- Imperfect capital markets, which includes failure to internalize all costs of conventional energy (e.g., effects of air pollution, risk of supply disruption) and failure

to internalize all benefits of renewable energy (e.g., cleaner air, energy security).

- Inadequate workforce skills and training, which includes lack of adequate scientific, technical, and manufacturing skills required for renewable energy production; lack of reliable installation, maintenance, and inspection services; and failure of the educational system to provide adequate training in new technologies.
- Lack of adequate codes, standards, utility interconnection, and net-metering guidelines.
- Poor public perception of renewable energy system aesthetics.
- Lack of stakeholder/community participation and co-operation in energy choices and renewable energy projects.

With such a wide range of non-technical barriers, there is no "silver bullet" solution to drive the transition to renewable energy. So ideally there is a need for several different types of policy instruments to complement each other and overcome different types of barriers.

A policy framework must be created that will level the playing field and redress the imbalance of traditional approaches associated with fossil fuels. The policy landscape must keep pace with broad trends within the energy sector, as well as reflecting specific social, economic and environmental priorities.

Public Policy Landscape

Public policy has a role to play in renewable energy commercialization because the free market system has some fundamental limitations. As the Stern Review points out:

In a liberalised energy market, investors, operators and consumers should face the full cost of their decisions. But this is not the case in many economies or energy sectors. Many policies distort the market in favour of existing fossil fuel technologies.

Lester Brown goes further and suggests that the market "does not incorporate the indirect costs of providing goods or services into prices, it does not value nature's services adequately, and it does not respect the sustainable-yield thresholds of natural systems". It also Favours the near term over the long term, thereby showing limited concern for future generations. Tax and subsidy shifting can help overcome these problems.

Shifting Taxes

Tax shifting involves lowering income taxes while raising levies on environmentally destructive activities, in order to create a more responsive market. It has been widely discussed and endorsed by economists. For example, a tax on coal that included the increased health care costs associated with breathing polluted air, the costs of acid rain damage, and the costs of climate disruption would encourage investment in renewable technologies. Several Western European countries are already shifting taxes in a process known there as environmental tax reform, to achieve environmental goals.

A four-year plan adopted in Germany in 1999 gradually shifted taxes from labour to energy and, by 2001, this plan had lowered fuel use by 5 percent. It had also increased growth in the renewable energy sector, creating some 45,400 jobs by 2003 in the wind industry alone, a number that is projected to rise to 103,000 by 2010. In 2001, Sweden launched a new 10-year environmental tax shift designed to convert 30 billion kroner ($3.9 billion) of taxes on income to taxes on environmentally destructive activities. Other European countries with significant tax reform efforts are France, Italy, Norway, Spain, and the United Kingdom. Asia's two leading economies, Japan and China, are considering the adoption of carbon taxes.

Shifting Subsidies

Subsidies are not inherently bad as many technologies and industries emerged through government subsidy schemes. The Stern Review explains that of 20 key innovations from the past 30 years, only one of the 14 they could source was funded entirely by the private sector and nine were totally funded by the public sector. In terms of specific examples, the Internet was the result of publicly funded links among computers in government laboratories and research institutes. And the combination of the federal tax deduction and a robust state tax deduction in California helped to create the modern wind power industry.

But just as there is a need for tax shifting, there is also a need for subsidy shifting. Lester Brown has argued that "a world facing the prospect of economically disruptive climate change can no longer justify subsidies to expand the burning of coal and oil. Shifting these subsidies to the development of climate-benign energy sources such as wind, solar, biomass, and geothermal power is the key to stabilizing the earth's climate."

Some countries are eliminating or reducing climate disrupting subsidies and Belgium, France, and Japan have phased out all subsidies for coal. Germany reduced its coal subsidy from $5.4 billion in 1989 to $2.8 billion in 2002, and in the process lowered its coal use by 46 percent. Germany plans to phase out this support entirely by 2010. China cut its coal subsidy from $750 million in 1993 to $240 million in 1995 and more recently has imposed a tax on high-sulfur coals.

While some leading industrial countries have been reducing subsidies to fossil fuels, most notably coal, the United States has been increasing its support for the fossil fuel and nuclear industries.

Renewable Energy Targets

Setting national renewable energy targets can be an important part of a renewable energy policy and these targets are usually defined as a percentage of the primary energy and/or electricity generation mix. For example, the European Union has prescribed an indicative renewable energy target of 12 per cent of the total EU energy mix and 22 per cent of electricity consumption by 2010. National targets for individual EU Member States have also been set to meet the overall target. Other developed countries with defined national or regional targets include Australia, Canada, Japan, New Zealand, Norway, Switzerland, and some US States.

National targets are also an important component of renewable energy strategies in some developing countries. Developing countries with renewable energy targets include

China, India, Korea, Indonesia, Malaysia, the Philippines, Singapore, Thailand, Brazil, Israel, Egypt, Mali, and South Africa. The targets set by many developing countries are quite modest when compared with those in some industrialized countries.

Renewable energy targets in most countries are indicative and nonbinding but they have assisted government actions and regulatory frameworks. The United Nations Environment Programme has suggested that making renewable energy targets legally binding could be an important policy tool to achieve higher renewable energy market penetration.

Employment

Current employment in the renewable energy sector and supplier industries is estimated at 2.3 million worldwide. The wind power industry employs some 300,000 people, the photovoltaics sector an estimated 170,000, and the solar thermal industry more than 600,000. Over 1 million jobs are found in the biofuels industry, associated with growing and processing a variety of feedstocks into ethanol and biodiesel.

Recent Developments

A number of events in 2006 pushed renewable energy up the political agenda, including the US mid-term elections in November, which confirmed clean energy as a mainstream issue. Also in 2006, the Stern Review made a strong economic case for investing in low carbon technologies now, and argued that economic growth need not be incompatible with cutting energy consumption. According to a trend analysis from the United Nations Environment Programme, climate change concerns coupled with recent high oil prices and increasing government support are driving increasing rates of investment in the renewable energy and energy efficiency industries.

Investment capital flowing into renewable energy reached a record US$77 billion in 2007, with the upward trend continuing in 2008. The OECD still dominates, but there is now increasing activity from companies in China, India and Brazil. Chinese companies were the second largest recipient of venture capital in 2006 after the United States. In the same year, India was the largest net buyer of companies abroad, mainly in the more established European markets.

Global revenues for solar photovoltaics, wind power, and biofuels expanded from $75.8 billion in 2007 to $115.9 billion in 2008. For the first time, one sector alone, wind, had revenues exceeding $50 billion. New global investments in clean energy technologies—including venture capital, project finance, public markets, and research and development—expanded by 4.7 percent from $148.4 billion in 2007 to $155.4 billion in 2008.

Continued growth for the renewable energy sector is expected in the mid-to long-term, but 2009 will be a year of refocus, consolidation, or retrenchment for many companies. At the same time, new government spending, regulation, and policies should help the industry weather the current economic crisis better than many other sectors. Most notably, U.S. President Barack Obama's American Recovery and Reinvestment Act of 2009 includes more than $70 billion in direct spending and tax credits for clean energy and associated transportation Programmes. This policy-stimulus combination represents the largest federal commitment in U.S. history for renewables, advanced transportation, and energy conservation initiatives. Based on these new rules, many more utilities are expected to strengthen their clean-energy Programmes.

Clean Edge suggests that the commercialization of clean energy will help countries around the world pull out of the current economic malaise.

Sustainable Energy

Moving towards energy sustainability will require changes not only in the way energy is supplied, but in the way it is used, and reducing the amount of energy required to deliver various goods or services is essential. Opportunities for improvement on the demand side of the energy equation are as rich and diverse as those on the supply side, and often offer significant economic benefits.

Renewable energy and energy efficiency are said to be the "twin pillars" of sustainable energy policy. Any serious vision of a sustainable energy economy requires commitments to both renewables and efficiency. The American Council for an Energy-Efficient Economy has explained that both resources must be developed in order to stabilize and reduce carbon dioxide emissions:

Efficiency is essential to slowing the energy demand growth so that rising clean energy supplies can make deep cuts in fossil fuel use. If energy use grows too fast, renewable energy development will chase a receding target. Likewise, unless clean energy supplies come online rapidly, slowing demand growth will only begin to reduce total emissions; reducing the carbon content of energy sources is also needed.

CASE STUDY: RENEWABLE FUELS ASSOCIATION OF USA

The **Renewable Fuels Association** (RFA) is an American lobbying organization which promotes policies, regulations, and research and development initiatives that will lead to the increased production and use of ethanol fuel. First organized in 1981, RFA serves as a voice of advocacy for the ethanol industry, providing research data and industry analysis to its members, to the public via the media, to the United States Congress, as well as to related federal and state agencies.

UNDERSTANDING BIO FUEL SYSTEMS

Bio Fuel Systems is a wholly Spanish owned firm that has developed a method of breeding plankton and turning the marine plants into oil, providing a potentially inexhaustible source of clean fuel. It was formed in 2006 in eastern Spain after three years of research by scientists and engineers connected with the University of Alicante.

The process it has developed converts energy, based on three elements: solar energy, photosynthesis and an electromagnetic field. Its new fuel is said to reduce CO_2, is free of other contaminants like sulphur dioxide and would be cheaper than standard fossil fuel that is available now. Their system of bioconversion is about 400 times more productive than any other plant-based system producing oil or ethanol.

CASE STUDY: FUEL BIO

Fuel Bio Holdings, LLC	
Type	Private Company
Founded	December, 2006
Founder (s)	Marty Borruso
Headquarters	Elizabeth, New Jersey, USA
Area served	Domestic and International
Key people	John Borruso
	(Government Affairs Liaison)
	Dave Johnson
	(CFO)
Industry	Biofuels
Products	Biodiesel, Glycerine
Website	FuelBio.com

Fuel Bio Holdings, LLC, is an American commercial producer of Biodiesel. Using proprietary technology and chemistry to produce ASTM D6751 product from multiple virgin feed stock sources, Fuel Bio's production plant is recognized as the largest of its kind in the northeast United States. Located at New York Terminals in Elizabeth New Jersey, Fuel Bio's operation is capable of producing a name plate capacity of 50,000,000 gallons of fuel each year.

Association with Green Movement

As a producer of biodiesel, an alternative fuel. Fuel Bio's efforts help further the cause of the green movement

In May 2007, The New Jersey Star Ledger reported that Fuel Bio "is trying to plant a green footprint on New Jersey's notorious Chemical Coast." Fuel Bio is the first alternative energy company located in this area.

CASE STUDY: AQUAFLOW BIONOMIC CORPORATION

Aquaflow Bionomic Corporation (ABC) is a startup that has set itself the objective to be the first company in the world to economically produce biofuel from wild algae harvested from open-air environments, to market it, and meet the challenge of increasing demand.

The company's office is in Nelson, New Zealand and laboratory and field development work is done in Malborough.

The world's first wild algae bio-diesel (a 5 per cent mixture with 95 per cent normal diesel), produced in New Zealand by Aquaflow Bionomic Corporation, was successfully test driven in Wellington on December 15, 2006 by the Minister for Energy and Climate Change Issues, David Parker.

The fuel was a mixture of 95 per cent normal diesel and 5 per cent laboratory derived algae based oil. Total algae based oil was less than 2 liters.

Company Directors

The directors of the company are Vicki Buck, Nicholas Gerritsen, Barrie Leay and David Milroy.

Technology

ABC harvests algae directly from the settling ponds of standard Effluent Management (EM) Systems and other nutrient-rich water. The process can be used in many industries that produce a waste stream, including the transport, dairy, meat and paper industries.

It uses dissolved air flotation (DAF) plus the addition of floculant to lift the algae to the surface followed by a belt press system to extract the algae. The extract is a 8 to 10 per cent algae concentration the rest is water.

Leay: The two-step process firstly optimises the ponds' productive capacity, and secondly, determines the most efficient and economic way of harvesting the pond algae. Algae are provided with full opportunity to exploit the nutrients available in the settling ponds, thereby cleaning up the water. The algae are then harvested to remove the remaining contaminant. A last stage of bio-remediation, still in development, will ensure that the water discharge from the process exceeds acceptable quality standards.

Leay: The water and sludge treatment process is an elegant clean-up and management service to councils responsible for sewage treatment systems while also generating a low-cost feedstock for conversion to fuel.

Leay: The result is an algae-based extract that will ultimately be converted to an alternative fuel source. This has not yet been achieved.

Experimental processing involves adding chemicals and processing the 8 to 10 per cent algae—water mixture at super critical water pressures and temperatures, approximately 230 bar (3500 psi) and 400 C. (750F) This process has proved problematical—the company is now looking elsewhere for technology to undertake the conversion process. The small output of sludge from the existing plant will need extensive further processing to turn it into a viable product. This has not been achieved. Only small laboratory scale samples have been produced. Most samples have been produced by super critical CO_2 extraction by a third party company. The whole process is at present energy negative by a large margin.

Leay: "In the Aquaflow process there is no conflict with land use or with the production of food crops as is occurring in America and Europe, which is becoming an increasing world problem."

Leay explains that the essence of Aquaflow's process is to use algae to capture 'current sunlight' through photosynthesis, which is the identical process used millions of years ago, when the world's oil and gas deposits were laid down, using 'ancient sunlight'.

"But as we all know those sources of captured 'ancient sunlight' are now being rapidly depleted as we reach Peak Oil.

"An extraordinarily beneficial by-product of the Aquaflow process is potentially releasing a clean water resource of millions of litres of clean water—to be recycled and available for use in irrigation, industrial washing, cooling, and so on."

The present installed harvest process only removes algae, the discharge water still needs a great deal of further treatment.

Marlborough Site

The Marlborough sewage pond details are:

Leay: 60ha of open oxidation ponds, serving a population of 27,000 with a mix of municipal and agro-industrial waste—including a significant wine industry—and with an annual water flow of 5 billion litres.

If the above figures are correct and with present pond algae levels of about 100g per cu meter this would yield approximately 500 cu meters of algae sludge. A 20 per cent conversion to biocrude would yield 100,000 liters, worth about $35,000 at today's oil price of US$70 per barrel. This is any before extraction costs are taken into account.

Leay: "We have now achieved commercial scale continuous harvesting of tonnes of wild algae at the Marlborough oxidation ponds so we can take the step up to commercial scale production of biocrude," says Aquaflow chairman, Barrie Leay.

The present installed equipment can harvest 300 to 400 Kg of processable material per day of which 8-10 per cent is algae—i.e. approximately 30-35kg of algae per day.

Media

In December 2006, Aquaflow Bionomic Corporation demonstrated how its algae-based fuel additive works in the standard diesel engine of a production vehicle, outside Parliament buildings in Wellington.

The fuel was a mixture of 95 per cent diesel and 5 per cent laboratory derived algae based oil. Total algae based oil was less than 2 liters.

As of July 2009 company has spent approximately NZ$3,000,000 trying to develop a process.

CASE STUDY: EAST TENNESSEE CLEAN FUELS COALITION

The East Tennessee Clean Fuels Coalition (ETCFC) was founded in February 2002 by Jonathan Overly with funding from the Tennessee Department of Economic and Community Development, Energy Division, and the University of Tennessee.

The ETCFC is a non-profit organization [501(c)3 corporation] that serves the entire eastern portion of the state of Tennessee and is a designated member of the [U.S. DOE "Clean Cities" Programme]

The main focus of the ETCFC is to get alternative fuels in use in that region to do two things:

1.   reduce regional and national oil dependence, and
2.   improve air quality.

The ETCFC directly engages regional fleets to help them move to using alternative fuels, holds public workshop and other events that works to do the same, and works on the education front for alternative fuels at many levels including mainly fleets and in K-12 schools.

The main alternative fuels used in East Tennessee and other parts of the U.S. include biodiesel, electricity, ethanol, natural gas and propane (or LPG). The most prominent alternative fuels used in East Tennessee currently (2007) are biodiesel and ethanol.

Much of the ETCFC's efforts are focused on putting in public and private alt-fuel refueling stations so that fleets and individuals alike can use these fuels. One such effort is the Biodiesel Brigade, a group of individuals encouraged by the ETCFC to prominently display their use of biodiesel through vehicle magnets and decals.

The ETCFC maintains a website that holds much information about their meetings and activities. The organization holds open, monthly meetings in Knoxville and every-other-month meetings in Chattanooga and Kingsport in attempts to provide the regional community plenty of opportunities to get involved and find out how they can begin using these fuels.

ETCFC Partner Organizations

UTK

- ORNL
- Great Smoky Mountains National Park
- Eastman Chemical Company
- East Tennessee cities: Maryville, Sevierville, Alcoa,

Gatlinburg, Knoxville, Chattanooga, Athens, Oak Ridge, Kingsport, Johnson City
- East Tennessee counties: Sevier County, Blount County, Knox County, Hamilton County, Anderson County, McMinn County, Sullivan County, Loudon County, Hawkins County, Cumberland County
- Alcoa
- Pilot Travel Centers
- Y-12 National Security Complex
- BBandT
- BAE Systems
- Nuclear Fuel Services
- Tate and Lyle

CASE STUDY: BIOENERGY PROGRAMME

The **Bioenergy Programme** was an initiative of the United States Department of Agriculture Farm Service Agency. The Programme started on October 1, 2002, and was terminated on June 30, 2006.

The Programme made payments to ethanol and biodiesel producers who expand their production capacity from eligible commodities. In the year of the expansion, the Programme payments help offset the cost of the additional commodity feedstocks (usually corn for ethanol and soybeans for biodiesel) needed for the expansion. In addition to corn and soyabeans, also barley, grain sorghum, oats, rice, wheat, sunflower seed, canola, crambe, rapeseed, safflower, sesame seed, flaxseed, mustard seed, and cellulosic crops were defined as eligible commodities if grown on farms for production of ethanol or biodiesel.

The Programme was codified into law by the United States 2002 farm bill (P.L. 107-171, Sec. 9010). It was funded through the Commodity Credit Corporation. Spending for the Programme was capped at $150 million annually between 2002 and 2006. Payments to a single producer were limited by 5 per cent of available funding. The Congressional Budget Office estimated that $204 million total was spent between 2002 and 2006.

The new **Bioenergy Programme for Advanced Biofuels** is planed to be implemented for fiscal year 2009.

CASE STUDY: US BIOENERGY ACTION PLAN

On April 25, 2006, California Governor Arnold Schwarzenegger issued Executive Order S-06-06, establishing targets for the use and production of biofuels and biopower and directing state agencies to work together to advance biomass Programmes in California while providing environmental protection and mitigation.

The Bioenergy Action Plan (Plan) provides the specific actions and timelines that the agencies have agreed to take to implement the Executive Order.

UNDERSTANDING SCENARIO OF BIOENERGY IN CHINA

China has set the goal of attaining one per cent of its renewable energy generation through *bioenergy* in 2020. The development of bioenergy in China is needed to meet the rising energy demand. Several institutions are involved in this development, most notably the Asian Development Bank and China's Ministry of Agriculture. There is also an added incentive to develop the bioenergy sector which is to increase the development of the rural agricultural sector. As of 2005, bioenergy use has reached more than 20 million households in the rural areas, with methane gas as the main biofuel. Also more than 4000 bioenergy facilities produce 8 billion cubic metres every year of methane gas. As of 2010, electricity generation by bioenergy is expected to reach 5 GW, and 30 GW by 2020. The annual use of methane gas is expected to be 19 cubic kilometers by 2010, and 40 cubic kilometers by 2020.

(*i*)   China is the world's third-largest producer of ethanol, after Brazil and the United States.(RFA)

(*ii*)   As of 2006, 20 per cent of "gasoline" consumed is actually a 10 per cent ethanol-gasoline blend.

(*iii*)   Although only 0.71 per cent of the country's grain yield

(3.366 million tons of grain) in 2006 was used for production of ethanol, concern has been expressed over potential conflicts between demands for food and fuel, as crop prices rose in late 2006.

Major Developments

(i) Chinese Enterprise Wins Award for Energy Efficiency, 23 June 2007 from chinagate.com.cn. Daxu wins an Ashden Award for producing over 25,000 efficient stoves that can burn crop waste for cooking and hot water. More details available here.

(ii) CASP agreement to benefit biofuel producers in Mekong, 11 April 2007 from Biofuelreview.com. Agriculture ministers from 6 countries, Cambodia, China, Laos, Myanmar, Thailand, and Vietnam have endorsed the Core Agricultural Support Programme, which will work toward increasing trade and investment in agriculture in the Greater Mekong Subregion. A major focus will be helping farmers reap the benefits of new energy crops and related technologies.

(iii) Chinese Biofuels Expansion Threatens Ecological Balance, March 27, 2007 from Renewable Energy Access. A recent agreement between China's top forestry authority and one of the nation's biggest energy giants to develop biofuels plantations in the southwest may come at great environmental loss to the region's forests and biological diversity.

(iv) China plans to plant an area the size of England with biofuel trees 8 February 2007 from China Daily. China will plant 130,000 square kilometres, an area the size of England, with Jatropha trees to produce oil amounting to nearly 6 million tons of biodiesel every year. The jatropha trees can also provide wood fuel for a power plant with an installed capacity of 12 million kilowatts, will account for 30 per cent of the country's renewable energy by 2010.

(v) Ethanol fuels hopes of China's small farmers 29 January

2007 from The Standard. Beijing's push to create more ethanol from cassava and sugar cane may benefit farmers in Guanxi, but with China already a net-importer of tapioca and sugar it is not clear that there will be enough feedstocks to go around.

(*vi*) Biodiesel Sweeps China in Controversy 23 January 2007 from Renewable Energy Access. China is looking at new biodiesel feedstocks including a new variety of rapeseed, Chinese pistachio and jatropha. However, standards and regulations are lacking and concerns over food vs. fuel are growing.

(*vii*) Biofuels eat into China's food stocks—21 December 2006 from Asia Times Online. China has clamped down on the use of corn and other edible grains for producing biofuels due to concerns that it will impact on food security.

(*ix*) China Clean Energy outlines plan to expand biodiesel capacity using palm oil leavings as a feedstock (go to story)—18 December 2006 from Biofuel Review.

(*ix*) China halts expansion of corn-based ethanol industry to arrest food price rise (go to story)—20 December 2006 from newKerala.com.

(*xi*) Shaanxi Mothers win top environmental award, 16 June 2006 from blueskieschina.com. Shaanxi Mothers wins an Ashden Award for the fitting of almost 1,300 biogas systems in farming households across China's Shaanxi Province. More details available here.

Major Policy Initiative

(*i*) The Renewable Energy Law of the People's Republic of China—English translation of the law, which took effect 1 January 2006.

Major Targets

(*i*) Target of 10 per cent renewable energy of the country's total energy consumption by 2010.

(*ii*) Alternative fuels: 6 million tons by 2010 and 15 million tons by 2020.

(*iii*)   Target of 50 per cent use of ethanol-blended gasoline by 2010.

(*iv*)   China has an annual production capacity of 1.02 million tons of ethanol. (source: People's Daily Online)

Issues in Biofuel production

(*i*)   "Corn accounted for 76 percent of the 1.02 million tons of ethanol produced" in 2005.

(*ii*)   "Non-grain crops in China could eventually produce as much as 300 million tonnes of ethanol a year, according to a report on the National Development and Reform Commission's website.

Regional, Government and Non-Government Organization (NGOs) Involved

(*i*)   Core Agriculture Support Programme—A programme that includes southern China and the countries of the Mekong Subregion in South-East Asia, that provides support for biofuel feedstock and other agricultural programmes.

(*ii*)   english.gov.net is the main English language portal for the Chinese Government. Many agencies do not yet have English language pages.

(*iii*)   China's circulars on bioenergy policy have been co-released by the following agencies:

(*iv*)   National Development and Reform Commission. English overview of the NDRC, which is "a macro-economic regulatory department, with a mandate to develop national economic strategies". It deals with China's targets for biofuels.

(*v*)   NDRC (Chinese only)

(*vi*)   State Environmental Protection Agency (SEPA)

(*vii*)   Ministry of Finance. The Ministry of Finance helps regulate subsidies and tax policy for bioenergy.

(*viii*)   MOF (Chinese)

(*ix*)   Ministry of Agriculture (Chinese)

(*x*)   State Forestry Administration (Chinese)

(*xi*)   State Administration of Taxation

(*xii*)   China National Petroleum Corporation

(*xiii*)   CNPC, through its subsidiary Jilin Fuel Ethanol Ltd. Co, built China's first ethanol plant using corn as a feedstock and now runs several other ethanol projects. Press Release: China's First fuel ethanol line into production in Jilin 27 November, 2003.

(*xiv*)   China National Cereals, Oils and Foodstuffs Corporation (Chinese only)

(*xv*)   Plans to invest more than US$1 billion in ethanol projects to increase production capacity to 3 million tons.

(*xvi*)   Currently owns an ethanol plant in Heilongjiang Province and has a 20 percent stake in another plant in Jilin Province, both using corn as a feedstock

(*xvii*)   The company is constructing an ethanol plant, which will use cassava as a feedstock, in the Guangxi Zhuang Autonomous Region.

(*xviii*)   Is awaiting Government permission to build two 300,000-ton-per-year ethanol plants in Hebei Province, using corn and sweet potatoes, and Liaoning Province, using only sweet potatoes.

(*xix*)   BBCA (Mostly Chinese) Large scale ethanol and biomass producer, using corn and cassava. Also doing research into cellulosic ethanol.

Scenario of Bioenergy in New Zealand

Bioenergy comes from a wide range of sources, including trees or crops grown specifically for their energy content. They also include indirect sources of biomass derived from waste products from industrial, commercial, agricultural and domestic activities. These wastes include straw, animal manure, animal fat and municipal solid waste. Biomass can be used to supply either electricity, heat, or transport fuels. Biogas is a gas produced during the biological breakdown of organic matter which can be used to provide energy. Biogas is also called Landfill gas. Wood biomass provides about five per cent (35PJ) of New Zealand's main energy supply. Wood biomass is mainly used by the timber industry, which burns residue wood to provide heat energy. Residential wood burners and open fires account for about 5PJ of our biomass use. There are opportunities to significantly increase our use of biomass as an energy source. EECA is administering the Wood Energy Grant Scheme (WEGS), which aims to increase the use of wood residues as an energy source.

Biofuels for Transport

Some biofuels such as biodiesel are superior fuels for transport, even though they are currently more expensive than some other fossil fuel alternatives.

Wood Processing Strategy

EECA contributed to the joint forest industry/government Wood Processing Strategy by co-chairing the Energy Working Group with the Forest Industries Council. The project recognised that energy is as important to Wood Processing Industry's infrastructure as transport and skilled labour. The project set out to review:

(*i*)   the supply and demand for energy in new wood processing regions,

(*ii*) the effects of energy market structures on the development of the sector,

(*iii*) likely changes in energy costs,

(*iv*) opportunities for the development of Bioenergy in the sector.

Bioenergy Industry Association

The Bioenergy Association of New Zealand (BANZ) represents the commercial bioenergy sector and provides a single point of contact on the bioenergy market. BANZ provides members with opportunities for networking, the latest technical and market information, opportunities for brand exposure and an avenue to influence market policies and issues. EECA works with BANZ to:

(*i*)   Identify and further opportunities for bioenergy in New Zealand.

(*ii*) Support industry workshops.

(*iii*) Help with feasibility and case studies.

TOWARDS A BIOENERGY VILLAGE

A bio-energy village is a regional orientated concept for the use of renewable energy sources in rural areas. The system uses biomass from local agriculture and forestry in a Biogas power plant in order to supply the energy demand of a village preferably complete, as electricity and district heating.

These villages tend to be self-powered and independent from external grids; despite being connected to overland grids for feeding surplus energy. The term bio-energy village describes only the energy dependency on fresh biologic material, whereas an ecovillage includes much more differentiated networks.

Examples of such villages are Jühnde near Göttingen and Mauenheim near Tuttlingen in Germany.

Energy Production

Liquid manure, grass, silage and other raw materials from agriculture are fermented in a biological gas facility. The biogas produced fuels a combined heat and power plant (CHP). The heat is distributed over a district heating system, while power is fed into a local electricity grid. In winter additional heat requirements can be supplied by an additional heating plant, in which wood chips or straw are burned.

Existing Projects

Jühnde

The first bio-energy village in Germany is Jühnde in the district of Göttingen. A project of the Interdisciplinary Centre For Sustainable Development (IZNE) at the University of Göttingen, and completed in January 2006, the project supplies the heat requirement of the village, and produces twice as much electricity as is used. It has been estimated that the participating households save €750 per year in energy costs.

Mauenheim

In Mauenheim, Baden-Württemberg, a bio-energy village has been developed in Immendingen in the district of Tuttlingen, with approximately 400 inhabitants and 148 buildings. The biogas facility and wood chip heating system are supplemented by a solar energy system. The project started operation in 2006. It has been calculated that about 1900 tonnes of CO_2 per year will be saved.

Rai Breitenbach

The Breuberger village of Rai Breitenbach in the Odenwald (approximately 890 inhabitants) is in the process of becoming a bio-energy village. At present the project is still in the planning stage. A feasibility study has been completed and a co-operative created to carry out the project, which is expected to be completed in 2008.

Freiamt

The village Freiamt in the Black Forest with 4300 inhabitants is using all forms of renewable energy. A biogas plant, Solar power, wind and water energy produce about 14 million kwh energy annually, about 3 million more than needed. Around 150 solar collectors are used for water heating.

Considerations

Advantages

- No climatically harmful waste gases are released. The risks and waste disposal problem of nuclear energy are avoided.
- The energy produced is often cheaper for consumers than conventional energy.
- Local resources are used, saving transportation energy costs.
- Energy costs are locally spent, strengthening the local economy and creating jobs.

Disadvantages

- Capital outlays are high.
- The system works only if a large majority of the inhabitants participate, and can be attached to the local heating supply network.

3

Biomass and Bioenergy:
With Special Focus on Heating System
and Power Plants

BIOENERGY FROM BIOMASS

Biomass, as a renewable energy source, refers to living and recently dead biological material that can be used as fuel or for industrial production. In this context, biomass refers to plant matter grown to generate electricity or produce for example trash such as dead trees and branches, yard clippings and wood chips biofuel, and it also includes plant or animal matter used for production of fibers, chemicals or heat. Biomass may also include biodegradable wastes that can be burnt as fuel. It excludes organic material which has been transformed by geological processes into substances such as coal or petroleum. Industrial biomass can be grown from numerous types of plants, including miscanthus, switchgrass, hemp, corn, poplar, willow, sorghum, sugarcane, and a variety of tree species, ranging from eucalyptus to oil palm (palm oil). The particular plant used is usually not very important to the end products, but it does affect the processing of the raw material. Production of biomass is a growing industry as interest in sustainable fuel sources is growing. Although fossil fuels have their origin in ancient biomass, they are not considered biomass by the generally accepted definition because they contain carbon that has been "out" of the carbon cycle for a very long time. Their combustion therefore disturbs the carbon dioxide content in the atmosphere. Plastics from biomass, like some recently developed to dissolve in seawater, are made the same way as petroleum-based plastics, are actually cheaper to manufacture and meet or exceed most performance standards. But they lack the same water resistance or longevity as conventional plastics.

Assessing Environmental Impact of Biomass Energy

Biomass is part of the carbon cycle. Carbon from the atmosphere is converted into biological matter by photosynthesis. On death or combustion the carbon goes back into the atmosphere as carbon dioxide (CO_2). This happens over a relatively short timescale and plant matter used as a fuel can be constantly replaced by planting for new growth. Therefore a reasonably stable level of atmospheric carbon results from its use as a fuel. It is accepted that the amount of carbon stored in dry wood is approximately 50 per cent by weight. Though biomass is a renewable fuel, and is sometimes called a "carbon neutral" fuel, its use can still contribute to global warming. This happens when the natural carbon equilibrium is disturbed; for example by deforestation or urbanization of green sites. When biomass is used as a fuel, as a replacement for fossil fuels, it still puts the same amount of CO_2 into the atmosphere. However, when biomass is used for energy production it is widely considered carbon neutral, or a net reducer of greenhouse gasses because of the offset of methane that would have otherwise entered the atmosphere. The carbon in biomass material, which makes up approximately fifty percent of its dry-matter content, is already part of the atmospheric carbon cycle. Biomass absorbs CO_2 from the atmosphere during its growing lifetime, after which its carbon reverts to the atmosphere as a mixture of CO_2 and methane (CH_4), depending on the ultimate fate of the biomass material. CH_4 converts to CO_2 in the atmosphere, completing the cycle. In contrast to biomass carbon, the car from long-term storage, and adds it to the stock of carbon in the atmospheric cycle. Energy produced from crap residues displaces the production of an equivalent amount of energy from fossil fuels, leaving the fossil carbon in storage. It also shifts the composition of the recycled carbon emissions associated with the disposal of the biomass residues from a mixture of CO_2 and CH_4, to almost exclusively CO_2. In the absence of energy production applications, biomass residue carbon would be recycled to the atmosphere through some combination of rotting (biodegradation) and open burning. Rotting produces a mixture of up to fifty percent CH_4, while open burning produces five to ten percent CH_4. Controlled combustion in a power plant converts virtually all of the carbon in the biomass to CO_2. Because CH_4 is a much stronger greenhouse gas than CO_2, shifting CH_4 emissions to CO_2 by converting biomass residues to energy significantly reduces the greenhouse warming potential of the recycled carbon associated with other fates or disposal of the biomass residues. The existing commercial biomass power generating industry in the United States, which consists of approximately 1,700 MW (megawatts) of operating capacity actively supplying power to the grid, produces about 0.5 percent of the U.S. electricity supply. This level of biomass power generation avoids approximately 11 million tons per year of CO_2 emissions from fossil fuel combustion. It also avoids approximately two million tons per year of CH_4 emissions from the biomass residues that, in the absence of energy production, would otherwise be disposed of by burial (in landfills, in disposal piles, or by the plowing under of agricultural residues), by spreading, and by open burning. The avoided CH_4 emissions associated with biomass energy production have a greenhouse warming potential that is more than 20 times greater than that of the avoided fossil-fuel CO_2 emissions. Biomass power production is at least five times more effective in reducing greenhouse gas emissions than any other greenhouse-gas-neutral power-production technology, such as other renewable and nuclear. Currently, the New Hope Power Partnership, is the largest biomass power plant in North America. The 140 MWH facility uses sugar cane fiber (bagasse) and recycled urban wood as fuel to generate enough power for its large milling and refining operations as well as to supply renewable electricity for nearly 60,000 homes. The facility reduces dependence on oil by more than one million barrels per year, and by recycling sugar cane and wood waste, preserves landfill space in urban communities in Florida. Anyway, most

Table 3.1: Global Biomass Energy Analysis

Biome Ecosystem Type	Area (Million Km²)	Mean Net Primary Production (Gram DryC/m²/Year)	World Primary Production (Billion Tonnes/Year)	Mean Biomass (kg DryC/m²)	World Biomass (Billion tonnes)	Minimum Replacement Rate (Years)
Tropical rain forest	17.00	2,200.00	37.40	45.00	765.00	20.50
Tropical monsoon forest	7.50	1,600.00	12.00	35.00	262.50	21.88
Temperate evergreen forest	5.00	1,320.00	6.60	35.00	175.00	26.52
Temperate deciduous forest	7.00	1,200.00	8.40	30.00	210.00	25.00
Boreal forest	12.00	800.00	9.60	20.00	240.00	25.00
Merranean open forest	2.80	750.00	2.10	18.00	50.40	24.00
Desert and semidesert scrub	18.00	90.00	1.62	0.70	12.60	7.78
Extreme desert, rock, sand or ice sheets	24.00	3.00	0.07	0.02	0.48	6.67
Cultivated land	14.00	650.00	9.10	1.00	14.00	1.54
Swamp and marsh	2.00	2,000.00	4.00	15.00	30.00	7.50
Lakes and streams	2.00	250.00	0.50	0.02	0.04	0.08
Total Continental	149.00	774.51	115.40	12.57	1,873.42	16.23
Open ocean	332.00	125.00	41.50	0.003	1.00	0.02
Upwelling zones	0.40	500.00	0.20	0.02	0.01	0.04
Continental shelf	26.60	360.00	9.58	0.01	0.27	0.03
Algal beds and reefs	0.60	2,500.00	1.50	2.00	1.20	0.80
Estuaries and mangroves	1.40	1,500.00	2.10	1.00	1.40	0.67
Total Marine	361.00	152.01	54.88	0.01	3.87	0.07
Grand Total	**510.00**	**333.87**	**170.28**	**3.68**	**1,877.29**	**11.02**

of the time the amount of biomass available is not as big as stated in the example above. Many times, especially in Europe where such huge agricultural developments like in the USA are not usual, the cost for transporting the biomass overcomes its actual value and therefore the gathering ground has to be limited to a certain small area. This fact leads to only small possible power outputs around 1 MW_{el}. To make an economic operation possible those power plants have to be equipped with the ORC technology, a cycle similar to the water steam power process just with an organic working medium. Such small power plants can be found in Europe. Despite harvesting, biomass crops may sequester (trap) carbon. So for example soil organic carbon has been observed to be greater in switchgrass stands than in cultivated cropland soil, especially at depths below 12 inches. The grass sequesters the carbon in its increased root biomass. But the perennial grass may need to be allowed to grow for several years before increases are measurable. Such small power plants can be found in Europe.

US Biomass Programme

Biomass is a clean, renewable energy source that can help to significantly diversify transportation fuels in the United States. The Biomass Programme is helping transform the nation's renewable and abundant biomass resources into cost competitive, high performance biofuels, bioproducts, and biopower.

BIOMASS HEATING SYSTEMS

Biomass heating systems refers to the various methods used to generate heat from biomass. The systems fall under the categories of direct combustion, gasification, combined heat and power (CHP), anaerobic and aerobic digestion.

Benefits of Biomass Heating Systems

The use of biomass in heating systems is beneficial because it uses agricultural, forest, urban and industrial residues and waste to produce heat and electricity with a very limited effect on the

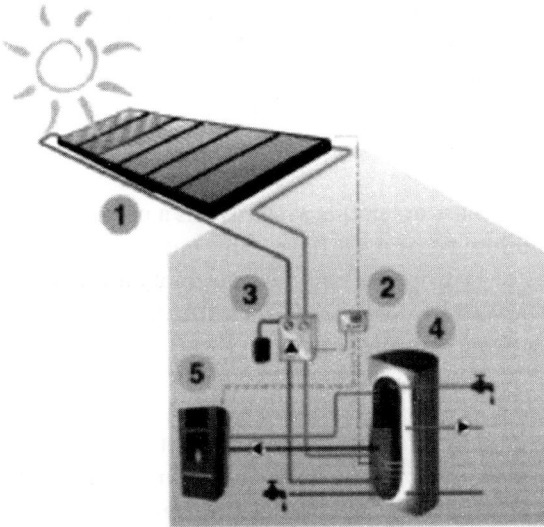

environment. This type of energy production has a very limited effect on the environment because the carbon in biomass is part of the natural carbon cycle, while the carbon in fossil fuels is not, and adds carbon to the environment when burned for fuel. Historically, before the use of fossil fuels in significant quantities, biomass in the form of wood fuel provided most of humanity's heating, as well as providing our first renewable energy resource.

Biomass Heating in Our World

The oil price increases since 2003 and consequent price increases for natural gas and coal have increased the value of biomass for heat generation. Forest renderings, agricultural waste, and crops grown specifically for energy production become competitive as the prices of energy dense fossil fuels rise. Efforts to develop this potential may have the effect of regenerating mismanaged croplands and be a cog in the wheel of a decentralized, multi-dimensional renewable energy industry. Efforts to promote and advance these methods became common throughout the European Union through the 2000s. In other areas of the world, inefficient and polluting means to generate heat from biomass coupled with poor forest practices have significantly added to environmental degradation.

Types of Biomass Heating Systems

The use of Biomass in heating systems has a use in many different types of buildings, and all have different uses. There are four main types of heating systems that use biomass to heat a boiler. The types are Fully Automated, Semi-Automated, Pellet-Fired, and Combined Heat and Power.

Fully Automated

Fully automated systems operate exactly how they sound. Chipped or ground up waste wood is brought to the site by delivery trucks and dropped into a holding tank. A system of conveyors then transports the wood from the holding tank to the boiler at a certain managed rate. This rate is managed by computer controls and a laser that measures the load of fuel the conveyor is bringing in. The system automatically goes on and off to maintain the pressure and temperature within the boiler. Fully automated systems offer a great deal of ease in their operation because they only require the operator of the system to control the computer, and not the transport of wood.

Semi-Automated or "Surge Bin"

Semi-Automated or "Surge Bin" systems are very similar to fully automated systems except they require more manpower to keep operational. They have smaller holding tanks, and a much simpler conveyor systems which will require personal to maintain the systems operation. The reasoning for the changes from the fully automated system is the efficiency of the system. Wood fire fueled boilers are most efficient when they are running at their highest capacity, and the heat required most days of the year will not be the peak heat requirement for the year. Considering that the system will only need to run at a high capacity a few days of the year, it is made to meet the requirements for the majority of the year to maintain its high efficiency.

Pellet-Fired

The third main type of biomass heating systems are pellet-fired systems. Pellets are a processed form of wood, which make them more expensive. Although they are more expensive, they are much more condensed and uniform, and therefore are more efficient. In these systems, the pellets are stored in a grain-type storage silo, and gravity is used to move them to the boiler. The storage requirements are much smaller for pellet-fired systems because of their condensed nature, which also helps cut down costs. these systems are used for a wide variety of facilities, but they are most efficient and cost effective for places where space for storage and conveyor systems is limited, and where the pellets are made fairly close to the facility.

Combined Heat and Power

Combined heat and power systems are very useful systems in which wood waste is used to generate power, and heat is created as a byproduct of the power generation system. They have a very high cost because of the high pressure operation. Because of this high pressure operation, the need for a highly trained operator is mandatory, and will raise the cost of operation. Another drawback is that while they produce electricity they will produce heat, and if producing heat is not desirable for certain parts of the year, the addition of a cooling tower is necessary, and will also raise the cost.

There are certain situations where CHP is a good option. Wood product manufacturers would use a combined heat and power system because they have a large supply of waste wood, and a need for both heat and power. Other places where these systems would be optimal are hospitals and prisons, which need energy, and heat for hot water. These systems are sized so that they will produce enough heat to match the average heat load so that no additional heat is needed, and a cooling tower is not needed.

SELECT EXAMPLES OF VARIED BIOMASS-BASED POWER PLANTS

Rauhalahti Power Station

Location	Jyväskylä
Owner	Jyväskylän Energiantuotanto
Employees	50 ·
Status	baseload
Fuel	firewood, peat, coal, oil
Commissioned	1986

Rauhalahti power station is a combined heat and power station built in 1986 in Jyväskylä, Finland. It is the main provider of district heat in Jyväskylä and it produces the steam used by Kangas Paper Mill. The construction cost of the plant was 84 million euros.

The plant is operated by Jyväskylän Energiantuotanto, which is owned by Jyväskylän Energia. Fortum owned 60 per cent of the company but Jyväskylän Energia bought the share for 40 million euros.

Power station's main sources of energy are firewood and peat; use of coal is rare. The peat is mainly provided by Vapo, while the firewood is provided by sawmills in Central Finland and nearby wood-processing industry.

Boiler	Year of Completion	District heat	Steam	Electricity	Total
Main boiler	1986	140 MW	40 MW	87 MW	267 MW
2-boiler	1992		65 MW		65 MW
District heat boiler	2004	40 MW			40 MW
Total		**180 MW**	**105 MW**	**87 MW**	**372 MW**

Dunaújváros Power Plant

Location	Dunaújváros, Fejér County
Owner	SWECO
Status	Under construction
Fuel	biomass
Maximum Capacity	50 MW
Commissioned	2010

The **Dunaújváros Power Plant** will be one of Hungary's largest biomass power plant having an installed heat capacity of 160 MW and electric capacity of 50 MW.

Gellénháza Power Plant

Location	Gellénháza, Zala County
Status	Under construction
Fuel	biomass
Maximum Capacity	142 MW
Commissioned	2008

The **Gellénháza Power Plant** will be one of Hungary's largest biomass power plant having an installed electric capacity of 142 MW.

Kalocsa Power Plant

Location	Kalocsa, Bács-Kiskun County
Owner	Kalocsa Hoeromu
Status	Under construction
Fuel	biomass
Maximum Capacity	50 MW
Commissioned	2009

The **Kalocsa Power Plant** will be one of Hungary's largest biomass power plant having an installed electric capacity of 50 MW.

Pécs Power Plant

Location	Pécs, Baranya County
Owner	Veolia Environment
Status	Active
Fuel	biomass
Maximum Capacity	182 MW

The **Pécs Power Plant** is one of Hungary's largest biomass power plant having an installed heat capacity of 313 MW and electric capacity of 182 MW.

Szakoly Power Plant

Location	Szakoly, Szabolcs-Szatmár-Bereg County
Owner	Tohuko Electric Power
Status	Under construction
Fuel	biomass
Maximum Capacity	20 MW
Commissioned	2008

The **Szakoly Power Plant** is one of Hungary's largest biomass power plant having an installed electric capacity of 20 MW.

Szerencs Power Plant

Location	Szerencs, Borsod-Abaúj-Zemplén County
Owner	BHD Hőerőmű
Status	Under construction
Fuel	biomass
Maximum Capacity	50 MW
Commissioned	2009

The **Szerencs Power Plant** will be one of Hungary's largest biomass power plant having an installed electric capacity of 50 MW.

Valdivia Pulp Mill

The **Valdivia Pulp Mill** or **Planta Valdivia** is a pulp mill and biomass-fueled electrical generating station in San José de la Mariquina, Los Ríos Region, Chile. Although the main activity is wood pulp production it generates 61 MW of electricity from the burning of volatiles and black liquor. The plant was built in 2004 and is owned by Celulosa Arauco y Constitución.

Río Cruces Contamination Controversy

In 2004 and 2005 thousands of Black-necked Swans in the Carlos Anwandter Nature Sanctuary in Chile died or migrated away following major contamination by a newly opened CELCO pulp mill located near the city of Mariquina and Cruces River which feeds the wetlands. By August 2005, the birds in the Sanctuary had been "wiped out"; only 4 birds could be observed from a population formerly estimated at 5,000 birds. Autopsies on dead swans attributed the deaths to high levels of iron and other metals polluting the water. The company had been dumping dioxins and heavy metals into the river illegally from a wastetube that had not been approved by the authorities. The plant was closed in 2005 after the company lawyers reportedly produced a misleading environmental study regarding pollution on the Cruces River. The scandal prompted Celco's chief executive to resign in June 2005 and the company to pledge to adopt cleaner technologies. The plant reopened two months later at limited production capacity. Even in 2006 the Latin American water tribunal recommended to close down the mill. In July of 2007 CELCO agreed to pay $614 millions Chilean Pesos to Valdivian tourism companies to avoid legal actions for supposed loses of the tourism sector of Valdivia due to contamination of Carlos Anwandter Nature Sanctuary. In a document signed the tourism companies CELCO was exempted from all responsibility involving the contamination of Cruces River. CELCO also promised to pay $2 million monthly each of the coming 3 years to promote tourism.

Blyth Biomass Power Station

The **Blyth Biomass Power Station** is a proposed biomass-fired power station, that if built would be situated at Blyth, Northumberland, on the south bank of the River Blyth near its tidal estuary.

The power station would be constructed on the redundant Bates Colliery site to the north of the town. Between 150 and 300 full time jobs would be created at the station once completed, 95 per cent of which, it is claimed, would go to local people. The station would generate 100 megawatts (MW) of electricity through burning a mixture of biomass fuels, including woodchips, wood pellets, wood briquettes and recycled chipped wood. This fuel would be delivered to the station via the nearby port. The electricity generated would be enough to power 150,000 homes, and would cut carbon dioxide emissions by over 300,000 tonnes per year.

In May 2009, Port of Blyth selected renewable energy company Renewable Energy Systems as the developer for the power station. The station is expected to be fully operational by 2015, if planning permission is granted.

The station is expected to feature a 100 m (330 ft) tall chimney, 60 m (200 ft) tall boiler house and a 20 m (66 ft) tall fuel storage warehouse.

Didcot Power Station

Didcot Power Station refers to two coal and natural-gas power plants that supply the National Grid. They are situated immediately adjoining one another in the civil parish of Sutton Courtenay, next to the town of Didcot in Oxfordshire (formerly in Berkshire), in the UK.

Didcot A

Didicot A Power Station is a coal and gas fired power station, designed by architect Frederick Gibberd. A vote was held in Didcot and surrounding villages on whether the power station should be built. There was strong opposition from Sutton Courtenay but the yes vote was carried, due to the number of jobs that would be created in the area. Building was started on the 2,000 MWe power station for the CEGB during 1964, and was completed in 1968 at a cost of £104m, with up to 2400 workers being employed at peak times. It is located on a 300 acres (1.2 km) site formerly part of the Ministry of Defence Central Ordnance Depot. The main chimney is 650 ft (200 m) tall with the six cooling towers 325 ft (99 m) each. The station uses four 500MWe generating units. In 2003 Didcot A burnt 3.7Mt of coal.

The station burns mostly pulverised coal, but also co-fires with natural gas. Didcot was the first large power station to be converted to have this function. In addition, a small amount of biomass, such as sawdust, is now burnt at the plant. This was introduced to try to depend more on renewable sources following the introduction of the Kyoto Protocol and, in April 2002, the Renewables Obligation. It is hoped that biomass could replace

2 per cent of coal burnt. In 1996 and 1997, Thales UK was awarded contracts by Innogy (now npower) to implement the APMS supervisory and control system on all of the four units, then allowing to have optimised emissions monitoring and reporting.

Some ash from Didcot A is used to manufacture building blocks at a factory on the adjacent Milton Parkand transported to Thacham near Newbury Berkshire for the manufacture of Thermalite aerated breeze blocks using both decabonized fly and raw ash but most is mixed with water and pumped via a pipeline to former quarries in Radley.

Environmental Protests

On the morning of Thursday 2 November 2006, 30 Greenpeace volunteers invaded the power station. One group chained themselves to a broken coal-carrying conveyor belt. A second group scaled the 200 metre high chimney, and set up a 'climate camp'. They proceeded to paint "Blair's Legacy" on the side of the chimney overlooking the town. Greenpeace claim Didcot Power Station is the second most polluting in Britain after Drax in Yorkshire, whilst Friends of the Earth describe it as the ninth worst in the UK.

Didcot A has opted out of the Large Combustion Plants Directive which means it will only be allowed to run for up to 20,000 hours after 1 January 2008 and must close by 31 December 2015. However, due to the amount of running hours the station is currently using, it will more than likely close before then.

Didcot B

History

Didcot B is the newer sibling initially owned by National Power, constructed from 1994-7 by Siemens and Atlantic Projects, and uses a (CCGT) type power plant to generate up to 1,360MWe of electricity. It opened in July 1997. There has been some controversy locally that the access for the site was originally agreed to be via the site entrance for Didcot A on *Basil Hill*

Road', however the 'temporary' access using the former National Grid stores access road is still in use.

Specification

It consists of two 680MWe modules, each with two 230MW SGT5-4000F (former V94.3A) Siemens gas turbines and two heat recovery steam generators, built by International Combustion Ltd (since 1997 known as ABB Combustion Services Ltd), and a steam turbine.

Ownership

Following privatisation of the CEGB in the early 1990s, Didcot A passed into the control of what became National Power, who also started construction of Didcot B. Successive demergers and mergers have meant the site passed through Innogy (in 2001) and now by npower (UK).

Tours

Tours of Didcot A are available and are free for educational institutions and community groups. Tours last 1.5 or 2 hours for the junior tour and adult tour respectively.

Architectual Reception

Didcot Power Station blending into the landscape—viewed from Wittenham Clumps

- It was voted Britain's third worst eyesore in 2003 by *Country Life* readers, although Didcot A won architectural awards for how well it blended into the landscape, following its construction. Radio Oxford received votes for the station when they conducted a survey of the worst building in Oxfordshire, with some listeners referring to it as looking like *somewhere up north*.
- British poet Kit Wright has written an "Ode to Didcot Power Station" using a parodic style akin to that of the early romantic poets.

Drax Ouse Renewable Energy Plant

Drax Ouse Renewable Energy Plant is one of the three proposed 300 MW biomass-fired renewable energy plants in the United Kingdom, which is developed by Drax Power Limited and Siemens Power Ventures GmbH. A site of the Drax Ouse Renewable Energy Plant adjacent to the current Drax Power Station on the River Ouse near Selby in North Yorkshire.

Description

The site has been selected as a potential location due to its proximity to the existing Drax Power Station infrastructure, including rail and road off-loading facilities, water supply and discharge capability, and existing connection to the electricity transmission network. In addition, the adjacent buildings and structures of the existing Drax Power Station will minimise the visual impact of the proposed plant.

The plant will consume approximately 1.4 million tonnes of biomass per year (dependent upon the actual types of biomass material used). Although imported biomass will initially made up much of the fuel source, Drax is keen to develop the use of indigenous biomass fuels where available and are encouraging the development of local energy crops and other renewable supplies. The Drax Ouse Renewable Energy Plant is predicted to result in a saving of 1,800,000 tonnes of carbon dioxide (CO_2) from fossil fuels each year, and to provide enough electricity to power 512,000 homes.

The plant will be water-cooled, using the water from the nearby River Ouse, and biomass fuel will be transferred to the site via conveyors from off-loading facilities. Electricity will be exported from the plant on to the national electricity transmission network.

It is anticipated that the stack will be 100 metres tall, the boiler house approximately 60 metres in height and the cooling towers approximately 20-25 metres tall.

Local Benefits

The total investment in the three renewable plants is worth

around £2 billion and will bring significant economic benefits to the local communities. For the area surrounding the development at Drax Power Station it will mean:

- Up to 850 jobs during a three-year construction period
- Up to 150 new jobs including permanent and contract opportunities during operations
- Contract opportunities for local businesses during the construction and operation period
- Local supply and maintenance contact opportunities during the planned 25-year operational phase

The Planning Process

The proposal requires consent from the Department of Energy and Climate Change (DECC). Drax will submit a planning application to DECC, accompanied by a comprehensive Environmental Statement on all aspects of the development. Drax intends to also consult with Selby District Council, North Yorkshire County Council, the Environment Agency, Natural England, members of the public and a range of other organisations.

Drax Power Station

Drax Power Station is a large coal-fired power station in North Yorkshire, England, capable of co-firing biomass and petcoke. It is situated near the River Ouse between Selby and Goole, and its name comes from the nearby village of Drax. Its generating capacity of 3,960 megawatts is the highest of any power station in the United Kingdom and Western Europe, and the station provides about 7 per cent of the UK's electricity demand.

Opened in 1974 and extended in the mid-1980s, the station was initially operated by the Central Electricity Generating Board, but since privatisation in 1990 the station has changed owner several times. It is currently operated by Drax Group plc and is now one of the cleanest and most efficient coal-fired power stations in the UK.

History

After the Selby Coalfield was discovered in 1967 the Central Electricity Generating Board built three large power stations to utilise its coal. These were an expansion of the station at Ferrybridge, a new station at Eggborough, and the power station at Drax. The station at Drax, near Selby was constructed on the site of Wood House.

Construction

The delivery of one of the station's Babcock boilers during construction in 1974.

Drax power station was constructed in two similar phases, each of three generating units. The first phase of construction, was begun in from 1973. Costain constructed the station's foundations and cable tunnels; Sir Robert McAlpine laid the roads in and about the station, as well as building the ancillary buildings; Mowlem laid the deep foundations; Alfred McAlpine built the administration and control buildings; Balfour Beatty undertook general building works and James Scott installed cabling. Although the first phase was not completed until 1975, the station's first generating set began generating electricity in 1974.

The second phase of construction began several years later in 1985. Tarmac Construction undertook the civil engineering works; Holst Civil Engineers built the chimney; N.G. Bailey installed cabling; Reyrolle, English Electric and South Wales Switchgear produced and installed the station's switchgear, English Electric also manufactured the generator cooling water pumps; T.W. Broadbent maintained the site's temporary electrical supplies, and the Sulzer Brothers manufactured the boiler feed pumps. This second and final phase was completed in 1986. In both stages the boilers were made by Babcock Power Ltd and the generators by C.A. Parsons and Company. Following the completion of the station, Mitsui Babcock fitted Flue Gas Desulfurization (FGD) equipment at the station between 1988 to 1995.

Post-privatisation

On privatisation of the UK's electric supply industry in 1990, the operation of Drax Power Station was transferred from the Central Electricity Generating Board to the privatised generating company National Power. They sold it on to the AES Corporation in November 1999 for £1.87 billion (US$3 billion). AES relinquished ownership of the station in August 2003, after falling into £1.3 billion of debt. Independent directors continued the operation of the station to ensure security of supply. In December 2005, after refinancing, ownership passed to the Drax Group. On 15 May 2009, the company lost its investment grade status and was downgraded to 'junk' status by Standard and Poor's.

Design and Specification

The station's turbine hall, inside and out, giving an idea of the immense scale of the plant. The taller light coloured building behind is the boiler house.

The station's main buildings are of steel frame and metal clad construction. The main features of the station consists of a turbine hall, a boiler house, a chimney and twelve cooling towers. The station's boiler house is 250 ft (76 m) high, and the turbine hall is 400 m (1,300 ft) long. The reinforced concrete chimney stands at 259 metres (850 ft) high, with a diameter of 9.1 metres (30 ft), and weighs 44,000 tonnes (43,310 LT; 48,500 ST). It consists of three flues, each serving two of the station's six boilers. When finished, the chimney was the largest industrial chimney in the world, and is still the tallest in the United Kingdom. The twelve 114 metres (370 ft) high natural draft cooling towers stand in two groups of six to the north and south of the station. They are made of reinforced concrete, in the typical hyperbolic design, and each have a base diameter of 92 m (300 ft). Other facilities on the site include a coal storage area, flue gas desulphurisation plant and gypsum handling facilities.

Drax power station is the second largest coal-fired power station in Europe, after Be[3]chatów Power Station in Poland. Drax produces around 24 terawatt-hours (TWh) (86.4 petajoules) of electricity annually. Although it generates around 1,500,000 tonnes of ash and 22,800,000 tonnes of carbon dioxide each year, Drax is the most carbon-efficient coal-fired powerplant in the United Kingdom.

Operations

Coal Supply and Transport

The station has a maximum potential consumption of 36,000 tonnes of coal a day. Per year, this equates to around 9 million tonnes. This coal comes from a mixture of both domestic and international sources, with domestic coal coming from mines in Yorkshire, the Midlands and Scotland, and foreign supplies coming from Australia, Colombia, Poland, Russia and South Africa.

When the station first opened, the majority of the coal burned there was taken from various local collieries in Yorkshire. These collieries included: Kellingley, Prince of Wales, Ackton Hall, Sharlston, Fryston, Askern and Bentley. However, since the miners' strike in the mid-1980s, all but one of these mines have shut, with the pit at Kellingley being the only one of these still open. UK Coal currently have a 5 year contract to supply the station with coal. This contract ends at the end of 2009. They supply the station with coal from Kellingley, Maltby and, until its closure in 2007, Rossington. Coal was also brought to the station from Harworth Colliery until it was mothballed, and is still supplied by Daw Mill in Warwickshire.

The foreign coal is brought to the station via various ports in the UK, and it is taken from these ports to the power station by railway. First GBRf have a contract with Drax Group to move coal brought to Port of Tyne to the power station. This contract has been celebrated by First GBRf naming one of their locomotives *"Drax Power Station"*. English Welsh and Scottish Railways haul coal to the station from the nearby ports of Hull and Immingham, and from Hunterston Terminal on the west coast of Scotland. Freightliner move coal imported through Redcar.

All of the coal is delivered to the station by train. Trains reach the station using a 4.5 mi (7.2 km) long freight only section of the closed Hull and Barnsley Railway, which branches away from the Pontefract Line at Hensall Junction. A balloon loop rail layout is used at the station so that wagons of coal do not need to be shunted after being unloaded. Merry-go-round trains are used, so that wagons can be unloaded without the train stopping, as it passes through an unloading house. On average, there are 35 deliveries a day, 6 days a week.

Electricity Generation

Coal is fed into one of thirty coal bunkers, each with a capacity of 1,000 tonnes (984.2 LT; 1,102 ST) of coal. Each bunker feeds two of the station's sixty pulverisers, each of which can crush 36 tonnes (35.43 LT; 39.68 ST) of coal an hour. The station has six Babcock Power boilers, each weighing 4,000 tonnes (3,937 LT; 4,409 ST). The powdered coal from ten pulverizers is blasted into each boiler through burners, which are ignited by propane. In 2003 the original burners were replaced by low nitrogen oxide burners. Each boiler feeds steam to one of six steam turbines. Each steam turbine consists of one High Pressure (HP) turbine, one Intermediate Pressure (IP) turbine and three Low Pressure (LP) turbines. The HP turbines generate at 140 megawatts (MW). Exhaust steam from them is fed back to the boiler and reheated, then fed to the 250 MW IP turbines and finally passes through the 90 MW LP turbines. This gives each generating set a generating capacity of 660 MW. This gives the station a capacity of 3,960 MW. Each of the generating units is equipped with the APMS control system, a system developed by RWE npower and Thales, and implemented by Capula.

The station also has six gas turbines installed. These standby turbines provide backup for breakdowns or shut downs in the National Grid, and their annual output is generally low. Three of these have been mothballed, and are out of operation, but they could be refurbished. Emissions from these units are released through the stations second, smaller chimney, to the south of the main stack.

Cooling System

Water is essential to a thermal power station, as water is heated to create steam to turn the steam turbines. Water used in the boilers is taken from two licensed boreholes on-site. Once this water has been through the turbines it is cooled using condensers. Water for these condensers is taken from the nearby River Ouse. Water is pumped from the river to the power station by a pumphouse on the river, north of the station. Once it has been through the condenser, the water is cooled by one of the station's natural draft cooling towers, with two towers serving each generating set. Once cooled, the water is discharged back into the river.

Flue Gas Desulphurisation

All six units are served by independent wet limestone-gypsum flue gas desulphurisation (FGD) plant, which was installed between 1988 and 1996. This diverts gasses from the boilers and passes them through a limestone slurry, which removes at least 90 per cent of the sulphur dioxide (SO_2) in the gasses. This is equivalent to removing over 250,000 tonnes of SO_2 from the station's emissions each year. The process requires 10,000 tonnes of limestone a week. This limestone is sourced from Tunstead Quarry in Derbyshire. A byproduct of the process is gypsum, and 15,000 tonnes of it is produced by the station each week. This goes to be used in the manufacture of plasterboard. The gypsum is sold exclusively to British Gypsum, and it is transported by rail to their plants at Kirkby Thore (on the Settle-Carlisle Railway), East Leake (on the Great Central Main Line) and occasionally to Robertsbridge (on the Hastings Line). English, Welsh and Scottish Railways transport the gypsum to the plants.

Ash Use and Disposal

Pulverised fuel ash (PFA) and furnace bottom ash (FBA) are two byproducts made through the burning of coal. Each year, the station produces about 1,000,000 tonnes of PFA and around 220,000 tonnes of FBA. Most of this ash is sold on, with all FBA and 85 per cent of PFA being sold. Under the trade name Drax Ash Products, the sold ash is sold to the local building industry, where it is used in the manufacture of blocks, cement products, grouting and the laying of roads. The ash is also used in other parts of the country. Between 2005 and 2007, PFA was used as an infill at four disused salt mines in Northwich in Cheshire. 1,100,000 tonnes of PFA was used in the project, which was to avoid a future risk of subsidence in the town. Ash was delivered by EWS in ten trains a week, each carrying 1,100 tonnes of PFA.

The unsold PFA is sent by conveyor belt to the Barlow ash mound, which is used for disposal and temporary stockpile. Three conveyors feed the mound, with a total capacity of delivering 750 tonnes of PFA an hour. Some FGD gypsum is disposed of on the mound, if it is not of a high enough grade to be sold on. The mound itself is notable as it has won a number of awards for its nature conservation work.

Co-firing

Co-firing is the process of burning two or more types of fuel together at the same time. As well as burning coal, Drax power station also co-fires biomass and petcoke. The station tested co-firing biomass in the summer of 2004, and in doing so was the first power station in the UK to be fueled by wood. The initial trial of 14,100 tonnes of willow was locally sourced from nearby Eggborough.

Since the trial, the station's use of biomass has continued. The station uses direct injection for firing the biomass, whereby it bypasses the station's pulverising mills and is either injected dirctly into the boiler or the fuel line, for greater throughput of biomass. The station's use of biomass has continued to increase and they have currently set a target for 12.5 per cent of the station's energy to be sourced from biomass. This will contribute to the station's aim to cut its CO_2 emissions by 15 per cent. The station burns a large range of biomass fuels, but among them, the most used are wood pellets, sunflower pellets, olive, peanut shell husk and rape meal. The majority of the station's biomass comes from overseas.

The station started to trial the co-firing of petcoke in one of its boilers in June 2005. The trial ended in June 2007. Over this period the boiler burned 15 per cent petcoke and 85 per cent coal. Petcoke was burned in the station to make the price of the station's electricity more competitive as the price of running the station's FGD equipment was making the station's electricity more expensive. The Environment Agency (EA) granted permission for the trial in June 2004, despite the plans being opposed by Friends of the Earth and Selby Council. To meet their concerns, the station's emissions were constantly monitored through the trial, and they were not allowed to burn petcoke without operating the FGD plant to remove the high sulphur content of the fuel's emissions. The trial proved that there were no significant negative effects on the environment, and so in late 2007, Drax Group applied to move from trial conditions to commercial burn. The EA granted permission in early 2008. The station can now burn up to 300,000 tonnes of the fuel a year, and stock anything up to 6,000 tonnes of the material on site.

Environmental Effects

The environmental effects of coal burning are well documented, the most significant of which is global warming, caused by the release of carbon dioxide (CO_2) into the earth's atmosphere. Coal is considered to be 'easily the most carbon-intensive and polluting form of energy generation available'. In 2007 Drax produced 22,160,000 tonnes of CO_2, making it the largest single source of CO_2 in the UK. Between 2000 and 2007, there has been a net increase in carbon dioxide CO_2 of over 3,000,000 tonnes. Drax power station also has the highest estimated emissions of Nitrogen oxide (NOx) in the European Union.

In 2007, in a move to try and lower the station's CO_2 emissions, Drax Group signed a £100 million contract with Siemens Power Generation to re-blade the station's steam turbines over the course of four years. This is the largest steam turbine modernisation ever undertaken in the UK, and will increase the station's efficiency. This, coupled with the co-firing of biomass, is part of a target to reduce the station's CO_2 emissions by 15 per cent by 2011.

Protests and Industrial Action

Climate Camp—August 2006

On 31 August 2006, over 600 people attended a protest against the power station's high carbon emissions. It was coordinated by the Camp for Climate Action group. At least 3,000 police officers from 12 forces were reported to have been drafted in for the duration of the protest, to safeguard electricity supplies and prevent the protesters from shutting the station down. Thirty nine people were arrested during the protest after after trying illegally to gain access to the plant.

Train Protest—June 2008

At 8:00 am on 13 June 2008, more than 30 climate change campaigners halted a EWS coal train en-route to Drax power station by disguising themselves as rail workers by wearing high-visibility clothing and waving red flags. Stopping the train on a bridge crossing the River Aire, they scaled the wagons with the aid of the bridge's girders. They then mounted a banner reading "Leave it in the ground" to the side of the wagon and tied the train to the bridge, preventing it moving. They then shoveled more than 20 tonnes of coal on to the railway line. The protest lasted the whole day, until several protesters were removed from the train by police that night. The station's management said that the protest had no effect on power station's output. The action was also coordinated by Camp for Climate Action.

Worker Strike—June 2009

On 18 June 2009, less than 200 workers and contractors walked out of or failed to show up at Drax power station in a wildcat strike, showing solidarity with workers at the Lindsey Oil Refinery in Lincolnshire where 51 workers had been laid off while another employer on the site was employing. A spokeswoman said the strike did not affect the station's electricity output.

Future of Station

Biomass Power Station

Drax Group have applied for planning permission to build a new 300 MW power station, fueled entirely by biomass, to the north of the current power station site. The Ouse Renewable energy plant is expected to burn 1,400,000 tonnes of biomass each year, saving 1,850,000 tonnes of CO_2 emissions per annum.

If the plans go ahead, 850 construction jobs will be created and 150 permanent jobs created once opened through direct and contract employment. Plans were submitted for review by the Department of Energy and Climate Change in July 2009. If planning permission is granted, construction may begin in late 2010 and is expected to last up to three and a half years. Two other similar plants are planned by Drax at the ports of Hull and Immingham.

Carbon Capture and Storage

On 17 June 2009 the Secretary of State for Energy and Climate Change, Ed Miliband, announced that all UK coal-fired power stations may be fitted with Carbon Capture and Storage (CCS) technology by the early 2020s or face closure. Drax currently has made no statement on the viability of CCS technology at the power station. If it was necessary to install CCS technology at Drax, it would require the construction of new turbines and boilers, and a secure way of transporting CO_2 emissions 40 mi (64 km) to the Yorkshire coast. Drax Power Limited are sponsoring development studies into the technology and its application.

Ferrybridge Power Station

Ferrybridge Power Station refers to a series of three coal-fired power stations situated on the River Aire in West Yorkshire, England. The first station on the site (Ferrybridge A) was constructed in the mid-1920s, and was closed as the second, B station was brought into operation in the 1950s. The A station has been retained since it closed. In the 1960s, Ferrybridge C was opened with a generating capacity of 2000 megawatts, which at the time was the largest of any power station in the UK. The B and C stations operated together until the B station's closure in the 1990s. The B station has since been demolished.

Ferrybridge C power station is currently the only power station operating on the site. Since 2004 it has been operated by Scottish and Southern Energy plc. It is capable of co-firing biomass and is currently being fitted with Flue Gas

Desulphurisation (FGD) plant. There are plans to build a fourth,. D station on the site.

History

Ferrybridge A

Ferrybridge A Power Station was a small coal-fired power station, and the first power station constructed on the site. Construction began in 1926 and the station began operating in 1927. The station closed on 25 October 1976, with a generating capacity of 125 MW. Its boiler and turbine hall still stands today, and the building is used as offices and workshops by the RWE npower Technical Support Group, who are responsible for maintenance and repairs of power station plant from around the country.

Ferrybridge B

Ferrybridge B Power Station was a 300 megawatt (MW) coal-fired power station constructed in the 1950s. Its three 100 MW generating sets were commissioned between 1957 and 1959. The station originally had a total generating capacity of 300 MW, but by the 1990s this was recorded as 285 MW. After the UK's electric supply industry was privatised in 1989, the station was operated by PowerGen. The station closed in 1992 and has since been completely demolished.

Ferrybridge C

Ferrybridge C Power Station is the current power station on the site. It is a 2000 MW coal-fired power station owned by Scottish and Southern Energy plc. It was the first 2000 MW power station in Europe and opened to begin supplying power to the National Grid in 1966.

Accident

On 1 November 1965 the station was the site of a collapse, when three of the cooling towers collapsed due to vibrations in 85mph winds. Although the structures had been built to withstand higher wind speeds, the design only considered average wind speeds over one minute and neglected shorter gusts. Furthermore, the grouped shape of the cooling towers meant that westerly winds were funnelled into the towers themselves, creating a vortex. Three out of the original eight cooling towers were destroyed and the remaining five were severely damaged. The towers were rebuilt and all eight cooling towers were strengthened to tolerate adverse weather conditions.

The power station was originally operated by the Central Electricity Generating Board, but ownership was passed to Powergen after privatisation of the UK's electrical supply industry. Ferrybridge Power Station, along with Fiddlers Ferry Power Station in Cheshire, was then sold to Edison Mission Energy in 1999. They were then sold on to AEP Energy Services Ltd in 2001, before both were sold again to Scottish and Southern Energy in July 2004 for £136 million.

Specification

Ferrybridge C Power Station comprises four 500 MW generating sets, using 800 tonnes of coal and 218 million litres of coolant water per hour. As well as the four main units, the station has two gas turbines which produce an extra 34 MW combined. These are used for extra generating capacity and in dead starts, as large power stations often need external electricity to start up if the main units are off line. Coal is delivered to the station by railway and road transport and until the late 1990's, by barge. The station has two 198 m (650 ft) high chimneys and eight 115 m (380 ft) high cooling towers, which are the largest of their kind in Europe.

Environmental Impact

Ferrybridge C has now had an operating life of over 40 years. Since 2003, the station has established itself as a market leader in the effective co-firing of biomass. In the 2002-2003 tax year, the station was responsible for 80 per cent of all co-fired renewable energy in the UK, resulting in a 3.5 per cent net reduction of the plant's greenhouse gas emissions.

Work is currently being completed on a new Flue Gas Desulphurisation plant, servicing units 3 and 4. In 2007, Scottish and Southern Energy announced plans to conduct a feasibility study to retrofit unit 1 with a 'supercritical' boiler. According to the 2007 Annual Report, the decision was taken to discontinue this scheme. As of June 2007 however, a separate study assessed the feasibility of building a new 800 MW supercritical coal-fired station at Ferrybridge, potentially with carbon capture technology. These developments could ensure that a power station will exist at Ferrybridge well into the 21st Century.

Fiddlers Ferry Power Station

Fiddlers Ferry Power Station is a 1989MW coal fired electricity generating power station established in 1971 and is located between Widnes and Warrington, England. The station is owned and operated by Scottish and Southern Energy plc.

History

The station was built by the CEGB but was transferred to Powergen PLC after privatisation. The station, along with Ferrybridge Power Station, a 1995MW coal-fired station in Yorkshire, was then sold to Edison Mission Energy in 1999, sold on to AEP Energy Services Ltd in 2001 and both were sold again in July 2004 to Scottish and Southern Energy for £136m. One of its cooling towers collapsed on Friday 13 January 1984, due to the freak high winds of that winter, but has since been rebuilt.

Operations

The station consumes 195 million litres of water daily from the River Mersey. Since the deep mines in the Lancashire coalfield closed, all its coal is either imported (largely by train from Liverpool docks), or supplied from mines in Yorkshire. 16,000 tonnes of coal are burned each day. It also burns biofuels together with the coal.

Fiddlers Ferry has been fitted with Flue Gas Desulphurisation (FGD) plant to reduce the emissions of sulphur by 94 per cent, meeting the European Large Combustion Plant directive. This work commenced in 2006 and was completed in 2008.

With its eight 114-metre (370 ft) high cooling towers and 200-metre (660 ft) high chimney the station is a prominent landmark and can be seen from as far away as the Peak District.

The station is seen in the title sequence to the BBC Three Programme, *Two Pints of Lager and a Packet of Crisps.*

Glanford Power Station

Glanford Power Station is an electricity generating plant located on the Flixborough Industrial Estate near Scunthorpe in North Lincolnshire. It generates around 13.5 megawatts (MW) of electricity, which is enough to provide power to about 32,000 homes. It was designed to generate electricity by the burning of poultry litter, and was only the second of this kind of power station in the world to have ever been built when it went into operation in 1993. The station is owned by Energy Power Resources (EPR) and operated by its subsidiary Fibrogen.

After the BSE crisis in the 1980s, millions of cattle were slaughtered, and almost half a million tons of dried meat and bone meal (MBM), the cause of the disease, was stockpiled in secure sheds. Glanford Power Station was re-commissioned in May 2000 to burn them. It charges a gate fee for the fuel it burns, which would have otherwise been disposed of using conventional landfills. MBM has around two thirds the energy value of fossil fuels such as coal, and has been labelled carbon neutral. Despite producing "green energy", Glanford Power Station is listed as the 27th largest arsenic air emitter in England and Wales in an air quality report published in February 2000.

In January 2004 Glandford Power Plant received planning permission for the site to be extended to allow other sources of biomass to be burned. The plant technology is based on a conventional moving grate boiler and steam cycle.

In accordance with the United Kingdom Governments new Renewables Obligation incentive mechanism, a premium is paid for renewable electricity generation. Each renewable generator is issued with Renewables Obligation Certificates (ROC), which they may sell to other electricity supply companies. This trade allows them to meet their obligation for the proportion of supplied electricity generated from renewable sources. The power output from the Glanford plant qualifies for ROC trading.

The combustion ashes produced by the station are then disposed of via landfill (21% of fuel by mass). Before the switch to MBM, they used to be sold as agricultural fertiliser.

Lynemouth Power Station

Lynemouth Power Station (also known as **Alcan Power Station**) is a coal-fired power station which co-fires biomass, located in North East England. The station is situated on the coast of Northumberland, 2 miles (3.2 km) north east of the town of Ashington. The station has stood as a landmark on the Northumberland coast since its construction in the 1970s, and has been privately owned by aluminium company Rio Tinto Alcan throughout its operation. The station is used to provide electricity for their nearby aluminium smelter, with surplus electricity being sold to the National Grid.

The station is one of the most recently built coal-fired power stations in the United Kingdom, but with a generating capacity of only 420 megawatts (MW), is now one of the smallest operating. Despite its small capacity, it is the most thermally efficient coal-fired power station in the UK. In 2009, plans were granted to construct a 13 turbine wind farm to the north of the station, and in the same year Alcan announced that they hope to fit the station with carbon capture and storage technology.

History

Under government direction, Alcan applied for planning permission to build a new aluminium smelter in Northumberland at Lynemouth in 1968. Permission was granted later that year and site preparation began shortly after. To meet the high electrical demands of the new smelter, a power station needed to be built. Lynemouth power station was to be constructed only 800 m (2,600 ft) from the new Alcan Lynemouth Aluminium Smelter. Both buildings were designed by architects Yorke,

Rosenberg and Mardall, with engineering consultation from Engineering and Power Consultants Ltd. The power station was constructed by Tarmac Construction and the smelter by M.J. Gleeson Company. Both the power station and smelter were brought into operation in March 1972.

The smelter and power station were constructed in south east area of Northumberland as part of an incentive to resolve the area's high unemployment. The site was also specifically chosen because of the close proximity of a plentiful coal supply for the power station. Nearby were the Ellington and Lynemouth collieries. Ellington Colliery was sunk in 1909, and Lynemouth in 1927. The two collieries were eventually joined underground in 1968 by the Bewick Drift near Lynemouth. This brought coal out of the ground which, because the drift had no rail connection, was sent to the washery at Lynemouth by conveyor belt. The power station was constructed near to the end of this conveyor belt.

Design and Specifications

The power station's main structures include a boiler house, turbine hall and a single chimney. The boiler house and turbine hall are of a steel frame construction with aluminium cladding. The chimney is of reinforced concrete construction and stands at 114 m (370 ft) tall. Other facilities on the site include coal delivery and sorting plant.

The station's boiler house houses three 380 MWth International Combustion Ltd boilers, which are fueled by pulverised bituminous coal. Each of these provide steam for on of three 140 megawatt (MW) Parsons turbo-alternators, situated in the station's turbine hall. These give the station a total generating capacity of 420 MW. The electricity generated is fed at 24 kilovolt (kV) to a substation which powers the smelter. The substation also has a 132 kV connection to the National Grid, where electricity is distributed to homes and other industries by Northern Electric Distribution Limited. The smelter's two pot lines only require 310 MW of the 420 MW that the power station produces, so it is the excess 110 MW which is fed into the national grid.

At the end of 2000 the power station was given a turbine upgrade, which saw an increase in the station's MWh production.

Operations

Coal Supply and Transport

The power station is the leading local customer of coal, burning 1,200,000 tonnes of coal a year, with a weekly coal consumption between 25,000 and 27,000 tonnes. The station has relatively limited coal storage facilities, and is only able to hold three to four weeks worth of its fuel.

The station was designed specifically to burn coal from the Northumberland coalfields. The neighbouring Ellington Colliery originally fed coal directly to the power station using a conveyor belt from its Bewick Drift Mine, situated 970 metres (3,200 ft) from the station. In 1994, Ellington Colliery connected underground with Lynemouth Colliery, but coal continued to be taken straight to the power station's coal sorting area using conveyor belts. This supply was supplemented by coal from local opencast mines. However, Ellington Colliery was forced to close when it flooded in January 2005. The station burned the colliery's remaining stock after it closed, and since then coal has been provided by local opencast mines and from mines in Scotland, but now a small amount of import is necessary.

Coal is now delivered to the station entirely using rail transport and is unloaded at the station using a merry-go-round system. Trains supplying the station use the Newbiggin and Lynemouth branch line of the Blyth and Tyne Railway, which also serves the smelter. This line was originally used to export coal from Easington, and also had passenger services. These passenger services ceased in 1964, and now the line is only used to serve the power station and smelter. Coal from the local opencast mines is brought to the station by road using heavy goods vehicles. Coal is graded and washed at the station prior to being burned.

With only one significant opencast in the local area mining past 2008, along with another smaller opencast at Stony Heap, there is a need for more local supplies of coal for the station because of the risks in depending upon overseas sources of coal.

Long distance supplies of coal can see sharp fluctuations in price, as well as the flexibility and security of the supply, whereas local sources aren't as vulnerable to interruptions and have fixed, contracted prices. The station is not an established importer of coal, having only imported since 2005. It is situated a long way from the major coal unloading ports of Teesside, Hull and Immingham, which have been booked by power stations closer to them. This means that coal for the power station needs to be imported via Blyth or the Port of Tyne. However, because of the small sizes of these docks, they only receive ships from Poland and Russia. Due to high production costs and industry restructuring in Poland though, the only realistic source of imported coal for the station is Russia. The environmental impact of shipping 1,000,000 tonnes of coal from Russia to Lynemouth is the production of 12,812 tonnes (12,610 LT; 14,120 ST) of CO_2, whereas hauling coal from local mines to the station would produce only 703 tonnes (691.9 LT; 774.9 ST) of CO_2. There are currently two local opencast mines for which planning approval have been granted, one at Shotton near Cramlington approved in 2007, the other at Potland Burn near Ashington approved in October 2008. However, the coal mined from Potland Burn would have too high a sulphur content to meet the station's environmental requirements, meaning it wouldn't be an immediate choice of coal for the station. Coal had been provided by the Delhi surface mine at Blagdon, owned by Banks Developments, since 2002. It finished extracting coal in March 2009, following the permission of extension proposals to its original plans in May 2007.

Cooling System

Water is essential to a thermal power station, to create the steam to turn the steam turbines and generate electricity. Water used in the Lynemouth station is taken from the North Sea, which the station is built alongside, through a system of shafts and tunnels. Once used in the station, the heated water needs to be cooled before it can be discharged into the sea.

Fishing bait company Seabait operate next door to the power station. They use some of the station's waste hot water to grow worms used as bait. This way, they are able to grow worms four times quicker than in the wild. As well as providing bait for fishing purposes, their worms are frozen and exported to shrimp and fish farms. The company was set up in 1985 and has won two Queen's Awards for Enterprise.

Ash Removal

Fly ash and bottom ash are two byproducts made through the burning of coal in power stations. Ash is normally dumped in the station's Ash Lagoons landfill site, which is located on site. Since 2006, ash produced at Lynemouth Power Station has been recycled and used as a sub-fill material in the construction industry and in the production of grout. In 2007, 63,000 tonnes (62,010 LT; 69,450 ST) of ash from the station, along with 100,000 tonnes (98,420 LT; 110,200 ST) of ash from the Ash ·Lagoons, was taken and recycled. In September 2007, Pulverised Fuel Ash was utilised as a filling material in the capping of Woodhorn Landfill, which had been used for the disposal of spent potlining from the smelter.

Biomass Usage

In December 2003 the Environment Agency granted permission for the plant to co-fire biomass fuels in the station. Since 2004 three different types of biomass fuel been in use at Lynemouth; Sawdust and Wood pellets from FSC certified forests and Olive residues. These fuels are mixed with the coal on the conveyor belt into the power station. In 2004 11,000 tonnes (10,830 LT; 12,130 ST) of biomass fuel were used in the station. Because biomass is considered 'Carbon neutral', this represents the equivalent to saving approximately 18,500 tonnes (18,210 LT; 20,390 ST) of CO_2 by burning this material. Alcan aimed to increase the amount of biomass used in future.

The station earned the world class OHSAS 18001 health and safety certificate in 2003, ahead of Alcan's global targets. All of the station's staff were required to take place in safety audits to improve working practice at the station. The certificate was presented to the station's manager by Wansbeck MP Denis Murphy on 15 March 2003. The station's attention to health and safety was further recognised on 6 June 2007 when they were honoured by the Royal Society for the Prevention of Accidents (RoSPA) with a RoSPA Occupational Health and Safety Award at the Hilton Birmingham Metropole Hotel. Workers at the station had been audited by RoSPA for 10 years before receiving the award.

Coastal Defence

In late 1994, the power station was flooded to a foot deep of sea water, after a freak high tide and strong winds. This led to a sea defence system being constructed to protect the building. The problems came about because of the temporary closure of Ellington Colliery. Tipped waste from the colliery had been used as a coastal defence measure, but as the colliery had closed, waste was no longer being tipped. The colliery was reopened by RJB Mining, and in July 1999 the station ensured the future of the colliery by signing a contract with RJB Mining to provided with 3,000,000 tonnes (2,953,000 LT; 3,307,000 ST) of coal from Ellington Colliery and opencast mines in Northumberland, over the course of three years. The colliery closed again in 2005, leading to problems with coastal defence again, threatening the station's coal stocking area. This required to a £2.5 million new coastal defence scheme be put in place, involving the use of large rocks as a defence wall.

Environmental Impact and Future

The power station's use of biomass since 2004 has been part of an attempt to reduce its carbon dioxide (CO_2) output. In 2002 and 2004 the station met its targets for reduction in greenhouse gas emissions. Despite this, in 2006 the power station was revealed as having the fourth highest CO_2 emissions in the north of England, for producing 2,685,512 tonnes (2,643,000 LT; 2,960,000 ST) of CO_2 per year.

In 2006 a proposal was made to construct three 110 m (360 ft) tall wind turbines on a currently unused part of the station's coal sorting area, north of the power station. ScottishPower Renewables were refused their initial planning application submitted in November 2006, as the site is spread over two council boundaries. Wansbeck Council approved the scheme, but Castle Morpeth refused. Following an appeal hearing in April 2008, permission was granted in January 2009 for the construction of up to 13 turbines. These will produce 30 MW of electricity, enough to supply over 16,000 homes with power.

Following a visit to the station by Prime Minister Gordon Brown on 3 July 2009, it became apparent that Rio Tinto Alcan were hoping to be able to demonstrate Carbon Capture and Storage (CCS) technology at the station in the future. However due to the economic climate, Rio Tinto are not willing to commit the funding for the project themselves, and so are hoping to secure European Union funding available for demonstration of CCS technology. The investment required could be up to £1 billion, but if funding is secured, "pre-combustion" CCS technology would be installed. This would involve treating the coal prior to burning so that less CO_2 was produced, with any remaining CO_2 being pumped under the North Sea into an aquifier.

Cultural Use and Visual Impact

Since its construction, the station has made appearances in a small number of films shot locally. These include:

- Seacoal—a movie made by Amber Films in 1985. The station is features heavily as a backdrop in the beach scenes, where the characters are working, collecting seacoal. Photographer Mik Critchlow (who would later become involved with Amber Films' sister company Side Gallery) also documented the seacoalers at Lynemouth, between 1981 and 1983. He also used the power station as an industrial backdrop to some of his images.
- Billy Elliot—a 2000 film directed by Stephen Daldry. The power station and the smelter both feature as an industrial backdrop in the film's cemetery scenes. The power station's coal sorting area is used to represent a colliery.

The chimneys of both the power station and the smelter are strong local landmarks on the south east Northumberland landscape, and can be seen over a 8.2 miles (13.2 km) stretch of coast, from Cresswell down to North Blyth.

Teesport Renewable Energy Plant

Teesport Renewable Energy Plant is a proposed biomass fueled power station situated on the River Tees at Teesport in Redcar and Cleveland, North East England.

The station would generate 300 megawatts (MW) of electricity through burning 2.5 million tonnes of sustainably sourced wood chip per year. If constructed the plant would be the largest dedicated wood chip biomass power station in the United Kingdom.

Tilbury Power Station

Tilbury Power Station is a coal-and-biofuel-fired power station located on the River Thames in Tilbury, Essex.

Tilbury 'A' station was commissioned in 1956 by the CEGB. Tilbury 'B' was opened in 1968 and assigned to National Power on privatisation in 1990, but is now operated by RWE npower.

The station contains four generating units with a combined capacity of 1428 MW, enough power for 1.4 million people, approximately 80 per cent of the population of Essex. Cooling water is drawn from the Thames. Fuel is delivered by ship to dedicated unloading jetties. The station connects to the National Grid at the nearby 275 kV substation.

Incidents

A fire broke out at the power station on 29th July 2009. The fire started in the turbine hall at the power station. The power station was offline for routine maintenance when the fire started. Fire crews were called shortly after 3pm, and the fire was reported to be under control by 5:30pm.

Wilton Power Station

Wilton Power Station (or **SembCorp Power Station**) refers to a series of coal, oil, gas and biomass fired CHP power stations on the Wilton International complex in Redcar and Cleveland, North East England. The site is currently owned and operated by SembCorp Industries. It comprises two individual power stations; Wilton Power Station, a coal, oil, gas and biomass co-fired power station which has been in operation since 1952, and Wilton 10, the UK's first ever large scale biomass power station to fire only wood.

History

Wilton Power Station has been generating electricity and providing steam to the plants on the Wilton International complex since March 1952. A secondary turbine commissioned at the station in November 1951 is still operating there today. For a long time the station was operated by Imperial Chemical Industries (ICI), but they sold the station to American company Enron in November 1998. In early 2002 Enron went bankrupt, and the station was sold to SembCorp.

The station has been approved as a "Good quality" combined heat and power plant under the Combined Heat and Power Quality Assurance registration scheme. This means that despite burning fossil fuels like coal and oil, the station is exempt from the Climate Change Levy, a tax which otherwise would have been put on the fuels supplied to the station and the electricity generated by it.

Specification

Wilton Power Station is a 197 megawatt (MW) fossil fuel power station, which is currently in constant operation. Over the years, nine boilers have been in operation at the station. In the 1970s the station had a generating capacity of 300 MW and produced 1,200 tonnes of steam per hour. In this period the station was burning waste from the chemical site, byproducts such as: liquid petroleum gas, tars, waste lubricating oils, emulsion residues, waste from aromatic and olefin plants and hydrogen gas from the nylon plant. There are currently three generating sets in operation at the station, two of which are fired by coal, the other of which is a gas turbine fueled by natural gas. Each of the two coal-fired generating sets are powered by a high pressure boiler, which are each capable of producing a continuous maximum rate of 280 tonnes per hour of steam, at 1700 psig. Each boiler is then connected to a 33 MW turbine generator. Intermediate pressure steam is then either diverted and fed to other plants in the complex, or sent to three secondary generating sets. After the secondary generating sets, low pressure steam can then be distributed to plants in the complex. Boiler Number 6, one of the coal boilers, was mothballed in 1997, however in 2001 it was refurbished and restored to operation. In 2002, one of the secondary turbines was replaced by a newer unit. In the same year, an automation project involving the boilers, turbines and ancillary equipment was completed.

Operations

Coal is brought to the power station by rail to the Wilton rail delivery terminal and unloaded at the station using a merry-go-round system. The coal is provided by a small number of opencast mines in Northumberland.

Electricity is generated for the major plants on the Wilton complex at 11 kilovolts (kV) and distributed throughout the complex using one of the largest privately owned distribution systems in the world. This system is connected to the National Grid so that surplus electricity can be distributed using four National Grid owned transformers at 66 kV and 275 kV. On top of this, the station produces around 4,000,000 tonnes of steam per year for the plants on the complex.

In late 2003, plans to install a 40 MW gas turbine at the station were finally fulfilled, after they were put off because of then owner Enron's bankruptcy. The gas turbine installation was completed in 2004, and it replaced an oil-fired boiler at the station, resulting in reduced emissions from the station. However, the station has retained the flexibility to fire oil as a backup to coal and gas.

In October 2003, SembCorp applied to the Environment Agency for permission to burn 110,000 tonnes of cow fat (tallow) from the carcasses of animals slaughtered during the BSE Crisis of 1996. The tallow bought was a large portion of the 200,000 tonne stockpile stored on farms in Merseyside and near London.

The tallow was brought to the station by road tanker from Merseyside. At the power station it was stored in a tanker, awaiting burning. Following consultation and a 10,000 tonne trial burn between March and May 2004, permission for the burning of the tallow was granted in August 2004. Following a £55 million boiler overhaul in 2005, the station began co-firing biomass.

Wilton 10

SembCorp announced plans to build the UK's first wood-only burning power station in March 2005. The Wilton 10 Power Station (so called because it was the tenth boiler constructed at the Wilton Power Station) cost £30 million to construct and was built alongside the other Wilton Power Station units. It began generating electricity in September 2007, but was officially opened on 12 November 2007 by Energy Minister Malcolm Wicks. 400 people were employed in the station's construction and there are 15 permanent jobs at the station. The station burns 300,000 tonnes of a combination of sustainable wood, sawmill waste and otherwise unusable wood offcuts a year to produce 30 megawatts (MW) of electricity, as well as 10 MW of thermal energy in the form of steam, which is piped for use in the rest of the Wilton complex. It operates separately from the fossil fuel power station.

Aberthaw Power Station

Aberthaw Power Station refers to a series of two coal-fired power stations situated on the coast of South Wales, near Barry in the Vale of Glamorgan. Although it shares its name with the village of Aberthaw, it is actually located on the waterfront of the nearby village of Gileston. The current power station on the site, **Aberthaw B Power Station**, co-fires biomass and as of May 2007, its generating capacity is 1455 megawatts (MW).

History

The site of the stations was originally a golf course prior to the construction of the first station. Aberthaw A Power Station opened in February 1966, and at the time was the most advanced power station in the world. Aberthaw B Power Station opened in the early 1970s. The A Station operated until 1995. It was subsequently demolished. Its two 425 feet (130 m) chimneys were the last section to be demolished, and were cleared on 25 July 1997.

Operations

The station takes its entire coal feed stock in by rail, under contract to EWS from the Vale of Glamorgan Line. Until its closure, it was in part powered by coal from the Tower Colliery in Hirwaun. Coal now comes from Cwmbargoed opencast mine in Merthyr Tydfyl and other opencasts and mines. Most coal is still taken to Tower Colliery for blending, before being taken to the station by rail. There is a possibility that the site could reopen as an opencast mine.

In response to the government's renewable energy obligation that came into effect in April 2002, the station is currently firing a range of biomass materials to replace some of the coal burned.

Flue Gas Desulfurization

Aberthaw B was due for closure, but in June 2005 station owners npower agreed to install new technology to reduce sulphur dioxide emissions by installing Flue Gas Desulfurization (FGD) equipment. This was to reduce sulphur dioxide levels by 90 per cent by 2008, when new European environmental regulations came into place. Construction of the equipment started on 21 June 2006, with a tree-planting ceremony attended by the Welsh Minister for Enterprise, Innovation and Networks, Andrew Davies. The desulfurization FGD project is being carried out by a consortium of ALSTOM and AMEC, which will employ 500 workers on site at the peak of construction.

Nuclear Proposal

It has recently been proposed that the plant, along with every other power station in the UK it should be noted, would be a suitable location for a power station using nuclear power based on the existing infrastructure and logistics. However, no political considerations were taken into account for the UK-wide survey, no plans made in spite of the sensational headlines and Aberthaw was not singled out or "earmarked", simply named alongside many other sites. It is generally held that nuclear stations would only be built in remote areas and on existing nuclear sites. Aberthaw is generally considered too close to Cardiff and is in a conservation area, not to mention the number of influential people who live in the Vale of Glamorgan.

Uskmouth Power Station

Uskmouth Power Station (also known as **Fifoots Point Power Station**) is a coal-fired power station at the mouth of the River Usk in the south-east of Newport, Wales. Two stations have stood on the site since the 1950s; **Uskmouth A Power Station** was built in the 1950s and demolished in the 1990s, and the current **Uskmouth B Power Station** was built in the 1960s. Both stations were coal fired.

Uskmouth A

Uskmouth A Power Station was built and opened in the early 1950s. It was closed on 26 October 1981 with a generating capacity of 228 megawatts. The station was demolished in 2002

Uskmouth B

Uskmouth B Power Station (or **Uskmouth Power** as it is now known) was built in 1959. It has a generating capacity of 363 MW, which is enough to power 360,000 homes, or the surrounding area of Newport. The electricity is provided by three generating sets. It is situated in an essential position for the National Grid, as there are very few power stations situated in the south of Wales. The station was closed in 1995, but in 1998 it was purchased by AES. The station was given a £120 million refurbishment to bring it up to date with legislative requirements. New environmental equipment was installed and it was given a refurbishment which is thought to have extended the station's life by 25 years. The station's generating capacity was also increased to 393 MW. In 2001 the work was completed and the station was reopened. However, only a year later the plant passed into receivership, but had a brief period of operating in the winter between 2003 and 2004. In June 2004 the station was put back into full operation, when it was bought by Welsh Power, who were then known as Carron Energy.

The station is currently one of the cleanest coal-fired power stations in the United Kingdom, and is fitted with Flue Gas Desulphurisation equipment and low NO_x burners. It also burns biomass, as well as coal, for its emissions to be considered closer to being carbon neutral. The station does not take water from or dump waste water into the River Usk. It instead uses secondary treated sewage water in its cooling system.

The station employs 90 people. It has been awarded RoSPA Gold Award for Occupational Health and Safety for its efforts to ensure station safety. The station's owners have participated in many local community projects—they donated land to the Newport Wetlands Reserve, and sponsor Welsh swimmer David Davies and the Newport Gwent Dragons.

Gas-Fired Station

There is currently an 800 megawatt (MW) gas-fired power station being built on the site of the A station. It will cost £400 million to build, creating 650 construction jobs. Once completed in 2010, it will create 60 full time jobs.

Uses in Culture

In 2006, the station was used as a location for two episodes of *Doctor Who*. In the episodes "Rise of the Cybermen" and "The Age of Steel", the station was used as the setting for the Cybermen Factory.

Fibrominn

Location	Benson, Minnesota
Owner	Fibrowatt LLC
Status	Operational
Fuel	poultry litter
Maximum Capacity	55-megawatt
Commissioned	October 2007

Fibrominn, located in Benson, Minnesota, is the first power plant under construction in the United States designed to burn poultry litter as its main source of fuel. It will produce a projected 55 megawatts of energy and will burn turkey manure combined with wood chips. Most of the energy will be purchased by Xcel Energy. Construction began in 2005 and began operating in mid-2007, making it the first biomass power plant in the United States. Grand opening ceremonies were held October 12-13, 2007.

Milltown Biomass-to-Energy Power Station

The Milltown Biomass-to-Energy Power Station is a 28mw Biomass to electricity power station under construction near Milltown, Indiana in Crawford County. It is being developed by Liberty Green Renewables (LGR), based in Houston, Texas.

This plant will utilize woody biomass material from a number of local forest product industrial sources, including residues from logging, land clearing activities, pallet manufacturing, furniture and cabinet manufacturing, sawmills, tree trimming and storm damage. This can be beneficial in the event of damaging ice storms like the January 2009 ice storm that hit Southern Indiana, and southern neighbour Kentucky, dropping thousands of limbs and wooden debris.

According to the company, the facility will utilize a fluidized bed boiler technology to enable utilization of a wide variety of woody biomass materials in an efficient, environmentally friendly manner.

Southeast Steam Plant

The Southeast Steam Plant, also known as the Twin City Rapid Transit Company Steam Power Plant, is a combined heat and power plant on the Mississippi River in the city of Minneapolis, Minnesota in the United States owned by the University of Minnesota.

History

The plant was constructed in 1903 to provide electricity for the

Twin City Rapid Transit street railway system. It supported the area's major form of public transportation for 50 years.

Minneapolis converted to buses in 1949—1954, and in the early 1950s, Northern States Power Company (now Xcel Energy) acquired the building. The University of Minnesota purchased the plant in 1976 for $1.

Operation

The facility heats 94 buildings—nearly all of the university's Minneapolis campus, cools 19 of those buildings, and provides steam to the University of Minnesota Medical Center, Minnesota State Board of Health and Cedar Riverside People's Center. Captured as the steam leaves the plant, pressure powers the plant and provides 20 per cent of the university's electricity. The plant's steam is transported through an 18 mile (29 km) network of tunnels to the campus buildings and would be enough to heat 55,000 homes. Each student pays about $200 for energy and those in residence halls pay $375 a year for heat and air conditioning, water heating and dining services.

The plant is university building 059. The university's Energy Management department, part of Facilities Management, oversees the plant. Foster Wheeler Twin Cities has contracted with the U of M to operate it since 1992.

Just upstream is the Hennepin Island Hydroelectric Plant operated by Xcel Energy. The university's Saint Paul, Minnesota campus three miles (5 km) away has its own plant. In addition the university has generators, pumps and boilers powered by diesel and natural gas, most used only in emergencies, with 11 used as peak shaving units.

Rehabilitation

Before pipes were reinsulated, employees needed breaks once an hour to work in the tunnels which reached 115°F (46°C). Insulation reduced the ambient temperature to 80°F (27°C), and the loss of energy from 10 per cent to 4 per cent, and over time resulted in a 25 per cent campus-wide decrease in energy consumption.

The university closed the Southeast plant to gut and rebuild the interior, and in 2000, reopened it and closed down its old coal-burning power plant.

Completed in 2005, exterior rehabilitation won a local historic preservation award, presented to the university and Miller Dunwiddie Architects, McGough Construction, Hess Roise Historical Consultants, Meyer Borgman Johnson, Michaud Cooley Erickson, INSPEC, Akiba Architects, and Kimley Horn.

Boilers, Fuel and Emissions

Among the "cleanest burning power plants in the country," the high temperature fires almost completely consume its fuels— natural gas, coal and wood waste. The plant has tested and been approved for oat hull biofuel, a renewable resource that would reduce each student's fees by about $21.

Five boilers are operational. A new fluidized bed boiler (CFB) is six stories high and capable of burning coal, wood, oat hulls or natural gas. There are two natural gas boilers, one pulverized coal boiler that can also fire fuel oil, and a spreader stoker coal boiler, also capable of burning fuel oil and possibly oat hulls. During May and October, the periods of lowest demand, the CFB boiler is not in use.

The CFB controls emissions of the acid gases sulfur dioxide, hydrogen chloride and hydrogen fluoride and particulate matter (PM) with limestone injection and a fabric filter. The pulverized coal and spreader stoker boilers are equipped with dry gas scrubbers (spray dryers). Two boilers have no control equipment but have flue gas recirculation to limit nitrogen oxide emission. The plant emits almost zero sulfur and mercury.

The unloading terminal for rail cars and its conveyors are enclosed and equipped with baghouse filters. The outdoor coal bunker is shielded from the wind by concrete retaining walls. Storage silos for ash have fabric filters.

Criticism

Environmental groups including the Save Our Riverfront Coalition and Friends of the Mississippi Inc. attempted and failed to move the plant off the river in 1996. Elected officials Phyllis Kahn and Larry Pogemiller, Arne Carlson who was governor of

Minnesota, and Sharon Sayles Belton who was mayor of Minneapolis supported the move. Concerned about potential emissions and noise from deliveries, some neighbourhood associations and a condominium developer at the nearby Pillsbury "A" Mill criticized the plant's 2005 application to amend its permit to allow tests of alternative fuels.

4

Global Biofuel Energy: Production, Technology and State of Regional Shares

FACTS RELATED TO BIOFUELS DATA

Bioenergy markets are dynamic and evolving, requiring constant monitoring of product streams. The Biomass Programme tracks market prices for commercially-available biofuels in cooperation with the National Renewable Energy Laboratory's Clean Cities programme, its own Energy Information Administration, the ethanol industry's Renewable Fuels Association, and the OPIS fuels information service, which obtains data from the Chicago Board of Trade and the Chicago Mercantile Exchange.

Table 4.1: Biofuels Data Table

E85 Retail Price/Gallon	$1.78 Mar 23
Gasoline Retail Price/Gallon	$1.96 Mar 23
Biodiesel Rack Price/Gallon	$3.12 Mar 23
Diesel Rack Price/Gallon	$1.38 Mar 23
E85 Station Count	1,871 Mar 23
New E85 Stations Opened	265 since Sept 1
Nameplate Ethanol Refineries	193 as of Mar 23
Nameplate Ethanol Production Capacity	12,375 million gallons Mar 23

EVALUATING ENERGY CONTENT OF BIOFUEL

Energy is the ability to do work. Per kilogram of mass, different substances can do different amounts of work. In other words they have different energy contents. Of course to do work we usually use a machine of some type. These machines vary in efficiency, or useful work done, and none are 100 per cent efficient. Thus the amount of useful work actually performed by these substances will never totally match these results. However this table gives us a relative measure of the amounts of these substances which could be equivalent in producing the required result (moving a car, heating a home, etc.).

In the example of the first two entries, Bagasse (Cane Stalks) has 9.6 MJ/kg (Mega Joules per kilogram) and Chaff (Seed Casings) has an energy content of 14.6 MJ/kg. In other words 1 kg of Chaff as a fuel would have 14.6-9.6, or 5 MJ more energy per kilogram energy content and potential work output than Bagasse.

Energy in MJ per kg CO_2 produced (MJ/kg) lists the energy produced per Kg of CO_2 produced. This is a measure of the potential environmental impact of the use of the substance as a fuel with respect to the release of CO_2. The more CO_2 released the worse it is for the environment. Thus a higher number in this column is better for the environment because we get more energy per kg of CO_2 produced. For example gasoline produces 13.64-14.64 MJ per kg of CO_2 but methane produces 20.05-20.30 MJ of energy, or nearly 50 per cent more energy for the same CO_2 production.

UNDERSTANDING SECOND GENERATION BIOFUELS

Biofuel technologies are competent to manufacture biofuels from biomass. Biomass is a wide-ranging term meaning any source of organic carbon that is renewed rapidly as part of the carbon cycle. Biomass is all derived from plant materials but can include animal materials.

Second generation biofuel technologies have been developed because first generation biofuels manufacture has important limitations . First generation biofuel processes are useful, but limited: there is a threshold above which they cannot produce enough biofuel without threatening food supplies and biodiversity. They are not cost competitive with existing fossil fuels such as oil, and some of them produce only limited greenhouse gas emissions savings. When taking emissions from production and transport into account, life_cycle_assessment from first-generation biofuels frequently exceed those of traditional fossil fuels.

Second generation biofuels can help solve these problems and can supply a larger proportion of our fuel supply sustainably, affordably, and with greater environmental benefits.

First generation bioethanol is produced by fermenting plant-derived sugars to ethanol, using a similar process to that used in beer and wine-making. This requires the use of 'food' crops such as sugar cane, corn, wheat, and sugar beet. These crops are required for food, so if too much biofuel is made from them, food prices could rise and shortages might be experienced in some countries. Corn, wheat and sugar beet also require high agricultural inputs in the form of fertilizers, which limit the greenhouse gas reductions that can be achieved.

The goal of second generation biofuel processes is to extend the amount of biofuel that can be produced sustainably by using biomass consisting of the residual non-food parts of current crops, such as stems, leaves and husks that are left behind once the food crop has been extracted, as well as other crops that are not used for food purposes (non food crops), such as switch grass, jatropha and cereals that bear little grain, and also industry waste such as wood chips, skins and pulp from fruit pressing, etc.

The problem that second generation biofuel processes are addressing is to extract useful feedstocks from this woody or fibrous biomass, where the useful sugars are locked in by lignin and cellulose. All plants contain cellulose and lignin. These are complex carbohydrates (molecules based on sugar). Lignocellulosic ethanol is made by freeing the sugar molecules from cellulose using enzymes, steam heating, or other pre-treatments. These sugars can then be fermented to produce ethanol in the same way as first generation bioethanol production. The by-product of this process is lignin. Lignin can be burned as a carbon neutral fuel to produce heat and power for the processing plant and possibly for surrounding homes and businesses.

The greenhouse gas emissions savings for lignocellulosic

ethanol are greater than those obtained by first generation biofuels. Lignocellulosic ethanol can reduce greenhouse gas emissions by around 90 per cent when compared with fossil petroleum

An operating lignocellulosic ethanol production plant is located in Canada, run by IOGEN Corporation . The demonstration-scale plant produces around 700,000 litres of bioethanol each year. A commercial plant is under construction. Many further lignocellulosic ethanol plants have been proposed in North America and around the world.

In the future, bio-synthetic liquid fuel might be available; such a fuel can be produced by the Fischer-Tropsch process a Gas-to-Liquid (GtL) process. When biomass is the source of the gas production the process is also referred to as Biomass-To-Liquids (BTL).

The following second generation biofuels are under development:

- *Biohydrogen*: Biohydrogen is the same as hydrogen except it is produced from a biomass feedstock. This is done using gasification of the biomass and then reforming the methane produced, or alternatively, this might be accomplished with some organisms that produce hydrogen directly under certain conditions. BioHydrogen can be used in fuel cells to produce electricity.
- *BioDME*: Bio-DME, Fischer-Tropsch, BioHydrogen diesel, Biomethanol and Mixed Alcohols all use syngas for production. This syngas is produced by gasification of biomass, however, it can be produced much easier from coal or natural gas, which is done on very large scales in power plants and in gas-to-liquid processes. HTU (High Temperature Upgrading) diesel is produced from particularly wet biomass stocks using high temperature and pressure to produce an oil.is the same as DME but is produced from a bio-sources. Bio-DME can be produced from Biomethanol using catalytic dehydration or it can be produced from syngas using

DME synthesis. DME can be used in the compression ignition engine.

- *Biomethanol*: Biomethanol is the same as methanol but it is produced from biomass. Biomethanol can be blended with petrol up to 10-20 per cent without any infrastructure changes.

- *Butanol and Isobutanol*: Via recombinant pathways expressed in hosts such as E. coli and yeast, butanol and isobutanol may be significant products of fermentation using glucose as a carbon and energy source.

- *DMF*: Recent advances in producing DMF from fructose and glucose using catalytic biomass-to-liquid process have increased its attractiveness.

- *HTU Diesel*: HTU diesel is produced from wet biomass. It can be mixed with fossil diesel in any percentage without need for infrastructure.

- *Fischer-Tropsch (FT) Fuels*: FT diesel is produced using the Fischer-Tropsch gas-to-liquids technology. In case biomass is used to produce hydrogen and CO, the reactants in the FT process, the carbon stays in a closed cycle. Disadvantage of this process is the high energy investment for the FT synthesis and consequently, the process is not yet economic. FT diesel can be mixed with fossil diesel at any percentage without need for infrastructure change and moreover, synthetic kerosene can be produced.

- *Mixed Alcohols (i.e., mixture of mostly ethanol, propanol and butanol, with some pentanol, hexanol, heptanol and octanol)*: Mixed alcohols are produced from syngas with catalysts similar to those used for methanol. Most RandD in this area is concentrated in producing mostly ethanol. However, some fuels are marketed as mixed alcohols. Mixed alcohols are superior to pure methanol or ethanol, in that the higher alcohols have higher energy content. Also, when blending, the higher alcohols increase compatibility of and decreases evaporative emissions. In addition, higher alcohols have also lower heat of vaporization than ethanol, which is important for cold starts.

- *Wood Diesel*: A new biofuel was developed by the University of Georgia from woodchips. The oil is extracted and then added to unmodified diesel engines. Either new plants are used or planted to replace the old plants. The charcoal byproduct is put back into the soil as a fertilizer. According to the director Tom Adams since carbon is put back into the soil, this biofuel can actually be carbon negative not just carbon neutral. Carbon negative decreases carbon dioxide in the air reversing the greenhouse effect not just reducing it.

Non-Food Crops

On the other hand, Biofuels from non-food energy crops that can be grown on marginal land (as Jatropha) are considered second generation biofuels and are nowadays available in mass production. Unfortunately these plants cannot be fed salt water and have a significant sweet water footprint as researchers by the University of Twente have established.

DOE Projects

USDOE has announced that it has selected six university-led advanced biofuels projects to receive up to $4.4 million, subject to annual appropriations. The awardees—Georgia Tech Research Corporation, the University of Georgia, the University of Maine, Montana State University, Steven's Institute of Technology in New Jersey, and the University of Toledo in Ohio—will all receive funding to conduct research and development of cost-effective, environmentally friendly biomass conversion technologies for turning non-food feedstocks into advanced biofuels. Combined with a university cost share of 20 per cent, more than $5.7 million is slated for investment in these projects.

Most of the projects will involve microbiology, including the University of Georgia and Montana State University projects, which are both focused on producing oils from algae. The University of Georgia will investigate the use of poultry litter to produce low-cost nutrients for algae, while Montana State, in partnership with Utah State University, will research the oil content, growth, and oil production of algae cultures in open ponds. Applying microbiology to biomass conversion, the University of Maine will study the use of bacteria to create biofuels from regionally available feedstocks, such as seaweed sludge and paper mill waste streams, while the University of Toledo will attempt to use pellets containing enzymes to efficiently convert cellulosic biomass into ethanol.

In contrast, Georgia Tech Research Corporation and Steven's Institute of Technology are both investigating the gasification of biomass. Georgia Tech will evaluate two experimental gasifiers run on forest residues, while Steven's Institute will test a novel microchannel reactor that gasifies pyrolysis oil, a petroleum-like oil produced by exposing biomass sources such as wood chips to high temperatures in the absence of oxygen. Gasified biomass can be used as a gaseous fuel or passed through a catalyst to produce a wide range of liquid fuels and chemicals.

CASE STUDY: VERASUN ENERGY

VeraSun Energy Corporation was a leading producer of renewable fuel. Founded in 2001, the company has a fleet of 16 production facilities in eight states, of which one is still under construction. VeraSun Energy is scheduled to have an annual production capacity of approximately 1.64 billion gallons (BGY) of ethanol and more than five million tons of distillers grains by the end of 2008. The company also has begun construction at its Aurora, South Dakota facility to extract oil from dried distillers grains, a co-product of the ethanol process, for use in biodiesel production.

VeraSun markets E85, a blend of 85 percent ethanol and 15 percent gasoline for use in Flexible-Fuel Vehicles (FFVs), directly to fuel retailers under the brand VE85. VeraSun Energy now has approximately 150 VE85 retail locations under contract in more than fifteen states and Washington, D.C..

History

VeraSun Energy was founded in 2001 with the goal of providing a renewable, home-grown energy source while boosting domestic rural economy and creating a future that includes renewable energy to help benefit the environment and reduce the nation's demand for foreign oil.

The company has been credited for a number of "industry firsts"—the first 100 million-gallon-per-year (MMGY) dry-grind production facility, the country's first branded E85, VE85; the first ethanol producer to form strategic alliances with Ford Motor Company, General Motors, Enterprise Rent-A-Car and Kroger to increase awareness and availability of E85, and the first company to place an E85 retail station in the Washington D.C. metro area.

VeraSun began producing ethanol in December 2003 when its Aurora, S.D., production facility came on-line. Less than two years later, Fort Dodge, Iowa, became the second VeraSun facility to begin production, increasing the company's production capacity to more than 200MMGY. Not stopping with two facilities, VeraSun began other Greenfield site developments in Iowa at Charles City and Hartley, in addition to a third location in Welcome, Minnesota. A fourth greenfield location is currently under development in Reynolds, Indiana—also referred to as BioTown USA. VeraSun Charles City began operation in April 2007, three months ahead of schedule.

On June 14, 2006, VeraSun became the first "pure play" ethanol producer to take its stock public when it listed on the New York Stock Exchange (NYSE). The public listing was the first of several major announcements for the company over the next 18 months. In July 2007, VeraSun announced the first major acquisition in the industry when the company purchased three 110MMGY production facilities from AS Alliances Biofuels, LLC. The facilities, located in Linden, Indiana; Albion, Nebraska, and Bloomingburg, Ohio, doubled the company's production capacity to more than 650MMGY.

VeraSun's second major acquisition came less than five months later when it was announced that VeraSun and US

BioEnergy would merge, creating a company with 16 biorefineries and a production capacity by the end of 2008 of more than 1.6 billion gallons per year. The merger closed on April 1, 2008 and positions VeraSun as the largest ethanol producer in the United States with plants located in eight different states.

Biorefinery Locations

VeraSun Energy has a fleet of 16 production facilities in eight states. They are listed below in alphabetical order.

Bankruptcy

As of 31 October 2008, VeraSun and twenty-four of its subsidiaries have filed for Chapter 11 bankruptcy protection to enhance liquidity while it reorganizes. The Sioux Falls-based company says the move was voluntary in order to maintain business as usual. South Dakota Senator John Thune says he is hopeful VeraSun Energy will emerge a stronger company after the chapter 11 filing.

TOWARDS BREAKTHROUGH IN BIOFUEL PRODUCTION PROCESS

Reporting in the April 7, 2008 issue of Chemistry and Sustainability, Energy and Materials (ChemSusChem), chemical engineer and National Science Foundation (NSF) CAREER awardee George Huber of the University of Massachusetts-Amherst (UMass) and his graduate students Torren Carlson and Tushar Vispute announced the first direct conversion of plant cellulose into gasoline components.

In the same issue, James Dumesic and colleagues from the University of Wisconsin-Madison announce an integrated process for creating chemical components of jet fuel using a green gasoline approach. While Dumesic's group had previously demonstrated the production of jet-fuel components using separate steps, their current work shows that the steps can be integrated and run sequentially, without complex separation and purification processes between reactors.

While it may be five to 10 years before green gasoline arrives at the pump or finds its way into a fighter jet, these breakthroughs have bypassed significant hurdles to bringing green gasoline biofuels to market.

"It is likely that the future consumer will not even know that they are putting biofuels into their car," said Huber. "Biofuels in the future will most likely be similar in chemical composition to gasoline and diesel fuel used today. The challenge for chemical engineers is to efficiently produce liquid fuels from biomass while fitting into the existing infrastructure today."

For their new approach, the UMass researchers rapidly heated cellulose in the presence of solid catalysts, materials that speed up reactions without sacrificing themselves in the process. They then rapidly cooled the products to create a liquid that contains many of the compounds found in gasoline.

The entire process was completed in under two minutes using relatively moderate amounts of heat. The compounds that formed in that single step, like naphthalene and toluene, make up one fourth of the suite of chemicals found in gasoline. The liquid can be further treated to form the remaining fuel components or can be used "as is" for a high octane gasoline blend.

"Green gasoline is an attractive alternative to bioethanol since it can be used in existing engines and does not incur the 30 percent gas mileage penalty of ethanol-based flex fuel," said John Regalbuto, who directs the Catalysis and Biocatalysis Programme at NSF and supported this research.

"In theory it requires much less energy to make than ethanol, giving it a smaller carbon footprint and making it cheaper to produce," Regalbuto said. "Making it from cellulose sources such as switchgrass or poplar trees grown as energy crops, or forest or agricultural residues such as wood chips or corn stover, solves the lifecycle greenhouse gas problem that has recently surfaced with corn ethanol and soy biodiesel."

Beyond academic laboratories, both small businesses and Fortune 500 petroleum refiners are pursuing green gasoline. Companies are designing ways to hybridize their existing refineries to enable petroleum products including fuels, textiles, and plastics to be made from either crude oil or biomass and the military community has shown strong interest in making jet fuel and diesel from the same sources.

"Huber's new process for the direct conversion of cellulose to gasoline aromatics is at the leading edge of the new 'Green Gasoline' alternate energy paradigm that NSF, along with other federal agencies, is helping to promote," states Regalbuto.

Not only is the method a compact way to treat a great deal of biomass in a short time, Regalbuto emphasized that the process, in principle, does not require any external energy. "In fact, from the extra heat that will be released, you can generate electricity in addition to the biofuel," he said. "There will not be just a small carbon footprint for the process; by recovering heat and generating electricity, there won't be any footprint."

The latest pathways to produce green gasoline, green diesel and green jet fuel are found in a report sponsored by NSF, the Department of Energy and the American Chemical Society entitled "Breaking the Chemical and Engineering Barriers to Lignocellulosic Biofuels: Next Generation Hydrocarbon Biorefineries" released April 1. In the report, Huber and a host of leaders from academia, industry and government present a plan for making green gasoline a practical solution for the impending fuel crisis.

"We are currently working on understanding the chemistry of this process and designing new catalysts and reactors for this single step technique. This fundamental chemical understanding will allow us to design more efficient processes that will accelerate the commercialization of green gasoline," Huber said.

UNDERSTANDING FISCHER—TROPSCH PROCESS

The **Fischer—Tropsch process** (or Fischer—Tropsch Synthesis) is a catalyzed chemical reaction in which synthesis gas, a mixture of carbon monoxide and hydrogen, is converted into liquid hydrocarbons of various forms. The most common catalysts are based on iron and cobalt, although nickel and ruthenium have also been used. The principal purpose of this process is to produce a synthetic petroleum substitute, typically from coal, natural gas or biomass, for use as synthetic lubrication oil or as synthetic fuel. This synthetic fuel runs trucks, cars, and some aircraft engines. (Refer to Sasol.) The use of diesel is increasing in recent years.

Combination of biomass gasification (BG) and Fischer-Tropsch (FT) synthesis is a possible route to produce renewable transportation fuels (biofuels).

Process Chemistry

The Fischer—Tropsch process involves a variety of competing chemical reactions, which lead to a series of desirable products and undesirable byproducts. The most important reactions are those resulting in the formation of alkanes. These can be described by chemical equations of the form:

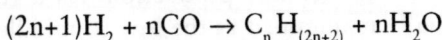

$$(2n+1)H_2 + nCO \rightarrow C_nH_{(2n+2)} + nH_2O$$

where 'n' is a positive integer. The simplest of these (n=1), results in formation of methane, which is generally considered an unwanted byproduct (particularly when methane is the primary feedstock used to produce the synthesis gas). Process conditions and catalyst composition are usually chosen to Favour higher order reactions (n>1) and thus minimize methane formation. Most of the alkanes produced tend to be straight-chained, although some branched alkanes are also formed. In addition to alkane formation, competing reactions result in the formation of alkenes, as well as alcohols and other oxygenated hydrocarbons. Usually, only relatively small quantities of these non-alkane products are formed, although catalysts Favouring some of these products have been developed.

Another important reaction is the water gas shift reaction:

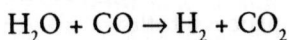

$$H_2O + CO \rightarrow H_2 + CO_2$$

Although this reaction results in formation of unwanted CO_2, it can be used to shift the H_2 : CO ratio of the incoming Synthesis gas. This is especially important for synthesis gas derived from coal, which tends to have a ratio of ~0.7 compared to the ideal ratio of ~2.

It should be noted that, according to published data on the current commercial implementations of the coal-based Fischer—Tropsch process, these plants can produce as much as 7 tonnes of CO_2 per tonne of liquid hydrocarbon products (excluding the reaction water product). This is due in part to the high energy demands required by the gasification process, and in part by the design of the process as implemented.

Process Conditions

Generally, the Fischer—Tropsch process is operated in the temperature range of 150-300°C (302-572°F). Higher temperatures lead to faster reactions and higher conversion rates, but also tend to Favour methane production. As a result the temperature is usually maintained at the low to middle part of the range. Increasing the pressure leads to higher conversion rates and also Favours formation of long-chained alkanes both of which are desirable. Typical pressures are in the range of one to several tens of atmospheres. Chemically, even higher pressures would be Favourable, but the benefits may not justify the additional costs of high-pressure equipment.

A variety of synthesis gas compositions can be used. For cobalt-based catalysts the optimal $H_2 : CO$ ratio is around 1.8-2.1. Iron-based catalysts promote the water-gas-shift reaction and thus can tolerate significantly lower ratios. This can be important for synthesis gas derived from coal or biomass, which tend to have relatively low $H_2 : CO$ ratios (<1).

Product Distribution

In general the product distribution of hydrocarbons formed during the Fischer—Tropsch process follows an Anderson-Schulz-Flory distribution, which can be expressed as:

$$W_n/n = (1-\alpha)^2\alpha^{n-1}$$

Where W_n is the weight fraction of hydrocarbon molecules containing n carbon atoms. á is the chain growth probability or the probability that a molecule will continue reacting to form a longer chain. In general, á is largely determined by the catalyst and the specific process conditions.

Examination of the above equation reveals that methane will always be the largest single product; however by increasing á close to one, the total amount of methane formed can be minimized compared to the sum of all of the various long-chained products. Increasing á increases the formation of long-chained hydrocarbons. The very long-chained hydrocarbons are waxes, which are solid at room temperature. Therefore, for production of liquid transportation fuels it may be necessary to crack some of the Fischer-Tropsch products. In order to avoid this, some researchers have proposed using zeolites or other catalyst substrates with fixed sized pores that can restrict the formation of hydrocarbons longer than some characteristic size (usually n<10). This way they can drive the reaction so as to minimize methane formation without producing lots of long-chained hydrocarbons. So far, such efforts have had only limited success.

Fischer-Tropsch Catalysts

A variety of catalysts can be used for the Fischer—Tropsch process, but the most common are the transition metals cobalt, iron, and ruthenium. Nickel can also be used, but tends to Favour methane formation. Cobalt seems to be the most active catalyst, although iron also performs well and can be more suitable for low-hydrogen-content synthesis gases such as those derived from coal due to its promotion of the water-gas-shift reaction. In addition to the active metal the catalysts typically contain a number of promoters, including potassium and copper, as well as high-surface-area binders/supports such as silica, alumina, or zeolites.

Unlike the other metals used for this process (Co, Ni, Ru) which remain in the metallic state during synthesis, iron catalysts tend to form a number of chemical phases, including various iron oxides and iron carbides during the reaction. Control of these phase transformations can be important in maintaining catalytic activity and preventing breakdown of the catalyst particles.

The Fischer-Tropsch catalysts are notoriously sensitive to the presence of sulfur-containing compounds among other poisons. The sensitivity of the catalyst to sulfur is higher for cobalt-based catalysts than for their iron counterparts.

Cobalt catalysts are preferred for Fischer-Tropsch synthesis when the feedstock is natural gas due to the higher activity of the cobalt catalyst. Natural gas has a high hydrogen to carbon ratio, so the water-gas-shift is not needed for cobalt catalysts. Iron catalysts are preferred for lower quality feedstocks such as coal or biomass. While iron catalysts are also susceptible to sulfur poisoning from coal with high sulfur content, the lower cost of iron makes sacrificial catalyst at the front of a reactor bed economical. Also, as mentioned earlier, iron can catalyze the water-gas-shift to increase the hydrogen to carbon ratio to make the reaction more Favourably selective.

Synthesis Gas Production

The initial reactants (synthesis gases) used in the Fischer—Tropsch process are hydrogen gas (H_2) and carbon monoxide (CO). These chemicals are usually produced by one of two methods:

The partial combustion of a hydrocarbon:

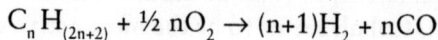

$$C_n H_{(2n+2)} + \tfrac{1}{2}\, nO_2 \rightarrow (n+1)H_2 + nCO$$

When n=1 (methane), the equation becomes:
$$2CH_4 + O_2 \rightarrow 4H_2 + 2CO$$

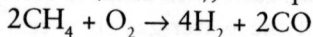

The gasification of coal, biomass, or natural gas:

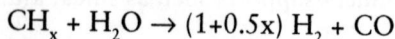

$$CH_x + H_2O \rightarrow (1+0.5x)\, H_2 + CO$$

The value of "x" depends on the type of fuel. For example, natural gas has a greater hydrogen content (from x=4 to x~2.5) than coal (x<1).

The energy needed for this endothermic reaction is usually provided by the (exothermic) combustion of the hydrocarbon source with oxygen.

The mixture of carbon monoxide and hydrogen is called synthesis gas or syngas. The resulting hydrocarbon products are refined to produce the desired synthetic fuel.

The carbon dioxide and carbon monoxide is generated by partial oxidation of coal and wood-based fuels. The utility of the process is primarily in its role in producing fluid hydrocarbons from a solid feedstock, such as coal or solid carbon-containing wastes of various types. Non-oxidative pyrolysis of the solid material produces syngas which can be used directly as a fuel without being taken through Fischer-Tropsch transformations. If liquid petroleum-like fuel, lubricant, or wax is required, the Fischer—Tropsch process can be applied.

History

Since the invention of the original process by the German researchers Franz Fischer and Hans Tropsch, working at the Kaiser Wilhelm Institute in the 1920s, many refinements and adjustments have been made, and the term "Fischer-Tropsch" now applies to a wide variety of similar processes (**Fischer-Tropsch synthesis** or **Fischer-Tropsch chemistry**). Fischer and Tropsch filed a number of patents, e.g. US patent no. 1,746,464, applied 1926, published 1930.

The process was invented in petroleum-poor but coal-rich Germany in the 1920s, to produce liquid fuels. It was used by Nazi Germany and Japan during World War II to produce *ersatz* (German: *substitute*) fuels. By early 1944 production reached more than 124,000 barrels per day (19,700 m³/d) from 25 plants ~ 6.5 million tons per year. However, the bombing of German oil facilities during World War II paralyzed much of Germany's synthetic fuel production.

The United States Bureau of Mines, in a Programme initiated by the Synthetic Liquid Fuels Act, employed seven Operation Paperclip synthetic fuel scientists in a Fischer-Tropsch chemical plant in Louisiana, Missouri in 1946.

In Britain, Alfred August Aicher obtained several patents for improvements to the process in the 1930s and 1940s, e.g. British patent no. 573,982, applied 1941, published 1945. Aicher's company was named *Synthetic Oils Ltd.* (There is no connection with the Canadian company of the same name.)

Use

Currently, only a handful of companies have commercialised their FT technology.

1. Shell in Bintulu, Malaysia, uses natural gas as a feedstock, and produces primarily low-sulfur diesel fuels and food-grade wax.
2. Sasol in South Africa uses coal and natural gas as a feedstock, and produces a variety of synthetic petroleum products. Sasol produces most of the country's diesel fuel.

The process was used in South Africa to meet its energy needs during its isolation under Apartheid. This process has received renewed attention in the quest to produce low-sulfur diesel fuel in order to minimize environmental degradation from the use of diesel engines.

A small US-based company, Rentech, is currently focusing on converting nitrogen-fertiliser plants from using a natural gas feedstock to using coal or coke, and producing liquid hydrocarbons as a co-product.

Also Choren Industries has built an FT plant in Germany.

The FT process is an established technology and already applied on a large scale in some industrial sectors, although its popularity is hampered by high capital costs, high operation and maintenance costs, the uncertain and volatile price of crude oil, and environmental concerns. In particular, the use of natural gas as a feedstock only becomes practical when using "stranded gas", i.e. sources of natural gas far from major cities which are impractical to exploit with conventional gas pipelines and LNG technology; otherwise, the direct sale of natural gas to consumers would become much more profitable. There are several companies developing the process to enable practical exploitation of so-called stranded gas reserves.

This technology has been proposed as a way to create transportation fuel from coal if conventional oil were to become more expensive. In Sept. 2005, Pennsylvania governor Edward Rendell announced a venture with Waste Management and Processors Inc.—using technology licensed from Shell and Sasol—to build an FT plant that will convert so-called waste coal (leftovers from the mining process) into low-sulfur diesel fuel at a site outside of Mahanoy City, northwest of Philadelphia. The state of Pennsylvania has committed to buy a significant percentage of the plant's output and, together with the U.S. Dept. of Energy, has offered over $140 million in tax incentives. Other coal-producing states are exploring similar plans. Governor Brian Schweitzer of Montana has proposed developing a plant that would use the FT process to turn his state's coal reserves into fuel in order to help alleviate the United States' dependence on foreign oil.

In Oct. 2006, Finnish paper and pulp manufacturer UPM announced its plans to produce biodiesel by Fischer—Tropsch process alongside the manufacturing processes at its European paper and pulp plants, using waste biomass resulted by paper and pulp manufacturing processes as source material.

In August 2007, Louisiana State University announced they had received funding from the US Department of Energy and Conoco Phillips for development of new nanotechnologies for catalysis of coal syngas to ethanol conversion.

U.S. Air Force Certification

Syntroleum, a publicly traded US company (Nasdaq: SYNM) has produced over 400,000 gallons of diesel and jet fuel from the Fischer—Tropsch process using natural gas and coal at its demonstration plant near Tulsa, Oklahoma. Syntroleum is working to commercialize its licensed Fischer-Tropsch technology via coal-to-liquid plants in the US, China, and Germany, as well as gas-to-liquid plants internationally. Using natural gas as a feedstock, the ultra-clean, low sulfur fuel has been tested extensively by the US Department of Energy, the Department of

Transportation, and most recently, Syntroleum has been working with the U.S. Air Force to develop a synthetic jet fuel blend that will help the Air Force to reduce its dependence on imported petroleum. The Air Force, which is the U.S. military's largest user of fuel, began exploring alternative fuel sources in 1999. On December 15, 2006, a B-52 took off from Edwards AFB, California for the first time powered solely by a 50-50 blend of JP-8 and Syntroleum's FT fuel. The seven-hour flight test was considered a success. The goal of the flight test Programme is to qualify the fuel blend for fleet use on the service's B-52s, and then flight test and qualification on other aircraft.

On August 8, 2007, Air Force Secretary Michael Wynne certified the B-52H as fully approved to use the FT blend, marking the formal conclusion of the test Programme.

This Programme is part of the Department of Defense Assured Fuel Initiative, an effort to develop secure domestic sources for the military energy needs. The Pentagon hopes to reduce its use of crude oil from foreign producers and obtain about half of its aviation fuel from alternative sources by 2016. With the B-52 now approved to use the FT blend, the USAF will use the test protocols developed during the Programme to certify the C-17 Globemaster III and then the B-1B to use the fuel. The Air Force intends to test and certify every airframe in its inventory to use the fuel by 2011.

Demonstration testing of the C-17 burning Fischer-Tropsch fuel was completed on October 22, 2007 at Edwards Air Force Base. Testing consisted of a ground test and two flights which demonstrated engine performance throughout the C-17 flight envelope and during some operationally representative manoeuvres. Test data is still being reviewed by the 418th FLTS to validate the subjective results of the test. On December 17, 2007 a C-17 Globemaster III using the synthetic fuel blend lifted off shortly before dawn from McChord Air Force Base, Washington, and flew to McGuire Air Force Base, New Jersey, where it was greeted by politicians and by officials from the airline and energy industries. Based on the two successful tests, the Air Force hopes to certify all of its C-17 fleet for the synthetic fuel mixture early in 2008.

US Navy

In 2009, the US Navy experimented with making jet fuel from seawater, using a variation of the Fischer—Tropsch process. To alleviate problems with global warming and potential oil shortages, Navy chemists tried to create fuel from seawater by splitting the molecules using electricity in order to extract the carbon dioxide. When combined with water using a cobalt-based catalyst this produces mostly methane gas, but by changing to an iron catalyst the process produced only 30 per cent methane with the rest being short-chain hydrocarbons. Further refining of the hydrocarbons produced could potentially lead to the production of kerosene-based jet fuel.

The abundance of CO_2 makes seawater an attractive alternative fuel source. Robert Dorner, a chemist at the Naval Research Laboratory, stated that, "although the gas forms only a small proportion of air—around 0.04 per cent—ocean water contains about 140 times that concentration". Doemer presented the findings to the American Chemical Society on 16 August 2009, at the Marriott Metro Center in Washington DC.

CO_2 Reuse

There are investigations underway to reduce CO_2 emissions by using solar power to convert waste CO_2 into CO from where the FT process can then convert it to hydrocarbons.

UNDERSTANDING ENERGY RETURN ON ENERGY INVESTED (EROEI)

In physics, energy economics and ecological energetics, EROEI (Energy Returned on Energy Invested), ERoEI, EROI (Energy Return On Investment) or less frequently, eMergy, is the ratio of the amount of usable energy acquired from a particular energy resource to the amount of energy expended to obtain that energy resource. When the EROEI of a resource is equal to or lower than 1, that energy source becomes an "energy sink", and can no longer be used as a primary source of energy.

$$EROEI = \frac{\text{Usable Acquired Energy}}{\text{Energy Expended}}$$

Non-manmade Energy Inputs

The natural or original sources of energy are not usually included in the calculation of energy invested, only the human-applied sources. For example in the case of biofuels the solar insulation driving photosynthesis is not included, and the energy used in the stellar synthesis of fissile elements is not included for nuclear fission. The energy returned includes any usable energy and not wasted heat for example.

Relationship to Net Energy Gain

EROEI and *Net energy (gain)* measure the same quality of an energy source or sink in numerically different ways. Net energy describes the amounts, while EROEI measures the ratio or efficiency of the process. They are related simply by:

$$\text{(Net Energy} + \text{Energy Expended)} \div \text{Energy Expended} = EROEI$$

or

$$\text{(Net Energy} \div \text{Energy Expended)} + 1 = EROEI$$

For example given a process with an EROEI of 5, expending 1 unit of energy yields a net energy gain of 4 units. The break-even point happens with an EROEI of 1 or a net energy gain of 0.

The Economic Influence of EROEI

High per-capita energy use is considered desirable as it is associated with a high standard of living based on energy-intensive machines. A society will generally exploit the highest available EROEI energy sources first, as these provide the most energy for the least effort. With non-renewable sources, progressively lower EROEI sources are then used as the higher-quality ones are exhausted.

For example, when oil was originally discovered, it took on average one barrel of oil to find, extract, and process about 100 barrels of oil. That ratio has declined steadily over the last century to about three barrels gained for one barrel used up in the U.S. (and about ten for one in Saudi Arabia). Currently (2006) the EROEI of wind energy in North America and Europe is about 20:1. which has driven its adoption.

Although many qualities of an energy source matter (for example oil is energy-dense and transportable, while wind is variable), when the EROEI of the main sources of energy for an economy fall energy becomes more difficult to obtain and its value rises relative to other resources and goods. Therefore the EROEI gains importance when comparing energy alternatives. Since expenditure of energy to obtain energy requires productive effort, as the EROEI falls an increasing proportion of the economy has to be devoted to obtaining the same amount of net energy.

Since the discovery of fire, humans have increasingly used exogenous sources of energy to multiply human muscle-power and improve living standards. Some historians have attributed our improved quality of life since then largely to more easily exploited (i.e. higher EROEI) energy sources, which is related to the concept of energy slaves. Thomas Homer-Dixon demonstrates that a falling EROEI in the Later Roman Empire was one of the reasons for the collapse of the Western Empire in the fifth century CE. In "The Upside of Down" he suggests that EROEI analysis provides a basis for the analysis of the rise and fall of civilisations. Looking at the maximum extent of the Roman Empire, (60 million) and its technological base the agrarian base of Rome was about 1:12 per hectare for wheat and 1:27 for alfalfa (giving a 1:2.7 production for oxen). One can then use this to calculate the population of the Roman Empire required at its height, on the basis of about 2,500-3,000 calories per day per person. It comes out roughly equal to the area of food production at its height. But ecological damage (deforestation, soil fertility loss particularly in southern Spain, southern Italy, Sicily and especially north Africa) saw a collapse in the system beginning in the 2nd century, as EROEI began to fall. It bottomed in 1084 when Rome's population, which had peaked under Trajan at 1.5 million, was only 15,000. Evidence also fits the cycle of Mayan and Cambodian collapse too. Joseph Tainter suggests that diminishing returns of the EROEI is a chief cause of the collapse of complex societies. Falling EROEI due to depletion of non-renewable resources also poses a difficult challenge for industrial economies.

Criticism of EROEI

Measuring the EROEI of a single physical process is unambiguous, but there is no agreed standard on which activities should be included in measuring the EROEI of an economic process. In addition, the form of energy of the input can be completely different from the output. For example, energy in the form of coal could be used in the production of ethanol. This might have an EROEI of less than one, but could still be desirable due to the benefits of liquid fuels.

How deep should the probing in the supply chain of the tools being used to generate energy go? For example, if steel is being used to drill for oil or construct a nuclear power plant, should the energy input of the steel be taken into account, should the energy input into building the factory being used to construct the steel be taken into account and amortized? Should the energy input of the roads which are used to ferry the goods be taken into account? What about the energy used to cook the steelworker's breakfasts? These are complex questions evading simple answers. A full accounting would require considerations of opportunity costs and comparing total energy expenditures in the presence and absence of this economic activity.

However, when comparing two energy sources a standard practice for the supply chain energy input can be adopted. For example, consider the steel, but don't consider the energy invested in factories deeper than the first level in the supply chain.

Energy return on Energy invested does not take into account

the factor of time. Energy invested in creating a solar panel may have consumed energy from a high power source like coal, but the return happens very slowly, i.e. over many years. If energy is increasing in relative value this should favour delayed returns. Some believe this means the EROEI measure should be refined further.

Conventional economic analysis has no formal accounting rules for the consideration of waste products that are created in the production of the ultimate output. For example, differing economic and energy values placed on the waste products generated in the production of ethanol makes the calculation of this fuel's true EROEI extremely difficult.

EROEI is only one consideration and may not be the most important one in energy policy. Energy independence (reducing international competition for limited natural resources), freedom from pollution (including carbon dioxide and other green house gases), and affordability could be more important, particularly when considering secondary energy sources. While a nation's primary energy source is not sustainable unless it uses less energy than it creates, the same is not true for secondary energy supplies. Some of the energy surplus from the primary energy source can be used to create the fuel for secondary energy sources, such as for transportation.

EROEI under Rapid Growth

A related recent concern is energy cannibalism where energy technologies can have a limited growth rate if climate neutrality is demanded. Many energy technologies, are capable of replacing significant volumes of fossil fuels and concomitant green house gas emissions. Unfortunately, neither the enormous scale of the current fossil fuel energy system nor the necessary growth rate of these technologies is well understood within the limits imposed by the net energy produced for a growing industry. This technical limitation is known as energy cannibalism and refers to an effect where rapid growth of an entire energy producing or energy efficiency industry creates a need for energy that uses (or cannibalizes) the energy of existing power plants or production plants. The solar breeder overcomes some of these problems. A solar breeder is a photovoltaic panel manufacturing plant which can be made energy-independent by using energy derived from its own roof using its own panels. Such a plant becomes not only energy self-sufficient but a major supplier of new energy, hence the name solar breeder. The reported investigation establishes certain mathematical relationships for the solar breeder which clearly indicate that a vast amount of net energy is available from such a plant for the indefinite future. BP Solar originally intended its plant in Frederick, Maryland to be such a Solar Breeder, but the project did not develop. Theoretically breeders of any kind can be developed.

CASE STUDY: BIOFIELDS
(RENEWABLE AND SUSTAINABLE ENERGY)

BioFields is a Mexican industrial group that produces biofuels from blue-green algae (cyanobacteria).

The company was founded in 2006, it has a commercial agreement with Algenol Biofuels the company that owns the technology to Direct Ethanol, which allows the production of blue-green algae based biofuels. BioFields got the first license to use such technology in Mexico, specifically in Puerto Libertad, in Pitiquito, Sonora, because the conditions of the place are ideal for the development of this technology (approximately 328 days of sunshine a year), it also has a near-by source of carbon dioxide, it is close to the sea and has great extent of dry, non-arable land.

BioFields' technology Direct to Ethanol uses non-toxic blue-green algae to produce biofuels in a fully enclosed secure system. This technology optimizes the metabolism of blue-green algae creating a new pathway for the use and fixation of carbon resulting in the synthesis of ethanol. The blue-green algae proliferates rapidly and uses efficiently sunlight, carbon dioxide and inorganic elements, for the production of ethanol or other biofuels.

The capture of carbon dioxide allows to accelerate the process of photosynthesis using the blue-green algae to convert the sun's energy in the form of intracellular sugar, which provides vital energy to grow and reproduce as well as to double its population in a few hours and it can be grown on farms allowing the biofuels production at an industrial level.

Their first project is Sonora Fields, consisting of 22,000 hectares of land in the desert of Sonora, Mexico. Their estimated production for the year 2012 is approximately 1 billion gallons.

UNDERSTANDING BIOFUELS BY REGION

The use of **biofuels** varies by **region** and with increasing oil prices there is a renewed interest in it as an energy source.

Recognizing the importance of implementing bioenergy, there are international organizations such as IEA Bioenergy, established in 1978 by the OECD International Energy Agency (IEA), with the aim of improving cooperation and information exchange between countries that have national Programmes in bioenergy research, development and deployment. The U.N. International Biofuels Forum is formed by Brazil, China, India, South Africa, the United States and the European Commission. The world leaders in biofuel development and use are Brazil, United States, France, Sweden and Germany.

Biofuels by Region

Brazil

Typical Brazilian "flex" models from several car makers, that run on any blend of ethanol and gasoline.

The government of Brazil hopes to build on the success of the Proálcool ethanol Programme by expanding the production of biodiesel which must contain 2 per cent biodiesel by 2008, and 5 per cent by 2013.

Canada

The government of Canada aims for 45 per cent of the country's gasoline consumption to contain 10 per cent ethanol by 2010.

China

In China, the government is making E10 blends mandatory in

five provinces that account for 16 per cent of the nation's passenger cars.

Colombia

Colombia mandates the use of 10 per cent ethanol in all gasoline sold in cities with populations exceeding 500,000. In Venezuela, the state oil company is supporting the construction of 15 sugar cane distilleries over the next five years, as the government introduces a E10 (10 per cent ethanol) blending mandate.

India

In India, a bioethanol Programme calls for E5 blends throughout most of the country targeting to raise this requirement to E10 and then E20.

Israel

IC Green Energy, a subsidiary of Israel Corp., aims by 2012 to process 4-5 per cent of the global biofuel market (~4 million tons). It is focused solely on non-edible feedstock such as jatropha, castor, cellulosic biomass and algae. In June 2008, Tel Aviv-based Seambiotic and Seattle-based Inventure Chemical announced a joint venture to use CO2 emissions-fed algae to make ethanol and biodiesel at a biofuel plant in Israel.

The European Union in its biofuels directive (updated 2006) has set the goal that for 2010 that each member state should achieve at least 5.75 per cent biofuel usage of all used traffic fuel. By 2020 the figure should be 10 per cent. As of January 2008 these aims are being reconsidered in light of certain environmental and social concerns associated with biofuels such as rising food prices and deforestation.

France

France is the second largest biofuel consumer among the EU States in 2006. According to the Ministry of Industry, France's consumption increased by 62.7 per cent to reach 682,000 toe (i.e. 1.6 per cent of French fuel consumption). Biodiesel represents the largest share of this (78 per cent, far ahead of bioethanol with 22 per cent). The unquestionable biodiesel leader in Europe is the French company Diester Industrie. In bioethanol, the French agro-industrial group Téréos is increasing its production capacities.

Germany

Germany remained the largest European biofuel consumer in 2006, with a consumption estimate of 2.8 million tons of biodiesel (equivalent to 2,408,000 toe), 0.71 million ton of vegetable oil (628.492 toe) and 0.48 million ton of bioethanol (307,200 toe).

The biggest German biodiesel company is ADM Ölmühle Hamburg AG, subsidiary of the American group Archer Daniels Midland Company. Among the other large German producers, MUW (Mitteldeutsche Umesterungswerke GmbH and Co KG) and EOP Biodiesel AG. A major contender in terms of bioethanol production is the German sugar corporation, Südzucker.

Spain

The Spanish group Abengoa, via its American subsidiary Abengoa Bioenergy, is the European leader in production of bioethanol.

Sweden

The government of Sweden and the national association of auto makers, BIL Sweden, have started work to end oil dependency. One-fifth of cars in Stockholm can run on alternative fuels, mostly ethanol fuel. Stockholm is to introduce a fleet of Swedish-made hybrid ethanol-electric buses. Plans for oil phase-out in Sweden by 2020 was announced in 2005.

United Kingdom

In the United Kingdom, the Renewable Transport Fuel Obligation (RTFO) (announced 2005) is the requirement that by 2010 5 per cent of all road vehicle fuel is renewable. In 2008 a critical report by the Royal Society stated *that biofuels risk failing to deliver significant reductions in greenhouse gas emissions from transport and could even be environmentally damaging unless the Government puts the right policies in place.*

The two types of **biofuel** supplied for use in motor vehicles in the UK are bioethanol, a denatured alcohol that can be blended with petrol, and biodiesel, which is oil-based and can be used as a replacement for normal diesel.

Bioethanol

Bioethanol is derived from plant material containing sugars and starches. The source is normally agricultural crops grown specifically for conversion to biofuel, but it could also be waste matter – crop residues or organic household waste.

This product can be made from grains or fruit, and also wood. Sugar beet and wheat are used in the UK, whilst around the world common alternatives include corn and potato. Sugar cane, used extensively in Brazil, has an extra advantage in that the fibrous waste, or bagasse, can be used to provide the thermal energy required to evaporate the cane juice. In the UK, manufacture of bioethanol from sugar beet began in 2007.

Bioethanol burns to produce very significantly lower levels of CO_2 than petrol. For use in motor vehicles, bioethanol is always blended with petrol. Two different petrol/ethanol mixtures are available in the UK. One is a blend of 5 per cent; this is the maximum percentage of ethanol that petrol suppliers are allowed to include in their ordinary unleaded petrol under current regulations, although there is speculation that the limit may soon be increased to 10 per cent. A blend of 5 per cent has no noticeable effect on vehicle performance, and a large proportion of the petrol sold on UK forecourts contains 5 per cent bioethanol. The alternative is an 85 per cent blend, marketed as E85, and this is only suitable for flexfuel vehicles.

Flexfuel Cars

Ethanol is higher octane and more corrosive than petrol. In order to run on an 85 per cent mixture, cars must be adapted to prevent engine damage. Because E85 has limited availability, it would be impractical to make vehicles that use only E85; they need to be able to use petrol as well. A number of car manufacturers have designed flexible fuel cars, known as 'flexfuel', that run equally well on either fuel. Potentially, the higher octane rating of E85 could be maximised to achieve enhanced performance, but this entails making advanced modifications to the engine settings, and at the time of writing only one manufacturer markets a flexfuel car that actually performs better on E85 than on standard petrol.

Biodiesel

Like bioethanol, biodiesel can be produced either from purpose-grown crops or from waste. In Europe, rapeseed oil is grown extensively as a biofuel crop. Vegetable oils and waste cooking oils and fats are also used, and kits are sold in the UK that enable individuals to convert waste oil into biodiesel.

Most older diesel engines can run on pure biodiesel, although if a vehicle is to be run for a long time on a 100 per cent blend some precautions might be necessary, such as replacing rubber hoses with corrosion-resistant plastic. For modern diesel vehicles, biodiesel is normally blended with normal diesel in varying proportions, according to manufacturer specifications.

Cars using crude-oil diesel produce less CO_2 that petrol cars, but more particulate emissions. Using biodiesel, CO_2 is cut down still further and particulates are also reduced; biodiesel contains more oxygen, so combustion is more complete. The higher the proportion of biodiesel in the fuel, the cleaner the emissions.

Major UK biofuel companies

1. Greenergy
2. Rix petroleum
3. Biofuels Corporation

Biofuel promotion

Virgos Co Ltd specializes in promoting BioFuels and OleoChemicals. The main products offered are:

1. Crude Palm Oil
2. Coco Oil
3. Palm/Coconut Fatty Acid Distillate
4. Refined Glycerine 99.5% Kosher
5. Fatty Alcohol
6. Soybean Acid Oil

7.   Refined Sunflower Oil
8.   Degummed SoyBean Oil

Biofuel in New Zealand

Ecodiesel, a company owned by a group of New Zealand farmers, plans to build a biodiesel plant by the end of 2008. The plant will be built in stages and cheaper than Argent's, and could produce 20 million litres of tallow-based biodiesel per year by April 2009.

In the effort to develop an aviation biofuel, Air New Zealand and Boeing are researching the jatropha plant to see if it can provide a renewable alternative to conventional fuel.

Bioethanol

Gull Force 10, a bioethanol blend, was introduced commercially in New Zealand for the first time by the company Gull on 1 August 2007. It contained 10% ethanol made from dairy by product by Anchor Ethanol, a subsidiary of Fonterra Ltd. On 8 August 2008, Gull introduced a 91-octane bioethanol blend in Albany. The blend, 'regular plus', contained 10% ethanol and included bioethanol made from whey. Gull planned to release the fuel to 33 stations, and marketed it as under $2 per litre. On release, the company said it would try to keep the price two cents less than its standard 91-octane fuel.

It was reported that British fuel producer Argent Energy would abandon plans to build a plant in Tauranga to produce tallow-based biodiesel. The plant would have cost over $100 million to build, and would have competed with cheaper sugar-based ethanol imports from Brazil. The plant could not proceed because a 42c/L tax break on bioethanol until 2010 had not been approved by the government.

Biomass

Firewood is used as a means of heating some homes in New Zealand and wood pellet fires are now becoming more common, especially in areas with high levels of air pollution.

Legislation and Government Funding

Biofuel Bill: The Labour-led government introduced a Biofuel Bill in October 2007. It passed its second reading in Parliament in September 2008. The Bill requires petrol and diesel to have a percentage of biofuels added with the amount increasing to 2.5% after five years.

In April 2008 the Parliamentary Commissioner for the Environment, independent from but funded by the government, recommended in a select committee briefing that the Biofuels Bill should not proceed. This conclusion was arrived at on a number of grounds. The Biofuels Bill did not restrict the importation of biofuels and this would lead to potential societal and environmental harm that may be greater than if biofuels were not used. It was also claimed that this may damage the "clean green" image of New Zealand. Waiting for second generation biofuels and curbing the growth in transport energy consumption were also seen as reasons for not supporting the Bill

After the National Party gained power it repealed parts of the Biofuel Bill with the Energy (Fuels, Levies, and References) Biofuel Obligation Repeal Act. It removed the mandatory requirement for all fuel to have a percentage of biofuel.

Sustainable Biofuels Bill: The Green Party tabled a Sustainable Biofuels Bill which passed its first reading in Parliament in July 2009. The Bill "is to ensure that biofuels that are supplied or sold in New Zealand from 1 May 2010 are sustainable biofuels".

Funding: In the 2009 Budget $36 million was made available over a three year period as grants for biofuel production. It is only available for producers who sell on the local market and are able to meet the quality specifications for engine fuels.

USA

The Energy Policy Act of 2005 was passed by the United States Congress on July 29, 2005 and signed into law by President George W. Bush on August 8 2005 at Sandia National Laboratories in Albuquerque, New Mexico. The Act, described by proponents as an attempt to combat growing energy problems, changed the energy policy of the United States by providing tax incentives and loan guarantees for energy production of various types.

In 2006, the United States president George W. Bush said in a State of the Union speech that the US is "addicted to oil" and should replace 75 per cent of imported oil by 2025 by alternative sources of energy including biofuels.

Essentially all ethanol fuel in the US is produced from corn. Corn is a very energy-intensive crop, which requires one unit of fossil-fuel energy to create just 0.9 to 1.3 energy units of ethanol. A senior member of the House Energy and Commerce Committee, Congressman Fred Upton introduced legislation to use at least E10 fuel by 2012 in all cars in the USA.

The 2007-12-19 US Energy Independence and Security Act of 2007 requires American "fuel producers to use at least 36 billion gallons of biofuel in 2022. This is nearly a fivefold increase over current levels." This is causing a significant shift of resources away from food production. American food exports have decreased (increasing grain prices worldwide), and US food imports have increased significantly.

Most biofuels are not currently cost-effective without significant subsidies. "America's ethanol Programme is a product of government subsidies. There are more than 200 different kinds, as well as a 54 cents-a-gallon tariff on imported ethanol. This prices Brazilian ethanol out of an otherwise competitive market. Brazil makes ethanol from sugarcane rather than corn (maize), which has a better EROEI. Federal subsidies alone cost $7 billion a year (equal to around $1.90 a gallon)."

General Motors is starting a project to produce E85 fuel from cellulose ethanol for a projected cost of $1 a gallon. This is optimistic, because $1/gal equates to $10/MBTU which is comparable to woodchips at $7/MBTU or cord wood at $6-$12/MBTU, and this does not account for conversion losses and plant operating and capital costs which are significant. The raw materials can be as simple as corn stalks and scrap petroleum-based vehicle tires, but used tires are an expensive feedstock with other more-valuable uses. GM has over 4 million E85 cars on the road now, and by 2012 half of the production cars for the US will be capable of running on E85 fuel. But by 2012, the supply of ethanol will not even be close to supplying this much E85. Coskata Inc. is building two new plants for the ethanol fuel. Theoretically, the process is claimed to be five times more energy efficient than corn based ethanol, but it is still in development and has not been proven to be cost effective in a free market.

The greenhouse gas emissions are reduced by 86 per cent for cellulose compared to corn's 29 per cent reduction.

The Food, Conservation, and Energy Act of 2008 is a $288 billion, five-year agricultural policy bill being considered by the United States Congress as a continuation of the 2002 Farm Bill. The bill continues the United States' long history of agricultural subsidy as well as pursuing areas such as energy, conservation, nutrition, and rural development. Some specific initiatives in the bill include increases in Food Stamp benefits, increased support for the production of cellulosic ethanol, and money for the research of pests, diseases and other agricultural problems.

Developing Countries

Biofuel industries are becoming established in many developing countries. Many developing countries have extensive biomass resources that are becoming more valuable as demand for biomass and biofuels increases. The approaches to biofuel development in different parts of the world varies. Countries such as India and China are developing both bioethanol and biodiesel Programmes. India is extending plantations of jatropha, an oil-producing tree that is used in biodiesel production. The Indian sugar ethanol Programme sets a target of 5 per cent bioethanol incorporation into transport fuel. China is a major bioethanol producer and aims to incorporate 15 per cent bioethanol into transport fuels by 2010. Costs of biofuel promotion Programmes can be very high, though.

In rural populations in developing countries, biomass provides the majority of fuel for heat and cooking. Wood, animal dung and crop residues are commonly burned. Figures from the International Energy Agency (IEA) show that biomass energy provides around 30 per cent of the total primary energy supply in developing countries; over 2 billion people depend on biomass fuels as their primary energy source.

The use of biomass fuels for cooking indoors is a source of

health problems and pollution. 1.3 million deaths were attributed to the use of biomass fuels with inadequate ventilation by the International Energy Agency in its World Energy Outlook 2006. Proposed solutions include improved stoves and alternative fuels. However, fuels are easily damaged, and alternative fuels tend to be expensive. Very low cost, fuel efficient, low pollution biomass stove designs have existed since 1980 or earlier. Issues are a lack of education, distribution, corruption, and very low levels of foreign aid. People in developing countries often cannot afford these solutions without assistance or financing such as microloans. Organizations such as Intermediate Technology Development Group work to make improved facilities for biofuel use and better alternatives accessible to those who cannot get them.

EVALUATING BIOFUEL IN THE UNITED STATES

The United States produces mainly biodiesel and ethanol fuel, which uses corn as the main feedstock. Since 2005 the US overtook Brazil as the world's largest ethanol producer. In 2006 the US produced 4.855 billion US gallons (18.38×10^6 m^3) of ethanol. The United States, together with Brazil accounted for 70 percent of all ethanol production, with total world production of 13.5 billion US gallons (51×10^6 m^3) (40 million tonnes). When accounting just for fuel ethanol production in 2007, the U.S. and Brazil are responsible for 88 per cent of the 13.1 billion US gallons (50×10^6 m^3) total world production. Biodiesel is commercially available in most oilseed-producing states. As of 2005, it was somewhat more expensive than fossil diesel, though it is still commonly produced in relatively small quantities (in comparison to petroleum products and ethanol fuel). Due to increasing pollution control and climate change requirements and tax relief, the U.S. market is expected to grow to 1 or 2 billion US gallons (3.8 or 7.6 m^3) by 2010.

Biofuels are mainly used mixed with fossil fuels. They are also used as additives. The largest biodiesel consumer is the U.S. Army. Most light vehicles on the road today in the US can run on blends of up to 10 per cent ethanol, and motor vehicle manufacturers already produce vehicles designed to run on much higher ethanol blends. The demand for bioethanol fuel in the United States was stimulated by the discovery in the late 90s that methyl tertiary butyl ether (MTBE), an oxygenate additive in gasoline, was contaminating groundwater. Cellulosic biofuels are under development, to avoid upward pressure on food prices and land use changes that would be expected to result from a major increase in use of food biofuels.

Biofuels are not just limited to liquid fuels. One of the often overlooked uses of biomass in the United States is in the gasification of biomass. There is a small, but growing number of people using woodgas to fuel cars and trucks all across America.

The challenge is to expand the market for biofuels beyond the farm states where they have been most popular to date. Flex-fuel vehicles are assisting in this transition because they allow drivers to choose different fuels based on price and availability.

It should also be noted that the growing ethanol and biodiesel industries are providing jobs in plant construction, operations, and maintenance, mostly in rural communities. According to the Renewable Fuels Association, the ethanol industry created almost 154,000 U.S. jobs in 2005 alone, boosting household income by $5.7 billion. It also contributed about $3.5 billion in tax revenues at the local, state, and federal levels.

Crane, an American based Green Holdings company, has an ethanol reformation system that uses the high octane of ethanol fuel. This increases the efficiency of an ethanol burning engine by 50 per cent. The inventor Russel Gehrke says he matches ethanol's flame speed to the engine's workload in a different and more efficient way than automakers have used to date.

"The United States leads the world in corn and soybean production, but even if 100 per cent of both crops were turned into fuel, it would be enough to offset just 20 per cent of on-road fuel consumption."

History

The United States used biofuel in the beginning of the 20th

century. For example, models of Ford T ran with ethanol fuel. Then the interest in biofuels declined until the first and second oil shock (1973 and 1979).

The Department of Energy established the National Renewable Energy Laboratory in 1974 and started to work in 1977. The NREL publish papers on biofuels. Congress also voted the Energy Policy Act in 1994 and a newer in 2005 to promote renewable fuels.

Legislation

Renewable Fuel Standard (RFS1)

The current National Renewable Fuel Standard Programme (RFS1) was established under the Energy Policy Act of 2005, which amended the Clean Air Act by establishing the first national renewable fuel standard. The U.S. Congress gave the U.S. Environmental Protection Agency (EPA) the responsibility to coordinate with the U.S. Department of Energy, the U.S. Department of Agriculture, and stakeholders to design and implement this new Programme.

The Renewable Fuel Standard called for 7.5 billion US gallons (28×10^6 m^3) of biofuels to be used annually by 2012, expanding the market for biofuels.

The EPA announced that the 2009 Renewable Fuel Standard will require most refiners, importers, and non-oxygenate blenders of gasoline to displace 10.21 per cent of their gasoline with renewable fuels such as ethanol. That requirement aims to ensure that at least 11.1 billion US gallons (42×10^6 m^3) of renewable fuels will be sold in 2009, in keeping with the targets established by the Energy Independence and Security Act of 2007 (EISA). While the RFS requirement is increasing by about 23 per cent—from 9 billion US gallons (34×10^6 m^3) in 2008 to 11.1 billion US gallons (42×10^6 m^3) in 2009—the percentage requirement is increasing by nearly one third, from 7.76 per cent in 2008 to 10.21 per cent in 2009.

The 2009 RFS is also pushing up against what is known as the "blend wall". To address the blend wall issue, DOE and others are studying the use of mid-range blends, such as E15 and E20, for use in standard gasoline-burning vehicles. Allowing all gasoline blends to contain up to 20 per cent ethanol would double the potential market for ethanol.

Renewable Fuel Standard (RFS2)

In May 2009, the EPA released proposed revisions to the National Renewable Fuel Standard Programme. These revisions addressed changes to the Renewable Fuel Standard Programme as required by EISA. The revised statutory requirements establish new specific volume standards for cellulosic biofuel, biomass-based diesel, advanced biofuel, and total renewable fuel that must be used in transportation fuel each year. The revised statutory requirements also include new definitions and criteria for both renewable fuels and the feedstocks used to produce them, including new greenhouse gas emission (GHG) thresholds for renewable fuels. The regulatory requirements for RFS will apply to domestic and foreign producers and importers of renewable fuel.

Of these modifications, several are significantly notable. First, the volume standard under RFS2 was increased beginning in 2008 from 5.4 to 9.0 billion US gallons (20–34 m^3). Thereafter, the required volume continues to increase under RFS2, eventually reaching 36 billion US gallons (140×10^6 m^3) by 2022.

Volume Standards

Table 4.2: Renewable Fuel Volume Requirements for RFS2

Year	Biomass-Based Diesel		Cellulosic Biofuel		Total Advanced Biofuel		Total Renewable Fuel	
	billion US gallons	million cubic meters	billion US gallons	million cubic meters	billion US gallons	million cubic meters	billion US gallons	million cubic meters
(1)	(2)	(3)	(4)	(5)	(6)	(7)	(8)	(9)
2009	0.5	1.9	—	—	0.6	2.3	11.1	42
2010	0.65	2.5	0.1	0.38	0.95	3.6	12.95	49.0
2011	0.8	3.0	0.25	0.95	1.35	5.1	13.95	52.8
2012	1.0	3.8	0.5	1.9	2.0	7.6	15.2	58
2013	1.0	3.8	1.0	3.8	2.75	10.4	16.55	62.6
2014	1.0	3.8	1.75	6.6	3.75	14.2	18.15	68.7
2015	1.0	3.8	3.0	11	5.5	21	20.5	78

(Contd.)

(1)	(2)	(3)	(4)	(5)	(6)	(7)	(8)	(9)
2016	1.0	3.8	4.25	16.1	7.25	27.4	22.25	84.2
2017	1.0	3.8	5.5	21	9.0	34	24.0	91
2018	1.0	3.8	5.5	21	9.0	34	24.0	91
2019	1.0	3.8	8.5	32	13.0	49	28.0	106
2020	1.0	3.8	10.5	40	15.0	57	30.0	114
2021	1.0	3.8	13.5	51	18.0	68	33.0	125
2022	1.0	3.8	16.0	61	21.0	79	36.0	136

Greenhouse Gas Reduction Thresholds

EISA established new renewable fuel categories and eligibility requirements, including setting the first ever mandatory greenhouse gas reduction thresholds for the various categories of fuels. For each renewable fuel pathway, greenhouse gas emissions are evaluated over the full life cycle, including production and transport of the feedstock; land use change; production, distribution, and blending of the renewable fuel; and end use of the renewable fuel. The GHG emissions are then compared to the life cycle emissions of 2005 petroleum baseline fuels (base year established as 2005 by EISA) displaced by the renewable fuel, such as gasoline or diesel. The life cycle GHG emissions performance reduction thresholds as established by EISA range from 20 to 60 percent reduction depending on the renewable fuel category.

Table 4.3: Lifecycle GHG Thresholds Specified in EISA. (percent reduction from 2005 baseline)

Conventional biofuel	20%
Advanced biofuels	50%
Biomass-based diesel	50%
Cellulosic biofuel	60%

Ethanol Fuel

The demand for ethanol fuel in the United States was stimulated by the discovery in the late 90s that methyl tertiary butyl ether (MTBE), an oxygenate additive in gasoline, was contaminating groundwater. Due to the risks of widespread and costly litigation, and because MTBE use in gasoline was banned in almost 20 states by 2006, the substitution of MTBE opened a new market for ethanol fuel. This demand shift for ethanol as an oxygenate additive took place at a time when oil prices were already significantly rising. This shift also contributed to a expansion in the use of gasohol E10 and to a sharp increase in the production and sale of E85 flex vehicles since 2002.

Low Ethanol Blends

Most cars on the road today in the U.S. can run on blends of up to 10 per cent ethanol (E10), and motor vehicle manufacturers already produce vehicles designed to run on much higher ethanol blends. Though E10 is mandatory only in 10 states, ethanol blends in the US are available in other states as optional or added on lower percentages as a subsititute to MTBE (used to oxygenate gasoline) without any labeling, making E blends present in two-thirds of the US gas supply.

Flexi-Fuel Vehicles

Ford, DaimlerChrysler, and GM are among the automobile companies that sell "flexible-fuel" cars, trucks, and minivans that can use gasoline and ethanol blends ranging from pure gasoline up to 85 per cent ethanol (E85). By mid-2008, there were approximately seven million E85-compatible vehicles on U.S. roads. However, a 2005 survey found that 68 per cent of American flex-fuel car owners were not aware they owned an E85 flex. This is due to the fact that the exterior of flex and non-flex vehicles look exactly the same; there is no sale price difference between them; the lack of consumer's awareness about E85s; and also the decision of American automakers of not putting any kind of exterior labeling, so buyers can be aware they are getting an E85 vehicle. Since 2006 many new FFV models in the US feature a bright yellow gas cap to remind drivers of the E85 capabilities, and GM is also using badging with the text "Flexfuel/E85 Ethanol" to clearly mark the car as an E85 FFV.

A major restriction hampering sales of E85 flex vehicles or fuelling with E85, is the limited infrastructure available to sell E85 to the public, as by October 2008 there were only 1,802 gasoline filling stations selling E85 to the public in the entire US, with a great concentration of E85 stations in the Corn Belt states, lead by Minnesota with 357 stations, the most that any other state, followed by Illinois with 189, Wisconsin with 118, and Missouri with 112. Only seven states do not have E85 available to the public, Alaska, Hawaii, Maine, New Hampshire, New Jersey, Rhode Island, and Vermont. The main constraint for a more rapid expansion of E85 availability is that it requires dedicated storage tanks at filling stations, at an estimated cost of USD 60,000 for each dedicated ethanol tank.

Chrysler, General Motors, and Ford have each pledged to manufacture 50 percent of their entire vehicle line as flexible fuel in model year 2012, if enough fueling infrastructure develops. Regarding energy policy, President-elect Barack Obama pledged during his electoral campaign to significantly reduce oil consumption, with measures that among others include mandating all new vehicles to have FFV capability by the end of 2013.

Biodiesel

GreenHunter Energy, Inc. has begun commercial operations at its biodiesel refinery in Houston, Texas, that can produce 105 million US gallons per year (400×10^3 m³/a) of biodiesel. That production capacity makes it the largest biodiesel refinery in the United States, barely beating out the 100 million US gallons per year (380×10^3 m³/a) biodiesel refinery built by Imperium Renewables in Washington.

For comparison, the total U.S. production capacity for biodiesel reached 2,240 million US gallons per year (8.5×10^6 m³/a) in 2007, although poor market conditions held 2007 production to about 450 million US gallons (1.7×10^6 m³), according to the National Biodiesel Board (NBB).

In 2006, Fuel Bio Opened the largest biodiesel manufacturing plant on the east coast of the United States in Elizabeth, New Jersey. Fuel Bio's operation is capable of producing a name plate capacity of 50 million US gallons (190×10^3 m³) of biodiesel per year.

Methanol Fuel

The 1996 Ford Taurus was the first flexible-fuel vehicle produced with versions capable of running with either ethanol (E85) or methanol (M85) blended with gasoline.

Methanol was first produced from pyrolysis of wood, resulting in its common English name of wood alcohol. Presently, methanol is usually produced using methane (the chief constituent of natural gas) as a raw material. It may also be produced by pyrolysis of many organic materials or by Fischer Tropsch from synthetic gas, so be called biomethanol. Production of methanol from synthesis gas using Biomass-To-Liquid can offer methanol production from biomass at efficiencies up to 75 per çent. Widespread production by this route has a postulated potential to offer methanol fuel at a low cost and with benefits to the environment. These production methods, however, are not suitable for small scale production.

Successful test Programmes in Europe and the US, mainly in California, were conducted with methanol flex-fuel vehicles, known as M85 flex-fuel vehicles. Ford began development of a flexible-fuel vehicle in 1982, and between 1985 and 1992, 705 experimental FFVs were built and delivered to California and Canada, including the 1.6L Ford Escort, the 3.0L Taurus, and the 5.0L LTD Crown Victoria. These vehicles could operate on either gasoline or methanol with only one fuel system. Legislation was passed to encourage the US auto industry to begin production, which started in 1993 for the M85 FFVs at Ford. In 1996, a new FFV Ford Taurus was developed, with models fully capable of running on either methanol or ethanol blended with gasoline. This ethanol version of the Taurus became the first commercial production of an E85 FFV. The momentum of the FFV production Programmes at the American car companies continued, although by the end of the 1990s, the emphasis shifted to the E85 version, as it is today. Ethanol was preferred over methanol because there is a large support from the farming community, and thanks to the government's incentive Programmes and corn-based ethanol subsidies.

In 2005, California's Governor, Arnold Schwarzenegger, terminated the use of methanol after 25 years and 200 million miles of successful operation, to join the expanding use of ethanol driven by producers of corn. In spite of this, he was optimistic about the future of the Programme, claiming "it will be back." Ethanol is currently (as of 2007) priced at 3 to 4 dollars per gallon, while methanol made from natural gas remains at 47 cents per gallon. Presently there are over 60 operating gas stations in California supplying methanol in their pumps.

Butanol Fuel

Questions have been raised about the amount of energy needed to produce fuel-grade ethanol compared to the amount of energy it releases upon combustion. Depending on who you ask, you may get a different answer as to whether or not ethanol from corn produces more energy than it consumes.

Therefore, companies like BP and DuPont have been looking at the next generation of biofuel, specifically investigating butanol.

Specific advantages to using butanol compared to ethanol include, higher energy content per kg, the ability to be blended at higher concentrations without further adaptation of vehicles, and potential to use existing storage/transport infrastructure.

However, using corn as a feedstock to produce either ethanol or butanol seems infeasible without significant technology improvements regarding yields. Currently, about 2.5 US gallons of butanol can be produced per bushel of corn (373 l/t). Meanwhile, about 2.75 US gallons of ethanol can be produced per bushel of corn (410 l/t). Our current production of ethanol is about 5 billion US gallons per year (19×10^6 m³/a), but it requires 20 per cent of the United States' corn crop and only replaces 1 per cent of its petroleum use. Reaching the 36 billion US gallons (140×10^6 m³) biofuel mandate by 2022, would be a difficult task if only using a corn grain feedstock.

By State

Oregon

Oregon Governor Ted Kulongoski signed legislation in July 2007 that will require all gasoline sold in the state to be blended with 10 per cent bioethanol (a blend known as BE10) and all diesel fuel sold in the state to be blended with 2 per cent biodiesel (a blend known as BD2).

Oregon currently has the only biofuels station in the country that can be used by any type of vehicle.

Michigan

Michigan State University researcher Bruce Dale says that 30 per cent of USA's energy can be achieved by 2030. The greenhouse emissions are reduced by 86 per cent for cellulose compared to corn's 29 per cent reduction. A plant is being built now in Georgia to make up to 100 million gallons per year.

Minnesota

Minnesota Governor Tim Pawlenty signed a bill on May 12, 2008 that will require all diesel fuel sold in the state for use in internal combustion engines to contain at least 20 per cent biodiesel by May 1, 2015.

Biofuel Companies

Unfortunately, costs of producing ethanol from cellulosic feedstock such as wood chips are still about 70 per cent higher than production from corn, because of an extra step in the production process, when compared to production of corn-derived ethanol. Until recently, the idea of extracting ethanol from farm waste and other sources was barely clinging to life in the recesses of university campuses and federal labs, because production problems, as well as the need to bring together a vast team of specialists. Consider: Finding a bacterium from a cow's intestinal tract or from elephant dung that has the correct enzyme to degrade cellulose, and then bringing in geneticists to modify that enzyme kept this discouraging feat from ever growing beyond its embryonic state. Now, that is all changing with a race by approximately thirty companies attempting to accomplish this alchemical feat, and in the process working directly or coordinating with: environmental groups, biotechnology firms, some major oil companies, chemical giants, auto makers, defense hawks and venture capitalists. The winner will be whoever can make cellulosic ethanol in mass quantities for as little money per gallon as possible.

With the majority of such biofuel-companies (Iogen Corporation, SunOpta's BioProcess Group, Genencor, Novozymes, Dyadic International, Inc. (AMEX: DIL), Kansas City-based Alternative Energy Sources, Inc. [Nasdaq: AENS], Flex Fuels USA based in Huntsville, Alabama (now owned by Alternative Energy Sources), or BRI Energy, LLC, Abengoa Bioenergy) located in North America, the United States is in a unique position to lead the way in the development, production, and sale of a new source of energy.

One notable company that deserves special mention is Archer-Daniels-Midland Company (ADM) which has already invested heavily into building approximately 100 corn-ethanol production plants, known as bio-refineries, and churns out about one-fifth of the country's ethanol supply. This occurred due to seasonal overcapacity in its corn syrup plants when surplus was available to produce ethanol. Moreover, ADM is in a unique position to utilize unused parts of the corn crop, and convert previously discarded waste into a viable product. The hull surrounding corn contains fiber that the Decatur, Illinois, grain-processing giant's ethanol-making microorganisms can not use. Figuring out how to convert the fiber into more sugar could increase the output of an existing corn-ethanol plant by 15 per cent. Consequently, ADM wouldn't have to figure out how to collect a new source of biomass but merely use the existing infrastructure for gathering corn—resulting in an advantage over its competitors. ADM executives want government help to build a plant that could cost between $50 million and $100 million. Prescient in their position in the quest for success, ADM recently hired the head of petroleum refining at Chevron, Patricia A Woertz, to metamorphasize ADM into the Exxon-Mobil of the ethanol industry. If ADM succeeds, it will catapult beyond the ethanol industry to compete with the larger, global energy industry. In essence, the old paradigm of processing a barrel of crude oil into gasoline will be replaced with processing a bushel of corn into ethanol.

Meanwhile DuPont, the chemical giant, is attempting to

figure out how to construct a bio-refinery fueled by corn stover—the stalk and leaves that are left in the field after farmers harvest their crop. The company's goal is to make ethanol from cellulose as cheaply as from corn kernels by 2009. If it works, the technology could double the amount of ethanol produced by a field of corn.

Diversa Corporation, a biotech company based in San Diego, examined how biomass is converted into energy in the natural environment. They have found that the enzymes inherent in the bacteria and protozoa that inhabit the digestive tracts of the household termite efficiently convert 95 per cent of cellulose into fermentable sugars. Using proprietary DNA extraction and cloning technologies, they were able to isolate the cellulose-degrading enzymes. By reenacting this natural process, the company created a cocktail of high-performance enzymes for industrial ethanol production enablers. Although still in the early stages of this work, the initial results are promising. Currently, these expensive enzymes cost about 25 cents per gallon of ethanol, although this price is very likely to decline by half in the coming years.

Construction of the first U.S. commercial plant producing cellulosic ethanol begins will commence in the State of Iowa in February 2007. The Voyager Ethanol plant in Emmetsburg, owned by Poet Energy, LLC, will be converted from a 50-million-US-gallon-per-year (190×10^3 m³/a) conventional corn dry mill facility into a 125-million-US-gallon-per-year (470×10^3 m³/a) commercial-scale biorefinery producing ethanol from not only corn but also the stalk, leaves and cobs of the corn plant. Most ethanol plants rely on natural gas to power their processing equipment. The process to be used at the Emmetsburg plant will enable the plant to make 11 per cent more ethanol by weight of corn and 27 per cent more by area of corn. The process cuts the need for fossil fuel power at the plant by 83 per cent by using some of its own byproduct for power. The $200 million plant is scheduled to begin in February and take about 30 months to complete. Project completion is contingent upon partial funding from a USDOE grant, which is likely as the U.S. Government views the renewable energy project as a full-blown national security issue.

EVALUATING BIOFUEL IN AUSTRALIA

Biofuel in Australia is available both as biodiesel and as ethanol fuel, which can be produced from sugarcane or grains. There are currently three commercial producers of fuel ethanol in Australia, all on the East Coast.

Legislation imposes a 10 per cent cap on the concentration of fuel ethanol blends. Blends of 90 per cent unleaded petrol and 10 per cent fuel ethanol are commonly referred to as E10, which is available through service stations operating under the BP, Caltex, Shell and United brands as well as those of a number of smaller independents. Not surprisingly, E10 is most widely available closer to the sources of production in Queensland and New South Wales. The Australian Government has set a target for the sale of 350 million litres of E10 fuel each year by 2010.

Recently BP Australia celebrated a milestone with over 100 million litres of the new BP Unleaded with renewable ethanol being sold to Queensland motorists. In partnership with the Queensland Government, the Canegrowers organisation launched a regional billboard campaign in March 2007 to promote the renewable fuels industry.

Australia is drafting new biodiesel legislation in 2008.

Caltex markets a B2 biodiesel blend suitable for all vehicles since 2006 and a B5 biodiesel blend following trials in 2005.

Production

There are currently three commercial producers of fuel ethanol in Australia, all on the East Coast. CSR's Sarina distillery and the Rocky Point distillery are located in Queensland and produce ethanol from molasses feedstock. The Manildra Group also produces fuel ethanol from waste starch and grain at a facility near Nowra, New South Wales. The combined capacity of these three producers has been estimated at less than 150 million litres per annum. A number of other prospective producers have projects at various stages of development.

Regulation

Legislation imposes a 10 per cent cap on the concentration of fuel ethanol blends. Blends of 90 per cent unleaded petrol and 10 per cent fuel ethanol are commonly referred to as E10. There is also a requirement that retailers label blends containing fuel ethanol on the dispenser.

Taxation

Domestically produced fuel ethanol is currently effectively exempt from excise tax until July 1, 2011 (an excise of 38.143 cents per litre is payable on petrol). From this date, excise will be increased at 2.5 cents per litre annually until it reaches 12.5 cents per litre in 2015.

Government Support

Federal Government support for fuel ethanol includes a voluntary industry biofuels target (encompassing ethanol, biodiesel, and other biofuels) of 350 million litres per annum by 2010, capital grants to current and prospective producers, fuel excise relief, and an effective tariff on imported ethanol until July 1, 2011.

In 2006, the Premiers of both New South Wales and Queensland proposed mandating the blending of ethanol into petrol.

Marketing

E10 is available through service stations operating under the BP, Caltex, Shell and United brands as well as those of a number of smaller independents. Not surprisingly, E10 is most widely available closer to the sources of production in Queensland and New South Wales. E10 is most commonly blended with 91 RON "regular unleaded" fuel.

BP has been marketing ethanol blended fuel in Queensland since 2001 and will soon commence the rollout to BP branded service stations in New South Wales, with the fuel to be available at about 50 additional locations by the end of the year. Over the next few years BP's planned expansion of ethanol blended fuel in New South Wales will see this number of service stations doubling.

E85 Vehicle

GM-Holden are committed to having locally built Holden Commodores running E85 in the market by 2010. The group says that biofuels will become a leading alternative fuel for the company.

EVALUATING BIOFUEL IN SWEDEN

Sweden has achieved the largest E85 flexible-fuel vehicle fleet in Europe, with a sharp growth from 717 vehicles in 2001 to 300 thousand by May 2009. Also, Sweden has one of the largest ethanol bus fleets in the world, with over 600 buses running on ED95, with almost 400 of them on Stockholm. Most ethanol fuel in Sweden is imported, mainly from Italy and Brazil.

The recent and accelerated growth of the Swedish fleet of E85 flexifuel vehicles is the result of the National Climate Policy in Global Cooperation Bill passed in 2005, which not only ratified the Kyoto Protocol but also sought to meet the 2003 EU Biofuels Directive regarding targets for use of biofuels, and also let to the 2006 government's commitment to eliminate oil imports by 2020, with the support of BIL Sweden, the national association for the automobile industry.

In order to achieve these goals several government incentives were implemented. Biofuels were exempted of both, the CO_2 and energy taxes until 2009, resulting in a 30 per cent price reduction at the pump of E85 fuel over gasoline and 40 per cent for biodiesel. Furthermore, other demand side incentives for flexifuel vehicle owners include a SEK 10,000 (USD 1,300 as of May, 2009) bonus to buyers of FFVs, exemption from the Stockholm congestion tax, up to 20 per cent discount on auto insurance, free parking spaces in most of the largest cities, lower annual registration taxes, and a 20 per cent tax reduction for flexifuel company cars. Also, a part of the Programme, the

Swedish Government ruled that 25 per cent of their vehicle purchases (excluding police, fire and ambulance vehicles) must be alternative fuel vehicles. By the first months of 2008, this package of incentives resulted in sales of flexible-fuel cars representing 25 per cent of new car sales.

On the supply side, since 2005 the gasoline fulling stations selling more than 3 million liters of fuel a year are required to sell at least one type of biofuel, resulting in more than 1,200 gas stations selling E85 by August 2008. Despite all the sharp growth of E85 flexifuel cars, by 2007 they represented just 2 per cent of the 4 million Swedish vehicle fleet. In addition, this law also mandated all new filling stations to offer alternative fuels, and stations with an annual volume of more than 1 million liters are required to have an alternative fuel pump by 31 December 2009. Therefore, the number of E85 pumps is expected to reach by 2009 nearly 60 per cent of Sweden's 4,000 filling stations.

History

Ethanol-powered ED95 buses were introduced in 1986 on a'trial basis as the fuel for two buses in Örnsköldsvik, and by 1989 30 ethanol-operated buses were in service in Stockholm. SEKAB provided the fuel, called ED95, consists of a blend of 95 per cent ethanol and 5 per cent ignition improver and it is used in modified diesel engines where high compression is used to ignite the fuel. Other countries have now this technology on trial under the auspicies of the BioEthanol for Sustainable Transport (BEST) project, which is coordinated by the city of Stockholm.

Flexible-fuel vehicles were introduced in Sweden as a demonstration test in 1994, when three Ford Taurus were imported to show the technology existed. Because of the existing interest, a project was started in 1995 with 50 Ford Taurus E85 flexifuel in different parts of Sweden: Umea, Örnsköldsvik, Härnösand, Stockholm, Karlstad, Linköping, and Växjö. Between 1997 to 1998 an additional 300 Taurus were imported, and the number of E85 fueling grew to 40. Then in 1998 the city of Stockholm placed an order for 2,000 of FFVs for any car manufacturer willing to produce them. The objective was to jump-start the FFV industry in Sweden. The two domestic car makers Volvo Group and Saab AB refused to participate arguing there were not in place any ethanol filling stations. However, Ford Motor Company took the offer and began importing the flexifuel version of its Focus model, delivering the first cars in 2001, and selling more than 15,000 FFV Focus by 2005, then representing an 80 per cent market share of the flexifuel market. In 2005 both Volvo and Saab introduced to the Swedish market their flexifuel models, and to the European market in the following years.

Current Situation

Most ethanol fuel in Sweden is imported, mainly from Italy and Brazil. During 2004 the government passed a law that said all bigger Swedish fuel stations were required to provide an alternative fuel option. From 2009 all small gas stations, that sell more than 1 000 000 L per year, have to provide this as well. The lower cost of building a station for ethanol compared with a station for petroleum makes it very common to see gas stations that sell ethanol.

One fifth of cars in Stockholm can run on alternative fuels, mostly ethanol fuel. As of December 2007, carmakers that offer ethanol-powered vehicles in Sweden are SAAB, Volvo, VW, Koenigsegg, Skoda, SEAT, Citroen, Peugeot, Renault and Ford.

Stockholm will introduce a fleet of Swedish-made hybrid electric buses in its public transport system on a trial basis in 2008. These buses will use ethanol-powered internal combustion engines and electric motors.

EVALUATING BIOFUEL IN INDIA

Biofuel development in India centers mainly around the cultivation and processing of Jatropha plant seeds which are very rich in oil (40%). The drivers for this are historic, functional, economic, environmental, moral and political. Jatropha oil has been used in India for several decades as biodiesel to cater to the diesel fuel requirements of remote rural and forest communities;

jatropha oil can be used directly after extraction (i.e. without refining) in diesel generators and engines. Jatropha provides immediate economic benefits at the local level since it grows well in dry marginal non-agricultural lands, thereby allowing villagers and farmers to leverage non-farm land for income generation. As well, increased Jatropha oil production delivers economic benefits to India on the macroeconomic or national level as it reduces the nation's fossil fuel import bill for diesel production (the main transportation fuel used in the country); minimizing the expenditure of India's foreign-currency reserves for fuel allowing India to increase its growing foreign currency reserves (which can be better spent on capital expenditures for industrial inputs and production). And since Jatropha oil is carbon-neutral, large-scale production will improve the country's carbon emissions profile. Finally, since no food producing farmland is required for producing this biofuel (unlike corn or sugar cane ethanol, or palm oil diesel), it is considered the most politically and morally acceptable choice among India's current biofuel options; it has no known negative impact on the production of the massive amounts grains and other vital agriculture goods India produces to meet the food requirements of its massive population (circa 1.1 Billion people as of 2008). Other biofuels which displace food crops from viable agricultural land such as corn ethanol or palm biodiesel have caused serious price increases for basic food grains and edible oils in other countries.

Analysis from Frost and Sullivan, *Strategic Analysis of the Indian Biofuels Industry*, reveals that the market is an emerging one and has a long way to go before it catches up with global competitors.

The Government is currently implementing an ethanol-blending Programme and considering initiatives in the form of mandates for biodiesel. Due to these strategies, the rising population, and the growing energy demand from the transport sector, biofuels can be assured of a significant market in India. On 12 September 2008, the Indian Government announced its 'National Biofuel Policy'. It aims to meet 20 per cent of India's diesel demand with fuel derived from plants. That will mean setting aside 140,000 square kilometres of land. Presently fuel yielding plants cover less than 5,000 square kilometres.

Jatropha Incentives in India

Jatropha oil is vegetable oil produced from the seeds of the *Jatropha curcas*, a plant that can grow in marginal lands and common lands. *Jatropha curcas* grows almost anywhere, even on gravelly, sandy and saline soils. It can also thrive on the poorest stony soil and grow in the crevices of rocks.

Jatropha incentives in India is a part of India's goal to achieve energy independence by the year 2012. Jatropha oil is produced from the seeds of the Jatropha curcas, a plant that can grow in across India, and the oil is considered to be an excellent source of bio-diesel. India is keen on reducing its dependence on coal and petroleum to meet its increasing energy demand and encouraging Jatropha cultivation is a crucial component of its energy policy.

Large plots of waste land have been selected for Jatropha cultivation and will provide much needed employment to the rural poor of India. Businesses are also seeing the planting of Jatropha as a good business opportunity. The Government of India has identified 400,000 square kilometres (98 million acres) of land where Jatropha can be grown, hoping it will replace 20 per cent of India's diesel consumption by 2011.

Implementation

The former President of India, Dr. Abdul Kalam, is one of the strong advocaters of jatropha cultivation for production of bio-diesel. In his recent speech, the Former President said that out of the 600,000 km² of wasteland that is available in India over 300,000 km² are suitable for Jatropha cultivation. Once this plant is grown the plant has a useful lifespan of several decades. During its life, Jatropha requires very little water when compared to other cash crops.

Recently, the State Bank of India provided a boost to the cultivation of Jatropha in India by signing a Memorandum of Understanding with D1 Mohan, a joint venture of D1 Oils plc, to give loans to the tune of 1.3 billion rupees to local farmers in India. Farmers will also be able to pay back the loan with the money that D1 Mohan pays for the Jatropha seeds.

Indian Railways

The Indian Railways has started to use the oil (blended with diesel fuel in various ratios) from the Jatropha plant to power its diesel engines with great success. Currently the diesel locomotives that run from Thanjavur to Nagore section and Tiruchirapalli to Lalgudi, Dindigul and Karur sections run on a blend of Jatropha and diesel oil.

Andhra Pradesh

Andhra Pradesh has entered into a formal agreement with Reliance Industries for Jatropha planting. The company has selected 200 acres (0.81 km²) of land at Kakinada to grow jatropha for high quality bio-diesel fuel. Kerala is planning a massive Jatropha planting campaign.

Chhattisgarh

Chhattisgarh has decided to plant 160 million saplings of jatropha in all its 16 districts during 2006 with the aim of becoming a bio-fuel self-reliant state by 2015. Chhattisgarh plans to earn Rs. 40 billion annually by selling seeds after 2010. The central government has provided Rs. 135 million to Chhattisgarh this year for developing jatropha nursery facilities.

In May 2005, Chief Minister Raman Singh became the first head of a state government to use jatropha diesel for his official vehicle. Chhattisgarh plans to replace with jatropha fuel all state-owned vehicles using diesel and petrol by 2007. Chattisgarh Biofuel Development Authority now oversees the production of the Jatropha curcas seed as a rich source of bio-diesel.

Karnataka

Farmers in semi-arid regions of Karnataka are planting Jatropha as it is well suited to those conditions.

Labland Biodiesel is a Mysore based Private Limited Company. Since the year 2002, the Company is active in Biodiesel and Jatropha curcas-based Research and Development activities headed by its Chairman and Managing Director, Dr. Sudheer Shetty.

Tamil Nadu

Tamil Nadu is aggressively promoting the plantation of Jatropha to help farmers over come the loss due to irregular rains during the past few years. The government has contracted the development of Jatropha in Tamilnadu in a large scale to four entrepreneurs. Namely M/s Mohan Breweries and Distilleries Limited. M/s Shiva Distilleries Limited, M/s Dharani Sugars and Chemicals Limited and M/s Riverway Agro Products Private Ltd. Currently the firms have cultivated the plant in about 3 square kilometres as against the goal of 50 km². The government of Tamilnadu has also abolished purchase tax on Jatropha., but presently government has announsed to 7.5 per cent tolgate charges is reduced to 2.5 per cent

Rajasthan

Jatropha is ideally suited for cultivation in Rajasthan as it needs very little water which is scarce in Rajasthan. Jatropa plantations have been undertaken in Udaipur, Kota, Sikar, Banswara, Chittor and Churu districts. In the Udaipur district, *Jatropha curcas* is planted in agroforestry formats with food or cash crops on marginal lands (in India often called waste lands). As its leaves are toxic and therefore non-palatable to livestock, they remain intact in their sapling stage, unlike most other tree saplings.

Maharashtra

In September 2007, the Hindustan Petroleum Corporation Limited (HPCL) joined hands with the Maharashtra State Farming Corporation Ltd (MSFCL) for a jatropha seed-based bio-diesel venture. As part of the project, a jatropha plant would be grown on 500 acres (2 km²) in Nashik and Aurangabad. In November 2005, the Maharashtra Government aimed to cultivate jatropha on 600 km² in the state, with half the land going to the public sector and the other half to the private sector. On July 1 2006, Pune Municipal Corporation took the lead among Indian cities in using bio-diesel from jatropha in over 100 public buses.

Ahmednagar

Gulabrao Kale studied the prospects of plantation in the Ahmednagar district in Maharashtra and under his guidance, Govind Gramin Vikas Pratishthan (GOGVIP), decided to plan under DPAP Programme of government. Initially, it was a very difficult task to make farmers ready for the Jatropha plantation. When 20-25 farmers were offered the plan, only 2-3 farmers were convinced to plant jatropha. Lack of literacy was a big hindrance in convincing the farmers. It was hard to convince them about the future benefits of the plant and its potential to produce bio-diesel, an equivalent of diesel. But after untiring and continuous efforts more than 1000 farmers are working with the GOGVIP for the Jatropha planting Programme now. For this task, under the watershed development Programme, GOGVIP took an area of 10.92 square kilometres for making CCT'S. To date, more than 2 million Jatropha plants have been planted in the target area of the five villages of Vankute, Dhoki, Dhotre, Dhavalpuri and Gajdipoor in the project. The villages are in the remote locations and that made connecting them with GOGVIP a difficult task.

Eastern India

[D1 Williamson Magor Bio Fuel Limited] is a joint venture company between D1 Oils plc, UK and Williamson Magor group. This biodiesel initiative was incorporated in July 2006. Advocating the creation of energy from renewable resources, the company promotes Jatropha Plantations on the wasteland possessed by the farmers in the North Eastern States, Orissa and Jharkhand. The Company has a comprehensive network to manufacture bio-diesel from the oilseeds harvested by the farmers.

Biodiesel initiative hopes to benefit local communities through commercial plantation of Jatropha. NGOs and self help groups are also involved.

Practices

The Project on Development of Agronomic practices for Jatropha curcas is being implemented, with the financial assistance of DBT, New Delhi. Dr. Panjabrao Deshmukh Krishi Vidyapeeth, Akola (MS) India has Planted Jatropha on 3 square kilometres, with the financial assistance of National Oilseeds and Vegetable oils development Board.

Use as Biodiesel

When jatropha seeds are crushed, the resulting jatropha oil can be processed to produce a high-quality biodiesel that can be used in a standard diesel car, while the residue (press cake) can also be processed and used as biomass feedstock to power electricity plants or used as fertilizer (it contains nitrogen, phosphorous and potassium).

The plant may yield more than four times as much fuel per hectare as soybean, and more than ten times that of maize (corn). A hectare of jatropha has been claimed to produce 1,892 litres of fuel. However, as it has not yet been domesticated or improved by plant breeders, yields are variable

Researchers at Daimler Chrysler Research explored the use of jatropha oil for automotive use, concluding that although jatropha oil as fuel "has not yet reached optimal quality, ...it already fulfills the EU norm for biodiesel quality". Archer Daniels Midland Company, Bayer CropScience and Daimler AG have a joint project to develop jatropha as a biofuel. Three Mercedes cars powered by Jatropha diesel have already put some 30,000 kilometres behind them. The project is supported by DaimlerChrysler and by the German Association for Investment and Development (Deutschen Investitions-und Entwicklungs gesellschaft, DEG).

Goldman Sachs recently cited *Jatropha curcas* as one of the best candidates for future biodiesel production. However, despite its abundance and use as an oil and reclamation plant, none of the *Jatropha* species has been properly domesticated and, as a result, its productivity is variable, and the long-term impact of its large-scale use on soil quality and the environment is unknown. However, because jatropha can grow in harsh climates, it can be planted in areas where it won't compete for resources needed to grow food.

Myanmar Biodiesel

Myanmar is also actively pursuing the use of jatropha oil. On 15 December 2005, head of state, Senior General Than Shwe, said "the States and Divisions concerned are to put 50,000 acres (200 km²) under the physic nut plants [Jatropha] each within three years totalling seven hundred thousand acres (2,800 km²) during the period". On the occasion of Myanmar's Peasant Day 2006, Chairman of the State Peace and Development Council Snr Gen Than Shwe described in his a message that "For energy sector which is an essential role in transforming industrial agriculture system, the Government is encouraging for cultivation of physic nut plants nationwide and the technical know how that can refine physic nuts to biodiesel has also identified." He would like to urge peasants to cultivate physic nut plants on a commercial scale with major aims for emergence of industrial agriculture system, for fulfilling rural electricity supply and energy needs, for supporting rural areas development and import substitute economy. (2005 from MRTV)

In 2006, the chief research officer at state-run Myanma Oil and Gas Enterprise said Myanmar hoped to completely replace the country's oil imports of 40,000 barrels a day with home-brewed, jatropha-derived biofuel. Other government officials declared Myanmar would soon start exporting jatropha oil. Despite the military's efforts, the jatropha campaign apparently has largely flopped in its goal of making Myanmar self-sufficient in fuel. (2006 from MyawaddyTV)

- Z.G.S. Bioenergy has started Jatropha Plantation Projects in Northern Shan State, the company has begun planting Jatropha plants during late June of 2007 and will start selling the seeds in large quantities by early 2010. The manager of the project site said that Z.G.S. will sell Jatropha seeds to both local and foreign markets; the company will also further research on Jatropha plants for higher quality seeds and better yields. (20 July 2007 from New Light of Myanmar)

A bizarre dimension of the Burmese junta's efforts appears to be related to the Burmese tradition of *yadaya*, the form of magic employed to ward off evil spirits, ensure good fortune and confound enemies.

Use as Jet Fuel

Aviation fuels may be more widely substituted with biofuels such as jatropha oil than fuels for other forms of transportation. There are fewer planes than cars or trucks and far fewer jet fueling stations to convert than gas stations. On December 30, 2008, Air New Zealand flew the first successful test flight with a Boeing 747 running one of its four Rolls-Royce engines on a 50:50 blend of jatropha oil and jet A-1 fuel. Subsequently, Air New Zealand and Houston based Continental Airlines have run tests in Jan. 2009, further demonstrating the viability of jatropha oil as a jet fuel. Japan Air also plans test flights in Jan. 2009 as well.

Light Hydrocarbon Fuel

In 2008-2009 a pair of student researchers at a public school in Connecticut tested Jatropha's ability to produce light Hydrocarbon fuels. Through the Gas Chromatographic analysis, the pair saw a significantly greater amount of Hydrocarbons present than diesel in a gas sample taken. The gas sample was taken through heating shell-less ground Jatropha at 800 degrees Celsius above a burner in a gas trap. The Gas Chromatographic analysis also showed that among the Hydrocarbons present was a significant amount of Methane along with Ethane and Propane. The sample taken also contained over 15 per cent per gram Carbon Dioxide. This research, along with other growing tests done by the researchers, may provide new light unto previous claims made regarding Jatropha's viability as a biodiesel. This may lead to research into Jatropha's potential as a solid fuel source in a solution.

Competition with Food Crops

Jatropha oil was lauded as being sustainable, and that its production would not compete with food production. However, after some years, it was obvious that the jatropha plant, when planted in soils not suitable for food growing, has very low yields. This has led some governments (of mostly third-world countries), to press farmers into converting their farms to produce jatropha, putting it in direct competition with food production, on countries where hunger is a reality. It is also worth noting that while the jatropha plant can grow on arid soils, when planted in fertile soils, it creates a huge pressure on them, consuming their natural nutrients at a much faster rate than other plants, and can even render the once fertile soils completely useless. Under these conditions, it also becomes a very agressive and invasive plant. Jatropha has toxic properties.

Many fuel business companies seem to be aware of this and are slowly abandoning jatropha oil production incentives.

UNDERSTANDING GLOBAL IMPACT OF FOOD VS. FUEL DEBATE

Food vs. fuel is the dilemma regarding the risk of diverting farmland or crops for biofuels production in detriment of the food supply on a global scale. The "food vs. fuel" or "food or fuel" debate is international in scope, with good and valid arguments on all sides of this issue. There is disagreement about how significant this is, what is causing it, what the impact is, and what can or should be done about it.

Biofuel production has increased in recent years. Some commodities like maize, sugar cane or vegetable oil can be used either as food, feed or to make biofuels. For example, since 2006, land that was also formerly used to grow other crops in the United States is now used to grow maize for biofuels, and a larger share of maize is destined to ethanol production, reaching 25 per cent in 2007. Since converting the entire grain harvest of the US would only produce 16 per cent of its auto fuel needs, some experts believe that placing energy markets in competition with food markets for scarce arable land will inevitably result in higher food prices. A lot of R&D efforts are currently being put into the production of second generation biofuels from non-food crops, crop residues and waste. Second generation biofuels could hence potentially combine farming for food and fuel and moreover, electricity could be generated simultaneously, which could be beneficial for developing countries and rural areas in developed countries. With global demand for biofuels on the increase due to the oil price increases taking place since 2003 and the desire to reduce oil dependency as well as reduce GHG emissions from transportation, there is also fear of the potential destruction of natural habitats by being converted into farmland. Environmental groups have raised concerns about this trade-off for several years, but now the debate reached a global scale due to the 2007-2008 world food price crisis. On the other hand, several studies do show that biofuel production can be significantly increased without increased acreage. Therefore stating that the crisis in hand relies on the food scarcity.

Brazil has been considered to have the world's first sustainable biofuels economy and its government claims Brazil's sugar cane based ethanol industry has not contributed to the 2008 food crises. A World Bank policy research working paper released in July 2008 concluded that "...large increases in biofuels production in the United States and Europe are the main reason behind the steep rise in global food prices", and also stated that "Brazil's sugar-based ethanol did not push food prices appreciably higher".

Food Price Inflation

From 1974 to 2005 real food prices (adjusted for inflation) dropped by 75 per cent. Food commodity prices were relatively stable after reaching lows in 2000 and 2001. Therefore, recent rapid food price increases are considered extraordinary. A World Bank policy research working paper published on July 2008 found that the increase in food commodities prices was led by grains, with sharp price increases in 2005 despite record crops worldwide. From January 2005 until June 2008, maize prices almost tripled, wheat increased 127 percent, and rice rose 170 percent. The increase in grain prices was followed by increases in fats and oil prices in mid-2006. On the other hand, the study found that sugar cane production has increased rapidly, and it was large enough to keep sugar price increases small except for 2005 and early 2006. The paper concluded that biofuels produced from grains have raised food prices in combination with other related factors between 70 to 75 percent, but ethanol produced from sugar cane has not contributed significantly to the recent increase in food commodities prices.

An economic assessment report published by the OECD in July 2008 found that *"...the impact of current biofuel policies on world crop prices, largely through increased demand for cereals and vegetable oils, is significant but should not be overestimated. Current biofuel support measures alone are estimated to increase average wheat prices by about 5 percent, maize by around 7 percent and vegetable oil by about 19 percent over the next 10 years."*

Corn is used to make ethanol and prices went up by a factor of three in less than 3 years (measured in US dollars). Reports in 2007 linked stories as diverse as food riots in Mexico due to rising prices of corn for tortillas, and reduced profits at Heineken the large international brewer, to the increasing use of corn (maize) grown in the US Midwest for ethanol production. (In the case of beer, the barley area was cut in order to increase corn production. Barley is not currently used to produce ethanol.) Wheat is up by almost a factor of 3 in 3 years, while soybeans are up by a factor of 2 in 2 years (both measured in US dollars).

As corn is commonly used as feed for livestock, higher corn prices lead to higher prices in Animal source foods. Vegetable oil is used to make biodiesel and has about doubled in price in the last couple years. The price is roughly tracking crude oil prices. The 2007-2008 world food price crisis is blamed partly on the increased demand for biofuels.

Rice prices have gone up by a factor of 3 even though rice is not directly used in biofuels.

The USDA expects the 2008/2009 wheat season to be a record crop and 8 per cent higher than the previous year. They also expect rice to have a record crop. Wheat prices have dropped from a high over $12/bushel in May 2008 to under $8/bushel in May. Rice has also dropped from its highs.

According to a new report from the World Bank, the production of biofuel is pushing up food prices. These conclusions were confirmed by the Union of Concerned Scientists in their September 2008 newsletter in which they remarked that the World Bank analysis "contradicts U.S. Secretary of Agriculture Ed Schaffer's assertion that biofuels account for only a small percentage of rising food prices."

According to the October Consumer Price Index released Nov. 19, 2008, food prices continued to rise in October 2008 and were 6.3 percent higher than October 2007. Since July of 2008 fuel costs dropped by nearly 60 percent.

Proposed Causes

Ethanol fuel as an Oxygenate Additive

The demand for ethanol fuel produced from field corn was spurred in the U.S. by the discovery that methyl tertiary butyl ether (MBTE) was contaminating groundwater. MBTE use as a oxygenate additive was widespread due to mandates of the Clean Air Act amendments of 1992 to reduce carbon monoxide emissions. As a result, by 2006 MTBE use in gasoline was banned in almost 20 states. There was also concern that widespread and costly litigation might be taken against the U.S. gasoline suppliers, and a 2005 decision refusing legal protection for MBTE, opened a new market for ethanol fuel, the primary substitute for MBTE. At a time when corn prices were around US$ 2 a bushel, corn growers recognized the potential of this new market and delivered accordingly. This demand shift took place at a time when oil prices were already significantly rising.

Other factors

That food prices went up at the same time fuel prices went up is not surprising and should not be entirely blamed on biofuels. Energy costs are a significant cost for fertilizer, farming, and food distribution. Also, China and other countries have had significant increases in their imports as their economies have grown. Sugar is one of the main feedstocks for ethanol and prices are down from 2 years ago. Part of the food price increase for international food commodities measured in US dollars is due to the dollar being devalued. Protectionism is also an important contributor to price increases. 36 per cent of world grain goes as fodder to feed animals, rather than people.

Over long time periods population growth and climate change could cause food prices to go up. However, these factors have been around for many years and food prices have jumped up in the last 3 years, so their contribution to the current problem is minimal.

Governments Distorting Food and Fuel Markets

France, Germany, The United Kingdom and The United States governments have supported biofuels with tax breaks, mandated use, and subsidies. These policies have the unintended consequence of diverting resources from food production and leading to surging food prices and the potential destruction of natural habitats. Current government policies cause distortions of supply and demand.

Fuel for agricultural use often does not have fuel taxes (farmers get duty-free petrol or diesel fuel). Biofuels may have subsidies and low/no retail fuel taxes. Biofuels compete with retail gasoline and diesel prices which have substantial taxes included. The net result is that it is possible for a farmer to use more than a gallon of fuel to make a gallon of biofuel and still make a profit. Some argue that this is a bad distortion of the market. There have been thousands of scholarly papers analyzing how much energy goes into making ethanol from corn and how that compares to the energy in the ethanol. Government distortions can make things happen that would not make sense in a free market.

A World Bank policy research working paper concluded that biofuels have raised food prices between 70 to 75 percent. The "month-by-month" five year analysis disputes that increases in global grain consumption and droughts were responsible for significant price increases, reporting that this had had only a marginal impact. Instead the report argues that the EU and US drive for biofuels has had by far the biggest impact on food supply and prices, as increased production of biofuels in the US and EU were supported by subsidies and tariffs on imports, and considers that without these policies, price increases would have been smaller. This research also concluded that Brazil's sugar cane based ethanol has not raised sugar prices significantly, and recommends removing tariffs on ethanol imports by both the US and EU, to allow more efficient producers such as Brazil and other developing countries, including many African countries, to produce ethanol profitably for export to meet the mandates in the EU and the US.

An economic assessment published by the OECD in July 2008 agrees with the World Bank report recommendations regarding the negative effects of subsidies and import tariffs, but found that the estimated impact of biofuels on food prices are much smaller. The OECD study found that trade restrictions, mainly through import tariffs, protect the domestic industry from foreign competitors but impose a cost burden on domestic biofuel users and limits alternative suppliers. The report is also critical of limited reduction of GHG emissions achieved from biofuels based on feedstocks used in Europe and North America, founding that the current biofuel support policies would reduce greenhouse gas emissions from transport fuel by no more than 0.8 percent by 2015, while Brazilian ethanol from sugar cane reduces greenhouse gas emissions by at least 80 percent compared to fossil fuels. The assessment calls for the need for more open markets in biofuels and feedstocks in order to improve efficiency and lower costs.

Oil Price Increases

Oil price increases since 2003 resulted in increased demand for biofuels. Transforming vegetable oil into biodiesel is not very hard or costly so there is a profitable arbitrage situation if vegetable oil is much cheaper than diesel. Diesel is also made from crude oil, so vegetable oil prices are partially linked to crude oil prices. Farmers can switch to growing vegetable oil crops if those are more profitable than food crops. So all food prices are linked to vegetable oil prices, and in turn to crude oil prices. A World Bank study concluded that oil prices and a weak dollar explain 25-30 per cent of total price rise between January 2002 until June 2008.

Demand for oil is outstripping the supply of oil and oil depletion is expected to cause crude oil prices to go up over the next 50 years. Record oil prices are inflating food prices worldwide, including those crops that have no relation to biofuels, such as rice and fish.

In Germany and Canada it is now much cheaper to heat a house by burning grain than by using fuel derived from crude oil. With oil at $120/barrel a savings of a factor of 3 on heating costs is possible. When crude oil was at $25/barrel there was no economic incentive to switch to a grain fed heater.

From 1971 to 1973, around the time of the 1973 oil crisis, corn and wheat prices went up by a factor of 3. There was no significant biofuel usage at that time.

US Government Policy

Some argue that the US government policy of encouraging ethanol from corn is the main cause for food price increases. US Federal government ethanol subsidizes total $7 billion per year, or $1.90 per gallon. Ethanol provides only 55 per cent as much energy as gasoline per gallon, realizing about a $3.45 per gallon gasoline trade off. Corn is used to feed chickens, cows, and pigs. So higher corn prices lead to higher prices for chicken, beef, pork, milk, cheese, etc.

U.S. Senators introduced the *BioFuels Security Act* in 2006. "It's time for Congress to realize what farmers in America's heartland have known all along—that we have the capacity and ingenuity to decrease our dependence on foreign oil by growing our own fuel," said U.S. Senator for Illinois Barack Obama.

Two-thirds of U.S. oil consumption is due to the transportation sector. The "*Energy Independence and Security Act of 2007*" has a significant impact on U.S. Energy Policy. With the high profitability of growing corn, more and more farmers switch to growing corn until the profitability of other crops goes up to match that of corn. So the ethanol/corn subsidies drive up the prices of other farm crops.

The US—an important export country for food stocks—will convert 18 per cent of its grain output to ethanol in 2008. Across the US, 25 per cent of the whole corn crop went to ethanol in 2007. The percentage of corn going to biofuel is expected to go up.

Since 2004 a US subsidy has been paid to companies that blend biofuel and regular fuel. The European biofuel subsidy is paid at the point of sale. Companies import biofuel to the US, blend 1 per cent or even 0.1 per cent regular fuel, and then ship the blended fuel to Europe, where it can get a second subsidy. These blends are called B99 or B99.9 fuel. The practice is called "splash and dash". The imported fuel may even come from Europe to the US, get 0.1 per cent regular fuel, and then go back to Europe. For B99.9 fuel the US blender gets a subsidy of $0.999 per gallon. The European biodiesel producers have urged the EU to impose punitive duties on these subsidized imports. US lawmakers are also looking at closing this loophole.

The US had arranged things so that Japan had to buy rice from US farmers even if they did not want it and they could not re-export that rice. This led to huge stockpiles of unused rice in Japan. This policy may be changing.

Proposed Action

Freeze on First Generation Biofuel Production

Environmental campaigner George Monbiot has argued for a 5-year freeze on biofuels while their impact on poor communities and the environment is assessed. It has been suggested that a problem with Monbiot's approach is that economic drivers may be required in order to push through the development of more sustainable second-generation biofuel processes: it is possible that these could be stalled if biofuel production decreases. Some environmentalists are suspicious that second-generation biofuels may not solve the problem of a potential clash with food as they also use significant agricultural resources such as water.

A recent UN report on biofuel also raises issues regarding food security and biofuel production. Jean Ziegler, then UN Special Rapporteur on food, concluded that while the argument for biofuels in terms of energy efficiency and climate change are legitimate, the effects for the world's hungry of transforming wheat and maize crops into biofuel are "absolutely catastrophic," and terms such use of arable land a "crime against humanity." Ziegler also calls for a 5-year moratorium on biofuel production. Ziegler's proposal for a five-year ban was rejected by the U.N. Secretary Ban Ki-moon, who called for a comprehensive review of the policies on biofuels, and said that "just criticising biofuel may not be a good solution".

Food surpluses exist in many developed countries. For example, the UK wheat surplus was around 2 million tonnes in 2005. This surplus alone could produce sufficient bioethanol to replace around 2.5 per cent of the UK's petroleum consumption, without requiring any increase in wheat cultivation or reduction in food supply or exports. However, above a few percent, there would be direct competition between first generation biofuel production and food production. This is one reason why many view second generation biofuels as increasingly important.

Non-food crops for biofuel

There are different types of biofuels and different feedstocks for them, and it has been proposed that only non-food crops be used for biofuel. This avoids direct competition for commodities like corn and edible vegetable oil. However, as long as farmers can make more money by switching to biofuels they will. The law of supply and demand predicts that if fewer farmers are producing food the price of food will rise.

Third generation biofuels (biofuel from algae) uses non-edible raw materials sources that can be used for biodiesel and bioethanol.

Biodiesel

Soybean oil, which only represents half of the domestic raw materials available for biodiesel production in the United States, is one of many raw materials that can be used to produce biodiesel.

Non-food crops like Camelina, Jatropha, seashore mallow and mustard, used for biodiesel, can thrive on marginal agricultural land where many trees and crops won't grow, or would produce only slow growth yields. Camelina is virtually 100 percent efficient. It can be harvested and crushed for oil and the remaining parts can be used to produce high quality omega-3 rich animal feed, fiberboard, and glycerin. Camelina does not take away from land currently being utilized for food production. Most camelina acres are grown in areas that were previously not utilized for farming. For example, areas that receive limited rainfall that can not sustain corn or soybeans without the addition of irrigation can grow camelina and add to their profitability.

Jatropha cultivation provides benefits for local communities:

Cultivation and fruit picking by hand is labour-intensive and needs around one person per hectare. In parts of rural India and Africa this provides much-needed jobs—about 200,000 people worldwide now find employment through jatropha. Moreover, villagers often find that they can grow other crops in the shade of the trees. Their communities will avoid importing expensive diesel and there will be some for export too.

NBB's Feedstock Development programme is addressing production of arid variety crops, algae, waste greases, and other feedstocks on the horizon to expand available material for biodiesel in a sustainable manner.

Bioalcohols

Cellulosic ethanol is a type of biofuel produced from lignocellulose, a material that comprises much of the mass of plants. Corn stover, switchgrass, miscanthus and woodchip are some of the more popular non-edible cellulosic materials for ethanol production. Commercial investment in such second-generation biofuels began in 2006/2007, and much of this investment went beyond pilot-scale plants. Cellulosic ethanol commercialization is moving forward rapidly. The world's first commercial wood-to-ethanol plant began operation in Japan in 2007, with a capacity of 1.4 million liters/year. The first wood-to-ethanol plant in the United States is planned for 2008 with an initial output of 75 million liters/year.

Other second generation biofuels may be commercialized in the future and compete less with food. Synthetic fuel can be made from coal or biomass and may be commercialized soon.

Biofuel from Food Byproducts and Coproducts

Biofuels can also be produced from the waste byproducts of food-based agriculture (such as citrus peels or used vegetable oil) to manufacture an environmentally sustainable fuel supply, and reduce waste disposal cost.

A growing percentage of U.S. biodiesel production is made from waste vegetable oil (recycled restaurant oils) and greases.

Collocation of a waste generator with a waste-to-ethanol plant can reduce the waste producer's operating cost, while creating a more-profitable ethanol production business. This innovative collocation concept is sometimes called holistic systems engineering. Collocation disposal elimination may be one of the few cost-effective, environmentally-sound, biofuel strategies, but its scalability is limited by availability of appropriate waste generation sources. For example, millions of tons of wet Florida-and-California citrus peels cannot supply billions of gallons of biofuels. Due to the higher cost of transporting ethanol, it is a local partial solution, at best.

More firms are investigating the potential of fractionating technology to remove corn germ (i.e. the portion of the corn kernel that contains oil) prior to the ethanol process. Furthermore, some ethanol plants have already announced their intention to employ technology to remove the remaining vegetable oil from dried distillers grains, a coproduct of the ethanol process. Both of these technologies would add to the biodiesel raw material supply.

End Unsustainable Biofuel Subsidies and Tariffs

Some people have claimed that ending subsidies and tariffs would enable sustainable development of a global biofuels market. Taxing biofuel imports while letting petroleum in duty-free does not fit with the goal of encouraging biofuels. Ending mandates, subsidies, and tariffs would end the distortions that current policy is causing. Some US senators advocate reducing subsidies for corn based ethanol. The US ethanol tariff and some US ethanol subsidies are currently set to expire over the next couple years. The EU is rethinking their biofuels directive due to environmental and social concerns. On January 18 2008 the UK House of

Commons Environmental Audit Committee raised similar concerns, and called for a moratorium on biofuel targets. Germany ended their subsidy of biodiesel on Jan 1 2008 and started taxing it.

Reduce Farmland Reserves and Set Asides

Some countries have programmes to hold farmland fallow in reserve. The current crisis has prompted proposals to bring some of the reserve farmland back into use.

The American Bakers Association has proposed reducing the amount of farmland held in the US Conservation Reserve Programme. Currently the US has 34,500,000 acres (140,000 km²) in the programme.

In Europe about 8 per cent of the farmland is in set aside programmes. Farmers have proposed freeing up all of this for farming. Two-thirds of the farmers who were on these programmes in the UK are not renewing when their term expires.

Sustainable Production of Biofuels

Second generation biofuels are now being produced from the cellulose in dedicated energy crops (such as perennial grasses), forestry materials, the co-products from food production, and domestic vegetable waste. Advances in the conversion processes will almost certainly improve the sustainability of biofuels, through better efficiencies and reduced environmental impact of producing biofuels, from both existing food crops and from cellulosic sources.

Lord Ron Oxburgh suggests that responsible production of biofuels has several advantages.

Produced responsibly they are a sustainable energy source that need not divert any land from growing food nor damage the environment; they can also help solve the problems of the waste generated by Western society; and they can create jobs for the poor where previously were none. Produced irresponsibly, they at best offer no climate benefit and, at worst, have detrimental social and environmental consequences. In other words, biofuels are pretty much like any other product.

Far from creating food shortages, responsible production

and distribution of biofuels represents the best opportunity for sustainable economic prospects in Africa, Latin America and impoverished Asia. Biofuels offer the prospect of real market competition and oil price moderation. According to the Wall Street Journal, crude oil would be trading 15 per cent higher and gasoline would be as much as 25 per cent more expensive, if it were not for biofuels. A healthy supply of alternative energy sources will help to combat gasoline price spikes.

Impact on Developing Countries

Demand for fuel in rich countries is now competing against demand for food in poor countries. Cars, not people, used most of the increase in world grain consumption in 2006. The grain required to fill a 25-gallon SUV gas tank with ethanol will feed one person for a year.

Several factors combine to make recent grain and oilseed price increases impact poor countries more:

- The World Bank estimated that in 2001 there were 2.7 billion people who lived in poverty on less than US$ (PPP) 2 per day. This was nearly half the 2001 world population of 6 billion.
- While rich people buy processed and packaged foods like Wheaties, where prices don't change much if wheat prices go up, poor people buy more grains like wheat and feel the full impact of grain price changes.
- Poor people spend a higher portion of their income on food, so higher food prices hurt them more, unless they are farmers. If a poor person spends 60 per cent of their money on food and then the food prices double, they will experience immediate hardship. So higher grain and oilseed prices will affect poorer countries more.
- Aid organizations that buy food and send it to poor countries are only able to send half as much food on the same budget if prices double. But the higher prices mean there are more people in need of aid.

The impact is not all negative. The Food and Agriculture Organization (FAO) recognizes the potential opportunities that the growing biofuel market offers to small farmers and aquaculturers around the world and has recommended small-scale financing to help farmers in poor countries produce local biofuel.

On the other hand, poor countries that do substantial farming have increased profits due to biofuels. If vegetable oil prices double, the profit margin could more than double. In the past rich countries have been dumping subsidized grains at below cost prices into poor countries and hurting the local farming industries. With biofuels using grains the rich countries no longer have grain surpluses to get rid of. Farming in poor countries is seeing healthier profit margins and expanding.

Interviews with local peasants in southern Ecuador provide strong anecdotal evidence that the high price of corn is encouraging the burning of tropical forests. The destruction of tropical forests now account for 20 per cent of all greenhouse gas emissions.

National Corn Growers Association

US government subsidies for making ethanol from corn have been attacked as the main cause of the food vs. fuel problem. To defend themselves, the US corn growers association has published their views on this issue. They consider the "food vs. fuel" argument to be a fallacy that is "fraught with misguided logic, hyperbole and scare tactics."

Claims made by the NCGA include:

- Corn growers have been and will continue to produce enough corn so that supply and demand meet and there is no shortage. Farmers make their planting decisions based on signals from the marketplace. If demand for corn is high and projected revenue-per-acre is strong relative to other crops, farmers will plant more corn. In 2007 US farmers planted 92,900,000 acres (376,000 km^2) with corn, 19 per cent more acres than they did in 2006.

- The U.S. has doubled corn yields over the last 40 years and expects to double them again in the next 20 years. With twice as much corn from each acre, corn can be put to new uses without taking food from the hungry or causing deforestation.

- US consumers buy things like corn flakes where the cost of the corn per box is around 5 cents. Most of the cost is packaging, advertising, shipping, etc. Only about 19 per cent of the US retail food prices can be attributed to the actual cost of food inputs like grains and oilseeds. So if the price of a bushel of corn goes up, there may be no noticeable impact on US retail food prices. The US retail food price index has gone up only a few percent per year and is expected to continue to have very small increases.

- Most of the corn produced in the US is field corn, not sweet corn, and not digestible by humans in its raw form. Most corn is used for livestock feed and not human food, even the portion that is exported.

- Only the starch portion of corn kernels is converted to ethanol. The rest (protein, fat, vitamins and minerals) is passed through to the feed coproducts or human food ingredients.

- One of the most significant and immediate benefits of higher grain prices is a dramatic reduction in federal farm support payments. According to the USDA, corn farmers received $8.8 billion in government support in 2006. Because of higher corn prices, payments are expected to drop to $2.1 billion in 2007, a 76 percent reduction.

- While the EROEI and economics of corn based ethanol are a bit weak, it paves the way for cellulosic ethanol which should have much better EROEI and economics.

- While basic nourishment is clearly important, fundamental societal needs of energy, mobility, and energy security are too. If farmers crops can help their country in these areas also, it seems right to do so.

Since reaching record high prices in June 2008, corn prices fell 50 per cent by October 2008, declining sharply together with other commodities, including oil. As ethanol production from corn has continue at the same levels, some have argued this trend shows the belief that the increased demand for corn to produce ethanol was mistaken. "*Analysts, including some in the ethanol sector, say ethanol demand adds about 75 cents to $1.00 per bushel to the price of corn, as a rule of thumb. Other analysts say it adds around 20 percent, or just under 80 cents per bushel at current prices. Those estimates hint that $4 per bushel corn might be priced at only $3 without demand for ethanol fuel.*". These industry sources consider that a speculative bubble in the commodity markets holding positions in corn futures was the main driver behind the observed hike in corn prices affecting food supply.

Controversy within the International System

The United States and Brazil lead the industrial world in global ethanol production, with Brazil as the world's largest exporter and biofuel industry leader. In 2006 the U.S. produced 18.4 billion liters (4.86 billion gallons), closely followed by Brazil with 16.3 billion liters (4.3 billion gallons), producing together 70 per cent of the world's ethanol market and nearly 90 per cent of ethanol used as fuel. These countries are followed by China with 7.5 per cent, and India with 3.7 per cent of the global market share.

Since 2007, the concerns, criticisms and controversy surrounding the food vs. biofuels issue has reached the international system, mainly heads of states, and inter-governmental organizations (IGOs), such as the United Nations and several of its agencies, particularly the Food and Agriculture Organization (FAO) and the World Food Programme (WFP); the International Monetary Fund; the World Bank; and agencies within the European Union.

The 2007 Controversy: Ethanol Diplomacy in the Americas

In March 2007, "ethanol diplomacy" was the focus of President George W. Bush's Latin American tour, in which he and Brazil's president, Luiz Inácio Lula da Silva, were seeking to promote the production and use of sugar cane based ethanol throughout Latin America and the Caribbean. The two countries also agreed to share technology and set international standards for biofuels. The Brazilian sugar cane technology transfer will permit various Central American countries, such as Honduras, Nicaragua, Costa Rica and Panama, several Caribbean countries, and various Andean Countries tariff-free trade with the U.S. thanks to existing concessionary trade agreements. Even though the U.S. imposes a USD 0.54 tariff on every gallon of imported ethanol, the Caribbean nations and countries in the Central American Free Trade Agreement are exempt from such duties if they produce ethanol from crops grown in their own countries. The expectation is that using Brazilian technology for refining sugar cane based ethanol, such countries could become exporters to the United States in the short-term. In August 2007, Brazil's President toured Mexico and several countries in Central America and the Caribbean to promote Brazilian ethanol technology.

This alliance between the U.S. and Brazil generated some negative reactions. While Bush was in São Paulo as part of the 2007 Latin American tour, Venezuela's President Hugo Chavez, from Buenos Aires, dismissed the ethanol plan as *"a crazy thing"* and accused the U.S. of trying *"to substitute the production of foodstuffs for animals and human beings with the production of foodstuffs for vehicles, to sustain the American way of life."* Chavez' complaints were quicky followed by then Cuban President Fidel Castro, who wrote that *"you will see how many people among the hungry masses of our planet will no longer consume corn."* *"Or even worse,"* he continued, *"by offering financing to poor countries to produce ethanol from corn or any other kind of food, no tree will be left to defend humanity from climate change."* Daniel Ortega, Nicaragua's President, and one of the preferential recipients of Brazil technical aid, said that *"we reject the gibberish of those who applaud Bush's totally absurd proposal, which attacks the food security rights of Latin Americans and Africans, who are major corn consumers"*, however, he voiced support for sugar cane based ethanol during Lula's visit to Nicaragua.

The 2008 Controversy: Global Food Prices

As a result of the international community's concerns regarding the steep increase in food prices, on April 14, 2008, Jean Ziegler, the United Nations Special Rapporteur on the Right to Food, at the Thirtieth Regional Conference of the Food and Agriculture Organization (FAO) in Brasília, called biofuels a *"crime against humanity"*, a claim he had previously made in October 2007, when he called for a 5-year ban for the conversion of land for the production of biofuels. The previous day, at their Annual IMF and World Bank Group meeting at Washington, D.C., the World Bank's President, Robert Zoellick, stated that *"While many worry about filling their gas tanks, many others around the world are struggling to fill their stomachs. And it's getting more and more difficult every day."*

Luiz Inácio Lula da Silva gave a strong rebuttal, calling both claims *"fallacies resulting from commercial interests"*, and putting the blame instead on U.S. and European agricultural subsidies, and a problem restricted to U.S. ethanol produced from maize. He also said that *"biofuels aren't the villain that threatens food security."* In the middle of this new wave of criticism, Hugo Chavez reaffirmed his opposition and said that he is concerned that *"so much U.S.-produced corn could be used to make biofuel, instead of feeding the world's poor"*, calling the U.S. initiative to boost ethanol production during a world food crisis a "crime."

German Chancellor Angela Merkel said the rise in food prices is due to poor agricultural policies and changing eating habits in developing nations, not biofuels as some critics claim. On the other hand, British Prime Minister Gordon Brown called for international action and said Britain had to be "selective" in supporting biofuels, and depending on the U.K.'s assessment of biofuels' impact on world food prices, *"we will also push for change in EU biofuels targets"*. Stavros Dimas, European Commissioner for the Environment said through a spokewoman that *"there is no question for now of suspending the target fixed for biofuels"*, though he acknowledged that the EU had underestimated problems caused by biofuels.

On April 29, 2008, U.S. President George W. Bush declared during a press conference that *"85 percent of the world's food prices are caused by weather, increased demand and energy prices"*, and recognized that *"15 percent has been caused by ethanol"*. He added that *"the high price of gasoline is going to spur more investment in ethanol as an alternative to gasoline. And the truth of the matter is it's in our national interests that our farmers grow energy, as opposed to us purchasing energy from parts of the world that are unstable or may not like us."* Regarding the effect of agricultural subsidies on rising food prices, Bush said that *"Congress is considering a massive, bloated farm bill that would do little to solve the problem. The bill Congress is now considering would fail to eliminate subsidy payments to multi-millionaire farmers"*, he continued, *"this is the right time to reform our nation's farm policies by reducing unnecessary subsidies"*.

Just a week before this new wave of international controversy began, U.N. Secretary General Ban Ki-moon had commented that several U.N. agencies were conducting a comprehensive review of the policy on biofuels, as the world food price crisis might trigger global instability. He said *"We need to be concerned about the possibility of taking land or replacing arable land because of these biofuels"*, then he added *"While I am very much conscious and aware of these problems, at the same time you need to constantly look at having creative sources of energy, including biofuels. Therefore, at this time, just criticising biofuel may not be a good solution. I would urge we need to address these issues in a comprehensive manner."* Regarding Jean Ziegler's proposal for a five-year ban, the U.N. Secretary rejected that proposal.

A report released by Oxfam in June 2008 criticized biofuel policies of rich countries as neither a solution to the climate crisis nor the oil crisis, while contributing to the food price crisis. The report concluded that from all biofuels available in the market, Brazilian sugarcane ethanol is far from perfect but it is the most favourable biofuel in the world in term of cost and GHG balance. The report discusses some existing problems and potential risks, and asks the Brazilian government for caution to avoid jeopardizing its environmental and social sustainability. The report also says that: "*Rich countries spent up to $15 billion last year supporting biofuels while blocking cheaper Brazilian ethanol, which is far less damaging for global food security.*"

A World Bank research report published on July 2008 found that from June 2002 to June 2008 "*biofuels and the related consequences of low grain stocks, large land use shifts, speculative activity and export bans*" pushed prices up by 70 percent to 75 percent. The study found that higher oil prices and a weak dollar explain 25-30 per cent of total price rise. The study said that "...*large increases in biofuels production in the United States and Europe are the main reason behind the steep rise in global food prices*" and also stated that "*Brazil's sugar-based ethanol did not push food prices appreciably higher*". The Renewable Fuel Association (RFA) published a rebuttal based on the version leaked before its formal release. The RFA critique considers that the analysis is highly subjective and that the author "*estimates the impact of global food prices from the weak dollar and the direct and indirect effect of high petroleum prices and attributes everything else to biofuels.*"

An economic assessment by the OECD also published on July 2008 agrees with the World Bank report regarding the negative effects of subsidies and trade restrictions, but found that the impact of biofuels on food prices are much smaller. The OECD study is also critical of the limited reduction of GHG emissions achieved from biofuels produced in Europe and North America, concluding that the current biofuel support policies would reduce greenhouse gas emissions from transport fuel by no more than 0.8 percent by 2015, while Brazilian ethanol from sugar cane reduces greenhouse gas emissions by at least 80 percent compared to fossil fuels. The assessment calls on governments for more open markets in biofuels and feedstocks in order to improve efficiency and lower costs. The OECD study concluded that "...*current biofuel support measures alone are estimated to increase average wheat prices by about 5 percent, maize by around 7 percent and vegetable oil by about 19 percent over the next 10 years.*"

Sustainable Biofuel

Biofuels—liquid fuels derived from plant materials—are entering the market, driven by factors such as oil price spikes and the need for increased energy security. However, many of the biofuels that are currently being supplied have been criticised for their adverse impacts on the natural environment, food security, and land use. The challenge is to support biofuel development, including the development of new cellulosic technologies, with responsible policies and economic instruments to help ensure that biofuel commercialization is sustainable. Responsible commercialization of biofuels represents an opportunity to enhance sustainable economic prospects in Africa, Latin America and Asia.

Biofuels offer the prospect of increased market competition and oil price moderation. A healthy supply of alternative energy sources will help to combat gasoline price spikes and reduce dependency on fossil fuels, especially in the transport sector. Using transportation fuels more efficiently is also an integral part of a sustainable transport strategy.

Biofuel Options

Biofuel development and use is a complex issue because there are many biofuel options which are available. Biofuels, such as ethanol and biodiesel, are currently produced from the products of conventional food crops such as the starch, sugar and oil feedstocks from crops that include wheat, maize, sugar cane, palm oil and oilseed rape. Any major switch to biofuels from such crops would create a direct competition with their use for food and animal feed, and in some parts of the world we are already seeing the economic consequences of such competition.

Second generation biofuels are now being produced from a much broader range of feedstocks including the cellulose in dedicated energy crops (perennial grasses such as switchgrass and Miscanthus giganteus), forestry materials, the co-products from food production, and domestic vegetable waste. Advances in the conversion processes will improve the sustainability of biofuels, through better efficiencies and reduced environmental impact of producing biofuels, from both existing food crops and from cellulosic sources.

Lord Ron Oxburgh suggests that responsible production of biofuels has several trade-offs:

Produced responsibly they are a sustainable energy source that need not divert any land from growing food nor damage the environment; they can also help solve the problems of the waste generated by Western society; and they can create jobs for the poor where previously were none. Produced irresponsibly, they at best offer no climate benefit and, at worst, have detrimental social and environmental consequences. In other words, biofuels are pretty much like any other product.

According to the Rocky Mountain Institute, sound biofuel production practices would not hamper food and fibre production, nor cause water or environmental problems, and would enhance soil fertility. The selection of land on which to grow the feedstocks is a critical component of the ability of biofuels to deliver sustainable solutions. A key consideration is the minimization of biofuel competition for prime cropland.

Plants Used as Sustainable Biofuel

Jatropha in India and Africa

Crops like Jatropha, used for biodiesel, can thrive on marginal agricultural land where many trees and crops won't grow, or would produce only slow growth yields. Jatropha cultivation provides benefits for local communities:

Cultivation and fruit picking by hand is labour-intensive and needs around one person per hectare. In parts of rural India and Africa this provides much-needed jobs—about 200,000 people worldwide now find employment through jatropha. Moreover, villagers often find that they can grow other crops in the shade of the trees. Their communities will avoid importing expensive diesel and there will be some for export too.

Jatropha in Cambodia

Cambodia has no proven fossil fuel reserves, and is almost completely dependent on imported diesel fuel for electricity production. Consequently Cambodians face an insecure supply and pay some of the highest energy prices in the world. The impacts of this are widespread and may hinder economic development.

Biofuels may provide a substitute for diesel fuel that can be manufactured locally for a lower price, independent of the international oil price. The local production and use of biofuel also offers other benefits such as improved energy security, rural development opportunities and environmental benefits. The Jatropha curcas species appears to be a particularly suitable source of biofuel as it already grows commonly in Cambodia. Local sustainable production of biofuel in Cambodia, based on the Jatropha or other sources, offers good potential benefits for the investors, the economy, rural communities and the environment.

Pongamia Pinnata in Australia and India

Pongamia pinnata is a legume native to Australia, India, Florida (USA) and most tropical regions, and is now being invested in as an alternative to Jatropha for areas such as Northern Australia, where Jatropha is classed as a noxious weed. Commonly known as simply 'Pongamia', this tree is currently being commercialised in Australia by Pacific Renewable Energy, for use as a Diesel replacement for running in modified Diesel engines or for conversion to Biodiesel using 1st or 2nd Generation Biodiesel techniques, for running in unmodified Diesel engines.

There are currently a number of Pongamia plantations in India and new trial plantations in Australia in Areas as diverse as Coastal South East Queensland, to Roma, Queensland.

Sweet Sorghum in India

Sweet sorghum overcomes many of the shortcomings of other biofuel crops. With sweet sorghum, only the stalks are used for biofuel production, while the grain is saved for food or livestock feed. It is not in high demand in the global food market, and thus has little impact on food prices and food security. Sweet sorghum is grown on already-farmed drylands that are low in carbon storage capacity, so concerns about the clearing of rainforest do not apply. Sweet sorghum is easier and cheaper to grow than other biofuel crops in India and does not require irrigation, an important consideration in dry areas.

International Collaboration on Sustainable Biofuels

Roundtable on Sustainable Biofuels

Public attitudes and the actions of stakeholders can play a crucial role in realising the potential of biofuels. Informed discussion and dialogue, based both on the scientific case and an understanding of public and stakeholder views, is important.

The Roundtable on Sustainable Biofuels is an international initiative which brings together farmers, companies, governments, non-governmental organizations, and scientists who are interested in the sustainability of biofuels production and distribution. During 2008, the Roundtable developed a series of principles and criteria for sustainable biofuels production through meetings, teleconferences, and online discussions.

The Roundtable for Sustainable Biofuels released "Version Zero" of its proposed standards for sustainable biofuels on August 13, 2008. This includes a dozen principles, each with several criteria developing the principle further. The 12 principles are:

- Biofuel production shall follow international treaties and national laws regarding such things as air quality, water resources, agricultural practices, labour conditions, and more.
- Biofuels projects shall be designed and operated in participatory processes that involve all relevant stakeholders in planning and monitoring.
- Biofuels shall significantly reduce greenhouse gas emissions as compared to fossil fuels. The principle seeks to establish a standard methodology for comparing greenhouse gases (GHG) benefits.
- Biofuel production shall not violate human rights or labour rights, and shall ensure decent work and the well-being of workers.
- Biofuel production shall contribute to the social and economic development of local, rural and indigenous peoples and communities.

- Biofuel production shall not impair food security.
- Biofuel production shall avoid negative impacts on biodiversity, ecosystems and areas of high conservation value.
- Biofuel production shall promote practices that improve soil health and minimize degradation.
- Surface and groundwater use will be optimized and contamination or depletion of water resources minimized.
- Air pollution shall be minimized along the supply chain.
- Biofuels shall be produced in the most cost-effective way, with a commitment to improve production efficiency and social and environmental performance in all stages of the biofuel value chain.
- Biofuel production shall not violate land rights.

Sustainable Biofuels Consensus

The Sustainable Biofuels Consensus is an international initiative which calls upon governments, the private sector, and other stakeholders to take concerted, collaborative and coordinated action to ensure the sustainable trade, use and production of biofuels. In this way biofuels may play a key role in the transformation of the energy sector, climate stabilization and resulting worldwide renaissance of rural areas, all of which are urgently needed.

The Sustainable Biofuels Consensus envisions a landscape that provides food, fodder, fiber, and energy, which offers opportunities for rural development; that diversifies energy supply, restores ecosystems, protects biodiversity, and sequesters carbon.

Oil Price Moderation

Biofuels offer the prospect of real market competition and oil price moderation. According to the Wall Street Journal, crude oil would be trading 15 per cent higher and gasoline would be as much as 25 per cent more expensive, if it were not for biofuels. A healthy supply of alternative energy sources will help to combat gasoline price spikes.

Sustainable Transport

Biofuels have a limited ability to replace fossil fuels and should not be regarded as a 'silver bullet' to deal with transport emissions. Biofuels on their own cannot deliver a sustainable transport system and so must be developed as part of an integrated approach, which promotes other renewable energy options and energy efficiency, as well as moderating the overall demand and need for transport. The development of hybrid and fuel cell vehicles, public transport, and better urban and rural planning all need to be considered.

In December 2008 an Air New Zealand jet completed the world's first commercial aviation test flight partially using jatropha-based fuel. More than a dozen performance tests were undertaken in the two hour test flight which departed from Auckland International Airport. A biofuel blend of 50:50 jatropha and Jet A1 fuel was used to power one of the Boeing 747-400's Rolls-Royce RB211 engines.

Air New Zealand set several criteria for its jatropha, requiring that "the land it came from was neither forest nor virgin grassland in the previous 20 years, that the soil and climate it came from is not suitable for the majority of food crops and that the farms are rain fed and not mechanically irrigated". The company has also set general sustainability criteria, saying that such biofuels must not compete with food resources, that they must be as good as traditional jet fuels, and that they should be cost competitive with existing fuels.

In January 2009, Continental Airlines used a sustainable biofuel to power a commercial aircraft for the first time in North America. This demonstration flight marks the first sustainable biofuel demonstration flight by a commercial carrier using a twin-engined aircraft, a Boeing 737-800, powered by CFM International CFM56-7B engines. The biofuel blend included components derived from algae and jatropha plants. The algae oil was provided by Sapphire Energy, and the jatropha oil by Terasol Energy.

INDIA APPROVES BIOFUEL RISE

SciDev.Net, 17 September 2008—India has approved a national

biofuel policy that aims to raise the proportion of biofuels from five to 20 per cent in petrol and diesel fuels over the coming decade, using non-edible plant sources.

The Indian government approved the policy last week (11 September). The policy states that by 2017, transport fuels in India need to contain 20 per cent biofuel.

Two main types of biofuels are envisaged: alcohol from plant wastes, chiefly sugarcane molasses, and biodiesel—oil produced from non-edible oilseed crops such as jatropha curcas, which can be blended with diesel.

The policy supports increasing biodiesel plantations on community, government-owned and forest wastelands, but not on fertile, irrigated lands. The government estimates 13.4 million hectares of barren land are available for jatropha cultivation, which could potentially yield 15 million tonnes of oil each year.

The policy also details incentives for growers of biofuel crops: removing taxes and duties on biodiesel, setting a minimum 'support' price for buying biodiesel oilseeds from growers and a minimum purchase price of bio-ethanol from oil marketing companies. These should ensure adequate returns to both crop growers and oil makers.

India imports over 70 per cent of its petroleum and the Indian Planning Commission estimates that, by 2017, the country's demand for petrol will rise to 16.40 million tonnes.

But biofuels will only substitute a little for fossil fuel use, yet lock huge land areas for crop cultivation, says Anumita Roychowdhury, associate director for research and advocacy at the Centre for Science and Environment, a Delhi-based nongovernmental organisation.

Instead of giving subsidies for biofuel production, the Indian government should invest in policies that reduce overall demand for fuels, such as encouraging use of public transport and restraining use of personal vehicles, Roychowdhury told SciDev.Net.

The Indian Express reported (16 September) on the experiences of Chhattisgarh state, where 400 million Jatropha saplings were planted on more than 155,000 hectares of fallow land in the last three years. However, until now, there has been no reported data on survival of saplings or seed production.

Farmers in many areas are in a fix as the trees have not yet borne fruits, while in places where they have, various departments and local agencies, are waiting for guidelines on collection and sale of seeds.

A 2006 analysis by the UN Conference on Trade and Development (UNCTAD) concluded that India cannot rely on sugarcane molasses as a reliable feedstock for alcohol, given the crop's dependence on monsoon and vagaries of the domestic sugar industry. Similarly, difficulties in procuring oilseeds and lack of infrastructure could obstruct substantial biodiesel production by 2011–12.

UNCTAD suggested that India might have to import both bio-ethanol and biodiesel to meet its targets.

BIOFUELS A SUSTAINABLE ANSWER TO FUTURE MOTORING, SAYS SWEDISH

Second generation biofuels based on sugar cane present a sustainable solution to the world's energy needs, a leading expert claimed last Friday. Speaking at the Green Power Forum at the Radisson St Helens Hotel in Dublin, Anders Fredriksson of leading Swedish biofuel company Sekab, warned delegates that dwindling oil reserves would allow for only 1 litre of fuel per week per adult by 2050. The Forum was hosted by the Irish Motoring Writers' Association and sponsored by Continental Tyres.

According to Fredriksson, only 200,000 hectares of land would be needed to fuel Sweden's car population with cellulosic biofuel, when considered in tandem with electric hybrid vehicles. On a world scale this would translate into 60 million hectares, of which Brazil could supply half without the need to impact on the rain forests. Africa presents a viable source for the remainder, he claimed.

This 'holy grail' of biofuels, which involves the use of low carbon soils, utilisation of biomass and ethanol-powered harvesting techniques, will be viable within 6 years, he said. The challenge in tackling the twin issues of dwindling oil reserves and rising emissions simultaneously—against the backdrop of rising populations—cannot be overestimated. "It's overwhelming, and requires a paradigm shift. The poorest stand to suffer most in the shake-up," Fredriksson warned.

Sekab is the largest provider of biofuels in northern Europe and a worldwide technology leader in developing the next generation of cellulose-based ethanol. The company has the world's first biofuel to be officially verified as sustainable on the basis of CO2 impact and social impact. Currently, 30 per cent of car sales in Sweden are of FlexiFuel Vehicles (FFV) and they account for the top 4 selling models on the market. 90 per cent of FFV drivers fill up primarily with bioethanol.

"If this can happen in Sweden, it can happen anywhere, "he claimed, citing the progress made in Ireland where several thousand FFVs have been sold. He also claimed the recent rise in diesel sales will peak, as EU directives will work against the fuel. A panel discussion featured contributions from Colin Roche, policy and advocacy coordinator, Oxfam, Norbert Krüger of Ford of Europe, Noel McMullan from the Maxol Group and Bernard Rice, biofuels researcher, Teagasc.

Rice maintained current production of biofuels in Ireland are a 'no-brainer' as they involve no land-use, and are "good on CO_2", being by-products of other production. He sees biofuels as an important outlet for maintaining cereal production here and emphasised the need to develop a biomass resource sooner rather than later. We need a first generation biofuel capability to pave the way for 2nd generation, he said.

"Biofuel is a topic that engenders emotive and sometimes ill-informed comment," comments Tony Toner, Chairman, Irish Motoring Writers Association. "In bringing a leading player in the sector to Ireland, I would like to think the motoring writers are making a contribution to the debate on sustainable mobility in this country.

"This a most pressing issue facing the motoring sector, and no doubt is a topic that will remain high on the agenda for motorists, the motor industry and motoring media for the next decade." Paddy Murphy, of sponsors Continental Tyres, commented: "All sectors of the motor industry share the goal of reducing CO_2 emissions. In the tyre sector this is translating into low rolling resistance tyres, for example. But it's the consumer, not the developer, who can contribute most to this goal, by keeping his or her tyres correctly inflated. As well as extending the life of the tyre, it also ensures they are safe."

Recent Continental research indicated that through increased fuel consumption, under-inflated tyres in Ireland could cost the equivalent in CO_2 emissions of 24,000 trips to New York each year.

The Forum was attended by policymakers from state bodies, local authorities, universities and representatives from the car and alternative energy sectors. Chairman was RTE's Rodney Rice.

5

Major Biofuel Crops: Types, Yields and Production Technologies

UNDERSTANDING VEGETABLE OIL USED AS FUEL

Vegetable oil is an alternative fuel for diesel engines and for heating oil burners. For engines designed to burn 2 diesel fuel, the viscosity of vegetable oil must be lowered to allow for proper atomization of the fuel, otherwise incomplete combustion and carbon build up will ultimately damage the engine. Many enthusiasts refer to vegetable oil used as fuel as **waste vegetable oil** (WVO) if it is oil that was discarded from a restaurant or **straight vegetable oil** (SVO) or **pure plant oil** (PPO) to distinguish it from biodiesel.

History

The first known use of vegetable oil as fuel in a diesel engine was a demonstration of an engine built by the Otto company and designed to burn mineral oil, which was run on pure peanut oil at the 1900 World's Fair. Late in his career, Rudolf Diesel investigated using vegetable oil to fuel engines of his design. In a 1912 presentation to the British Institute of Mechanical Engineers, he cited a number of efforts in this area and remarked, "The fact that fat oils from vegetable sources can be used may seem insignificant today, but such oils may perhaps become in course of time of the same importance as some natural mineral oils and the tar products are now."

Periodic petroleum shortages spurred research into vegetable oil as a diesel substitute during the 1930s and 1940s, and again in the 1970s and early 1980s when straight vegetable oil enjoyed its highest level of scientific interest. The 1970s also saw the formation of the first commercial enterprise to allow consumers to run straight vegetable oil in their automobiles, Elsbett of Germany. In the 1990s Bougainville conflict, islanders cut off from oil supplies due to a blockade used coconut oil to fuel their vehicles.

Academic research into straight vegetable oil fell off sharply in the 80s with falling petroleum prices and greater interest in biodiesel as an option that did not require extensive vehicle modifications.

Application and Usability

While engineers and enthusiasts have been experimenting with using vegetable oils as fuel for a diesel engine since at least 1900, it is only recently that the necessary fuel properties and engine parameters for reliable operation have become apparent. A number of peer reviewed studies exists that show reliable long term use of vegetable oil; the German Deutz AG F3l912W. and a high speed common rail engine fitted to a Mercedes-Benz 220 C Class

Most diesel car engines are suitable for the use of SVO, also commonly called Pure Plant Oil (PPO), with suitable modifications. Principally, the viscosity and surface tension of the SVO/PPO must be reduced by preheating it, typically by using waste heat from the engine or electricity, otherwise poor atomization, incomplete combustion and carbonization may result. One common solution is to add a heat exchanger, and an additional fuel tank for "normal" diesel fuel (petrodiesel or biodiesel) and a three way valve to switch between this additional tank and the main tank of SVO/PPO. (This aftermarket modification typically costs about $1200 USD.) The engine is started on diesel, switched over to vegetable oil as soon as it is warmed up and switched back to diesel shortly before being switched off to ensure that no vegetable oil remains in the engine or fuel lines when it is started from cold again. In colder climates it is often necessary to heat the vegetable oil fuel lines and tank as it can become very viscous and even solidify.

Single tank conversions have been developed, largely in Germany, which have been used throughout Europe. These conversions are designed to provide reliable operation with rapeseed oil that meets the German rapeseed oil fuel standard DIN 51605. Modifications to the engines cold start regime assist combustion on start up and during the engine warm up phase. Suitably modified indirect injection (IDI) engines have proven to be operable with 100 per cent PPO down to temperatures of -10°C. Direct injection (DI) engines generally have to be preheated with a block heater or diesel fired heater. The exception is the VW Tdi (Turbocharged Direct Injection) engine for which a number of German companies offer single tank conversions. For long term durability it has been found necessary to increase the oil change frequency and to pay increased attention to engine maintenance.

With unmodified engines the unfavourable effects may be reduced by blending, or "cutting", the SVO/PPO with diesel fuel; however, opinions vary as to the efficacy of this. Some WVO mechanics have found higher rates of wear and failure in fuel pumps and piston rings. This can generally be attributed to the use of oils with properties or contaminants that make them unsuitable for use in this type of application, poorly maintained engines, unsuitable engine modifications or operating regimes.

Many cars powered by indirect injection engines supplied by in-line injection pumps, or mechanical Bosch injection pumps are capable of running on pure SVO/PPO in all but winter temperatures. Indirect injection Mercedes-Benz vehicles with in-line injection pumps and cars featuring the PSA XUD engine tend to perform reasonably, especially as the latter is normally equipped with a coolant heated fuel filter. Engine reliability would depend on the condition of the engine. Attention to maintenance of the engine, particularly of the fuel injectors, cooling system and glow plugs will help to provide longevity. Ideally the engine would be converted.

Properties

The main form of SVO/PPO used in the UK is rapeseed oil (also known as canola oil, primarily in the United States and Canada) which has a freezing point of -10°C. However the use of sunflower oil, which freezes at -17°C, is currently being investigated as a means of improving cold weather starting. Unfortunately oils with lower gelling points tend to be less saturated (leading to a higher iodine number) and polymerize more easily in the presence of atmospheric oxygen.

Material Compatibility

Free fatty acids in WVO can have a detrimental effect on metals. Copper and its alloys, such as brass, are affected. Zinc and zinc-plating (galvanization) are stripped by FFA's and tin, lead, iron, and steel are affected too. Stainless steel and aluminium are generally unaffected.

Examples

Some Pacific island nations are using coconut oil as fuel to reduce their expenses and their dependence on imported fuels while helping stabilize the coconut oil market. Coconut oil is only usable where temperatures do not drop below 17 degrees Celsius (62 degrees Fahrenheit), unless two-tank SVO/PPO kits or other tank-heating accessories, etc. are used. Fortunately, the same techniques developed to use, for example, canola and other oils in cold climates can be implemented to make coconut oil usable in temperatures lower than 17 degrees Celsius.

Home Heating

With often minimal modification, most residential furnaces and boilers that are designed to burn No. 2 heating oil can be made to burn either biodiesel or filtered, preheated waste vegetable oil. These are generally not as clean-burning as petroleum fuel oil, but if processed at home, by the consumer, can result in considerable savings. Many restaurants will give away their used cooking oil either free or at minimal cost, and processing to biodiesel is fairly simple and inexpensive. Burning filtered WVO directly is somewhat more problematic, since it is much more viscous, but it can be accomplished with suitable preheating. WVO can thus be an economical heating option for those with the necessary mechanical and experimental aptitude, where fire regulations and insurance policy permit it.

Combined Heat and Power

A number of companies offer compressed ignition engine generators optimized to run on plant oils where the waste engine heat is recovered for heating.

Availability

Waste Vegetable Oil

As of 2000, the United States was producing in excess of 11 billion liters (2.9 billion U.S. gallons) of waste vegetable oil annually, mainly from industrial deep fryers in potato processing plants, snack food factories and fast food restaurants. If all those 11 billion liters could be collected and used to replace the energetically equivalent amount of petroleum (an ideal case), almost 1 per cent of US oil consumption could be offset. Use of waste vegetable oil as a fuel competes with some other uses of the commodity, which has effects on its price as a fuel and increases its cost as an input to the other uses as well.

Pure Vegetable Oil (Pure Plant Oil)

Pure plant oil (PPO) (or Straight Vegetable Oil (SVO)), in contrast to waste vegetable oil, is not a byproduct of other industries, and thus its prospects for use as fuel are not limited by the capacities of other industries. Production of vegetable oils for use as fuels is theoretically limited only by the agricultural capacity of a given economy.

Legal Implications

Taxation of Fuel

Taxation on SVO/PPO as a road fuel varies from country to country, and it is possible the revenue departments in many countries are even unaware of its use, or feel it insufficiently significant to legislate. Germany used to have 0 per cent taxation, resulting in it being a leader in most developments of the fuel use. However SVO/PPO as a road fuel began to be taxed at 0,09 €/liter from 1 January 2008 in Germany, with incremental rises up to 0,45 €/liter by 2012. However, in Australia it has become illegal to produce any fuel if it is to be sold unless a license to do so is granted by the Federal Government. This is a chargeable offence with a fine of up to 20,000 dollars but this bracket may alter circumstantially. Also a jail term may result if offenders are aware of the legality of selling the fuel.

In the USA

The legality of burning SVO in the United States of America is debated by many. The EPA clearly states it is illegal.

There seems to be no clear federal taxation system in the USA. Production of biodiesel in some US regions may require motor fuel taxes to be paid.

In Japan

The Japanese Government has also exempted the use of SVO as a fuel from road tax.

In Ireland

In Ireland a pilot scheme is currently running (as of April 2006) whereby eight suppliers have been approved to sell SVO/PPO for use as a fuel without the payment of excise duty (Value Added Tax at 21 per cent still applies, SVO from any other source still attracts excise duty at 36.8058 Euro cents per litre plus 21 per cent VAT).

In France

Despite its use being common in France, it would appear there has been no legislation to cover this.

In the UK

In the UK, it is legal once duty on the fuel is paid. In the UK, drivers using SVO/PPO have been prosecuted for failure to pay duty to Her Majesty's Revenue and Customs. The rate of taxation on SVO was originally set at a reduced rate of 27.1p per litre, but in late 2005, HMRC started to enforce the full diesel excise rate of 47.1p per litre.

Following a review in late 2006, HM Revenue and Customs has announced changes regarding the administration and collection of excise duty of biofuels and other fuel substitutes (Veg. Oil). The changes came into effect on June 30, 2007. There is no longer a requirement to register to pay duty on vegetable oil used as road fuel for those who "produce" or use less than 2,500 litres per year. For those producing over this threshold the biodiesel rate now applies.

HMRC argued that SVOs/PPOs on the market from small producers did not meet the official definition of "biodiesel" in Section 2AA of The Hydrocarbon Oil Duties Act 1979 (HODA), and consequently was merely a "fuel substitute" chargeable at the normal diesel rate. Such a policy seemed to contradict the UK Government's commitments to the Kyoto Protocol and to many EU directives and had many consequences, including an attempt to make the increase retroactive, with one organization being presented with a £16,000 back tax bill. This change in the rate of excise duty has effectively removed any commercial incentive to use SVO/PPO, regardless of its desirability on environmental grounds; unless waste vegetable oil can be obtained free of charge, the combined price of SVO/PPO and taxation for its use usually exceeds the price of mineral diesel. HMRC's interpretation is being widely challenged by the SVO/PPO industry and the UK pure Plant Oil Association (UKPPOA)] has been formed to represent the interests of people using vegetable oil as fuel and to lobby parliament.

CORN STOVE

A **corn stove** (also spelled as **CornStoves, Corn Fireplaces**), is a home heater or a small business heater that uses local renewable whole kernel shelled corn, wood pellets or multiple biomass as fuel. Local renewable whole kernel shelled corn is supplied by two million local farmers in the United States of America, or globally, in maize producing areas. Properly installed corn stoves can reduce home heating costs by up to 80 per cent. A cornstove can burn biomass fuel of any type that the auger will feed such as wood pellets, soybeans, cherry seeds, orange seeds, or screened wood chips. In the event of a malfunction, Corn as a fuel will self extinguish in less than 60 seconds. Other biomass fuels like wood will continue to burn until all the fuel is gone. A small amount of biomass or wood pellet stoves will burn corn or a corn pellet mix.

The cornstove has a variable controller for BTU output rate which reduces waste energy. Other types of heaters have Set Point Controls that continuously switch "off" and "on". Infinitely variable controls allow for cost effective control of relative humidity and temperature simultaneously. Here is why: A constant temperature is required to maintain a constant room relative humidity and avoid reheat cost of moist air and dry air each temperature cycle. Most heaters are actuated by the thermostat. Energy is wasted each cycle to reheat the cooled air and cool moisture in the air. Also the relative humidity is forced to swing 3 per cent for each degree of temperature change. The steady heat output of a corn stove cost effectively maintains a precise room temperature which allows the relative humidity to be controlled precisely and exactly at the desired per cent RH. The recommended healthy 50 per cent RH is more economical to maintain. Extra heat input is not required to raise "high BTU" moist air to room temperature. "High BTU" moist room air that escapes through a leak or doorway is overly expensive if the room RH is greater than 50 per cent. Dry air below 50 per cent RH is relatively inexpensive to heat and the loss thereof through a leak, window, or doorway is relatively inconsequential to the monthly heating or air conditioning bill no matter which type HVAC system is installed. This air quality issue can also be remedied by using a number of corn/biomass stoves that automatically cycle between a high and low setting to maintain temperature.

Overview

Some corn stoves need to stir the fuel during combustion. The requirement to stir corn during combustion is the one safety feature separating some corn stoves from solid fuel stoves. For example, a wood pellet stove can burn the house down if the hopper full of fuel catches ablaze. Corn in a container or stove hopper is completely safe from combustion. Corn must be stirred to release the 34 per cent hydrogen contained inside the kernel. A pellet stove does not stir the fuel as required for corn combustion. Some stoves will burn solid biomass fuel like whole kernel shelled corn, biomass pellet fuels, wood pellets, grass, or trash. Some corn stoves can burn 100 per cent shelled corn, or mixtures of wood pellets, cherry pits, soy beans, orange seeds, lemon seeds, grape seeds, rice, or screened crushed corn cobs. It is also a stove that will burn solid fuels. that require constant stirring or constant vibration to support combustion. Once the stirring or vibration of the burning fuel is eliminated, a corn flame will self extinguish within seconds. Locally renewable shelled corn is safe, clean, environmentally friendly, non-volatile, non-explosive. Corn produces no smoke (zero opacity), and very little (0.00x MMBTU) carbon dioxide. Corn combustion can be 98.6 per cent efficient releasing five to ten gallons of solid effluent per home per year. Unique to corn and other biomass stoves is the ability to safely preheat both the combustion air and fuel by safely storing the corn inside. Use inside combustion air. Vent the corn combustion exhaust effluents outside.

Corn has no [[Volatile organic compound|VOC). Corn is edible. Corn combustion will self-extinguish within 60 seconds even while applying the customary three "fat" elements required for combustion: fuel/air/temp. In addition to "FAT", (TTT) Time, Temperature and turbulence are also required for corn combustion. Corn will not burn in a bucket or in a pile but may parch, pop, or cook. Corn will not flame/explode/ propagate unless properly stirred. A corn stove will burn corn, soybeans, wood, pellets, trash. A wood pellet stove will not burn corn but may permit some ratio of corn to be mixed with the wood pellets which inadvertently provide the required turbulence for the corn as the wood size is reduced by combustion.

Fuel supply

Local farms in the US raise corn, supply commercial corn demand, and compete for the local and global corn market. Corn yield can be economically controlled from 50 bushel per acre to 300 bushel per acre within two months of harvest. Eighty to 100 million acres (400,000 km²) of corn are grown annually in the US alone. Over 125 million acres (510,000 km²) of corn were grown annually in the US to feed farm animals and people prior to modern mechanization methods of farming.

Styles

Corn stoves are manufactured in styles not limited to free standing, fireplace insert, HVAC connected home furnace heaters, home cooking corn stove grills, and outside corn burning furnaces.

Safety

Although corn stoves have been available for several years, there have been no recorded home fires resultant from a corn stove. Home insurance rates have no added cost for using corn as a heating fuel.

Environment

Corn has been estimated by the Sierra Club to reduce global warming by annually converting a net positive 484 pounds of carbon dioxide into oxygen.

The United States Environmental Protection Agencyý (EPA) test results for a Tennessee corn stove record airborne particulates of 0.001 pounds per million BTU as compared to 0.1 MMBTU for gas, 5.0 MMBTU for wood and 0.5 MMBTU for coal. A corn stove will produce approximately 5 gallons of solid potash particulate annually in a southern US city or 10 gal potash particulate annually in a 1500 sq foot home located in a northern city.

Availability

Corn stoves, corn fireplaces, corn fireplace inserts and corn grills are available in all 50 US states.

The following table shows the vegetable oil yields of common energy crops associated with biodiesel production. This is unrelated to ethanol production, which relies on starch, sugar and cellulose content instead of oil yields.

Table 5.1: Biofuel Crop Yields

Crop	kg oil/ha/yr	litres oil/ha	lbs oil/acre	US gal/acre
maize (corn)	145	172	129	18
cashew nut	148	176	132	19
oats	183	217	163	23

(Contd.)

Crop	kg oil/ha/yr	litres oil/ha	lbs oil/acre	US gal/acre
lupin (lupine)	195	232	175	25
kenaf	230	273	205	29
calendula	256	305	229	33
cotton	273	325	244	35
hemp	305	363	272	39
soybean	375	446	335	48
coffee	386	459	345	49
flax (linseed)	402	478	359	51
hazelnuts	405	482	362	51
euphorbia	440	524	393	56
pumpkin seed	449	534	401	57
coriander	450	536	402	57
mustard seed	481	572	430	61
camelina	490	583	438	62
sesame	585	696	522	74
safflower	655	779	585	83
rice	696	828	622	88
tung tree	790	940	705	100
sunflowers	800	952	714	102
cacao (cocoa)	863	1026	771	110
peanut	890	1059	795	113
opium poppy	978	1163	873	124
rapeseed	1000	1190	893	127
olives	1019	1212	910	129
castor beans	1188	1413	1061	151
pecan nuts	1505	1791	1344	191
jojoba	1528	1818	1365	194
jatropha	1590	1892	1420	202
macadamia nuts	1887	2246	1685	240
brazil nuts	2010	2392	1795	255
avocado	2217	2638	1980	282
coconut	2260	2689	2018	287
chinese tallow	3950	4700	3500	500
oil palm	5000	5950	4465	635
Copaifera langsdorffii		12000		1283
algae (open pond)	80000	95000	70000	10000

Note: Chinese Tallow (*Sapium sebiferum,* or *Tradica sebifera*) is also known as the "Popcorn Tree".

Source: Sourced here, published in Hill, Amanda, Al Kurki, and Mike Morris. 2006. "*Biodiesel: The Sustainability Dimensions.*" ATTRA Publication. Butte, MT: National Center for Appropriate Technology, pp. 4-5.

Table 5.2: Yields of Common Crops Associated with Biofuels Production

Crop	Oil (kg/ha)	Oil (L/ha)	Oil (lbs/acre)	Oil (US gal/acre)	Oil per seeds (kg/100 kg)	Melting Range (°C) Oil/Fat	Methyl Ester	Ethyl Ester	Iodine number	Cetane number
(1)	(2)	(3)	(4)	(5)	(6)	(7)	(8)	(9)	(10)	(11)
Groundnut					(Kernel) 42					
Copra					62					
Tallow						35-42	16	12	40-60	75
Lard						32-36	14	10	60-70	65
Corn (maize)	145	172	129	18		-5	-10	-12	115-124	53
Cashew nut	148	176	132	19						
Oats	183	217	163	23						
Lupine	195	232	175	25						
Kenaf	230	273	205	29						
Calendula	256	305	229	33	(Seed) 13					
Cotton	273	325	244	35		-1-0	-5	-8	100-115	55
Hemp	305	363	272	39						
Soybean	375	446	335	48	14	-16--12	-10	-12	125-140	53
Coffee	386	459	345	49						
Linseed (flax)	402	478	359	51		-24			178	
Hazelnuts	405	482	362	51						
Euphorbia	440	524	393	56						
Pumpkin seed	449	534	401	57						
Coriander	450	536	402	57	35					
Mustard seed	481	572	430	61						
Camelina	490	583	438	62						

(Contd.)

(1)	(2)	(3)	(4)	(5)	(6)	(7)	(8)	(9)	(10)	(11)
Sesame	585	696	522	74	50					
Safflower	655	779	585	83						
Rice	696	828	622	88						
Tung oil tree	790	940	705	100		−2.5			168	
Sunflowers	800	952	714	102	32	−18—17	−12	−14	125-135	52
Cocoa (cacao)	863	1,026	771	110						
Peanuts	890	1,059	795	113		3			93	
Opium poppy	978	1,163	873	124						
Rapeseed	1,000	1,190	893	127	37	−10-5	−10-0	−12-2	97-115	55-58
Olives	1,019	1,212	910	129		−12-6	−6	−8	77-94	60
Castor beans	1,188	1,413	1,061	151	50	−18			85	
Pecan nuts	1,505	1,791	1,344	191						
Jojoba	1,528	1,818	1,365	194						
Jatropha	1,590	1,892	1,420	202						
Macadamia nuts	1,887	2,246	1,685	240						
Brazil nuts	2,010	2,392	1,795	255						
Avocado	2,217	2,638	1,980	282						
Coconut	2,260	2,689	2,018	287		20-25	−9	−6	8-10	70
Chinese Tallow		4,700		500						
Oil palm	5,000	5,950	4,465	635	20-36	20-40	−8-21	−8-18	12-95	65-85
Algae		95,000		10,000						

Oil per seeds = Typical oil extraction from 100 kg. of oil seeds

Note: Chinese Tallow (Sapium sebiferum, or Tradica Sebifera) is also known as the "Popcorn Tree".

Source: Used with permission from The Global Petroleum Club

UNDERSTANDING TYPES OF FOOD BIOENERGY CROPS

Jatropha Curcas

Jatropha curcas, Barbados nut or Physic nut is a perennial poisonous shrub (normally up to 5 m high) belonging to the Euphorbiaceae or spurge family. It is an uncultivated non-food wild-species. The plant, originating in Central America, whereas it has been spread to other tropical and subtropical countries as well and is mainly grown in Asia and in Africa, where it is known as *Pourghère*. It is used as a living fence to protect gardens and fields from animals. It is resistant to a high degree of aridity (it can be planted even in the desert) and as such does not compete with food crops. The seeds contain 30 per cent oil that can be processed to produce a high-quality biodiesel fuel, usable in a standard diesel engine.

Main Botanical features

(*i*)   *Leaves*: large green to pale-green leaves.
(*ii*)   *Flowers*: more female flowers yield more seeds.
(*iii*)   *Fruits*: fruits are produced in winter, or there may be several crops during the year if soil moisture is good and temperatures are sufficiently high.
(*iv*)   *Seeds*: the seeds are mature when the capsule changes from green to yellow.

Pattern of Cultivation

Cultivation is uncomplicated. *Jatropha curcas* grows in tropical and subtropical regions. The plant can grow in wastelands and grows on almost any terrain, even on gravelly, sandy and saline soils. It can thrive on the poorest stony soil and grow in the crevices of rocks. Complete germination is achieved within 9 days. Adding manure during the germination has negative effects during that phase, but is favourable if applied after germination is achieved. It is usually propagated by cuttings as this yields faster results than multiplication by seeds. The flowers only develop terminally (at the end of a stem), so a good ramification (plants presenting many branches) produces the greatest amount of fruits. Another productivity factor is the ratio between female and male flowers within an inflorescence (usually about 1 female to 10 male flowers—more female flowers mean more fruits). *Jatropha curcas* thrives on a mere 250 mm (10 in) of rain a year, and only during its first two years does it need to be watered in the closing days of the dry season. Ploughing and planting are not needed regularly, as this shrub has a life expectancy of approximately forty years. The use of pesticides and other polluting substances are not necessary, due to the pesticidal and fungicidal properties of the plant. While *Jatropha curcas* starts yielding from 9-12 months time, the effective yield is obtained only after 2—3 years time. If planted in hedges, the reported productivity of *Jatropha* is from 0.8 kg. to 1.0 kg. of seed per meter of live fence. The seed production is around 3.5 tons/ hectare (Seed production ranges from about 0.4 tons per hectare in first year to over 5 tons per hectare after 3 years).

Patterns of Propagation

Jatropha curcas has limited potential for vegetative propagation and is usually propagated by seed. It is cross-pollinated and propagation through seed leads to a lot of genetic variability in terms of growth, biomass, seed yield and oil content. Besides, the problems of low viability and recalcitrant nature of oil seeds limits the sexual propagation. On the other hand, clonal techniques will help in overcoming the biological problems hindering mass propagation of this tree-borne oilseed species. *Jatropha curcas* can be propagated by seed as well as vegetatively. Propagation by seed. Vegetatively, this crop can be propagated by stem cuttings, grafting, budding as well as by air layering techniques. The investigation leads to the recommendation that cuttings should be taken preferably from juvenile plants and treated with 200 micro gram per litre of IBA (rooting hormone) to ensure the highest level of rooting in stem cuttings of *Jatropha curcas*. Thus stem cuttings, grafting, budding and air layering methods of propagation could be used as a potential protocol for commercial propagation of *Jatropha curcas*.

Methods of Processing

Seed extraction is made simple with the use of the Universal Nut

Sheller, an appropriate technology designed by the Full Belly Project. Oil content varies from 28 per cent to 30 per cent and 94 per cent extraction, *one hectare of plantation will give 1.6t (metric tonne) of oil if the soil is average.* The oily seeds are processed into oil, which may be directly used to fuel combustion engines or may be subjected to transesterification to produce biodiesel. Jatropha oil is not suitable for human consumption, as it induces strong vomiting and diarrhea. A colourant can also be derived from the seed.

Other Names

(i)   Pinhão manso in Brazil
(ii)   Tempate in Nicaragua
(iii)   kasla also tubatuba or tubang bakod in Philippines
(iv)   Purging nut
(v)   Jarak in Indonesia
(vi)   Mbono in Tanzania
(vii)   Pourghère in Francophone Africa
(viii) Lahong Kwang in Cambodia
(ix)   Cay Dau Lai in Việt Nam
(x)   Dang iu ciu in Taiwan

Chinese Tallow

Triadica sebifera, also referred to as *Sapium sebiferum*, is commonly known as the *Chinese tallow tree, Florida aspen and Gray Popcorn Tree*. Introduced to the United States, the tree is native to eastern Asia, and is most commonly associated with eastern China, Taiwan, and Japan. In these regions, the waxy coating of the seeds is used for candle and soap making, and the leaves are used as herbal medicine to treat boils. The plant sap and leaves are reputed to be toxic, and decaying leaves from the plant are toxic to other species of plant. The specific epithets *sebifera* and *sebiferum* mean "wax-bearing" and refer to the vegetable tallow that coats the seeds. It is useful in the production of biodiesel because it is the third most productive vegetable oil producing crop in the world, after algae and oil palm. This species is considered to be a noxious invader in the U.S.

Main Physical Characteristics

The simple, deciduous leaves of this tree are alternate, broad rhombic to ovate in shape and have smooth edges, heart shaped and sometimes with an extended tail often resembling the bo tree, Ficus religiosa. The leaves are bright green in colour and. slightly paler underneath. They become bright yellows, oranges, purples and reds in the autumn. The tree is monoécious, producing male and female flowers on the same plant. The waxy green leaves set off the clusters of greenish-yellow and white flowers at bloom time. The flowers occur in terminal spike-like inflorescences up to 20 cm long. Light green in colour, these flowers are very conspicuous in the spring. Each pistillate (female) flower is solitary and has a three-lobed ovary, three styles, and no petals. They are located on short branches at the base of the spike. The staminate (male) flowers occur in clusters at the upper nodes of the inflorescence.

Fruits are three-lobed, three-valved capsules. As the capsules mature, their colour changes from green to a brown-black. The capsule walls fall away and release three globose seeds with a white, tallow-containing covering. Seeds usually hang on the plants for several weeks. In North America, the flowers typically mature from April to June and the fruit ripens from September to October.

Fields Range and Habitat Types

The plant is found throughout the southern United States. It was introduced in colonial times and has become naturalized from South Carolina southward along the Atlantic and the entire Gulf coast, where it grows profusely along ditchbanks and dikes. It grows especially well in open fields and abandoned farmland, and along the edges of the Western Gulf coastal grasslands biome, sometimes forming pure stands. In the Houston area, Chinese tallowtrees account for a full 23 percent of all trees, more than any other tree species and is the only invasive tree species in the 14 most common species in the area. Herbivores and insects have a conditioned behavioural avoidance to eating the leaves of Chinese tallowtree, and this, rather than plant toxins, may be a reason for the success of the plant as an invasive. The plant is sold in nurseries as an ornamental tree. It is not choosy about soil types or drainage, but will not grow in deep shade. It commonly grows all over Japan, and is reasonably hardy. It is prized for its abundant and often spectacular autumn foliage.

Major Uses

The wax of the seeds is harvested by placing the seeds in hot water, and skimming the surface of the water. Though other parts of the plant are toxic, the wax is not, and it can be used as a substitute for vegetable oil in cooking.

The nectar is also non-toxic, and it has become a major honey plant for beekeepers. The honey is not of high quality, being sold as bakery grade, but is produced copiously at a barren time of year, after most of the other spring bloom is done. In the Gulf coast states, beekeepers migrate with their honey bees to good tallow locations near the sea. The tree is highly ornamental, fast growing and a good shade tree. It is especially noteworthy if grown in areas that have strong seasonal temperature ranges with the leaves becoming a multitude of colours rivalling maples in the autumn.

Hemp

Hemp is the common name for plants of the entire genus *Cannabis*, although the term is often used to refer only to *Cannabis* strains cultivated for industrial (non-drug) use. Industrial hemp has many uses, including paper, textiles, biodegradable plastics, construction, health food, and fuel. It is one of the fastest growing biomasses known, and one of the earliest domesticated plants known. It also runs parallel with the "Green Future" objectives that are becoming increasingly popular. Hemp requires little to no pesticides, no herbicides, controls erosion of the topsoil, and produces oxygen. Furthermore, hemp can be used to replace many potentially harmful products, such as tree paper (the processing of which uses chlorine bleach, which results in the waste product polychlorinated dibensodioxins, popularly known as dioxins, which are carcinogenic, and contribute to deforestation), cosmetics, and plastics, most of which are petroleum-based and do not decompose easily. The strongest chemical needed to whiten the already light hemp paper is non-toxic hydrogen peroxide. *Cannabis sativa* L. subsp. *sativa* var. *sativa* is the variety grown for industrial use in Europe, Canada, and elsewhere, while *C. sativa* subsp. *indica* generally has poor fiber quality and is primarily used for production of recreational and medicinal drugs. The major difference between the two types of plants is the appearance and the amount of $f¢^9$-tetrahydrocannabinol (THC) secreted in a resinous mixture by epidermal hairs called glandular trichomes. Strains of *Cannabis* approved for industrial hemp production produce only minute amounts of this psychoactive drug, not enough for any physical or psychological effects. Typically, Hemp contains below 0.3 per cent THC, while *Cannabis* grown for marijuana can contain anywhere from 6 or 7 per cent to 20 per cent or even more. Industrial Hemp is produced in many countries around the world. Major producers include Canada, France, and China. While more hemp is exported to the United States than to any other country, the United States Government does not consistently distinguish between marijuana and the non-psychoactive *Cannabis* used for industrial and commercial purposes.

Major Uses

Hemp is used for a wide variety of purposes, including the manufacture of cordage of varying tensile strength, clothing, and nutritional products. The *bast fibers* can be used in 100 per cent hemp products, but are commonly blended with other organic fibers such as flax, cotton or silk, for apparel and furnishings, most commonly at a 55 per cent/45 per cent hemp/cotton blend. The inner two fibers of hemp are more woody, and are more often used in non-woven items and other industrial applications, such as mulch, animal bedding and litter. The oil from the fruits ("seeds") dries on exposure to air (similar to linseed oil) and is sometimes used in the manufacture of oil-based paints, in creams as a moisturising agent, for cooking, and in plastics. Hemp seeds have been used in bird seed mix.

Uses as Material

In Europe and China, hemp fibers are increasingly used to strengthen concrete, and in other composite materials for many construction and manufacturing applications. Hempcrete is used as a construction material containing hemp hurds, especially in France. Mercedes-Benz uses a "biocomposite" composed principally of hemp fiber for the manufacture of interior panels in some of its automobiles.

Uses as Food

Hemp seeds contain all the essential amino acids and essential fatty acids necessary to maintain healthy human life. The seeds can be eaten raw, ground into a meal, sprouted, made into hemp milk (akin to soy milk), prepared as tea, and used in baking. The fresh leaves can also be eaten in salads. Products range from cereals to frozen waffles, hemp tofu to nut butters. A few companies produce value added hemp seed items that include the seed oils, whole hemp grain (which is sterilized as per international law), hulled hemp seed (the whole seed without the mineral rich outer shell), hemp flour, hemp cake (a by-product of pressing the seed for oil) and hemp protein powder. Hemp is also used in some organic cereals, for non-dairy milk somewhat similar to soy and nut milks, and for non-dairy hemp "ice cream." Within the UK, the Department for Environment, Food and Rural Affairs (Defra) has treated hemp as purely a non-food crop. Seed appears on the UK market as a legal food product, and cultivation licenses are available for this purpose. In North America, hemp seed food products are sold in large volumes, particularly from Canada to the USA, and typically in health food stores or through mail order.

Applied Nutritional Value

About 30-35 per cent of the weight of hempseed is hemp oil, an edible oil that contains about 80 per cent as essential fatty acids (EFAs); i.e., linoleic acid, omega-6 (LA, 55%), alpha-linolenic acid, omega-3 (ALA, 22%), in addition to gamma-linolenic acid, omega-6 (GLA, 1-4%) and stearidonic acid, omega-3 (SDA, 0-2%). Hempseed also contains about 20 per cent of a highly-

Table 5.3: Typical Nutritional Analysis of Hemp Nut (Hulled Hemp Seeds)

Calories/100 g	567
Protein	30.6
Carbohydrate	10.9
Dietary fiber	6
Fat	47.2
Saturated fat	5.2
Palmitic 16:0	3.4
Stearic 18:0	1.5
Monounsaturated fat	5.8
Oleic 18:1 (Omega-9)	5.8
Polyunsaturated fat	36.2
Linoleic 18:2 (Omega-6)	27.6
Linolenic 18:3 (Omega-3)	8.7
Linolenic 18:3 (Omega-6)	0.8
Cholesterol	0 mg
Moisture	5
Ash	6.6
Vitamin A (B-Carotene)	4 IU
Thiamine (Vit B1)	1 mg
Riboflavin (Vit B2)	1 mg
Vitamin B6	0 mg
Niacin (Vit B3)	0 mg
Vitamin C	1.0 mg
Vitamin D	0 IU
Vitamin E	9 IU
Sodium	9 mg
Calcium	74 mg
Iron	4.7 mg

digestible protein, where 1/3 is edestin and 2/3 are albumins. Its amino acid profile is close to "complete" when compared to more common sources of proteins such as meat, milk, eggs and soy. The proportions of linoleic acid and alpha-linolenic acid in one tablespoon (15 ml) per day of hempseed oil easily provides human daily requirements for EFAs. Unlike flaxseed oil, hempseed oil can be used continuously without developing a deficiency or other imbalance of EFAs. This has been demonstrated in a clinical study, where the daily ingestion of flaxseed oil decreased the endogenous production of GLA. Hempseed is an adequate source of calcium and iron. Whole, toasted hempseeds are also a good source of phosphorus, magnesium, zinc, copper and manganese. Hempseed contains no gluten and therefore would not trigger symptoms of celiac disease.

Methods of Storage

Hempseed oil is a highly unsaturated oil. It can spontaneously oxidize and turn rancid within a short period of time if not stored properly. Hempseed oil is best stored in a dark glass bottle, in a refrigerator or freezer (its freezing point is -20C) Preservatives (antioxidants) are not necessary for high quality oils that are stored properly. Highly unsaturated oils are not suitable for frying.

Use as Dietary Supplement

Hempseed oil has been shown to relieve the symptoms of eczema (atopic dermatitis).

Use as Medicine

Hemp seed oil has anti-inflammatory properties.

Use as Fiber

The fiber is one of the most valuable parts of the hemp plant. It is commonly called bast, which refers to the fibers that grow on the outside of the woody interior of the plant's stalk, and under the outer most part (the bark). Bast fibers give the plants strength. Hemp fibers can be 0.91 m (3 ft) to 4.6 m (15 ft) long, running the length of the plant. Depending on the processing used to remove the fiber from the stem, the hemp may naturally be creamy white, brown, gray, black or green. The use of hemp for fiber production has declined sharply over the last two centuries, but before the industrial revolution, hemp was a popular fiber because it is strong and grows quickly; it produces 250 per cent more fiber than cotton and 600 per cent more fiber than flax when grown on the same land. Hemp has been used to make paper. It was used to make canvas, and the word *canvas* derives from *cannabis*. Hemp was very popular as it had many uses. Abaca or Manila replaced its use for rope. Burlap, made from jute, took over the sacking market. The paper industry began using wood pulp. The carpet industry switched over to wool, sisal, and jute, then nylon. Netting and webbing applications were taken over by cotton and synthetics. The world hemp paper pulp production was believed to be around 120,000 tons per year in 1991 which was about 0.05 per cent of the world's annual pulp production volume. In 1916, US Department of Agriculture chief scientists Lyster H. Dewe, and Jason L. Merrill created paper made from hemp pulp, which they concluded was "favourable in comparison with those used with pulp wood." Jack Herer later summarized the findings of the bulletin in his book "The Emperor Wears No Clothes." Herer wrote:

In 1916, USDA Bulletin No. 404, reported that one acre of cannabis hemp, in annual rotation over a 20-year period, would produce as much pulp for paper as 4.1 acres (17,000 m^2) of trees being cut down over the same 20-year period. This process would use only 1/4 to 1/7 as much polluting sulfur-based acid chemicals to break down the glue-like lignin that binds the fibers of the pulp, or none at all using soda ash. The problem of dioxin contamination of rivers is avoided in the hemp paper making process, which does not need to use chlorine bleach (as the wood pulp paper making process requires) but instead safely substitutes hydrogen peroxide in the bleaching process.... If the new (1916) hemp pulp paper process were legal today, it would soon replace about 70 per cent of all wood pulp paper, including computer printout paper, corrugated boxes and paper bags.

New technology has allowed for more environmentally-friendly paper production from wood pulp, though the production of wood pulp paper still claims the one of the highest CO$_2$ emissions by industry (second only to concrete production). The recovery boiler was invented in the early 1930s. The first recovery boilers were commissioned to wood-pulp mills during the mid-1930s, ECF (Elemental Chlorine Free), or TCF (Total chlorine Free) bleaching, better fiber filters etc. has created less of a demand for alternative raw materials. Hemp is currently of little significance as raw material for paper; however, it is scarcely grown in the developed world. The long-term price for pulpwood has been low compared with any alternative except recycled paper. The decision of the United States Congress to pass the 1937 Marijuana Tax Act was based in part on testimony derived from articles in newspapers owned by William Randolph Hearst, who, some authors stress, had significant financial interests in the forest industry, which manufactured his newsprint. The background material also included that from 1880 to 1933 the hemp grown in the United States had declined from 15,000 to 1,200 acres (4.9 km^2), and that the price of line hemp had dropped from $12.50 per pound in 1914 to $9.00 per pound in 1933. In 1935, however, hemp would also make a significant rebound. Hearst began a campaign against hemp, and published stories in his newspapers associating hemp with marijuana and attacking marijuana usage. As a result of the act, the production and use of hemp was discontinued. Characteristics of hemp fibre are its superior strength and durability, resistance to ultraviolet light and mold, comfort and good absorbency (8%). The original Levi Strauss jeans were made from lightweight hemp canvas. Hemp rope is notorious for breaking due to rot as the capillary effect of the rope-woven fibres tended to hold liquid at the interior, while seeming dry from the outside. Hemp rope used in the age of sailing-ships was protected by tarring, a labour-intensive process (and source of the Jack Tar nickname for sailors). Hemp rope was phased out when Manila, which does not require tarring, became available. There is a niche market for hemp paper, but the cost of hemp pulp is approximately six times that of wood pulp, mostly due to the small size and outdated equipment of the few hemp processing plants in the Western world. Hemp pulp is bleached with hydrogen peroxide, which can also be used for wood pulp. Kenaf is another fast-growing plant which can be used as a replacement for wood pulp. Kenaf paper has been produced in commercial quantities. A modest hemp industry exists. Recent developments in processing have made it possible to soften coarse fibers to a wearable level. The Hempest is an example of a clothing store that specializes in hemp clothing.

Uses in Water Purification

Hemp can be used as a "mop crop" to clear impurities out of wastewater, such as sewage effluent, excessive phosphorus from chicken litter, or other unwanted substances or chemicals. Eco-technologist Dr. Keith Bolton from Southern Cross University in Lismore, New South Wales, Australia, is a leading researcher in this area. Hemp is being used to clean contaminants at Chernobyl nuclear disaster site.

Uses in Weed Control

Hemp, because of its height, dense foliage and its high planting density as a crop, is a very effective and long used method of killing tough weeds in farming. Using hemp this way can help farmers avoid the use of herbicides, to help gain organic certification and to gain the benefits of crop rotation *per se.*

Uses as Fuel

Biofuels such as biodiesel and alcohol fuel can be made from the oils in hemp seeds and stalks, and the fermentation of the plant as a whole, respectively. Filtered hemp oil can be used directly to power diesels. The inventor of the diesel engine, Rudolf Diesel, used hemp oil to power his new engine, and assumed that it would be run in the future on hemp oil. Henry Ford grew industrial hemp on his estate after 1937, possibly to prove the cheapness of methanol production at Iron Mountain. He made plastic cars with wheat straw, hemp and sisal. (Popular Mechanics, Dec. 1941, "Pinch Hitters for Defense.") In 1892, Rudolph Diesel invented the diesel engine, which he intended to fuel "by a variety of fuels, especially vegetable and seed oils."

Pattern of Cultivation

Millennia of selective breeding have resulted in varieties that look quite different. Also, breeding since circa 1930 has focused quite specifically on producing strains which would perform very poorly as sources of drug material. Hemp grown for fibre is planted closely, resulting in tall, slender plants with long fibres. Ideally, according to Britain's Department for Environment, Food and Rural Affairs, the herb should be harvested before it flowers. This early cropping is done because fibre quality declines if flowering is allowed and, incidentally, this cropping also pre-empts the herb's maturity as a potential source of drug material. However, in these strains of industrial hemp the tetrahydrocannabinol (THC) content would have been very low regardless. The name *Cannabis* is the genus and was the name favoured by the 19th century medical practitioners who helped to introduce the herb's drug potential to modern English-speaking consciousness. *Cannabis* for non-drug purposes (especially ropes and textiles) was then already well known as hemp. The name "marijuana" is Spanish in origin and associated almost exclusively with the herb's drug potential.

Details of Historical cultivation

While the fibre has been grown for millennia in Asia and the Middle East, commercial production of hemp in the West took off in the eighteenth century. Due to colonial and naval expansion of the era, economies needed large quantities of hemp for rope and oakum. The endless European Wars, and ever expanding naval fleets, all used the material. To this end, the young United States of America became a large hemp producer. The Gulf and Carolina states had very large hemp industries. In fact the market was second only to cotton fibre. Machinery was invented in the United States for producing hemp fibre. An unpleasant task performed by prison labour was the manufacture of rope and boat caulking. Before the age of nylon rope, hemp rope had a short lifetime and was ever in need of replacement. In the 19th century it was cultivated by binders. The soils most suited to the culture of this plant are those of the deep, black, putrid vegetable kind, that are low, and rather inclined to moisture, and those of the deep mellow, loamy, or sandy descriptions. The quantity of produce is generally much greater on the former than on the latter; but it is said to be greatly inferior in quality. It may, however, be grown with success on lands of a less rich and fertile kind by proper care and attention in their culture and preparation. In order to render the grounds proper for the reception of the crop, they should be reduced into a fine mellow state of mould, and be perfectly cleared from weeds, by repeated ploughings. When it succeeds grain crops, the work is mostly accomplished by three ploughings, and as many harrowings: the first being given immediately after the preceding crop is removed, the second early in the spring, and the last, or seed earth, just before the seed is to be put in. In the last ploughing, well rotted manure, in the proportion of fifteen or twenty, or good compost, in the quantity of twenty-five or thirty-three horse-cart loads, should be turned into the land; as without this it is seldom that good crops can be produced. The surface of the ground being left perfectly flat, and as free from furrows as possible; as by these means the moisture is more effectually retained, and the growth of the plants more fully promoted. It is of much importance in the cultivation of hemp crops that the seed be new, and of a good quality, which may in some measure be known by its feeling heavy in the hand, and being of a bright shining colour. The proportion of seed that is most commonly employed, is from two to three bushels [per acre], according to the quality of the land; but, as the crops are greatly injured by the plants standing too closely together, two bushels, or two bushels and a half may be a more advantageous quantity. As the hemp plant is extremely tender in its early growth, care should be taken not to put the seed into the ground at so early a period, as that it may be liable to be injured by the effects of frost; nor to protract the sowing to so late a season as that the quality of the produce may be effected. The best season, on the drier sorts of land in the southern districts, is as soon as possible after the frosts are over in April; and, on the same descriptions of soil, in the more northern ones, towards the close of the same month or early in the ensuing one. The most general method of putting crops of this sort into the soil is the broadcast, the seed being dispersed over the surface of the land in as even a manner as possible, and afterwards covered in by means of a very light harrowing. In many cases, however, especially when the crops are to stand for seed, the drill method in rows, at small distances, might be had recourse to with advantage; as, in this way, the early growth of the plants would be more effectually promoted, and the land be kept in a more clean and perfect state of mould, which are circumstances of importance in such crops. In whatever method the seed is put in, care must constantly be taken to keep the birds from it for some time afterwards. This sort of crop is frequently cultivated on the same piece of ground for a great number of years, without any other kind intervening; but, in such cases, manure must be applied with almost every crop, in pretty large proportions, to prevent the exhaustion that must otherwise take place. It may be sown after most sorts of grain crops, especially where the land possesses sufficient fertility, and is in a proper state of tillage. As hemp, from its tall growth and thick foliage, soon covers the surface of the land, and prevents the rising of weeds, little attention is necessary after the seed has been put into the ground, especially where the broadcast method of sowing is practised; but, when put in by the drill machine, a hoeing or two may be had recourse to with advantage in the early growth of the crop. In the culture of this plant, it is particularly necessary that the same piece of land grows both male and female, or what is sometimes denominated simple hemp. The latter kind contains the seed. When the grain is ripe (which is known by its becoming of a whitish-yellow colour, and a few of the leaves beginning to drop from the stems); this happens commonly about thirteen or fourteen weeks from the period of its being sown, according as the season may be dry or wet (the first sort being mostly ripe some weeks before the latter), the next operation is that of taking it from the ground; which is effected by pulling it up by the roots, in small parcels at a time, by the hand, taking care to shake off the mould well from them before the handsful are laid down. In some districts, the whole crop is pulled together, without any distinction being made between the different kinds of hemp; while, in others, it is the practice to separate and pull them at different times, according to their ripeness. The latter is obviously the better practice; as by pulling a large proportion of the crop before it is in a proper state of maturity, the quantity of produce must not only be considerably lessened, but its quality greatly injured by being rendered less durable. After being thus pulled, it is tied up in small parcels, or what are sometimes termed baits. Where crops of this kind are intended for seeding, they should be suffered to stand till the seed becomes in a perfect state of maturity, which is easily known by the appearance of it on inspection. The stems are then pulled and bound up, as in the other case, the bundles being set up in the same manner as grain, until the seed becomes so dry and firm as to shed freely. It is then either immediately threshed out upon large cloths for the purpose in the field, or taken home to have the operation afterwards performed. The hemp, as soon as pulled, is tied up in small bundles, frequently at both ends. It is then conveyed to pits, or ponds of stagnant water, about six or eight feet in depth, such as have a clayey soil being in general preferred, and deposited in beds, according to their size, and depth, the small bundles being laid both in a straight direction and crosswise of each other, so as to bind perfectly together; the whole, being loaded with timber, or other materials, so as to keep the beds of hemp just below the surface of the water. It is not usual to water more than four or five times in the same pit, until it has been filled with water. Where the ponds are not sufficiently large to contain the whole of the produce at once, it is the practice to pull the hemp only as it can be admitted into them, it being thought disadvantageous to leave the hemp upon the ground after being pulled. It is left in these pits four, five, or six days, or even more, according to the warmth of the season and the judgment of the operator, on his examining whether the hempy material readily separates from the reed or stem; and then taken up and conveyed to a pasture field which is clean and even, the bundles being loosened and spread out thinly, stem by stem, turning it every second or third day, especially in damp weather, to prevent its being injured by worms or other insects. It should remain in this situation for two, three, four, or more weeks, according to circumstances, and be then collected together when in a perfectly dry state, tied up into large bundles, and placed in some secure building until an opportunity is afforded for breaking it, in order to separate the hemp. By this means the process of grassing is not only shortened, but the more expensive ones of breaking, scutching, and bleaching the yarn, rendered less violent and troublesome. After the hemp has been removed from the field it is in a state to be broken and swingled, operations that are mostly performed by common labourers, by means of machinery for the purpose, the produce being tied up in stones. The refuse collected in the latter process is denominated sheaves, and is in some districts employed for the purposes of fuel. After having undergone these different operations, it is ready for the purposes of the manufacturer.

Pattern of Harvesting

Smallholder plots are usually harvested by hand. The plants are cut at 2 to 3 cm above the soil and left on the ground to dry. Mechanical harvesting is now common, using specially adapted cutter-binders or simpler cutters. The cut hemp is laid in swathes to dry for up to four days. This was traditionally followed by *retting*, either water retting (the bundled hemp floats in water) or dew retting (the hemp remains on the ground and is affected by the moisture in dew moisture, and by molds and bacterial action). Modern processes use steam and machinery to separate the fibre, a process known as thermo-mechanical pulping.

Listening Varieties

There are broadly three groups of *Cannabis* varieties being cultivated today:

- Varieties primarily cultivated for their fibre, characterized by long stems and little branching, extreme red, yellow, blue or purple colouration, or thickness of stem and solid core, such as hemp cannabis oglalas, and more generally called industrial hemp.
- Varieties grown for seed from which hemp oil is extracted or which can be dehulled.
- Varieties grown for medicinal, spiritual development or recreational purposes. A nominal if not legal distinction is often made between hemp, with concentrations of the psychoactive chemical THC far too low to be useful as a drug, and *Cannabis* used for medical, recreational, or spiritual purposes.

Major Diseases

Hemp plants can be vulnerable to various pathogens including bacteria, fungi, nematodes, viruses and other miscellaneous pathogens. Such diseases often lead to problems such as reduced fiber quality, stunted growth, and (eventually) death of the plant.

Time Frame

Hemp use dates back to the Stone Age, with hemp fibre imprints found in pottery shards in China and Taiwan over 10,000 years old. These ancient Asians also used the same fibres to make clothes, shoes, ropes, and an early form of paper. Contrary to the traditional view that Cai Lun invented paper in around 105 AD, specimens of hemp paper were found in the Great Wall of

China dating back 200 years earlier. Hemp cloth was more common than linen until the mid 14th century. The use of hemp as a cloth was centered largely in the countryside, with higher quality textiles being available in the towns. Virtually every small town had access to a hemp field. In late medieval Germany and Italy, hemp was employed in cooked dishes, as filing in pies and tortes, or boiled in a soup. The traditional *European hemp* was by tradition and due to its low narcotic effect not used as a drug in Europe. It was cultivated for its fibers and for example used by Christopher Columbus for ropes on his ships. The Spaniards first brought hemp to the Western Hemisphere and cultivated it in Chile starting about 1545. However, in May 1607, "hempe" (*sic*) was among the crops Gabriel Archer observed being cultivated by the natives at the main Powhatan village, where Richmond, Virginia is now situated; and in 1613, Samuell Argall reported wild hemp "better than that in England" growing along the shores of the upper Potomac. As early as 1619, the first Virginia House of Burgesses passed an Act requiring all planters in Virginia to sow "both English and Indian" hemp on their plantations. The Puritans are first known to have cultivated hemp in New England in 1645. George Washington and Thomas Jefferson both cultivated hemp on their farms. Benjamin Franklin started the first American paper mill, which made paper exclusively from hemp. In the Napoleonic era, many military uniforms were made of hemp. While hemp linens were coarser than those made of flax, the added strength and durability of hemp, as well as the lower cost, meant that hemp uniforms were preferred. Hemp was used extensively by the United States during WWII. Uniforms, canvas, and rope were among the main textiles created from the hemp plant at this time. Much of the hemp used was planted in the Midwest and Kentucky. Historically, hemp production made up a significant portion of Kentucky's economy and many slave plantations located there focused on producing hemp. By the early twentieth century, the advent of the steam engine and the diesel engine ended the reign of the sailing ship. The advent of iron and steel for cable and ships' hulls further eliminated natural fibers in marine use, although hemp had long since fallen out of favour in the sailing industry in preference to Manila hemp. The invention of artificial fibers in the late thirties by DuPont further put strain on the market.

Producing Hemp Countries

Every industrialized country in the world, excluding the United States, produces industrial hemp including Australia, Austria, Canada, China, Great Britain, France, Russia and Spain. From the 1950s to the 1980s the Soviet Union was the world's largest producer (3,000 km2 in 1970). The main production areas were in Ukraine, the Kursk and Orel regions of Russia, and near the Polish border. Since its inception in 1931, the Hemp Breeding Department at the Institute of Bast Crops. Other important producing countries were China, North Korea, Hungary, the former Yugoslavia, Romania, Poland, France and Italy. In Japan, hemp was historically used as paper and a fiber crop; it was restricted as a narcotic drug in 1948. The ban on marijuana imposed by the US authorities was alien to Japanese culture, as the drug had never been widely used in Japan before. There is archaeological evidence that cannabis was used for clothing and the seeds were eaten in Japan right back to the Jomon period (10,000 to 300 BCE). Many Kimono designs portray hemp, or "Asa" (Japanese: *–f*), as a beautiful plant. France is Europe's biggest producer, with 8,000 hectares cultivated. Canada (9,725 ha in 2004), the United Kingdom, and Germany all resumed commercial production in the 1990s. British production is mostly used as bedding for horses; other uses are under development. The largest outlet for German fibre is composite automotive panels. Companies in Canada, UK, US and Germany among many others process hemp seed into a growing range of food products and cosmetics; many traditional growing countries still continue to produce textile grade fibre. Hemp is illegal to grow in the U.S. under federal law due to its relation to marijuana. Some states have defied federal law and made the cultivation of industrial hemp legal. These states, North Dakota, Hawaii, Kentucky, Maine, Maryland, Montana and West Virginia, have not yet begun to grow hemp due to resistance from the federal Drug Enforcement Administration.

Patterns of Industrial Growth under Licence

Licenses for hemp cultivation are issued in the European Union, Canada, in three states of Australia, and two of the United States. In the United Kingdom, these licenses are issued by the Home Office under the Misuse of Drugs Act 1971. When grown for non-drug purposes hemp is referred to as industrial hemp, and a common product is fiber for use in a wide variety of products, as well as the seed for nutritional aspects as well as for the oil. Feral hemp or ditch weed is usually a naturalized fiber or oilseed strain of Cannabis that has escaped from cultivation and is self-seeding. In Australia the states of Victoria, Queensland and most recently New South Wales issue licences to grow hemp for industrial use. Victoria was an early adopter in 1998, and has reissued the regulation in 2008. Queensland has allowed industrial production under licence since 2002 where the issuance is controlled under the Drugs Misuse Act 1986. Most recently New South Wales now issues licences under a law that came into effect as at the 6th of November 2008, the Hemp Industry Regulations Act 2008 (No 58). Vermont and North Dakota have passed laws enabling hemp licensure. Both states are waiting for permission to grow hemp from the Drug Enforcement Agency (DEA). Currently, North Dakota representatives are pursuing legal measures to force DEA approval.

Evaluating Socioeconomic and Legal Future of Hemp

Hemp is becoming increasingly popular as a health food in America and abroad. Canadian hemp growth and exports have also been increasing, though growth is focused on seed and oil, not fiber production. However, prices remain too high for any significant consumer use. The increased demand for health food has stimulated the trade of shelled hemp seed, hemp protein powder and hemp oil as well as finished and ready-to-eat food products including waffles, granola bars, ice cream, and hemp milk using these derivatives as ingredients. Hemp oil has also been used in the manufacture of body care products. As a natural fiber, hemp is involved in the International Year of Natural Fibres, 2009. Hemp is also receiving attention as a possible fiber for production of clothes. New uses for hemp are receiving greater media attention as environmental awareness becomes more prevalent and environmentally-friendly alternatives to conventional materials. With help from Vote Hemp, laws relating to the growth of industrial hemp are being passed around the at least twelve US states including Kentucky, Vermont and North Dakota. North Dakota has at least five laws relating to the growth of industrial hemp. Under one new North Dakota law, farmers no longer need permission from the DEA to grow industrial hemp, which now is distinguished from "marijuana" by THC content, though legal action to establish the feasibility of farmers following the North Dakota law is ongoing. Hemp Plastic is a new technology based on 20-100 per cent hemp fiber-based plastics that can be molded or injection molded. Demand for fiber-reinforced composites and other natural plastics could become more popular as oil prices rise and environmental awareness increases.

Chemical Called THC in Hemp

Hemp contains *delta-9-tetrahydrocannabinol* (THC), which is the psychoactive ingredient found in hashish and marijuana. While THC is present in all Cannabis plant varieties to some extent, industrial hemp does not contain an amount to produce any intoxicating effect, even in significant quantities. In varieties grown for use as a drug, where males are removed in order to prevent fertilization, THC levels can reach as high as 24 per cent in the unfertilized females which are given ample room to flower.

In hemp varieties grown for seed or fibre use, the plants are grown very closely together and a very dense biomass product is obtained, rich in oil from the seeds and fibre from the stalks and low in THC content. EU and Canadian regulations limit THC content to 0.3 per cent in industrial hemp. On October 9, 2001, the US Drug Enforcement Administration (DEA) ruled that even traces of THC in products intended for food use would be illegal as of February 6, 2002. This *Interpretive Rule* would have ruled out the production or use of hempseed or hempseed oil in food use in the USA, but after the Hemp Industries Association (HIA) filed suit the rule was stayed by the United States Court of Appeals for the Ninth Circuit on March 7, 2002. On March 21, 2003, the DEA issued a nearly identical *Final Rule* which was also stayed by the Ninth Circuit Court of Appeals on April 16, 2003. On February 6, 2004, the Ninth Circuit Court of Appeals issued a unanimous decision in favour of the HIA in which Judge Betty Fletcher wrote, "[T]hey (DEA) cannot regulate naturally-occurring THC not contained within or derived from marijuana—i.e. non-psychoactive hemp is not included in Schedule I. The DEA has no authority to regulate drugs that are not scheduled, and it has not followed procedures required to schedule a substance. The DEA's definition of "THC" contravenes the unambiguously expressed intent of Congress in the Controlled Substances Act (CSA) and cannot be upheld." On September 28, 2004, the HIA claimed victory after DEA declined to appeal to the Supreme Court of the United States the ruling from the Ninth Circuit Court of Appeals protecting the sale of hemp-containing foods. Industrial hemp remains legal for import and sale in the U.S., but U.S. farmers still are not permitted to grow it. Strong opposition to trace amounts of THC, a chemical shown by scientific research to be less harmful and less addictive than nicotine or alcohol, leads some of its critics, like Jack Herer in The Emperor Wears No Clothes, to charge ulterior motives such as protection of the synthetic-fibre, wood pulp, petrochemical, and pharmochemical industries. The US government's position has not been completely constant, as shown by the wide-spread cultivation of industrial hemp in Kentucky and Wisconsin during World War II. Critics of the HIA, however, argue that the necessities of the war and the unavailability of adequate synthetic substitutes outweighed the social, health, and public safety risks of producing hemp. The presence of THC in hemp varieties and the fear that fields with hemp can hide cultivation of cannabis with more THC has hampered the development of hemp in many countries, most notably, the United States. Regulations in certain countries in EU demand approved variety of the seed and registration of the field in advance every year. Marijuana is often female only, and kept completely isolated from any males, to keep the THC production up and seed production low. There are specially developed strains that require a very specific growing operation, and there is much care put into increasing THC production. Hiding marijuana in a hemp field would create a variety of problems. One is, the dense hemp would most likely "choke out" the marijuana, taking valuable and necessary nutrients and sunlight that the marijuana needs to produce THC. Even more, the male hemp plants would fertilize the marijuana plants, which would have several side effects. First, the marijuana would produce seeds, quickly lowering its value. Energy growing seeds quickly diminishes THC content. More importantly, the fertilization essentially crossbreeds the hemp and marijuana (only in the THC potent females). While the hemp will not produce any more THC, the marijuana, once "tainted" by the Hemp, will produce significantly less THC, depending on how long and how close the contact is with the Hemp. If marijuana was successfully hidden and grown in a hemp field, the resulting plant matter would be of very little street value. It would be full of seeds and stems (because of fertilization), be malnourished (because hemp, like a strong weed, sucks up nutrients and grows taller, taking the available sun), and above all, have a very low THC content, making it undesirable to even the indiscriminate marijuana users. There is a consensus between experts and marijuana growers alike; the risk to reward ratio is too far out of proportion for it to even be considered.

Maize

Maize, known as corn in some countries, is a cereal grain domesticated in Mesoamerica and subsequently spread throughout the American continents. After European contact with the Americas in the late 15th and early 16th century, maize spread to the rest of the world. Maize is the most widely grown crop in the Americas (332 million tonnes annually in the United States alone). Hybrid maize, due to its high grain yield as a result of heterosis ("hybrid vigor"), is preferred by farmers over conventional varieties. While some maize varieties grow up to 7 metres (23 ft) tall, most commercially grown maize has been bred for a standardized height of 2.5 metres (8 ft). Sweet corn is usually shorter than field-corn varieties.

The term *maize* derives from the Spanish form (*maíz*) of the indigenous Taino term for the plant, and was the form most commonly heard in the United Kingdom. In the United States, Canada (maïs in French speaking Canadian regions) and Australia, the usual term is *corn*, which originally referred to any grain, but which now refers exclusively to maize, having been shortened from the form "Indian corn" (which currently, at least in the U.S. and Canada, is often used to refer specifically to multi-coloured "field corn" cultivars).

Understanding Physiology

Maize stems superficially resemble bamboo canes and the internodes can reach 20-30 centimetres (8-12 in). Maize has a very distinct growth form; the lower leaves being like broad flags, 50-100 centimetres long and 5-10 centimetres wide (2-4 ft by 2-4 in); the stems are erect, conventionally 2-3 metres (7-10 ft) in height, with many nodes, casting off flag-leaves at every node. Under these leaves and close to the stem grow the ears. They grow about 3 milimetres a day. The ears are female inflorescences, tightly covered over by several layers of leaves, and so closed-in by them to the stem that they do not show themselves easily until the emergence of the pale yellow silks from the leaf whorl at the end of the ear. The silks are elongated stigmas that look like tufts of hair, at first green, and later red or yellow. Plantings for silage are even denser, and achieve an even lower percentage of ears and more plant matter. Certain varieties of maize have been bred to produce many additional developed ears, and these are the source of the "baby corn" that is used as a vegetable in Asian cuisine. Maize is a facultative long-night plant and flowers in a certain number of growing degree days > 50 °F (10°C) in the environment to which it is adapted. The magnitude of the influence that long nights have on the number of days that must pass before maize flowers is genetically prescribed and regulated by the phytochrome system. Photoperiodicity can be eccentric in tropical cultivars, while the long days characteristic of higher latitudes allow the plants to grow so tall that they do not have enough time to produce seed before being killed by frost. These attributes, however, may prove useful in using tropical maize for biofuels. The apex of the stem ends in the tassel, an inflorescence of male flowers. Each silk may become pollinated to produce one kernel of corn. Young ears can be consumed raw, with the cob and silk, but as the plant matures (usually during the summer months) the cob becomes tougher and the silk dries to inedibility. By the end of the growing season, the kernels dry out and become difficult to chew without cooking them tender first in boiling water. Modern farming techniques in developed countries usually rely on dense planting, which produces on average only about 0.9 ears per stalk because it stresses the plants.

Properties of Seeds

The kernel of corn has a pericarp of the fruit fused with the seed coat, typical of the grasses. It is close to a multiple fruit in structure, except that the individual fruits (the kernels) never fuse into a single mass. The grains are about the size of peas, and adhere in regular rows round a white pithy substance, which forms the ear. An ear contains from 200 to 400 kernels, and is from 10-25 centimetres (4-10 inches) in length. They are of various colours: blackish, bluish-gray, red, white and yellow. When ground into flour, maize yields more flour, with much less bran, than wheat does. However, it lacks the protein gluten of wheat and, therefore, makes baked goods with poor rising capability and coherence. A genetic variation that accumulates more sugar and less starch in the ear is consumed as a vegetable and is called sweet corn. Immature maize shoots accumulate a powerful antibiotic substance, DIMBOA (2,4-dihydroxy-7-methoxy-1,4-benzoxazin-3-one). DIMBOA is a member of a group of hydroxamic acids (also known as benzoxazinoids) that serve as a natural defense against a wide range of pests including insects, pathogenic fungi and bacteria. DIMBOA is also found in related grasses, particularly wheat. A maize mutant (bx) lacking DIMBOA is highly susceptible to be attacked by aphids and fungi. DIMBOA is also responsible for the relative resistance of immature maize to the European corn borer (family Crambidae). As maize matures, DIMBOA levels and resistance to the corn borer decline. Due to its shallow roots of only one to two inches deep, maize is susceptible to droughts, intolerant of nutrient-deficient soils, and prone to be uprooted by severe winds.

Major Allergy

Maize contains lipid transfer protein, an undigestable protein which survives cooking. This protein has been linked to a rare and understudied allergy to maize in humans. The allergic reaction can cause skin rash, swelling or itching of mucus membranes, diarrhoea, vomiting, asthma and, in severe cases, anaphylactic shock. It has been noted that those with corn allergy almost always have peach allergy as well. It is unclear how common this allergy is in the general populace.

Understanding Genetics

Many forms of maize are used for food, sometimes classified as various subspecies:

(*i*)   Flour corn—*Zea mays var. amylacea*
(*ii*)   Popcorn—*Zea mays var. everta*
(*iii*)   Dent corn—*Zea mays var. indentata*
(*iv*)   Flint corn—*Zea mays var. indurata*
(*v*)   Sweet corn—*Zea mays var. saccharata* and *Zea mays var. rugosa*
(*vi*)   Waxy corn—*Zea mays var. ceratina*
(*vii*)   Amylomaize—*Zea mays*
(*viii*) Pod corn—*Zea mays var. tunicata* Larrañaga ex A. St. Hil.
(*ix*)   Striped maize—*Zea mays var. japonica*

This system has been replaced (though not entirely displaced) over the last 60 years by multi-variable classifications based on ever more data. Agronomic data were supplemented by botanical traits for a robust initial classification, then genetic, cytological, protein and DNA evidence was added. Now the categories are forms (little used), races, racial complexes, and recently branches. Maize has 10 chromosomes (n = 10). The combined length of the chromosomes is 1500 cM. Some of the maize chromosomes have what are known as "chromosomal knobs": highly repetitive heterochromatic domains that stain darkly. Individual knobs are polymorphic among strains of both maize and teosinte. Barbara McClintock used these knob markers to prove her transposon theory of "jumping genes", for which she won the 1983 Nobel Prize in Physiology or Medicine. Maize is still an important model organism for genetics and developmental biology today. There is a stock center of maize mutants, *The Maize Genetics Cooperation Stock Center*, funded by the USDA Agricultural Research Service and located in the Department of Crop Sciences at the University of Illinois at Urbana-Champaign. The total collection has nearly 80,000 samples. The bulk of the collection consists of several hundred named genes, plus additional gene combinations and other heritable variants. There are about 1,000 chromosomal aberrations (e.g., translocations and inversions) and stocks with abnormal chromosome numbers (e.g., tetraploids). Genetic data describing the maize mutant stocks as well as myriad other data about maize genetics can be accessed at MaizeGDB, the Maize Genetics and Genomics Database. In 2005, the U.S. National Science Foundation (NSF), Department of Agriculture (USDA) and the Department of Energy (DOE) formed a consortium to sequence the maize genome. The resulting DNA sequence data will be deposited immediately into GenBank, a public repository for genome-sequence data. Sequencing the corn genome has been considered difficult because of its large size and complex genetic arrangements. The genome has 50,000-60,000 genes scattered among the 2.5 billion bases—molecules that form DNA—that make up its 10 chromosomes. (By comparison, the human genome contains about 2.9 billion bases and 26,000 genes.) On February 26, 2008, researchers announced that they had sequenced the entire genome of maize.

Tracing Origin and Evolution

There are several theories about the specific origin of maize in Mesoamerica:

(i) It is a direct domestication of a Mexican annual teosinte, *Zea mays* ssp. *parviglumis*, native to the Balsas River valley of southern Mexico, with up to 12 per cent of its genetic material obtained from *Zea mays* ssp. *mexicana* through introgression.

(ii)   It derives from hybridization between a small domesticated maize (a slightly changed form of a wild maize) and a teosinte of section *Luxuriantes*, either *Z. luxurians* or *Z. diploperennis*.

(iii) It underwent two or more domestications either of a wild maize or of a teosinte.

(iv)   It evolved from a hybridization of *Z. diploperennis* by *Tripsacum dactyloides*. (The term "teosinte" describes all species and subspecies in the genus *Zea*, excluding *Zea mays* ssp. *mays*.) In the late 1930s, Paul Mangelsdorf suggested that domesticated maize was the result of a hybridization event between an unknown wild maize and a species of *Tripsacum*, a related genus. However, the proposed role of tripsacum (gama grass) in the origins of maize has been refuted by modern genetic testing, refuting Mangelsdorf's model and the fourth listed above.

The first model was proposed by Nobel Prize winner George Beadle in 1939. Though it has experimental support, it has not explained a number of problems, among them:

(i)   how the immense diversity of the species of sect. *Zea* originated,

(ii)   how the tiny archaeological specimens of 3500-2700 BC (uncorrected) could have been selected from a teosinte, and

(iii) how domestication could have proceeded without leaving remains of teosinte or maize with teosintoid traits until ca. 1100 BC.

The domestication of maize is of particular interest to researchers—archaeologists, geneticists, ethnobotanists, geographers, etc. The process is thought by some to have started 7,500 to 12,000 years ago (corrected for solar variations). Recent genetic evidence suggests that maize domestication occurred 9,000 years ago in central Mexico, perhaps in the highlands between Oaxaca and Jalisco. More recent evidence now points to the lowlands of the Balsas River valley. The crop wild relative teosinte most similar to modern maize grows in the area of the Balsas River. Archaeological remains of early maize ears, found at Guila Naquitz Cave in the Oaxaca Valley, date back roughly 6,250 years (corrected; 3450 BC, uncorrected); the oldest ears from caves near Tehuacan, Puebla, date ca. 2750 BC. Little change occurred in ear form until ca. 1100 BC when great changes appeared in ears from Mexican caves: maize diversity rapidly increased and archaeological teosinte was first deposited. Perhaps as early as 1500 BC, maize began to spread widely and rapidly. As it was introduced to new cultures, new uses were developed and new varieties selected to better serve in those preparations. Maize was the staple food, or a major staple, of most the pre-Columbian North American, Mesoamerican, South American, and Caribbean cultures. The Mesoamerican civilization was strengthened upon the field crop of maize; through harvesting it, its religious and spiritual importance and how it impacted their diet. Maize formed the Mesoamerican people's identity. During the 1st millennium AD, maize cultivation spread from Mexico into the U.S. Southwest and a millennium later into U.S. Northeast and southeastern Canada, transforming the landscape as Native Americans cleared large forest and grassland areas for the new crop. It is unknown what precipitated its domestication, because the edible portion of the wild variety is too small and hard to obtain to be eaten directly, as each kernel is enclosed in a very hard bi-valve shell. However, George Beadle demonstrated that the kernels of teosinte are readily "popped" for human consumption, like modern popcorn. Some have argued that it would have taken too many generations of selective breeding in order to produce large compressed ears for efficient cultivation. However, studies of the hybrids readily made by intercrossing teosinte and modern maize suggest that this objection is not well founded. In 2005, research by the USDA Forest Service indicated that the rise in maize cultivation 500 to 1,000 years ago in what is now the southeastern United States contributed to the decline of freshwater mussels, which are very sensitive to environmental changes.

Major Production Quantities and Methods

Maize is widely cultivated throughout the world, and a greater weight of maize is produced each year than any other grain. While the United States produces almost half of the world's harvest (~42.5%), other top producing countries include China, Brazil, Mexico, Argentina, India and France. Worldwide production was around 800 million tonnes in 2007—just slightly more than rice (~650 million tonnes) or wheat (~600 million tonnes). In 2007, over 150 million hectares of maize were planted worldwide, with a yield of 4970.9 kilogram/hectare. Because it is cold-intolerant, in the temperate zones maize must be planted in the spring. Its root system is generally shallow, so the plant is dependent on soil moisture. As a C4 plant (a plant that uses C4 carbon fixation), maize is a considerably more water-efficient crop than C3 plants (plants that use C3 carbon fixation) like the small grains, alfalfa and soybeans. Maize is most sensitive to drought at the time of silk emergence, when the flowers are ready for pollination. In the United States, a good harvest was traditionally predicted if the corn was "knee-high by the Fourth of July," although modern hybrids generally exceed this growth rate. Maize used for silage is harvested while the plant is green and the fruit immature. Sweet corn is harvested in the "milk stage," after pollination but before starch has formed, between

Table 5.4: Top Ten Maize Producers in 2007

Country	Production (Tonnes)
United States	332,092,180
People's Republic of China	151,970,000
Brazil	51,589,721
Mexico	22,500,000
Argentina	21,755,364
India	16,780,000
France	13,107,000
Indonesia	12,381,561
Canada	10,554,500
Italy	9,891,362
World	784,786,580

Source: Food and Agricultural Organization of United Nations: Economic and Social Department: The Statistical Division.

late summer and early to mid-autumn. Field corn is left in the field very late in the autumn in order to thoroughly dry the grain, and may, in fact, sometimes not be harvested until winter or even early spring. The importance of sufficient soil moisture is shown in many parts of Africa, where periodic drought regularly causes famine by causing maize crop failure.

Maize was planted by the Native Americans in hills, in a complex system known to some as the Three Sisters: beans used the corn plant for support and in turn provided nitrogen from nitrogen-fixing bacteria which live on the roots of beans and other legumes; and squashes provided ground cover to stop weeds and inhibit evaporation by providing shade over the soil. This method was replaced by single species hill planting where each hill 60-120 cm (2-4 ft) apart was planted with 3 or 4 seeds, a method still used by home gardeners. A later technique was *checked corn* where hills were placed 40 inches apart in each direction, allowing cultivators to run through the field in two directions. In more arid lands this was altered and seeds were planted in the bottom of 10-12 cm (4-5 in) deep furrows to collect water. Modern technique plants maize in rows which allows for cultivation while the plant is young, although the hill technique is still used in the cornfields of some Native American reservations. In North America, fields are often planted in a two-crop rotation with a nitrogen-fixing crop, often alfalfa in cooler climates and soybeans in regions with longer summers. Sometimes a third crop, winter wheat, is added to the rotation. Fields are usually ploughed each year, although no-till farming is increasing in use. Many of the maize varieties grown in the United States and Canada are hybrids. Over half of the corn area planted in the United States has been genetically modified using biotechnology to express agronomic traits such as pest resistance or herbicide resistance. Before about World War II, most maize in North America was harvested by hand (as it still is in most of the other countries where it is grown). This often involved large numbers of workers and associated social events. Some one-and two-row mechanical pickers were in use but the corn combine was not adopted until after the War. By hand or mechanical picker, the entire ear is harvested which then requires a separate operation of a corn sheller to remove the kernels from the ear. Whole ears of corn were often stored in *corn cribs* and these whole ears are a sufficient form for some livestock feeding use. Few modern farms store maize in this manner. Most harvest the grain from the field and store it in bins. The combine with a corn head (with points and snap rolls instead of a reel) does not cut the stalk; it simply pulls the stalk down. The stalk continues downward and is crumpled in to a mangled pile on the ground. The ear of corn is too large to pass through a slit in a plate and the snap rolls pull the ear of corn from the stalk so that only the ear and husk enter the machinery. The combine separates out the husk and the cob, keeping only the kernels.

Pellagra

When maize was first introduced into other farming systems than those used by traditional native-American peoples, it was generally welcomed with enthusiasm for its productivity. However, a widespread problem of malnutrition soon arose wherever maize was introduced as a staple. This was a mystery since these types of malnutrition were not normally seen among the indigenous Americans, to whom maize was the principal staple food. It was eventually discovered that the indigenous Americans learned long ago to add alkali—in the form of ashes among North Americans and lime (calcium carbonate) among Mesoamericans—to corn meal, which liberates the B-vitamin niacin, the lack of which was the underlying cause of the condition known as pellagra. This alkali process is known by its Nahuatl (Aztec)-derived name: nixtamalization. Besides the lack of niacin, pellagra was also characterized by protein deficiency, a result of the inherent lack of two key amino acids in pre-modern maize, lysine and tryptophan. Nixtamalisation was also found to increase the lysine and tryptophan content of maize to some extent, but more importantly, the indigenous Americans had learned long ago to balance their consumption of maize with beans and other protein sources such as amaranth and chia, as well as meat and fish, in order to acquire the complete range of amino acids for normal protein synthesis. Since maize had been introduced into the diet of non-indigenous Americans without the necessary cultural knowledge acquired over thousands of years in the Americas, the reliance on maize in other cultures was often tragic. In the late 19th century pellagra reached endemic proportions in parts of the deep southern U.S., as medical researchers debated two theories for its origin: the deficiency theory (eventually shown to be true) posited that pellagra was due to a deficiency of some nutrient, and the germ theory posited that pellagra was caused by a germ transmitted by stable flies. In 1914 the U.S. government officially endorsed the germ theory of pellagra, but rescinded this endorsement several years later as evidence grew against it. By the mid-1920s the deficiency theory of pellagra was becoming scientific consensus, and the theory was proved in 1932 when niacin deficiency was determined to be the cause of the illness. Once alkali processing and dietary variety was understood and applied, pellagra disappeared. The development of high lysine maize and the promotion of a more balanced diet has also contributed to its demise.

Pests

Insect pests

(i) Corn earworm (*Helicoverpa zea*)
(ii) Fall armyworm (*Spodoptera frugiperda*)
(iii) Common armyworm (*Pseudaletia unipuncta*)
(iv) Stalk borer (*Papaipema nebris*)
(v) Corn leaf aphid (*Rhopalosiphum maidis*)
(vi) European corn borer (*Ostrinia nubilalis*) (ECB)
(vii) Corn silkfly (*Euxesta stigmatis*)
(viii) Lesser cornstalk borer (*Elasmopalpus lignosellus*)
(ix) Corn delphacid (*Peregrinus maidis*)
(x) Western corn rootworm (*Diabrotica virgifera virgifera* LeConte)

The susceptibility of maize to the European corn borer, and the resulting large crop losses, led to the development of transgenic expressing the *Bacillus thuringiensis* toxin. "Bt corn" is widely grown in the United States and has been approved for release in Europe.

Diseases

(i)   Corn smut or common smut (*Ustilago maydis*): a fungal disease, known in Mexico as *huitlacoche*, which is prized by some as a gourmet delicacy in itself.

(ii)   Maize dwarf mosaic virus

(iii)   Stewart's Wilt (*Pantoea stewartii*)

(iv)   Common Rust (*Puccinia sorghi*)

(v)   Goss's Wilt (*Clavibacter michiganese*)

(vi)   Grey Leaf Spot

(vii)   Mal de Río Cuarto Virus (MRCV)

(viii) Stalk and Kernal Rot

Major Uses

Use as Food

Corn and cornmeal (corn flour) constitutes a staple food in many regions of the world. Corn meal is made into a thick porridge in many cultures: from the polenta of Italy, the angu of Brazil, the mamaliga of Romania, to mush in the U.S. or the food called sadza, nshima, ugali, and mealie pap in Africa. Corn meal is also used as a replacement for wheat flour, to make cornbread and other baked products. Masa (cornmeal treated with lime water) is the main ingredient for tortillas, atole and many other dishes of Mexican food. Popcorn is kernels of certain varieties that explode when heated, forming fluffy pieces that are eaten as a snack. Chicha and "chicha morada"(purple chicha) are drinks made usually from particular types of maize. The first one is fermented and alcoholic, the second one is a soft drink commonly drunk in Peru. Corn flakes are a common breakfast staple in the United States, and are increasingly popular all over the world. Maize can also be prepared as hominy, in which the kernels are soaked with lye; or grits, which are coarsely ground hominy. These are commonly eaten in the Southeastern United States, foods handed down from Native Americans. The Brazilian dessert canjica is made by boiling maize kernels in sweetened milk. Roasted dried corn cobs with semi-hardened kernels, coated with a seasoning mixture of fried chopped spring onions with salt added to the oil, is a popular snack food in Vietnam. Maize can also be harvested and consumed in the unripe state, when the kernels are fully grown but still soft. Unripe corn must usually be cooked to become palatable; this may be done by simply boiling or roasting the whole ears and eating the kernels right off the cob. Such corn on the cob is a common dish in the United States, United Kingdom and some parts of South America, but virtually unheard of in some European countries. The cooked unripe kernels may also be shaved off the cob and served as a vegetable in side dishes, salads, garnishes, etc. Alternatively, the raw unripe kernels may also be grated off the cobs and processed into a variety of cooked dishes, such as corn purée, tamales, pamonhas, curau, cakes, ice creams, etc. Sweetcorn, a genetic variety that is high in sugars and low in starch, is usually consumed in the unripe state. Maize is a major source of starch, a major ingredient in home cooking and in many industrialized food products. It is also a major source of cooking oil (corn oil) and of corn gluten. Maize starch can be hydrolyzed and enzymatically treated to produce syrups, particularly high fructose corn syrup, a sweetener; and also fermented and distilled to produce grain alcohol. Grain alcohol from maize is traditionally the source of bourbon whiskey. Maize is used to make chicha, a fermented beverage of Central and South America; and sometimes as the starch source for beer. In the United States and Canada maize is also widely grown to feed for livestock, as forage, silage (made by fermentation of chopped green cornstalks), or grain. Corn meal is also a significant ingredient of some commercial animal food products, such as dog food. Maize is also used as a fish bait, called "dough balls". It is particularly popular in Europe for coarse fishing.

Use as Chemicals and Medicines

Starch from maize can also be made into plastics, fabrics, adhesives, and many other chemical products. Stigmas from female corn flowers, known popularly as corn silk, are sold as herbal supplements. The corn steep liquor, a plentiful watery byproduct of maize wet milling process, is widely used in the biochemical industry and research as a culture medium to grow many kinds of microorganisms.

Use as Biofuel

"Feed corn" is being used increasingly for heating; specialized corn stoves (similar to wood stoves) are available and use either feed corn or wood pellets to generate heat. Corncobs are also used as a biomass fuel source. Maize is relatively cheap and home-heating furnaces have been developed which use maize kernels as a fuel. They feature a large hopper that feeds the uniformly sized corn kernels (or wood pellets or cherry pits) into the fire. Maize is increasingly used as a biomass fuel, such as ethanol, which as researchers search for innovative ways to reduce fuel costs, has unintentionally caused a rapid rise in food costs. This has led to the 2007 harvest being one of the most profitable corn crops in modern history for farmers. Maize is widely used in Germany as a feedstock for biogas plants. Here the maize is harvested, shredded then placed in silage clamps from which it is fed into the biogas plants. A biomass gasification power plant in Strem near Güssing, Burgenland, Austria was begun in 2005. Research is being done to make diesel out of the biogas by the Fischer Tropsch method. Increasingly ethanol is being used at low concentrations (10% or less) as an additive in gasoline (gasohol) for motor fuels to increase the octane rating, lower pollutants, and reduce petroleum use (what is nowadays also known as "biofuels" and has been generating an intense debate regarding the human beings' necessity of new sources of energy, on the one hand, and the need to maintain, in regions such as Latin America, the food habits and culture which has been the essence of civilizations such as the one originated in Mesoamerica; the entry, January 2008, of maize among the commercial agreements of NAFTA has increased this debate, considering the bad labour conditions of workers in the fields, and mainly the fact that NAFTA "opened the doors to the import of corn from the United States, where the farmers who grow it receive multi-million dollar subsidies and other government supports. (...) According to OXFAM UK, after NAFTA went into effect, the price of maize in Mexico fell 70 per cent between

1994 and 2001. The number of farm jobs dropped as well: from 8.1 million in 1993 to 6.8 million in 2002. Many of those who found themselves without work were small-scale maize growers."). However, introduction in the northern latitudes of the U.S. of tropical maize for biofuels, and not for human or animal consumption, may potentially alleviate this. As a result of the U.S. federal government announcing its production target of 35 billion gallons of biofuels by 2017, ethanol production will grow to 7 billion gallons by 2010, up from 4.5 billion in 2006, boosting ethanol's share of corn demand in the U.S. from 22.6 percent to 36.1 percent.

Major Ornamental and Other Uses

Some forms of the plant are occasionally grown for ornamental use in the garden. For this purpose, variegated and coloured leaf forms as well as those with colourful ears are used. Additionally, size-superlative varieties, having reached 31 ft (9.4m) tall, or with ears 24 inches (60 cm) long, have been popular for at least a century.

Corncobs can be hollowed out and treated to make inexpensive smoking pipes, first manufactured in the United States in 1869. An unusual use for maize is to create a *maize maze* as a tourist attraction. This is a maze cut into a field of maize. The idea of a maize maze was introduced by Adrian Fisher, one of the most prolific designers of modern mazes, with The American Maze Company who created a maze in Pennsylvania in 1993. Traditional mazes are most commonly grown using yew hedges, but these take several years to mature. The rapid growth of a field of maize allows a maze to be laid out using GPS at the start of a growing season and for the maize to grow tall enough to obstruct a visitor's line of sight by the start of the summer. In Canada and the U.S., these are called "corn mazes" and are popular in many farming communities. Corn kernels can be used in place of sand in a sandbox-like enclosure for children's play.

Use as Fodder

Maize makes a greater quantity of epigeous mass than other

cereal plants, so can be used for fodder. Digestibility and palatability are higher when ensiled and fermented, rather than dried.

Use in Art

Maize has been an essential crop in the Andes since the pre-Columbian Era. The Moche culture from Northern Peru made ceramics from earth, water, and fire. This pottery was a sacred substance, formed in significant shapes and used to represent important themes. Maize represented anthropomorphically as well as naturally. In the United States, maize itself is sometimes used for temporary architectural detailing when the intent is to celebrate local agricultural productivity and culture. A well-known example of this use is the Corn Palace in Mitchell, South Dakota, which utilizes cobs of coloured maize to implement a design that is recycled annually.

Sorghum Bicolour

Sorghum bicolour (*Sorghum japonicum*), commonly called sorghum, is a plant species in the grass family Poaceae. It is the primary Sorghum species cultivated for grain for human consumption and for animal feed. It can be popped in a similar fashion to popcorn. The species originated in northern Africa and can grow in arid soils and withstand prolonged droughts. *S. bicolour* is typically an annual, but some cultivars are perennial. It grows in clumps which may reach over 4 meters high. The grain is small reaching about 3 to 4 mm in diameter. Sweet sorghums are sorghum cultivars that are primarily grown for foliage; they are shorter than those grown for grain.

Major Uses

A traditional food plant in Africa, this little-known grain has potential to improve nutrition, boost food security, foster rural development and support sustainable landcare. The species is source of ethanol fuel, and in some environments may be better than maize or sugarcane as it can grow under more harsh conditions. Used for making a traditional corn broom.

Soybean

The soybean (U.S.) or soya bean (UK) (*Glycine max*) is a species of legume native to East Asia. The plant is classed as an oilseed rather than a pulse. It is an annual plant that has been used in China for 5,000 years as a food and a component of drugs. Soy contains significant amounts of all the essential amino acids for humans, and so is a good source of protein. Soybeans are the primary ingredient in many processed foods, including dairy product substitutes. The plant is sometimes referred to as *greater bean* or *edamame,* though the latter is more commonly used in English when referring to a specific dish. Soybeans are an important source of vegetable oil and protein world wide. Soy products are the main ingredients in many meat and dairy substitutes. They are also used in to make soy sauce, and the oil is used in many industrial applications. The main producers of soy are the United States, Brazil, Argentina, and China. The beans contain significant amounts of alpha-linolenic acid, omega-6 fatty acid, and the isoflavones genistein and daidzein.

Major Classification

The genus name Glycine was originally introduced by Linnaeus (1737) in his first edition of Genera Plantarum. The word glycine is derived from the Greek-glykys (sweet) and very likely refers to the sweetness of the pear-shaped (*apios* in Greek) edible tubers produced by the native North American twining or climbing herbaceous legume, Glycine apios, now known as *Apios americana.* The cultivated soybean first appeared in the Species Plantarum, Linnaeus, under the name *Phaseolus max* L. The combination, *Glycine max*(L.) Merr., as proposed by Merrill in 1917, has become the valid name for this useful plant. Like some other crops of long domestication, the relationship of the modern soy bean to wild-growing species can no longer be traced with any degree of certainty. It is a cultural variety with a very large number of cultivars. The genus Glycine Wild. is divided into two subgenera (species), Glycine and Soja. The subgenus Soja (Moench) includes the cultivated soybean, G. max (L.) Merrill, and the wild soybean, G. soja Sieb.and Zucc. Both species are annual. The soybean grows only under cultivation while G. soja grows wild in China, Japan, Korea, Taiwan and Russia. Glycine soja is the wild ancestor of the soybean: the wild progenitor. At present, the subgenus Glycine consists of at least 16 wild perennial species: for example, Glycine canescens, and G. tomentella Hayata found in Australia, and Papua New Guinea.

Main Physical Characteristics

Soy varies in growth, habit, and height. It may grow prostrate, not higher than 20 cm (7.8 inches), or grow up to 2 meters (6.5 feet) high. The pods, stems, and leaves are covered with fine brown or gray hairs. The leaves are trifoliolate, having 3 to 4 leaflets per leaf, and the leaflets are 6-15 cm (2-6 inches) long and 2-7 cm (1-3 inches) broad. The leaves fall before the seeds are mature. The big, inconspicuous, self-fertile flowers are borne in the axil of the leaf and are white, pink or purple. The fruit is a hairy pod that grows in clusters of 3-5, each pod is 3-8 cm long (1-3 inches) and usually contains 2-4 (rarely more) seeds 5-11 mm in diameter. Soybeans occur in various sizes, and in many hull or seed coat colours, including black, brown, blue, yellow, green and mottled. The hull of the mature bean is hard, water resistant, and protects the cotyledon and hypocotyl (or "germ") from damage. If the seed coat is cracked, the seed will not germinate. The scar, visible on the seed coat, is called the hilum (colours include black, brown, buff, gray and yellow) and at one end of the hilum is the micropyle, or small opening in the seed coat which can allow the absorption of water for sprouting. Remarkably, seeds such as soybeans containing very high levels of protein can undergo desiccation yet survive and revive after water absorption. A. Carl Leopold, son of Aldo Leopold, began studying this capability at the Boyce Thompson Institute for Plant Research at Cornell University in the mid 1980s. He found soybeans and corn to have a range of soluble carbohydrates protecting the seed's cell viability. Patents were awarded to him in the early 1990s on techniques for protecting "biological membranes" and proteins in the dry state. Compare to tardigrades.

Understanding the Chemical Composition of the Seed

Together, oil and protein content account for about 60 per cent of dry soybeans by weight; protein at 40 per cent and oil at 20 per cent. The remainder consists of 35 per cent carbohydrate and about 5 per cent ash. Soybean cultivars comprise approximately 8 per cent seed coat or hull, 90 per cent cotyledons and 2 per cent hypocotyl axis or germ.

Carbohydrates	30.16 g
— Sugars 7.33 g	
— Dietary fiber 9.3 g	
Fat	19.94 g
— saturated 2.884 g	
— monounsaturated 4.404 g	
— polyunsaturated 11.255 g	
Protein	36.49 g
Water	8.54 g
Vitamin A equiv. 1 µg	0%
Vitamin B6 0.377 mg	29%
Vitamin B12 0 µg	0%
Vitamin C 6.0 mg	10%
Vitamin K 47 µg	45%
Calcium 277 mg	28%
Iron 15.70 mg	126%
Magnesium 280 mg	76%
Phosphorus 704 mg	101%
Potassium 1797 mg	38%
Sodium 2 mg	0%
Zinc 4.89 mg	49%

Source: USDA Nutrient database.

The majority of soy protein is a relatively heat-stable storage protein. This heat stability enables soy food products requiring high temperature cooking, such as tofu, soymilk and textured vegetable protein (soy flour) to be made. The principal soluble carbohydrates, saccharides, of mature soybeans are the disaccharide sucrose (range 2.5-8.2%), the trisaccharide raffinose (0.1-1.0%) composed of one sucrose molecule connected to one molecule of galactose, and the tetrasaccharide stachyose (1.4 to 4.1%) composed of one sucrose connected to two molecules of galactose. While the oligosaccharides raffinose and stachyose protect the viability of the soy bean seed from desiccation they are not digestible sugars and therefore contribute to flatulence and abdominal discomfort in humans and other monogastric animals; compare to the disaccharide trehalose. Undigested oligosaccharides are broken down in the intestine by native microbes producing gases such as carbon dioxide, hydrogen, methane, etc. Since soluble soy carbohydrates are found in the whey and are broken down during fermentation, soy concentrate, soy protein isolates, tofu, soy sauce, and sprouted soy beans are without flatus activity. On the other hand, there may be some beneficial effects to ingesting oligosaccharides such as raffinose and stachyose, namely, encouraging indigenous bifidobacteria in the colon against putrefactive bacteria. The insoluble carbohydrates in soybeans consist of the complex polysaccharides cellulose, hemicellulose, and pectin. The majority of soybean carbohydrates can be classed as belonging to dietary fiber.

Nutritional Value

Soybeans are considered by many agencies, including the US Food and Drug Administration, to be a source of complete protein. A complete protein is one that contains significant amounts of all the essential amino acids that must be provided to the human body because of the body's inability to synthesize them. For this reason, soy is a good source of protein, amongst many others, for many vegetarians and vegans or for people who cannot afford meat. According to the FDA, "Soy protein products can be good substitutes for animal products because, unlike some other beans, soy offers a "complete" protein profile. Soybeans contain all the amino acids essential to human nutrition, which must be supplied in the diet because they cannot be synthesized by the human body. Soy protein products can replace animal-based foods—which also have complete proteins but tend to contain more fat, especially saturated fat—without requiring major adjustments elsewhere in the diet." However, as with any dietary health claim, there are opposing viewpoints on the health benefits of soybeans. The gold standard for measuring protein quality, since 1990, is the

Protein Digestibility Corrected Amino Acid Score (PDCAAS) and by this criterion soy protein is the nutritional equivalent of meat and eggs for human growth and health. Soybean protein isolate has a Biological Value of 74, whole soybeans 96, soybean milk 91, and eggs 97. Soy protein is essentially identical to that of other legume pulses seeds. Moreover, it has the highest yield per square meter of growing area, and is the least expensive source of dietary protein. Consumption of soy may also reduce the risk of colon cancer, possibly due to the presence of sphingolipids.

Cultivation Pattern

Soybeans are an important global crop, providing oil and protein. The bulk of the crop is solvent-extracted for vegetable oil and then defatted Soymeal is used for animal feed. A small proportion of the crop is consumed directly by humans. Soybean products do appear in a large variety of processed foods. During World War II, soybeans became important in both North America and Europe chiefly as substitutes for other protein foods and as a source of edible oil. It was during World War II that the soybean was discovered as fertilizer by the Department of Agriculture. In the 1960-1 Dillion round of the GATT the United States secured tariff-free access for its soybeans in the European market. In the 1960s the United States exported over 90 per cent of the worlds soybeans. The soybean is now a leading crop in the United States. Brazil, Argentina, and Paraguay also are significant soybean-exporting nations. Cultivation is successful in climates with hot summers, with optimum growing conditions in mean temperatures of 20°C to 30°C (68°F to 86°F); temperatures of below 20°C and over 40°C (68°F, 104°F) retard growth significantly. They can grow in a wide range of soils, with optimum growth in moist alluvial soils with a good organic content. Soybeans, like most legumes, perform nitrogen fixation by establishing a symbiotic relationship with the bacterium *Bradyrhizobium japonicum* (syn. *Rhizobium japonicum*; Jordan 1982). However, for best results an inoculum of the correct strain of bacteria should be mixed with the Soy bean (or any legume) seed before planting.

Modern crop cultivars generally reach a height of around 1 m (3 ft), and take 80-120 days from sowing to harvesting. Soybeans are native to east Asia but only 45 percent of soybean production is located there. The other 55 percent of production is in the Americas. U.S.A. produced 75 million tons of soybeans in 2000, of which more than one-third was exported. Other leading producers are Brazil, Argentina, Paraguay, China, and India. Environmental groups, such as Greenpeace and the WWF, have reported that both soybean cultivation and the probability of increased soybean cultivation in Brazil, has destroyed huge areas of Amazon rainforest and is encouraging further deforestation. Growing soy also requires the use of harmful pesticides and herbicides, while alternative sources of plant protein such as hemp do not. American soil scientist Dr. Andrew McClung, who first showed that the ecologically biodiverse savannah of the Cerrado region of Brazil could grow profitable soybeans, was awarded the 2006 World Food Prize on October 19, 2006. The first research on soybeans in the United States was conducted by George Washington Carver at Tuskegee, Alabama, but he decided it was too exotic a crop for the poor black farmers of the South so he turned his attention to peanuts.

Time Frame

Soybeans were a crucial crop in eastern Asia long before written records, and they remain a major crop in China, Japan, and Korea. Prior to fermented products such as Soy sauce, tempeh, natto, and miso, soy was considered sacred for its use in crop rotation as a method of fixing nitrogen. The plants would be plowed under to clear the field for food crops. Soy was first introduced to Europe in the early 1700s and what is now the United States in 1765, where it was first grown for hay. Benjamin Franklin wrote a letter in 1770 mentioning sending soybeans home from England. Soybeans did not become an important crop outside of Asia until about 1910. In America, soy was considered an industrial product only and not used as a food prior to the 1920s. Soy was introduced to Africa from China in the late 19th Century and is now widespread across the continent.

Asia

The origins of the soybean plant are obscure, but many botanists believe it to have derived from *Glycine ussuriensis*, a legume native to central China. The soybean has been used in China for 5,000 years as a food and a component of drugs. According to the ancient Chinese, in 2853 BC the legendary Emperor Shennong of China proclaimed that five plants were sacred: soybeans, rice, wheat, barley, and millet. Cultivation of soybeans was long confined chiefly to China, but gradually spread to other countries. The earliest preserved soybeans were found in archaeological sites in Korea. Radiocarbon dating of soybean samples recovered through flotation during excavations at the Early Mumun period Okbang site in Korea indicates that soybean was cultivated as a food crop in ca. 1000-900 BC. From about the first century AD to the Age of Discovery (15-16th century), soybeans were introduced into several countries such as Japan, Indonesia, the Philippines, Vietnam, Thailand, Malaysia, Burma, Taiwan, Nepal and India. This spread was due to the establishment of sea and land trade routes. The best current evidence on the Japanese Archipelago suggests that soybean cultivation occurred in the early Yayoi period. The earliest Japanese textual reference to the soybean is in the classic *Kojiki* (Records of Ancient Matters) which was completed in 712 AD. Many people have claimed that soybeans in Asia were historically only used after a fermentation process, which lowers the high phytoestrogens content found in the raw plant. However, terms similar to "soy milk" have been in use since 82 AD, and there is evidence of tofu consumption that dates to 220.

United States of America

Soy took on a very important role in the United States after World War I. During the Great Depression, the drought stricken (Dust Bowl) regions of the United States were able to use soy to regenerate their soil because of its nitrogen-fixing properties. Farms were increasing production in order to meet with government demands, and Henry Ford was a great leader of the soybean industry. In 1932-33 the Ford Motor Company spent approximately $1,250,000 on soybean research. By 1935 every

Ford car had soy involved in its manufacture. For example, the soybean oil was used to paint the automobiles as well as fluid for shock absorbers. Ford scientists had developed a fiber from soy protein which was wool-like, very soft, and chosen to be used for the making of suits, felt hats, and overcoats. Ford himself wore a suit made entirely from soybeans, he was even said to have had dinner parties with nothing but soybean based foods on the menu. Ford's involvement with the soybean opened many doors for agriculture and industry to be linked more strongly than it ever had before. Henry Ford promoted the soybean, helping to develop uses for it both in food and in industrial products, even demonstrating auto body panels made of soy-based plastics. Ford's interest led to two bushels of soybeans being used in each Ford car as well as products like the first commercial soy milk, ice cream and all-vegetable non-dairy whipped topping. The Ford development of so-called soy-based plastics was based on the addition of soybean flour and wood flour to phenolformaldehyde plastics. In 1931, Ford hired chemists Robert Boyer and Frank Calvert to produce artificial silk. They succeeded in making a textile fiber of spun soy protein fibers, hardened or tanned in a formaldehyde bath which was given the name Azlon by the Federal Trade Commission. Pilot production of Azlon reached 5000 pounds per day in 1940, but never reached the commercial market.

Main Soybean Diseases

Role of Genetic Modification

Soybeans are one of the "biotech food" crops that have been genetically modified, and genetically modified soybeans are being used in an increasing number of products. In 1995 Monsanto introduced *Roundup Ready* (RR) soybeans that have been genetically modified to be resistant to the herbicide Roundup through substitution of the *Agrobacterium sp.* (strain CP4) gene EPSP (5-enolpyruvyl shikimic acid-3-phosphate) synthase. The substituted version is not sensitive to glyphosate. In 1997, about 8 per cent of all soybeans cultivated for the commercial market in the United States were genetically modified. In 2006, the figure was 89 per cent. As with other "Roundup Ready" crops, concern is expressed over damage to biodiversity. However, the RR gene has been bred into so many different soybean cultivars that the genetic modification itself has not resulted in any decline of genetic diversity, as demonstrated by a 2003 study on genetic diversity. The widespread use of such types of GM soybeans in the Americas has caused problems with exports to some regions. GM crops require extensive certification before they can be legally imported into the European Union, where there is considerable supplier and consumer reluctance to use GM products for consumer or animal use. Difficulties with coexistence and subsequent traces of cross-contamination of non-GM stocks have caused shipments to be rejected and have put a premium on non-GM soy.

Major Uses

Soybeans can be broadly classified as "vegetable" (garden) or field (oil) types. Vegetable types cook more easily, have a mild nutty flavor, better texture, are larger in size, higher in protein, and lower in oil than field types. Tofu and soymilk producers prefer the higher protein cultivars bred from vegetable soybeans originally brought to the United States in the late 1930s. The "garden" cultivars are generally not suitable for mechanical combine harvesting because they have a tendency for the pods to shatter on reaching maturity. Among the legumes, the soybean, also classed as an oilseed, is pre-eminent for its high (38-45%) protein content as well as its high (20%) oil content. Soybeans are the leading agricultural export in the United States. The bulk of the soybean crop is grown for oil production, with the high-protein defatted and "toasted" soy meal used as livestock feed. A smaller percentage of soybeans are used directly for human consumption. Immature soybeans may be boiled whole in their green pod and served with salt, under the Japanese name *edamame edamame*. Because of the proclaimed health benefits of soy, edamame has been featured as an ideal snack alternative in fitness and healthy living magazines such as Real Simple. Edamame is sold in the frozen vegetable section at some larger grocery stores, and as ready-to-eat snackfood in many Asian delis. The beans can be processed in a variety of ways. Common forms of soy (or *soya*) include soy meal, soy flour, soy milk, tofu, textured vegetable protein (TVP, which is made into a wide variety of vegetarian foods, some of them intended to imitate meat), tempeh, soy lecithin and soybean oil. Soybeans are also the primary ingredient involved in the production of soy sauce (or *shoyu*). Archer Daniels Midland (ADM) is among the largest processors of soybeans and soy products. ADM along with Dow Chemical Company, DuPont and Monsanto support the industry trade associations United Soybean Board (USB) and Soyfoods Association of North America (SANA). These trade associations have increased the consumption of soy products dramatically in recent years.

Use as Oil

In processing soybeans for oil extraction and subsequent soy flour production, selection of high quality, sound, clean, dehulled yellow soybeans are very important. Soybeans having a dark coloured seed coat, or even beans with a dark hilum will inadvertently leave dark specks in the flour, and are undesirable for use in commercial food products. All commercial soybeans in the United States are yellow or yellow brown. To produce soybean oil, the soybeans are cracked, adjusted for moisture content, rolled into flakes and solvent-extracted with commercial hexane. The oil is then refined, blended for different applications, and sometimes hydrogenated. Soybean oils, both liquid and partially hydrogenated, are exported abroad, sold as "vegetable oil," or end up in a wide variety of processed foods. The remaining soybean husks are used mainly as animal feed. The major unsaturated fatty acids in soybean oil triglycerides are 7 per cent linolenic acid (C18:3); 51 per cent linoleic acid (C-18:2); and 23 per cent oleic acid (C-18:1). It also contains the saturated fatty acids 4 per cent stearic acid and 10 per cent palmitic acid. Soybean oil has a relatively high proportion, 7-10 per cent, of oxidation prone linolenic acid, which is an undesirable property for continuous service, such as in a restaurant. In the early nineties, Iowa State University developed soybean oil with 1 per cent linolenic acid in the oil. Three companies, Monsanto, DuPont/Bunge, and Asoyia in 2004 introduced low linolenic,

(C18:3; cis-9, cis-12, cis-15 octadecatrienoic acid) Roundup Ready soybeans. In the past, hydrogenation was used to reduce the unsaturation in linolenic acid, but this produced the unnatural *trans*-fatty acid configuration, whereas in nature the configuration is *cis*. This external picture from North Dakota State University compares soybean oil fatty acid content with other oils. In the 2002-2003 growing season, 30.6 million tons of soybean oil were produced worldwide, constituting about half of worldwide edible vegetable oil production, and thirty percent of all fats and oils produced, including animal fats and oils derived from tropical plants. While soybean oil has no direct insect repellent activity, it is used as a fixative to extend the short duration of action of essential oils such as geranium oil in several commercial products.

Use as Meal

Soybean meal, the material remaining after solvent extraction of soybean flakes, with a 50 per cent soy protein content, toasted (a misnomer because the heat treatment is with moist steam) and ground in a hammer mill, provided the energy for the American production method, beginning in the 1930s, of growing farm animals such as poultry and swine on an industrial scale; and more recently the aquaculture of catfish.

Use as Flour

Soy flour refers to defatted soybeans where special care was taken during desolventizing (not toasted) in order to minimize denaturation of the protein to retain a high Nitrogen Solubility Index (NSI), for uses such as extruder texturizing (TVP). It is the starting material for production of soy concentrate and soy protein isolate. Defatted soy flour is obtained from solvent extracted flakes, and contains less than 1 per cent oil. Full-fat soy flour is made from unextracted, dehulled beans, and contains about 18 per cent to 20 per cent oil. Due to its high oil content a specialized Alpine Fine Impact Mill must be used for grinding rather than the more common hammer mill. Low-fat soy flour is made by adding back some oil to defatted soy flour. The lipid content varies according to specifications, usually between 4.5

per cent and 9 per cent. High-fat soy flour can also be produced by adding back soybean oil to defatted flour at the level of 15 per cent. Lecithinated soy flour is made by adding soybean lecithin to defatted, low-fat or high-fat soy flours to increase their dispersibility and impart emulsifying properties. The lecithin content varies up to 15 per cent.

Use as Infant Formula

Infant formulas based on soy (SBIF) are used by lactose-intolerant babies and for babies that are allergic to cow milk proteins. The formulas are sold in powdered, ready-to-feed, or concentrated liquid forms. Some reviews express the opinion that more research is needed to answer the question of what effect the phytoestrogens contained in soy formula may have on infants, but did not find any adverse effects. Diverse studies conclude there are no adverse effects in human growth, development, or reproduction as a result of the consumption of soy-based infant formula. One of these studies, published at the Journal of Nutrition, concludes that:

...there is no clinical concerns with respect to nutritional adequacy, sexual development, neurobehavioural development, immune development, or thyroid disease. SBIFs provide complete nutrition that adequately supports normal infant growth and development. FDA has accepted SBIFs as safe for use as the sole source of nutrition.

Use as Meat and Dairy Substitutes

Soybeans can be processed to produce a texture and appearance similar to many other foods. For example, soybeans are the primary ingredient in many dairy product substitutes (e.g., soy milk, margarine, soy ice cream, soy yogurt, soy cheese, and soy cream cheese) and meat substitutes (e.g. veggie burgers). These substitutes are readily available in most supermarkets. Although soy milk does not naturally contain significant amounts of digestable calcium (the high calcium content of soybeans is bound to the insoluble constituents and remains in the soy pulp), many manufacturers of soy milk sell calcium-enriched products as well. Soy products also are used as a low cost filler in meat and poultry products. Food service, retail and institutional (primarily school lunch and correctional) facilities regularly use such "extended" products. Extension may result diminished flavor, but fat and cholesterol are reduced. Vitamin and mineral fortification can be used to make soy products nutritionally equivalent to animal ·protein; the protein quality is already roughly equivalent.

Use as Other Products

Soybeans are the bean used in Chinese fermented black beans, douchi, not the sometimes confused black turtle beans. Soybeans are also used in industrial products including oils, soap, cosmetics, resins, plastics, inks, crayons, solvents, and clothing. Soybean oil is the primary source of biodiesel in the United States, accounting for 80 per cent of domestic biodiesel production. Soybeans have also been used since 2001 as fermenting stock in the manufacture of a brand of vodka.

Use as Soybean Meal

Soybean meal is a byproduct of milling soybeans and used in lower end dog foods.

Major Role of Soyfoods in Disease Prevention

(a) *Omega-3 Fatty Acids*: Omega-3 fatty acids, for example, alpha-linolenic acid C18-3, all cis, 9,12,15 octadecatrienoic acid (where the omega-3 refers to carbon number 3 counting from the hydrocarbon tail whereas C-15 refers to carbon number 15 counting from the carboxyl acid head) are special fat components that benefit many body functions. However, the effects which are beneficial to health are associated mainly with the longer-chain, more unsaturated fatty acids eicosapentaenoic (20:5n-3, EPA) and docosahexaenoic acid (22:6n-3, DHA) found in fish oil and oily fish. For instance, EPA and DHA, inhibit blood clotting, while there is no evidence that alpha-linolenic acid (aLNA) can do this. Soybean oil is one of the few common vegetable oils that contains a significant amount of aLNA; others include canola, walnut, hemp, and flax. However, soybean oil does not contain EPA or DHA. Soybean oil does contain significantly greater amount of omega-6 fatty acids in the oil: 100g of soybean oil contains 7g of omega-3 fatty acids to 51g of omega-6: a ratio of 1:7. Flaxseed, in comparison, has an omega-3: omega-6 ratio of 3:1.

(b) *Isoflavones*: Soybeans also contain the isoflavones genistein and daidzein, types of phytoestrogen, that are considered by some nutritionists and physicians to be useful in the prevention of cancer and by others to be carcinogenic and endocrine disruptive. Soy's content of isoflavones are as much as 3 mg/g dry weight. Isoflavones are polyphenol compounds, produced primarily by beans and other legumes, including peanuts and chickpeas. Isoflavones are closely related to the antioxidant flavonoids found in other plants, vegetables and flowers. Isoflavones such as genistein and daidzein are found in only some plant families, because most plants do not have an enzyme, chalcone isomerase which converts a flavone precursor into an isoflavone. In contradiction to well known benefits of isoflavones, Genistein acts as an oxidant (stimulating nitrate synthesis), as well as it blocks formation of new blood vessels (antiangiogenic effects). Some studies show Genistein to act as inhibitor of the activity of substances in the body that regulate cell division and cell survival (growth factors).

(c) *Cholesterol Reduction*: The dramatic increase in soyfood sales is largely credited to the Food and Drug Administration's (FDA) approval of soy as an official cholesterol-lowering food, along with other heart and health benefits. A 2001 literature review argued that these health benefits were poorly supported by the available evidence, and noted that disturbing data on soy's effect on the cognitive function of the elderly existed. In 2008, an epidemiological study of 719 Japanese men found that tofu intake was associated with worse memory, but tempeh (a fermented soy product) intake was associated with better memory. This study replicated other studies. From 1992 to 2003, sales have experienced a 15 per cent compound annual growth rate, increasing from $300 million to $3.9 billion over 11 years, as new soyfood categories have been introduced, soyfoods have been repositioned in the market place, thanks to a better emphasis on marketing nutrition. In 1995, the *New England Journal of Medicine* (Vol. 333, No. 5) published a meta-analysis financed by DuPont Protein Technologies International (PIT), which produces and markets soy through The Solae Company. The meta-analysis concluded that soy protein is correlated with significant decreases in serum cholesterol, LDL (bad cholesterol) and triglycerides. However, HDL (good cholesterol) did not increase by a significant amount. Soy phytoestrogens (isoflavones: genistein and daidzein) adsorbed onto the soy protein were suggested as the agent reducing serum cholesterol levels. On the basis of this research PTI filed a petition with FDA in 1998 for a health claim that soy protein may reduce cholesterol and the risk of heart disease. The FDA granted the following health claim for soy: "25 grams of soy protein a day, as part of a diet low in saturated fat and cholesterol, may reduce the risk of heart disease." One serving, (1 cup or 240 mL) of soy milk, for instance, contains 6 or 7 grams of soy protein. Solae resubmitted their original petition, asking for a more vague health claim, after their original was challenged and highly criticized. Solae also submitted a petition for a health claim that soy can help prevent cancer. They quickly withdrew the petition for lack of evidence and after more than 1,000 letters of protest were received. In February 18, 2008 Weston A. Price Foundation submitted a petition for removal of this health claim. In January 2006 an American Heart Association review (in the journal "Circulation") of a decade long study of soy protein benefits casts doubt on the FDA allowed "Heart Healthy" claim for soy protein and did not recommend isoflavone supplementation. The review panel also found that soy isoflavones have not been shown to reduce post menopause "hot flashes" in women and the efficacy and safety of isoflavones to help prevent cancers of the breast, uterus or prostate is in question.

Major Health Risks

(a) *Phytoestrogen*: Soybeans contain isoflavones called genistein and daidzein, which are one source of phytoestrogens in the human diet. Because most naturally occurring estrogenic substances show weak activity, normal consumption of foods that contain these phytoestrogens should not provide sufficient amounts to elicit a physiological response in humans. Plant lignans associated with high fiber foods such as cereal brans and beans are the principal precursor to mammalian lignans which have an ability to bind to human estrogen sites. Soybeans are a significant source of mammalian lignan precursor secoisolariciresinol containing 13-273 µg/100 g dry weight. Another phytoestrogen in the human diet with estrogen activity is coumestans, which are found in beans, split-peas, with the best sources being alfalfa, clover, and soybean sprouts. Coumestrol, an isoflavone coumarin derivative is the only coumestan in foods. Soybeans and processed soy foods do not contain the highest "total phytoestrogen" content of foods. A study in which data were presented on an as-is (wet) basis per 100 g and per serving found that food groups from highest to lowest levels of total phytoestrogens per 100 g are nuts and oilseeds, soy products, cereals and breads, legumes, meat products, various processed foods that may contain soy, vegetables, and fruits.

(b) *Allergy*: Allergy to soy is often said to be rather common, and the food is listed with other foods that commonly cause allergy, such as milk, eggs, peanuts, tree nuts, shellfish. However, a critical review of medical literature reveals surprisingly little solid information on the topic. The problem has been reported amongst younger children and the diagnosis of soy allergy is often based on symptoms reported by parents and/or results of skin tests or blood tests for allergy. Only a few reported studies have attempted to confirm allergy to soy by direct challenge with the food under controlled conditions. In these circumstances it is, clear that skin/blood tests considerably overestimate the problem, as do parental reports. It is very difficult to give a reliable estimate of the true prevalence of soy allergy in the general population. To the extent that it does exist, soy allergy may cause cases of urticaria (hives) and angioedema (swelling), usually within minutes to two hours of ingestion of the food. In rare, severe cases true anaphylaxis may occur, a condition that is much more common with allergy to foods such as peanut and shellfish. The reason for the discrepancy is likely that soy proteins, the causative factor in allergy, are far less potent at triggering allergy symptoms than the proteins of peanut and shellfish. An allergy test that is positive demonstrates that the immune system has formed IgE antibodies to soy proteins. However, when soy is ingested proteins must evade digestion and be absorbed in a form capable of triggering allergy and also in sufficient quantities to reach a threshold to provoke actual symptoms. The low potency of soy proteins as allergens may help explain why allergy skin/ blood tests suggest that soy allergy is common, yet few cases are confirmed when the food is eaten under observation. Soy can also trigger symptoms via food intolerance, a situation where no immunologic (allergic) mechanism can be proven. One scenario is seen in very young infants who have vomiting and diarrhoea when fed soy-based formula. The symptoms resolve when the formula is withdrawn and recur when it is re-administered. Older infants can suffer a more severe disorder with vomiting, diarrhoea that may be bloody, anemia, weight loss and failure to thrive. The commonest cause of this unusual disorder is a sensitivity to cow's milk, but there is no doubt that soy formulas can also be the trigger. The precise mechanism is unclear and it could be immunologic, although not through the IgE-type antibodies that have the leading role in urticaria and anaphylaxis. Fortunately it is also self-limiting and will often disappear in the toddler years.

(c) *Promotion as Health Food*: Soy consumption has been promoted by natural food companies and the soy industry's aggressive marketing campaign in various magazines, television ads and in health food markets. Research has been conducted examining the validity of the beneficial health claims with regard to the increase in consumption of soybeans which mimic hormonal activity. A practice guideline published in the journal *Circulation* questions the efficacy and safety of soy isoflavones for preventing or treating cancer of the breast, endometrium, and prostate (although the same study also concludes that soy in some foods should be beneficial to cardiovascular and overall health) and does not recommend usage of isoflavone supplements in food or pills. A review of the available studies by the United States' Health and Human Services' Agency for Healthcare Research and Quality (AHRQ) found little evidence of substantial health improvements and no adverse effects, but also noted that there was no long-term safety data on estrogenic effects from soy consumption.

(d) *Brain*: Estrogen helps protect and repair the brain during and after injury. The mimicry of estrogen by the phytoestrogens in soy has introduced a controversy over whether such a replacement is harmful or helpful to the brain. Several studies have found soy to be harmful for rats. Nevertheless the cited study was based on rats fed with concentrated phytoestrogens and not common soybeans. The common amounts of phytoestrogens in soy beans are not to be compared to concentrated estrogen. One study followed over 3000 Japanese men between 1965 and 1999, and that showed a positive correlation between brain atrophy and consumption of tofu. A study on elderly Indonesian men and women found that tempeh consumption was independently related to better memory.

(e) *Carcinogen*: Raw soy flour is known to cause pancreatic cancer in rats. Whether this is also true in humans is unknown because no studies comparing cases of pancreatic cancer and soy intake in humans have yet been conducted, and the doses used to induce pancreatic cancer in rats are said to be larger than humans would normally consume. Heated soy flour may not be carcinogenic in rats. Existing cancer patients are being warned to avoid foods rich in soy because they can accelerate the growth of tumours.

Major Soybean Futures

Soybean futures are traded on the Chicago Board of Trade and have delivery dates in January (F), March (H), May (K), July (N), August (Q), September (U), November (X). It is also traded on other commodity futures exchanges under different contract specifications:

(i) *SAFEX*: The South African Futures Exchange
(ii) *DC*: Dalian Commodity Exchange
(iii) *KEX*: Kansai Commodities Exchange in Japan
(iv) *TGE*: Tokyo Grain Exchange in Japan
(v) *KCX*: Fukuoka Commodity Exchange in Japan that was absorbed by the KEX

Stover

Stover consists of the leaves and stalks of corn (maize), sorghum or soybean plants that are left in a field after harvest. It can be directly grazed by cattle or dried for use as fodder (forage). It is similar to straw, the residue left after any cereal grain or grass has been harvested at maturity for its seed. Stover has attracted some attention as a potential alternative fuel source, and as biomass for fermentations.

Straw

Straw is an agricultural by-product, the dry stalk of a cereal plant, after the grain or seed has been removed. Straw makes up about half of the yield of cereal crops such as barley, oats, rice, rye and wheat. In times gone by, it was regarded as a useful by-product of the harvest, but with the advent of the combine harvester, straw has become more burdensome to agriculture. However, straw can be put to many uses, old and new.

Major Uses

(a) *Biofuels:* The use of straw as a carbon-neutral energy source is increasing rapidly, especially for biobutanol.

(b) *Biomass:* The use of straw in large-scale biomass power plants is becoming mainstream in the EU, with several facilities already online. The straw is either used directly in the form of bales, or densified into pellets which allows for the feedstock to be transported over longer distances. Finally, torrefaction of straw with pelletisation is gaining attention, because it increases the energy density of the resource, making it possible to transport it still further. This processing step also makes storage much easier, because torrefied straw pellets are hydrophobic. Torrefied straw in the form of pellets can be directly co-fired with coal or natural gas at very high rates and make use of the processing infrastructures at existing coal and gas plants. Because the torrefied straw pellets have superior structural, chemical and combustion properties to coal, they can replace all coal and turn a coal plant into an entirely biomass-fed power station. First generation pellets are limited to a co-firing rate of 15 per cent in modern IGCC plants.

(c) *Bedding Humans or Livestock:* The straw-filled mattress, also known as palliasse, is still used in many parts of the world.

It is commonly used as bedding for ruminants and

horses. It may be used as bedding and food for small animals, but this often leads to injuries to mouth, nose and eyes as straw is quite sharp.

(d) *Animal Feed*: Straw may be fed as part of the roughage component of the diet to cattle that are on a near maintenance level of energy requirement. It has a low digestible energy and nutrient content. The heat generated when microorganisms in a herbivore's gut digest straw can be useful in maintaining body temperature in cold climates. Due to the risk of impaction and its poor nutrient profile, it should always be restricted to part of the diet.

(e) *Hats*: There are several styles of straw hats that are made of woven straw.

Until about 100 years ago, thousands of women and children in England were employed in plaiting straw for making hats. These days the straw plait is imported.

(f) *Thatching*: Thatched roofs are becoming increasingly popular, and the skills of a master thatcher are once again in demand.

(g) *Packaging*: Straw is resistant to being crushed and therefore makes a good packing material. A company in France makes a straw mat sealed in thin plastic sheets. Straw envelopes for wine bottles have become rarer, but are still to be found at some wine merchants.

(h) *Paper*: Straw can be pulped to make paper.

(i) *Archery targets*: Heavy gauge straw rope is coiled and sewn tightly together. This is no longer done entirely by hand, but is partially mechanised.

(j) *Horse collars*: Working horses are making a comeback, and there is a need for horse collars stuffed with good quality rye straw. Being a "long straw filler" is a highly skilled job.

(k) *Construction material: bricks/cob*: In many parts of the world, straw is used to bind clay and concrete. This mixture of clay and straw, known as cob, can be used as a building material. There are many recipes for making cob.

When baled, straw has excellent insulation characteristics. It can be used, alone or in a post-and-beam construction, to build straw bale houses.
Enviroboard can be made from straw.

(*l*) *Rope:* Rope made from straw was used by thatchers, in the packaging industry and even in iron foundries.

(*m*) *Basketry:* Bee skeps and linen baskets are made from coiled and bound together continuous lengths of straw. The technique is known as lip work.

(*n*) *Sandals:* Koreans wear Jipsin, sandals made of straws.

(*o*) *Horticulture:* Straw is used in cucumber houses and for mushroom growing.

In Japan, certain trees are wrapped with straw to protect them from the effects of a hard winter as well as to use them as a trap for parasite insects. It is also used in ponds to reduce algae by changing the nutrient ratios in the water. The soil under strawberries is covered with straw to protect the ripe berries from dirt, and straw is also used to cover the plants during winter to prevent the cold from killing them.

Sugarcane

Sugarcane (*Saccharum*) is a genus of 6 to 37 species (depending on taxonomic interpretation) of tall perennial grasses (family Poaceae, tribe Andropogoneae), native to warm temperate to tropical regions of the Old World. They have stout, jointed, fibrous stalks that are rich in sugar and measure 2 to 6 meters tall. All of the sugar cane species interbreed, and the major commercial cultivars are complex hybrids.

Patterns in Cultivation and Uses

About 195 countries grow the crop to produce 1,324.6 million tons (more than six times the amount of sugar beet produced). As of the year 2005, the world's largest producer of sugar cane by far is Brazil followed by India. Uses of sugar cane include the production of sugar, Falernum, molasses, rum, soda, cachaça (the national spirit of Brazil) and ethanol for fuel. The bagasse that remains after sugar cane crushing may be burned to provide both heat—used in the mill—and electricity, typically sold to the consumer electricity grid. It may also, because of its high cellulose content, be used as raw material for paper, cardboard, and eating utensils branded as "environmentally friendly" as it is made from a by-product of sugar production.

Time Frame

Sugarcane was originally from tropical South Asia and Southeast Asia. Different species likely originated in different locations with *S. barberi* originating in India and *S. edule* and *S. officinarum* coming from New Guinea. The thick stalk stores energy as sucrose in the sap. From this juice, sugar is extracted by evaporating the water. Crystallized sugar was reported 5000 years ago in India. Around the eighth century A.D., Arabs introduced sugar to the Mediterranean, Mesopotamia, Egypt, North Africa, and Spain. By the tenth century, sources state, there was no village in Mesopotamia that didn't grow sugar cane. It was among the early crops brought to the Americas by Spaniards. Brazil is currently the biggest sugar cane producing country. A *boiling house* was used in the 17th through 19th centuries to make sugarcane juice into raw sugar. These houses were add-ons to the sugar plantations in the western colonies. This process was often conducted by the African slaves, under very poor conditions. The boiling house was made of cut stone. The furnaces were rectangular boxes of brick or stone with openings near to one side, and at the bottom to stoke the fire and pull out the ashes. At the top of each furnace were up to seven copper kettles or boilers, each one smaller than the previous one and hotter. The cane juice was placed in the first copper kettle which was the largest. The juice was then heated and a little lime added to remove impurities. The juice was then skimmed then channeled to the other copper kettles. The last kettle, which was called the 'teache' was where the cane juice became syrup. It was then put into cooling troughs where the sugar crystals hardened around a sticky core of molasses. The raw sugar was then shoveled from the cooling trough into hogsheads (wooden barrels) where they were put in the curing house. Sugarcane was, and still is, extensively grown in the Caribbean, where it was first brought by Christopher Columbus during his second voyage to The Americas, initially to the island of Hispaniola (modern day Haiti and the Dominican Republic). In colonial times, sugar was a major product of the triangular trade of New World raw materials, European manufactures, and African slaves. France found its sugarcane islands so valuable it effectively traded its portion of Canada, famously dubbed "a few acres of snow," to Britain for their return of Guadeloupe, Martinique and St. Lucia at the end of the Seven Years' War. The Dutch similarly kept Suriname, a sugar colony in South America, instead of seeking the return of the New Netherlands (New Amsterdam). Cuban sugarcane produced sugar that received price supports from and a guaranteed market in the USSR; the dissolution of that country forced the closure of most of Cuba's sugar industry. Sugarcane remains an important part of the economy of Belize, Barbados, Haiti along with the Dominican Republic, Guadeloupe, Jamaica, and other islands. The sugarcane industry is a major export for the Caribbean, but it is expected to collapse with the removal of European preferences by 2009. Sugarcane production greatly influenced many tropical Pacific islands, including Okinawa and most particularly Hawaii and Fiji. In these islands, sugar cane came to dominate the economic and political landscape after the arrival of powerful European and American agricultural business, which promoted immigration from various Asian countries for workers to tend and harvest the crop. Sugar-industry policies eventually established the ethnic makeup of the island populations that now exist, profoundly affecting modern politics and society in the islands. Brazil is a major grower of sugarcane, which is used to produce sugar and provide the ethanol used in making gasoline-ethanol blends (gasohol) for transportation fuel. In India, sugarcane is sold as jaggery and also refined into sugar, primarily for consumption in tea and sweets, and for the production of alcoholic beverages.

Cultivation Pattern

Sugarcane cultivation requires a tropical or subtropical climate, with a minimum of 600 mm (24 in) of annual moisture. It is one of the most efficient photosynthesizers in the plant kingdom. It is a C-4 plant, able to convert up to 2 percent of incident solar energy into biomass. In prime growing regions, such as Peru, Brazil, Bolivia, Colombia, Australia, Ecuador, Cuba, the Philippines and Hawaii, sugarcane can produce 20 kg for each square meter exposed to the sun. Sugarcane is propagated from cuttings, rather than from seeds; although certain types still produce seeds, modern methods of stem cuttings have become the most common method of reproduction. Each cutting must contain at least one bud, and the cuttings are usually planted by hand. Once planted, a stand of cane can be harvested several times; after each harvest, the cane sends up new stalks, called **ratoons**. Usually, each successive harvest gives a smaller yield, and eventually the declining yields justify replanting. Depending on agricultural practice, two to ten harvests may be possible between plantings. Sugarcane is harvested mostly by hand and sometimes mechanically. Hand harvesting accounts for more than half of the world's production, and is especially dominant in the developing world. When harvested by hand, the field is first set on fire. The fire spreads rapidly, burning away dry dead leaves, and killing any venomous snakes hiding in the crop, but leaving the water-rich stalks and roots unharmed. With cane knives or machetes, harvesters then cut the standing cane just above the ground. A skilled harvester can cut 500 kg of sugarcane in an hour. With mechanical harvesting, a sugarcane combine (or chopper harvester), a harvesting machine originally developed in Australia, is used. The Austoft 7000 series was the original design for the modern harvester and has now been copied by other companies including Cameco and John Deere. The machine cuts the cane at the base of the stalk, separates the cane from its leaves, and deposits the cane into a haulout transporter while blowing the thrash back onto the field. Such machines can harvest 100 tonnes of cane each hour, but cane harvested using these machines must be transported to the processing plant rapidly; once cut, sugarcane begins to lose its sugar content, and damage inflicted on the cane during mechanical harvesting accelerates this decay. Sugar cane is cultivated in almost all the world only for some months of the year.

Major Pests

The most important sugarcane pests are the larvae of some butterfly/moth species, including the turnip moth, the sugarcane borer (*Diatraea saccharalis*), the Mexican rice borer (*Eoreuma loftini*), leaf-cutting ants, termites; spittlebugs (especially *Mahanarva fimbriolata* and *Deois flavopicta*), and the beetle *Migdolus fryanus*. The planthopper *Eumetopina flavipes* is an insect which acts as a vector for the phytoplasma which causes the sugarcane disease ramu stunt.

Major Diseases

(a) *Processing*: Traditionally, sugarcane has been processed in two stages. Sugarcane mills, located in sugarcane-producing regions, extract sugar from freshly harvested sugarcane, resulting in raw sugar for later refining, and in "mill white" sugar for local consumption. Sugar refineries, often located in heavy sugar-consuming regions, such as North America, Europe, and Japan, then purify raw sugar to produce refined white sugar, a product that is more than 99 percent pure sucrose. These two stages are slowly becoming blurred. Increasing affluence in the sugar-producing tropics has led to an increase in demand for refined sugar products in those areas, where a trend toward combined milling and refining has developed.

(b) *Milling*: Sugarcane first has to be moved to a mill which is usually located close to the area of cultivation. Small rail networks are a common method of transporting the cane to a mill. Once the factories acquire the cane it will be subjected to the quality test. In Sri Lanka cane will be evaluated according to the brix and trash percentage. In a sugar mill, sugarcane is washed, chopped, and shredded by revolving knives. The shredded cane is repeatedly mixed with water and crushed between rollers; the collected juices (called garapa in Brazil) contain 10-15 per cent sucrose, and the remaining fibrous solids, called bagasse, are burned for fuel. Bagasse makes a sugar mill more than self-sufficient in energy; the surplus bagasse can be used for animal feed, in paper manufacture, or burned to generate electricity for the local power grid. The cane juice is next mixed with lime to adjust its pH to 7. This mixing arrests sucrose's decay into glucose and fructose, and precipitates out some impurities. The mixture then sits, allowing the lime and other suspended solids to settle out, and the clarified juice is concentrated in a multiple-effect evaporator to make a syrup about 60 percent by weight in sucrose. This syrup is further concentrated under vacuum until it becomes supersaturated, and then seeded with crystalline sugar. Upon cooling, sugar crystallizes out of the syrup. A centrifuge is used to separate the sugar from the remaining liquid, or molasses. Additional crystallizations may be performed to extract more sugar from the molasses; the molasses remaining after no more sugar can be extracted from it in a cost-effective fashion is called blackstrap. Raw sugar has a yellow to brown colour. If a white product is desired, sulfur dioxide may be bubbled through the cane juice before evaporation; this chemical bleaches many colour-forming impurities into colourless ones. Sugar bleached white by this *sulfitation* process is called "mill white", "plantation white", and "crystal sugar". This form of sugar is the form most commonly consumed in sugarcane-producing countries. Traditionally, sugarcane has been processed in two stages. Sugarcane mills, located in sugarcane-producing regions, extract sugar from freshly harvested sugarcane, resulting in raw sugar for later refining, and in "mill white" sugar for local consumption. Sugar refineries, often located in heavy sugar-consuming regions, such as North America, Europe, and Japan, then purify raw sugar to produce refined white sugar, a product that is more than 99 percent pure sucrose. These two stages are slowly becoming blurred. Increasing affluence in the sugar-producing tropics has led to an increase in demand for refined sugar products in those areas, where a trend toward combined milling and refining has developed.

(c)   *Refining*: In sugar refining, raw sugar is further purified. It is first mixed with heavy syrup and then centrifuged clean. This process is called 'affination'; its purpose is to wash away the outer coating of the raw sugar crystals, which is less pure than the crystal interior. The remaining sugar is then dissolved to make a syrup, about 70 percent by weight solids. The sugar solution is clarified by the addition of phosphoric acid and calcium hydroxide, which combine to precipitate calcium phosphate. The calcium phosphate particles entrap some impurities and absorb others, and then float to the top of the tank, where they can be skimmed off. An alternative to this "phosphatation" technique is 'carbonatation,' which is similar, but uses carbon dioxide and calcium hydroxide to produce a calcium carbonate precipitate. After any remaining solids are filtered out, the clarified syrup is decolourized by filtration through a bed of activated carbon; bone char was traditionally used in this role, but its use is no longer common. Some remaining colour-forming impurities adsorb to the carbon bed. The purified syrup is then concentrated to supersaturation and repeatedly crystallized under vacuum, to produce white refined sugar. As in a sugar mill, the sugar crystals are separated from the molasses by centrifuging. Additional sugar is recovered by blending the remaining syrup with the washings from affination and again crystallizing to produce brown sugar. When no more sugar can be economically recovered, the final molasses still contains 20-30 percent sucrose and 15-25 percent glucose and fructose. To produce granulated sugar, in which the individual sugar grains do not clump together, sugar must be dried. Drying is accomplished first by drying the sugar in a hot rotary dryer, and then by conditioning the sugar by blowing cool air through it for several days.

(d) *Ribbon Cane Syrup*: Ribbon cane is a subtropical type that was once widely grown in southern United States, as far north as coastal North Carolina. The juice was extracted with horse or mule-powered crushers; the juice was boiled, like maple syrup, in a flat pan, and then used in the syrup form as a sweetener for other foods. It is not a commercial crop nowadays, but a few growers try to keep alive the old traditions and find ready sales for their product. Most sugarcane production in the United States occurs in Florida and Louisiana, and to a lesser extent in Hawaii and Texas.

Patterns of Production

In India, the states of Uttar Pradesh (38.57 %), Maharashtra (17.76 %) and Karnataka (12.20 %) lead the nation in sugarcane production. In the United States, sugar cane is grown commercially in Florida, Hawaii, Louisiana, and Texas.

Table 5.5: Top Ten Sugarcane Producers, June 2008

Country	(Tonnes)
Brazil	514,079,729
India	355,520,000
People's Republic of China	106,316,000
Thailand	64,365,682
Pakistan	54,752,000
Mexico	50,680,000
Colombia	40,000,000
Australia	36,000,000
United States	27,750,600
Philippines	25,300,000
World	1,557,664,978

Note: P = official figure, F = FAO estimate, * = Unofficial/Semi-official/ mirror data, C = Calculated figure A = Aggregate (may include official, semi-official or estimates);

Source: Food And Agricultural Organization of United Nations: Economic And Social Department: The Statistical Devision.

Cane Ethanol

This is generally available as a by-product of sugar mills producing sugar. It can be used as a fuel, mainly as a biofuel alternative to gasoline, and is widely used in cars in Brazil. It is steadily becoming a promising alternative to gasoline throughout much of the world and thus instead of sugar may be produced as a primary product out of sugar canes processing. A textbook on renewable energy describes the energy transformation:

At present, 741 tons of raw sugar cane are produced annually per hectare in Brazil. The cane delivered to the processing plant is called burned and cropped (bandc) and represents 77 per cent of the mass of the raw cane. The reason for this reduction is that the stalks are separated from the leaves (which are burned and whose ashes are left in the field as fertilizer) and from the roots that remain in the ground to sprout for the next crop. Average cane production is, therefore, 58 tons of bandc per hectare per year. ·

Each ton of bandc yields 740 kg of juice (135 kg of sucrose and 605 kg of water) and 260 kg of moist bagasse (130 kg of dry bagasse). Since the higher heating value of sucrose is 16.5 MJ/kg, and that of the bagasse is 19.2 MJ/kg, the total heating value of a ton of bandc is 4.7 GJ of which 2.2 GJ come from the sucrose and 2.5 from the bagasse.

Per hectare per year, the biomass produced corresponds to 0.27 TJ. This is equivalent to 0.86 W per square meter. Assuming an average insolation of 225 W per square meter, the photosynthetic efficiency of sugar cane is 0.38 per cent.

The 135 kg of sucrose found in 1 ton of bandc are transformed into 70 liters of ethanol with a combustion energy of 1.7 GJ. The practical sucrose-ethanol conversion efficiency is, therefore, 76 per cent (compare with the theoretical 97%).

One hectare of sugar cane yields 4000 liters of ethanol per year (without any additional energy input because the bagasse produced exceeds the amount needed to distill the final product). This however does not include the energy used in tilling, transportation, and so on. Thus, the solar energy-to-ethanol conversion efficiency is 0.13 per cent.

Sugarcane as Food

In most countries where sugarcane is cultivated, there are several foods and popular dishes derived from it, such as:

(*i*) Direct consumption of raw sugarcane cylinders or cubes, which are chewed to extract the juice, and the bagasse is spat out

(*ii*) Freshly extracted juice (garapa, *guarab, guarapa, guarapo, papelón, 'aseer asab*, Ganna sharbat, mosto or *caldo de cana*) by hand or electrically operated small mills, with a touch of lemon and ice, makes a popular drink.

(*iii*) Molasses, used as a sweetener and as a syrup accompanying other foods, such as cheese or cookies

(*iv*) Rapadura, a candy made of flavored solid brown sugar in Brazil, which can be consumed in small hard blocks, or in pulverized form (flour), as an add-on to other desserts.

(*v*) Sugarcane is also used in rum production, especially in the Caribbean.

(*vi*) Cane sugar syrup was the traditional sweetener in soft drinks for many years, but has been largely supplanted (in the US at least) by high-fructose corn syrup, which is less expensive, but is considered by some to not taste quite like the sugar it replaces.

(*vii*) Hard rock candy is a confection that is enjoyed by people around the world.

(*viii*) Jaggery—solidified molasses of sugarcane, known as Gur or Gud in South Asia. Traditionally produced by heat evaporating sugarcane juice until it is a thick sludge and then letting it cool in buckets used as molds. Modern production methods make use of partial freeze drying to give reduce caramelization and give it lighter colour. It is used as sweetener in cooking traditional meal entrees as well as sweets and desserts.

Role in Nitrogen Fixation

Some sugarcane varieties are known to be capable of fixing atmospheric nitrogen in association with a bacterium,

Acetobacter diazotrophicus. Unlike legumes and other nitrogen fixing plants which form root nodules in the soil in association with bacteria, Acetobacter diazotrophicus lives within the intercellular spaces of the sugarcane's stem.

Sunflower

The sunflower (*Helianthus annuus*) is an annual plant in the family Asteraceae and native to the Americas, with a large flowering head (inflorescence). The stem can grow as high as 3 meters (9 3/4 ft), and the flower head can reach 30 cm (11.8 in) in diameter with the "large" seeds. The term "sunflower" is also used to refer to all plants of the genus *Helianthus*, many of which are perennial plants.

Characteristics

What is usually called the flower is actually a *head* (formally *composite flower*) of numerous florets (small flowers) crowded together. The outer florets are the sterile *ray florets* and can be yellow, maroon, orange, or other colours. The florets inside the circular head are called *disc florets*, which mature into what are traditionally called "sunflower seeds," but are actually the fruit (an *achene*) of the plant. The inedible husk is the wall of the fruit and the true seed lies within the kernel. The florets within the sunflower's cluster are arranged in a spiraling pattern. Typically each floret is oriented toward the next by approximately the golden angle, 137.5°, producing a pattern of interconnecting spirals where the number of left spirals and the number of right spirals are successive Fibonacci numbers. Typically, there are 34 spirals in one direction and 55 in the other; on a very large sunflower there could be 89 in one direction and 144 in the other.

Heliotropism

Sunflowers in the bud stage exhibit heliotropism. At sunrise, the faces of most sunflowers are turned towards the east. Over the course of the day, they follow the sun from east to west, while at night they return to an eastward orientation. This motion is performed by motor cells in the pulvinus, a flexible segment of the stem just below the bud. As the bud stage ends, the stem stiffens and the blooming stage is reached. Sunflowers in the blooming stage are not heliotropic anymore. The stem has "frozen", typically in an eastward orientation. The stem and leaves lose their green colour. The wild sunflower typically does not turn toward the sun; its flowering heads may face many directions when mature. However, the leaves typically exhibit some heliotropism.

Time Frame

Francisco Pizarro found it in Tahuantinsuyo, Peru, where the natives Incas worshipped the sunflower image as a symbol of the sun god. At the beginning of the 16th century, gold figures of this flower as well as its seeds were brought to Europe. The sunflower is native to the Americas. The evidence thus far is that the sunflower was first domesticated in Mexico, by at least 2600 BC. It may have been domesticated a second time in the middle Mississippi Valley, or been introduced there from Mexico at an early date, as corn (maize) was. The earliest known examples of a fully domesticated sunflower north of Mexico have been found in Tennessee and date to around 2300 BC. Many indigenous American peoples used the sunflower as the symbol of the sun deity, including the Aztecs and the Otomi of Mexico and the Incas in South America. Gold images of the flower, as well as seeds, were taken back to Spain early in the 16th century. Some researchers argue that the Spaniards tried to suppress cultivation of the sunflower because of its association with solar religion and warfare.

During the 18th Century, the use of sunflower oil became very popular in Europe, particularly with members of the Russian Orthodox Church because sunflower oil was one of the few oils that was not prohibited during Lent.

Cultivation and Uses

To grow well, sunflowers need full sun. They grow best in fertile, moist, well-drained soil with a lot of mulch. In commercial planting, seeds are planted 45 cm (1.5 ft) apart and 2.5 cm (1 in) deep. Sunflower "whole seed" (fruit) are sold as a snack food, after roasting in ovens, with or without salt added. Sunflowers can be processed into a peanut butter alternative, Sunbutter, especially in China, Russia, the United States, the Middle East and Europe. In Germany, it is mixed together with rye flour to make *Sonnenblumenkernbrot* (literally: sunflower whole seed bread), which is quite popular in German-speaking Europe. It is also sold as food for birds and can be used directly in cooking and salads. Sunflower oil, extracted from the seeds, is used for cooking, as a carrier oil and to produce margarine and biodiesel, as it is cheaper than olive oil. A range of sunflower varieties exist with differing fatty acid compositions; some 'high oleic' types contain a higher level of healthy monounsaturated fats in their oil than even olive oil.

The cake remaining after the seeds have been processed for oil is used as a livestock feed. Some recently developed cultivars have drooping heads. These cultivars are less attractive to gardeners growing the flowers as ornamental plants, but appeal to farmers, because they reduce bird damage and losses from some plant diseases. Sunflowers also produce latex and are the subject of experiments to improve their suitability as an alternative crop for producing hypoallergenic rubber. Traditionally, several Native American groups planted sunflowers on the north edges of their gardens as a "fourth sister" to the better known three sisters combination of corn, beans, and squash. Annual species are often planted for their allelopathic properties. However, for commercial farmers growing commodity crops, the sunflower, like any other unwanted plant, is often considered a weed. Especially in the midwestern USA, wild (perennial) species are often found in corn and soybean fields and can have a negative impact on yields. Sunflowers may also be used to extract toxic ingredients from soil, such as lead, arsenic and uranium. They were used to remove uranium, cesium-137, and strontium-90 from soil after the Chernobyl accident.

Major Varieties

The following are varieties of sunflowers (in alphabetical order):

(*i*)   American Giant Hybrid
(*ii*) Arikara

(*iii*)   Autumn Beauty
(*iv*)   Aztec Sun
(*v*)   Black Oil
(*vi*)   Dwarf Sunspot
(*vii*)   Evening Sun
(*viii*)   Giant Primrose
(*ix*)   Indian Blanket Hybrid
(*x*)   Irish Eyes
(*xi*)   Italian White
(*xii*)   Kong Hybrid
(*xiii*)   Large Grey Stripe
(*xiv*)   Lemon Queen
(*xv*)   Mammoth Sunflower
(*xvi*)   Mongolian Giant
(*xvii*)   Orange Sun
(*xviii*)   Red Sun
(*xix*)   Ring of Fire
(*xx*)   Rostov
(*xxi*)   Soraya
(*xxii*)   Strawberry Blonde
(*xxiii*)   Sunny Hybrid
(*xxiv*)   Taiyo
(*xxv*)   Tarahumara
(*xxvi*)   Teddy Bear
(*xxvii*)   Titan
(*xxviii*)   Valentine
(*xxix*)   Velvet Queen

Other Species

The Maximillian sunflower (*Helianthus maximillianii*) is one of 38 species of perennial sunflower native to North America. The Land Institute and other breeding programmes are currently exploring the potential for these as a perennial seed crop. The Jerusalem artichoke (*Helianthus tuberosa*) is related to the sunflower, another example of perennial sunflower. The Mexican sunflower is *Tithonia rotundifolia*. It is only very distantly related to North American sunflowers. False sunflower refers to plants of the genus *Heliopsis*.

UNDERSTANDING TYPES OF NON-FOOD BIOENERGY CROPS

An energy crop is a plant grown as a low cost and low maintenance harvest used to make biofuels, or directly exploited for its energy content. Commercial energy crops are typically densely planted, high yielding crop species where the energy crops will be burnt to generate power. Woody crops such as Miscanthus, Willow or Poplar are widely utilised and tropical grasses such as [Napier Grass] or [Elephant Grass] receiving more attention by emerging energy crop companies. If carbohydrate content is desired for the production of biogas, whole-crops such as maize, Sudan grass, millet, white sweet clover and many others, can be made into silage and then converted into biogas. Through genetic modification and application of biotechnology plants can be manipulated to create greater yields, reduce associated costs and require less water. However, high energy yield can be realized with existing crops, especially maize.

(a) Solid Biomass

The terms biofuel, biomass, and so on, are often used interchangeably. Energy generated by burning plants grown for the purpose, often after the dry matter is pelletized. Energy crops are used for firing power plants, either alone or co-fired with other fuels. Alternatively they may be used for heat or combined heat and power (CHP) production.

(b) Gas Biomass (Methane)

Anaerobic digesters or biogas plants can be directly supplemented with energy crops once they have been ensiled into silage. The fastest growing sector of German biofarming has been in the area of "Renewable Energy Crops" (Nachwachsender Rohstoff = "NaWaRo") on nearly 500,000 ha land (2006). Energy crops can also be grown to boost gas yields where feedstocks have a low energy content, such as manures and spoiled grain. It is estimated that the energy yield presently of bioenergy crops converted via silage to methane is about 2 GWh/km2. Small mixed cropping enterprises with animals can use a portion of their acreage to grow and convert energy crops and sustain the entire farms energy requirements with about 1/5 the acreage. In Europe and especially Germany, however, this rapid growth has occurred only with substantial government support, as in the German bonus system for renewable energy (ranging from €0.02-€0.16/kWh made from renewable sources). Similar developments of integrating crop farming and bioenergy production via silage-methane have been almost entirely overlooked in N. America, where political and structural issues and a huge continued push to centralize energy production has overshadowed positive developments.

(c) Liquid Biomass

(*i*) *Biodiesel:* European production of biodiesel from energy crops has grown steadily in the last decade, principally focused on rapeseed used for oil and energy. In North America rapeseed was renamed "Canada Oil = Canola". Production of oil/biodiesel from rape covers more than 12,000 km² in Germany alone, and has doubled in the past 15 years. Typical yield of oil as pure biodiesel may be is 100,000 L/km² or more, making biodiesel crops economically attractive, provided sustainable crop rotations exist that are nutrient-balanced and preventative of the spread of disease such as clubroot. Biodiesel yield of soybeans is significantly lower than that of rape.

Table 5.6: Typical Oil Extractable by Weight

Crop	Oil %
Copra	62
Castor seed	50
Sesame	50
Groundnut kernel	42
Jatropha	40
Rapeseed	37
Palm kernel	36
Mustard Seed	35
Sunflower	32
Palm fruit	20
Soybean	14
Cotton seed	13

(ii) *Bioethanol:* Energy crops for biobutanol are grasses. A non-food crop for the production of cellulosic bioethanol is switchgrass. There has been a preoccupation with cellulosic bioethanol in America as the agricultural structure supporting biomethane is absent in many regions, with no credits or bonus system in place. Consequently a lot of private money and investor hopes are being pinned on marketable and patentable innovations in enzyme hydrolysis and the like and therefore America is viewed by some technology planners as falling further behind Europe in real bioenenergy gains. Bioethanol also refers to the technology of using animal and human grains, principally corn (maize seed) to make ethanol directly through fermentation, a process that is widely reputed to consume as much energy as it produces, therefore being non-sustainable. New developments in converting grain stillage (referred to as distillers grain stillage or DGS) into biogas energy looks promising as a means to improve the poor energy ratio of this type of bioethanol process. 2007 saw a setback in the economics of building grain refineries in the USA while the shipment of grains and ethanol by rail car has prompted the train industries largest growth phase since 50 years.

(d) By dedication

Dedicated energy crops are non-food energy crops as switchgrass, jatropha, and algae. Also byproducts (green waste) of food and non-food energy crops can be used to produce biofuels.

Energy Forestry

Energy forestry is a form of forestry in which a fast-growing species of tree or woody shrub is grown specifically to provide biomass or biofuel for heating or power generation. The two forms of energy forestry are short rotation coppice and short rotation forestry:

(i) Short rotation coppice are crops of Poplar or Willow, grown for 2 to 5 years before harvest.

(ii) Short rotation forestry are crops of Alder, Ash, Birch, Eucalyptus, Poplar, and Sycamore, grown for 8 to 20 years before harvest.

Benefits

The main advantage of using "grown fuels", as opposed to "fossil fuels" such as coal, natural gas and oil, is that while they are growing they absorb the near-equivalent in carbon dioxide (an important greenhouse gas) to that which is later released in their burning. Whereas by burning fossil fuels we are increasing atmospheric carbon unsustainably, by using carbon that was added to the earths carbon sink millions of years ago in processes which took millions of years to complete, and this is a prime cause of global warming. According to the FAO, compared to other energy crops, wood is among the most efficient sources of bioenergy in terms of quantity of energy released by unit of carbon emitted. Another advantage of generating energy from trees, as opposed to agricultural crops, is that trees do not have to be harvested each year, the harvest can be delayed when market prices are down, and the products can fulfil a variety of end-uses. Yields of some varieties can be as high as 12 oven dry tonnes every year. These crops can also be used in bank stabilisation and phytoremediation.

Problems

Although in many areas of the world government funding is still required to support large scale development of energy forestry as an industry, it is seen as a valuable component of the renewable energy network and will be increasingly important in the future.

Growing trees are relatively water intensive.

The system of energy forestry has faced criticism over food vs. fuel, whereby it has become financially profitable to replace food crops with energy crops.

Miscanthus Giganteus

Miscanthus giganteus is a large perennial grass (it can grow up to 13 feet in height) used for energy production. It is currently used commercially in the UK with a rapidly growing market demand. In addition to providing clean and affordable electricity and heat, Miscanthus is an environmentally friendly crop. Its large root system captures nutrients, and stems provide wildlife cover. As a high yielding, low input perennial, Miscanthus is also excellent for carbon sequestration and soil building.

Major Uses

Research trials being conducted in the United States and Ireland are making strides towards developing Miscanthus giganteus as a source of biomass for the production of energy either for direct combustion or through cellulosic ethanol or other biofuel production.

Panicum Virgatum

Panicum virgatum, commonly known as switchgrass, is a perennial warm season grass native to North America, where it occurs naturally from 55° N latitude in Canada southwards into the United States and Mexico. Switchgrass is one of the dominant species of the central North American tallgrass prairie and can be found in remnant prairies, in native grass pastures, and naturalized along roadsides. It is used primarily for soil conservation, forage production, game cover, as an ornamental grass, and more recently as a biomass crop for ethanol, fibre, electricity, and heat production. Other common names for switchgrass include tall panic grass, Wobsqua grass, blackbent, tall prairiegrass, wild redtop and thatchgrass.

Main Properties

Switchgrass is a hardy, deep rooted, perennial rhizomatous grass that begins growth in late spring. It can grow up to 1.8-2.2 m high but is typically shorter than Big Bluestem grass or Indiangrass. The leaves are 30-90 cm long, with a prominent midrib. Switchgrass uses C_4 carbon fixation, giving it an advantage in conditions of drought and high temperature. Its flowers have a well-developed panicle, often up to 60 cm long, and it bears a good crop of seeds. The seeds are 3-6 mm long and up to 1.5 mm wide, and are developed from a single-flowered spikelet. Both glumes are present and well developed. When ripe, the seeds sometimes take on a pink or dull-purple tinge, and turn golden brown with the foliage of the plant in the fall. Switchgrass is both a perennial and self-seeding crop, which means farmers do not have to plant and re-seed after annual harvesting. Once established, a switchgrass stand can survive for ten years or longer. Also, unlike corn, switchgrass can grow on marginal lands and requires relatively modest levels of chemical fertilizers. Overall, it is considered a resource-efficient, low-input crop for producing bioenergy from farmland.

Background

Much of North America, especially the prairies of the Midwestern United States, was once prime habitat to vast swaths of native grasses, including Switchgrass (*Panicum virgatum*), Indiangrass (*Sorghastrum nutans*), Eastern Gamagrass (*Tripsacum dactyloides*), Big Bluestem (*Andropogon gerardii*), Little Bluestem (*Schizachyrium scoparium*), and others. As European settlers began spreading west across the continent, the native grasses were plowed up and the land converted to growing crops such as corn, wheat, and oats. Introduced grasses such as fescue, bluegrass, and orchardgrass also replaced the native grasses for use as hay and pasture for cattle.

Distribution

Switchgrass is a very versatile and adaptable plant. It can grow and even thrive in many weather conditions, lengths of growing seasons, soil types, and land conditions. Its distribution spans south of latitude 55°N from Saskatchewan to Nova Scotia, south over most of the United States east of the Rocky Mountains, and further south into Mexico. As a warm season perennial grass, most of its growth occurs from late spring through early fall, becoming dormant and unproductive during colder months. Thus, the productive season in its northern habitat can be as short as three months, but in the southern reaches of its habitat, the growing season may be as long as eight months, around the

Gulf Coast area. Switchgrass is a diverse species, with striking differences between plants. This diversity, which presumably reflects evolution and adaptation to new environments as the species spread across the continent, provides a range of valuable traits for breeding programmes. Switchgrass has two distinct forms, or "cytotypes": the lowland cultivars, which tend to produce more biomass, and the upland cultivars, which are generally of more northern origin, more cold tolerant, and therefore usually preferred in northern areas. Upland switchgrass types are generally shorter (£ 8 ft, or 2.4 m, tall) and less coarse than lowland types. Lowland cultivars may grow to ³ 9 ft, or 2.7 m, in favourable environments. Both upland and lowland cultivars are deeply rooted (> 6 ft, or 1.8 m, in favourable soils) and have short rhizomes. The upland types tend to have more vigorous rhizomes. Subsequently, the lowland cultivars may appear to have a bunchgrass habit, while the upland types tend to be more sod forming. Lowland cultivars appear more plastic in their morphology, and produce larger plants if stands become thin or when planted in wide rows. On the other hand, lowland types seem to be more sensitive to moisture stress than upland cultivars. In native prairies, switchgrass is historically found in association with several other important native tallgrass prairie plants, such as big bluestem, indiangrass, little bluestem, sideoats grama, eastern gamagrass, and various forbs (sunflowers, gayfeather, prairie clover, and prairie coneflower). These widely adapted tallgrass species once occupied millions of hectares.

Establishment and Management

Switchgrass can be grown on land that typically isn't well suited to row crop production, including land that is too erodible for corn production as well as sandy and gravelly soils in humid regions that typically produce low yields of other farm crops. No single method of establishing switchgrass can be suggested for all situations. The crop can be established both by no-till and conventional tillage. When seeded as part of diverse mixture, planting guidelines for warm season grass mixtures for conservation plantings should be followed. Regional guidelines for growing and managing switchgrass for bioenergy or conservation plantings are available. Several key factors can increase the likelihood of success for establishing switchgrass. These include:

(*i*) Planting switchgrass after the soil is well warmed during the spring.

(*ii*) Using seeds that are highly germinable and planting 1/4 inch to 1/2 inch deep, or up to 3/4 inch deep in sandy soils.

(*iii*) Packing or firming the soil both before and after seeding.

(*iv*) Providing no fertilization at planting to minimize competition.

(*v*) Controlling weeds with chemical and/or cultural control methods.

Mowing and properly labeled herbicides are recommended for weed control. Chemical weed control can be used in the fall prior to establishment, pre-plant and post-plant. Weeds should be mowed just above the height of the growing switchgrass. Hormone herbicides such as 2,4-D should be avoided as they are known to reduce development of switchgrass when applied early in the establishing year. Plantings that appear to have failed due to weed infestations are often wrongly assessed, as the failure is often more apparent than real. Switchgrass stands that are initially weedy commonly become well established with appropriate management in subsequent years. Once established, switchgrass can take up to three years to reach its full production potential. Depending on the region, it can typically produce 1/4 to 1/3 of its yield potential in its first year and 2/3 of its potential in the year after seeding. After establishment, switchgrass management will depend on the goal of the seeding. Historically, most switchgrass seedings have been managed for the Conservation Reserve Programme in the US. Disturbance such as periodic mowing, burning, or discing is required to optimize the stand's utility for encouraging biodiversity. Presently, increased attention is being placed on switchgrass management as an energy crop. Generally, the crop requires modest application of nitrogen fertilizer as it is not a heavy feeder. Typical nitrogen (N) content of senescent material in the fall is 0.5 per cent N. Fertilizer nitrogen applications of about 5 kg N/hectare (ha) applied for each tonne of biomass removed is a general guideline. More specific recommendations for fertilization are available regionally in North America. Herbicides are not often used on switchgrass after the seeding year, as the crop is generally quite competitive with weeds. Most bioenergy conversion processes for switchgrass, including those for cellulosic ethanol and pellet fuel production, can generally accept some alternative species in the harvested biomass. Stands of switchgrass should be harvested no more than twice per year, and one cutting often provides as much biomass as two. Switchgrass can be harvested with the same field equipment used for hay production and it is well-suited to baling or bulk field harvesting. If its biology is properly taken into consideration, switchgrass can offer great potential as an energy crop.

Uses

Switchgrass can be used as a biomass feedstock for energy production, as ground cover for soil conservation and to control erosion, for forages and grazing, and as game cover. It can be used by cattle farmers for hay and pasture and as a substitute for wheat straw in a diversity of applications, including livestock bedding, straw bale housing, and as a substrate for growing mushrooms. Additionally, switchgrass is grown as a drought resistant ornamental grass in average to wet soils and in full sun to part shade.

Bioenergy

Switchgrass has been researched as a bioenergy crop since the mid-1980s, because it is a native perennial warm season grass that has the ability to produce moderate to high yields on marginal farmlands. It is now being considered for use in a diversity of bioenergy conversion processes, including cellulosic ethanol production, biogas, and direct combustion for thermal energy applications. The main agronomic advantages of switchgrass as a bioenergy crop are its stand longevity, drought and flooding tolerance, relatively low herbicide and fertilizer input requirements, ease of management, hardiness in poor soil and climate conditions, and widespread adaptability in temperate climates. In some warm humid southern zones such as Alabama it has the ability to produce up to 25 ODT/ha. A summary of switchgrass yields across 13 research trial sites in the United States found the top two cultivars in each trial to yield 9.4 to 22.9 t/ha, with an average yield of 14.6 ODT/ha. However, these yields were recorded on small plot trials, and commercial field sites could be expected to be at least 20 per cent lower than these results. In the United States, switchgrass yields appear to be highest in warm humid regions with long growing seasons such as the US Southeast and lowest in the dry short season areas of the Northern Great Plains. The energy inputs required to grow switchgrass are favourable when compared with annual seed bearing crops such as corn, soybean, or canola, which can require relatively high energy inputs for field operations, crop drying, and fertilization. Whole plant herbaceous perennial C4 grass feedstocks are desirable biomass energy feedstocks, as they require fewer fossil energy inputs to grow and effectively capture solar energy because of their C4 photosynthetic system and perennial nature. One study cites that it takes from 0.97 to 1.34 GJ to produce 1 tonne of switchgrass, compared with 1.99 to 2.66 GJ to produce 1 tonne of corn. Another study found that switchgrass uses 0.8 GJ per oven dry tonne (ODT) of fossil energy compared to grain corn's 2.9 GJ/ODT. Given that switchgrass contains approximately 18.8 GJ/ODT of biomass, the energy output-to-input ratio for the crop can be up to 20:1. This highly favourable ratio is attributable to its relatively high energy output per hectare and low energy inputs for production. Presently, a great deal of effort is being put into developing switchgrass as a cellulosic ethanol crop in the USA. In his 2006 State of the Union address, President George W. Bush proposed the usage of switchgrass for ethanol; since then, over $100 million has been invested into researching switchgrass as a potential biofuel source. Switchgrass has the potential to produce up to 100 gallons (380 liters) of ethanol per metric ton harvested. However, current technology for herbaceous biomass conversion to ethanol is about 90 gallons (340 liters) per tonne. In contrast, corn ethanol yields about 106 gallons (400 liters) per tonne. The main advantage of using switchgrass over corn as an ethanol feedstock is that its cost of production is generally about 1/2 that of grain corn and more biomass energy per hectare can be captured in the field. Thus, switchgrass cellulosic ethanol should give a higher yield of ethanol per hectare at lower cost. However, this will depend on cellulosic ethanol plant construction and ethanol processing costs being reduced considerably. The switchgrass ethanol industry energy balance is also considered to be substantially better than that of corn ethanol. During the bioconversion process, the lignin fraction of switchgrass can be burned to provide sufficient steam and electricity to operate the biorefinery. Studies have found, for example, that for every unit of energy input needed to create a biofuel from switchgrass, four units of energy are yielded. In contrast, corn ethanol yields about 1.28 units of energy for each unit of energy input. A recent study from the Great Plains indicated that for ethanol production from switchgrass, this figure is 5.4, or alternatively, that 540 per cent more energy was contained in the ethanol produced than was used in growing the switchgrass and converting it to liquid fuel. However, there remain commercialization barriers to the development of cellulosic ethanol technology. Projections in the early 1990s for commercialization of cellulosic ethanol by the year 2000 have not been met. The commercialization of cellulosic ethanol is thus proving to be a significant challenge, despite noteworthy research efforts. Thermal energy applications for switchgrass appear to be closer to near-term scale-up than cellulosic ethanol for industrial or small-scale applications. For example, switchgrass can be pressed into fuel pellets that are subsequently burned in pellet stoves used to heat homes (which typically burn corn or wood pellets). Switchgrass has been widely tested as a suitable fuel for substituting for coal in power generation. The most widely studied project to date has been the Chariton Valley Project in Iowa. As well, the Show-Me-Energy Cooperative (SMEC) in Missouri is using switchgrass and other warm season grasses along with wood residues as feedstocks for pellets used for the firing of a coal-fired power plant. In Eastern Canada, switchgrass is being used on a pilot scale as a feedstock for commercial heating applications. Combustion studies have been undertaken and it appears to be well-suited as a commercial boiler fuel. Research is also being undertaken to develop switchgrass as a pellet fuel because of lack of surplus wood residues in Eastern Canada, as a slowdown in the forest products industry in 2009 is now resulting in wood pellet shortages throughout Eastern North America. Generally speaking, the direct firing of switchgrass for thermal applications can provide the highest net energy gain and energy output-to-input ratio of all switchgrass bioconversion processes. Research has found that switchgrass, when pelletized and used as a solid biofuel, is a highly efficient strategy for displacing fossil fuels. Switchgrass pellets were identified to have a 14.6:1 energy output to input ratio, which is substantially better than that for liquid biofuel options from farmland. As a greenhouse gas mitigation strategy, switchgrass pellets were found to be a highly effective means to use farmland to mitigate greenhouse gases. Using farmland to produce switchgrass pellets could mitigate 7.6-13 tonnes per hectare of CO_2. In contrast, switchgrass cellulosic ethanol and corn ethanol were found to mitigate 5.2 and 1.5 tonnes of CO_2 per hectare, respectively. Historically, the major constraint to the development of grasses for thermal energy applications has been the difficulty associated with burning grasses in conventional boilers, as biomass quality problems can be of particular concern in combustion applications. These technical problems now appear to have been largely resolved through crop management practices such as fall mowing and spring harvesting that allow for leaching to occur, which leads to fewer aerosol forming compounds (such as K, and Cl) and N in the grass. This reduces clinker formation and corrosion and enables switchgrass to be a clean combustion fuel source for use in smaller combustion appliances. Fall harvested grasses likely have more application for larger commercial and industrial boilers. In regions where the potassium and chlorine contents of switchgrass cannot be successfully leached out for thermal applications, it may be that biogas applications for switchgrass will prove more promising. Switchgrass has demonstrated some promise in biogas research as an alternative feedstock to whole plant corn silage for biogas digesters. Switchgrass is also currently being used to heat small industrial and farm buildings in Germany and China through a process used to make a low quality natural gas substitute.

Soil Conservation

Switchgrass is useful for soil conservation and amendment, particularly in the United States and Canada where switchgrass is endemic. Switchgrass has a deep fibrous root system—nearly as deep as the plant is tall. Since it, along with other native grasses and forbs, once covered the plains of the United States that are now the Corn Belt, the effects of the past switchgrass habitat have been beneficial, lending to the fertile farmland that exists today. The deep fibrous root systems of switchgrass left a very deep rich layer of organic matter in the soils of the Midwest, making those mollisol soils some of the most productive in the world. By returning switchgrass and other perennial prairie grasses as an agricultural crop, many marginal soils may benefit from increased levels of organic material, permeability, and fertility, due to the grass's deep root system. Soil erosion, both from wind and water, is of great concern in regions where switchgrass grows. Due to its height, switchgrass can form an effective wind erosion barrier. Its root system, also, is excellent for holding soil in place, which helps prevent erosion from flooding and runoff. Some highway departments (for example, KDOT) have used switchgrass in their seed mixes when re-establishing growth along roadways. It can also be used on strip mine sites, dikes, and pond dams. Conservation districts in many parts of the United States use it to control erosion in grass waterways because of its excellent ability to anchor soils while also doubling as native habitat for wildlife.

Forages and Grazing

Switchgrass is an excellent forage for cattle; however, it has shown toxicity in horses, sheep, and goats through chemical compounds known as saponins, which cause photosensitivity and liver damage in these animals. Researchers are continuing to learn more about the specific conditions under which switchgrass causes harm to these species, but until more is discovered, it is recommended that switchgrass not be fed to them. For cattle, however, it can be fed as hay, or grazed. Grazing switchgrass calls for watchful management practices to ensure survival of the stand. It is recommended that grazing begin when there is 18-22 inches of growth, to stop grazing when there are 8-12 inches of stubble left, and to rest the pasture 30-45 days between grazing periods. Switchgrass becomes very stemmy and unpalatable as it matures, but during the target grazing period, it is a highly favourable forage with a relative feed value (RFV) of 90-104. The grass' upright growth pattern places its growing point off the soil surface onto its stem, so leaving 8-12 inches of stubble is important for regrowth. When harvesting switchgrass for hay, the first cutting occurs at the late boot stage—around mid-June. This should allow for a second cutting in mid-August, leaving enough regrowth to survive the winter.

Game Cover

Switchgrass is well-known among wildlife conservationists as a favourite forage and habitat among upland game bird species such as pheasant, quail, grouse, wild turkey, and song birds, with its plentiful small seeds and tall cover. Depending on how thickly switchgrass is planted, and what it is partnered with, it also offers excellent forage and cover for a wide variety of other wildlife across the country. For those producers who have switchgrass stands on their farm, it is considered an environmental and aesthetic benefit due to the abundance of wildlife attracted by the switchgrass stands. Some members of Prairie Lands Bio-Products, Inc. in Iowa have even turned this benefit into a profitable business by leasing their switchgrass land for hunting during the proper seasons. The benefits to wildlife can be extended even in large scale agriculture through the process of strip harvesting, as recommended by The Wildlife Society, which suggests that rather than harvesting an entire field at once, strip harvesting could be practiced so that the entire habitat is not removed, thereby protecting the wildlife that has inhabited the switchgrass.

Wood Fuel

Wood fuel is wood used as fuel. The burning of wood is currently the largest use of energy derived from a solid fuel biomass. Wood fuel can be used for cooking and heating, and occasionally for fueling steam engines and steam turbines that generate electricity. Wood fuel may be available as firewood (eg. logs, bolts, blocks), charcoal, chips, sheets, pellets and sawdust. The particular form used depends upon factors such as source, quantity, quality and application. Sawmill waste and construction industry by-products also include various forms of lumber tailings. Wood may be burned in a furnace, stove, fireplace, or in a campfire, or used for a bonfire. Wood is the most easily available form of fuel, requiring no tools in the case of picking up dead wood, or little tools, although as in any industry, specialized tools, such as skidders and hydraulic wood splitters, have evolved to mechanize production. The discovery of how to make fire for the purpose of burning wood is regarded as one of humanity's most important advances.

Historical Development

The use of wood as a fuel source for heating is as old as civilization itself. Historically, it was limited in use only by the distribution of technology required to make a spark. **Wood heat** is still common throughout much of the world.

Early examples include the use of wood heat in tents. Fires

were constructed on the ground, and a smoke hole in the top of the tent allowed the smoke to escape by convection.

In permanent structures and in caves, hearths were constructed or established—surfaces of stone or another noncombustible material upon which a fire could be built. Smoke escaped through a smoke hole in the roof.

Wood has been used as fuel for millennia. The Greeks, Romans, Celts, Britons, and Gauls all had access to forests suitable for using as fuel. Over the centuries there was a partial deforestation of climax forests and the evolution of the remainder to coppice with standards woodland as the primary source of wood fuel. These woodlands involved a continuous cycle of new stems harvested from old stumps, on rotations between seven and thirty years. One of the earliest printed books in English was John Evelyn "Sylva, or a discourse on forest trees" (1664) advising landowners on the proper management of forest estates. H.L.Edlin, in "Woodland Crafts in Britain", 1949 outlines the extraordinary techniques employed, and range of wood products that have been produced from these managed forests since pre-roman times. And throughout this time the preferred form of wood fuel was the branches of cut coppice stems bundled into faggots. Larger, bent or deformed stems that were of no other use to the woodland craftsmen were converted to charcoal.

As with most of Europe, these managed woodlands continued to supply their markets right up to the end of World War two. Since then much of these woodlands have been converted to broadscale agriculture. Total demand for fuel increased considerably with the industrial revolution but most of this increased demand was met by the new fuel source, Coal, which was more compact and more suited to the larger scale of the new industries.

The development of the chimney and the fireplace allowed for more effective exhaustion of the smoke. Masonry heaters or stoves went a step further by capturing much of the heat of the fire and exhaust in a large thermal mass, becoming much more efficient than a fireplace alone.

The metal stove was a technological development concurrent with the industrial revolution. Stoves were manufactured or constructed pieces of equipment that contained the fire on all sides and provided a means for controlling the draft—the amount of air allowed to reach the fire. Stoves have been made of a variety of materials. Cast iron is among the more common. Soapstone (talc), tile, and steel have all been used. Metal stoves are often lined with refractory materials such as firebrick, since the hottest part of a woodburning fire will burn away steel over the course of several years' use.

The Franklin stove was developed in the United States by Benjamin Franklin. More a manufactured fireplace than a stove, it had an open front and a heat exchanger in the back that was designed to draw air from the cellar and heat it before releasing it out the sides. The heat exchanger was never a popular feature and was omitted in later versions. So-called "Franklin" stoves today are made in a great variety of styles, though none resembles the original design.

The 1800s became the high point of the cast iron stove. Each local foundry would make their own design, and stoves were built for myriads of purposes—parlour stoves, box stoves, camp stoves, railroad stoves, portable stoves, cooking stoves and so on. Elaborate nickel and chrome edged models took designs to the edge, with cast ornaments, feet and doors. Wood or coal could be burnt in the stoves and thus they were popular for over one hundred years. The action of the fire, combined with the causticity of the ash, ensured that the stove would eventually disintegrate or crack over time. Thus a steady supply of stoves was needed. The maintenance of stoves, needing to be blacked, their smokiness, and the need to split wood meant that oil or electric heat found favour.

The airtight stove, originally made of steel, allowed greater control of combustion, being more tightly fitted than other stoves of the day. Airtight stoves became common in the 19th century.

Use of wood heat declined in popularity with the growing availability of other, less labor-intensive fuels. Wood heat was gradually replaced by coal and later by fuel oil, natural gas and propane heating except in rural areas with available forests.

After the 1967 Oil Embargo, many in the United States used wood for the first time. The EPA provided information on clean stoves, which burned much more efficiently.

Origin and Evolution

The use of wood as a fuel source for heating is as old as civilization itself. Historically, it was limited in use only by the distribution of technology required to make a spark. Wood heat is still common throughout much of the world.

Early examples include the use of wood heat in tents. Fires were constructed on the ground, and a smoke hole in the top of the tent allowed the smoke to escape by convection. In permanent structures and in caves, hearths were constructed or established—surfaces of stone or another non-combustible material upon which a fire could be built. Smoke escaped through a smoke hole in the roof. Wood has been used as fuel for millennia. The Greeks, Romans, Celts, Britons, and Gauls all had access to forests suitable for using as fuel. Over the centuries there was a partial deforestation of climax forests and the evolution of the remainder to coppice with standards woodland as the primary source of wood fuel. These woodlands involved a continuous cycle of new stems harvested from old stumps, on rotations between seven and thirty years. One of the earliest printed books in English was John Evelyn "Sylva, or a discourse on forest trees" (1664) advising landowners on the proper management of forest estates. H.L. Edlin, in "Woodland Crafts in Britain", 1949 outlines the extraordinary techniques employed, and range of wood products that have been produced from these managed forests since pre-roman times. And throughout this time the preferred form of wood fuel was the branches of cut coppice stems bundled into faggots. Larger, bent or deformed stems that were of no other use to the woodland craftsmen were converted to charcoal. As with most of Europe, these managed woodlands continued to supply their markets right up to the end of World War two. Since then much of these woodlands have been converted to broad scale agriculture. Total demand for fuel increased considerably with the industrial revolution but most of this increased demand was met by the new fuel source, Coal, which was more compact and more suited to the larger scale of the new industries. The development of the chimney and the fireplace allowed for more effective exhaustion of the smoke. Masonry heaters or stoves went a step further by capturing much of the heat of the fire and exhaust in a large thermal mass, becoming much more efficient than a fireplace alone. The metal stove was a technological development concurrent with the industrial revolution. Stoves were manufactured or constructed pieces of equipment that contained the fire on all sides and provided a means for controlling the draft—the amount of air allowed to reach the fire. Stoves have been made of a variety of materials. Cast iron is among the more common. Soapstone (talc), tile, and steel have all been used. Metal stoves are often lined with refractory materials such as firebrick, since the hottest part of a woodburning fire will burn away steel over the course of several years' use. The Franklin stove was developed in the United States by Benjamin Franklin. More a manufactured fireplace than a stove, it had an open front and a heat exchanger in the back that was designed to draw air from the cellar and heat it before releasing it out the sides. The heat exchanger was never a popular feature and was omitted in later versions. So-called "Franklin" stoves today are made in a great variety of styles, though none resembles the original design. The 1800s became the high point of the cast iron stove. Each local foundry would make their own design, and stoves were built for myriads of purposes—parlour stoves, box stoves, camp stoves, railroad stoves, portable stoves, cooking stoves and so on. Elaborate nickel and chrome edged models took designs to the edge, with cast ornaments, feet and doors. Wood or coal could be burnt in the stoves and thus they were popular for over one hundred years. The action of the fire, combined with the causticity of the ash, ensured that the stove would eventually disintegrate or crack over time. Thus a steady supply of stoves was needed. The maintenance of stoves, needing to be blacked, their smokiness, and the need to split wood meant that oil or electric heat found favour. The airtight stove, originally made of steel, allowed greater control of combustion, being more tightly fitted than other stoves of the day. Airtight stoves became common in the 19th century. Use of wood heat declined in popularity with the growing availability of other, less labour-intensive fuels. Wood heat was gradually replaced by coal and later by fuel oil, natural gas and propane heating except in rural areas with available forests. After the 1967 Oil Embargo, many in the United States used wood for the first time. The EPA provided information on clean stoves, which burned much more efficiently.

Firewood

Some firewood is harvested in "woodlots" managed for that purpose, but in heavily wooded areas it is more usually harvested as a byproduct of natural forests. Deadfall that has not started to rot is preferred, since it is already partly seasoned. Standing dead timber is considered better still, as it is both seasoned, and has less rot. Harvesting this form of timber reduces the speed and intensity of bushfires. Harvesting timber for firewood is normally carried out by hand with chainsaws. Thus, longer pieces—requiring less manual labour, and less chainsaw fuel—are less expensive and only limited by the size of their firebox. Prices also vary considerably with the distance from wood lots, and quality of the wood.

Firewood usually relates to timber or trees unsuitable for building or construction. Firewood is a renewable resource provided the consumption rate is controlled to sustainable levels. The shortage of suitable firewood in some places has seen local populations damaging huge tracts of bush thus leading to further desertification.

Energy Content

A common hardwood, red oak, weighs 3757 pounds per cord, with an energy content of 24 million BTU per cord, and 16.8 million recoverable BTU if burned at 70 per cent efficiency. The Sustainable Energy Development Office (SEDO), part of the Government of Western Australia states that the energy content of wood is 4.5 kWh/kg or 16.2 gigajoules/tonne (GJ/t).

Measurement of firewood

In the metric system, firewood is normally sold by the stere (1 m3 = ~0.276 cords). In the United States, firewood is usually sold by the cord, 128 ft3 (3.62 m3), corresponding to a woodpile 8 ft wide × 4 ft high of 4 ft-long logs. The cord is legally defined by statute in most states. A "thrown cord" is firewood that has not been stacked and is defined as 4 feet wide × 4 feet tall × 10 feet long. The additional volume is to make it equivalent to a standard stacked cord, where there is less void space. It is also common to see wood sold by the "face cord", which is usually *not* legally defined, and varies from one area to another. For example, in one state a pile of wood 8 feet wide × 4 feet high of 16"-long logs will often be sold as a "face cord", though its volume is only one-third of a cord. In another state, or even another area of the same state, the volume of a face cord may be considerably different. Hence, it is risky to buy wood sold in this manner, as the transaction is not based on a legally enforceable unit of measure.

Combustion by-products

As with any fire, burning wood fuel creates numerous by-products, some of which may be useful (heat and steam), and others that are undesirable, irritating or dangerous. One by-product of wood burning is wood ash, which in moderate amounts is a fertilizer (mainly potash), contributing minerals, but is strongly alkaline as it contains potassium hydroxide (lye). Wood ash can also be used to manufacture soap. Smoke, containing water vapor, carbon dioxide and other chemicals and aerosol particulates, can be an irritating (and potentially dangerous) by-product of partially burnt wood fuel. A major component of wood smoke is fine particles that may account for a large portion of particulate air pollution in some regions. During cooler months, wood heating accounts for as much as 60 per cent of fine particles in Melbourne, Australia.

Slow combustion stoves increase efficiency of wood heaters burning logs, but also increase particulate production. Low pollution/slow combustion stoves are a current area of research. An alternative approach is to use pyrolysis to produce several

useful biochemical byproducts, and clean burning charcoal, or to burn fuel extremely quickly inside a large thermal mass, such as a masonry heater. This has the effect of allowing the fuel to burn completely without producing particulates while maintaining the efficiency of the system. In some of the most efficient burners, the temperature of the smoke is raised to a much higher temperature where the smoke will itself burn (e.g., 1,200 degrees for igniting carbon monoxide gas). This may result in significant reduction of smoke hazards while also providing additional heat from the process. By using a catalytic converter, the temperature for obtaining cleaner smoke can be reduced. Some U.S. jurisdictions prohibit sale or installation of stoves that do not incorporate catalytic converters.

Combustion by-product Effects on Human Health

Depending on population density, topography, climatic conditions and combustion equipment used, wood heating may substantially contribute to air pollution, particularly particulates. Wood combustion products can include toxic and carcinogenic substances. The conditions in which wood is burnt will greatly influence the content of the emission. Particulate air pollution can contribute to human health problems and increased hospital admissions for asthma and heart diseases. The technique of compressing wood pulp into pellets or artificial logs can reduce emissions. The combustion is cleaner, and the increased wood density and reduced water content can eliminate 3 to 7 per cent of the transport bulk. Thus the fossil energy consumed in transport is reduced (and in fact represents a tiny fraction of the fossil fuel consumed in producing and distributing heating oil or gas).

Environmental Impact

- *Harvesting Operations*: Much wood fuel comes from native forests around the world. Plantation wood is rarely used for firewood, as it is more valuable as timber or wood pulp. The collection or harvesting of this wood can have serious environmental implications for the collection area. The concerns are often specific to the particular area, but can include all the problems that regular logging create. The heavy removal of wood from forests can cause habitat destruction and soil erosion.
- *Greenhouse Gases*: Wood burning does *not* release more carbon dioxide than its biodegradation (rottening). Wood burning can therefore be called "carbon neutral". Of course, harvesting and transport operations can produce significant amounts of greenhouse gas pollution. Significant amount of the carbon stored in trees is released as CO_2, and is not carbon neutral when used as a clearcutting measure.

Wood Fuels Around the World

- *European Use of Wood Fuel*: Some countries produce a significant fraction of their electricity needs from wood or wood wastes. Sweden, for example produces 1,490 megawatts of electricity this way and Austria produces 747 megawatts. The Swedish figure corresponds circa 4.5 per cent of the nation's total installed production capacity (33 400 MW in 2003). In Finland, there is a growing interest in using wood waste as fuel for home and industrial heating, in the form of compacted pellets. In Scandinavian countries the costs of manual labour to process firewood is very high. Therefore it is common to import firewood from countries with cheap labour and natural resources. The main exporters to Scandinavia are the Baltic countries (Estonia, Lithuania, and Latvia).

- *Historic Japanese Use of Wood Fuel*: Wood, during the Edo period, was used for many purposes, and the consumption of wood led Japan to develop a forest management policy during that era. Demand for timber resources was on the rise not only for fuel, but also for construction of ships and buildings, and consequently deforestation was widespread. As a result, forest fires occurred, along with floods and soil erosion. Around 1666, the shogun made it a policy to reduce logging and increase the planting of trees. This policy decreed that only the shogun, and/or a daimyo, could authorize the use of wood. By the 18th century, Japan had developed detailed scientific knowledge about silviculture and plantation forestry.

- *Firewood Use in Australia*: About 1.5 million households in Australia use firewood as the main form of domestic heating. As of 1995, approximately 1.85 million cubic metres of firewood (1m3 equals approximately one car trailer load) was used in Victoria annually, with half being consumed in Melbourne. This amount is comparable to the wood consumed by all of Victoria's sawlog and pulplog forestry operations (1.9 million m3).

  Species used as sources of firewood include:

  (i)   Red Gum, from forests along the Murray River (the Mid-Murray Forest Management Area, including the Barmah and Gunbower forests, provides about 80% of Victoria's red gum timber).

  (ii) Box and Messmate Stringybark, in southern Australia.

  (iii) Sugar gum, a wood with high thermal efficiency that usually comes from small plantations.

- *Environmental Concerns*: In Victoria, red gum is the most popular and commonly used firewood. Although some consider that Victoria's red gum forests are being depleted wholesale as a direct result of firewood harvesting, much of the forest clearance occurred decades ago, when the wood was used in large-scale infrastructure projects such as railway construction. Victoria and NSW's remnant red gum forests, including the Murray River's Barmah-Millewa forest, are increasingly being clear-felled using mechanical harvesters, destroying ecologically significant and already endangered habitat. Macnally estimates that approximately 81 per cent of fallen timber has been removed from the southern Murray-Darling Basin. There are concerns with the extent to which firewood harvesting in red gum forests depletes the habitat of various animal species that would otherwise reside in the hollows which form in fallen timber (it is estimated that 37 per cent of mammals and 39 per cent of woodland bird species in Victoria reside in such hollows). In areas where there is extensive firewood harvesting (eg. Barmah), there are concerns that many species may be at risk. The Barmah forest contains 51 per cent of the threatened species found in eastern north Victoria. At least 37 threatened plants are found in Barmah-Millewa, four of which are found nowhere else in Victoria. The forest includes the only remaining Victorian breeding grounds of the Superb Parrot, a bird that is listed as endangered in that state, and as vulnerable on the IUCN Red List. Mammals listed as threatened in both Victoria and New South Wales reside in the forest, including the squirrel glider, the brush tailed phascogale, and the large footed myotis.

Efficiency and Sustainability

With appropriately certified and operated modern wood heaters, the use of good quality wood fuel is one of the most efficient and cheapest forms of heating in Australia. The replacement of existing national domestic heating needs supplied by wood with gas and electricity would result in a significant net increase in carbon dioxide emissions, while the application and enforcement of national standards for wood heaters and wood fuel would substantially reduce particulate emissions. The peak industry body, the Australian Home Heating Association Inc is a major financial supporter of Landcare Australia, sponsoring the planting of over 40,000 trees per year. Landcare groups have planted millions of trees in revegetation programmes to replace the estimated 20 billion trees removed since European settlement, laid thousands of kilometres of protective fencing, introduced sustainable farming techniques, removed hundreds of thousands of tonnes of weeds, and volunteered countless hours to the land care ethic. Firewood plantations also provide alternative financial opportunities for farmers and local government, with fuel being one of the multi-uses of tree plantations.

1973 Energy Crisis

A brief resurgence in popularity occurred during and after the 1973 energy crisis, when some believed that fossil fuels would become so expensive as to preclude their use. A period of innovation followed, with many small manufacturers producing stoves based on designs old and new. Notable innovations from that era include the Ashley heater, a thermostatically-controlled stove with an optional perforated steel enclosure that prevented accidental contact with hot surfaces. A number of dual-fuel furnaces and boilers were made, which utilized ductwork and piping to deliver heat throughout a house or other building. The growth in popularity of wood heat also led to the development and marketing of a greater variety of equipment for cutting and splitting wood. Consumer grade hydraulic log splitters were developed to be powered by electricity, gasoline, or PTO of farm tractors.

Pellet Stove

A pellet stove is an appliance that burns compressed wood or biomass pellets. Wood heat continues to be used in areas where firewood is abundant. For serious attempts at heating, rather than mere ambiance (open fireplaces), stoves, fireplace inserts, and furnaces are most commonly used today. In rural, forested parts of the U.S., freestanding boilers are increasingly common. They are installed outdoors, some distance from the house, and connected to a heat exchanger in the house using underground piping. The mess of wood, bark, smoke, and ashes is kept outside and the risk of fire is reduced. The boilers are large enough to hold a fire all night, and can burn larger pieces of wood, so that less cutting and splitting is required. There is no need to retrofit a chimney in the house. However, outdoor wood boilers emit more wood smoke and associated pollutants than other wood-burning appliances. This is due to design characteristics such as the water-filled jacket surrounding the firebox, which acts to cool the fire and leads to incomplete combustion. Outdoor wood boilers also typically have short stack heights in comparison to other wood-burning appliances, contributing to ambient levels of particulates at ground level. An alternative that is increasing in popularity are wood gasification boilers, which burn wood at very high efficiencies (85-91%) and can be placed indoors or in an outbuilding. Wood is still used today for cooking in many places, either in a stove or an open fire. It is also used as a fuel in many industrial processes, including smoking meat and making maple syrup. As a sustainable energy source, wood fuel also remains viable for generating electricity in areas with easy access to forest products and by-products.

Retail Cost

United States
    In 2008, wood for fuel cost $15.15 per 1 million BTUs.

Potential use in renewable energy technologies

(*i*)   Pellet stove,
(*ii*) Wood pellets,
(*iii*) Efficient stove for developing nations,
(*iv*) Sawdust can be pelletized.

6

Biodiesel: Current Production, Benefits and Future Prospects

BIODIESEL AS A FUTURE FUEL

Biodiesel refers to a vegetable oil-or animal fat-based diesel fuel consisting of long-chain alkyl (methyl, propyl or ethyl) esters. Biodiesel is typically made by chemically reacting lipids (e.g., vegetable oil, animal fat (tallow)) with an alcohol.

Biodiesel is meant to be used in standard diesel engines and is thus distinct from the vegetable and waste oils used to fuel *converted* diesel engines. Biodiesel can be used alone, or blended with petrodiesel.

The term "biodiesel" is standardized as mono-alkyl ester in the United States.

Blends

Blends of biodiesel and conventional hydrocarbon-based diesel are products most commonly distributed for use in the retail diesel fuel marketplace. Much of the world uses a system known as the "B" factor to state the amount of biodiesel in any fuel mix: fuel containing 20 per cent biodiesel is labeled B20, while pure biodiesel is referred to as B100. It is common in the USA to see B99.9 because a federal tax credit is awarded to the first entity which blends petroleum diesel with pure biodiesel. Blends of 20 percent biodiesel with 80 percent petroleum diesel (B20) can generally be used in unmodified diesel engines. Biodiesel can also be used in its pure form (B100), but may require certain engine modifications to avoid maintenance and performance problems. Blending B100 with petroleum diesel may be accomplished by:

- Mixing in tanks at manufacturing point prior to delivery to tanker truck
- Splash mixing in the tanker truck (adding specific percentages of Biodiesel and petroleum diesel)
- In-line mixing, two components arrive at tanker truck simultaneously.
- Metered pump mixing, petroleum diesel and Biodiesel meters are set to X total volume, transfer pump pulls from two points and mix is complete on leaving pump.

Applications

Biodiesel can be used in pure form (B100) or may be blended with petroleum diesel at any concentration in most modern diesel engines. Biodiesel has different solvent properties than petrodiesel, and will degrade natural rubber gaskets and hoses in vehicles (mostly vehicles manufactured before 1992), although these tend to wear out naturally and most likely will have already been replaced with FKM, which is nonreactive to biodiesel. Biodiesel has been known to break down deposits of residue in the fuel lines where petrodiesel has been used. As a result, fuel filters may become clogged with particulates if a quick transition to pure biodiesel is made. Therefore, it is recommended to change the fuel filters on engines and heaters shortly after first switching to a biodiesel blend.

Distribution

Since the passage of the Energy Policy Act of 2005 biodiesel use has been increasing in the United States. In Europe, the Renewable Transport Fuel Obligation obliges suppliers to include 5 per cent renewable fuel in all transport fuel sold in the EU by 2010. For road diesel, this effectively means 5 per cent biodiesel. A growing number of transport fleets use it as an additive in their fuel. Biodiesel is often more expensive to purchase than petroleum diesel but this is expected to diminish due to economies of scale and agricultural subsidies versus the rising cost of petroleum as reserves are depleted.

Vehicular Use and Manufacturer Acceptance

In 2005, Chrysler (then part of DaimlerChrysler) released the Jeep Liberty CRD diesels from the factory into the American market with 5 per cent biodiesel blends, indicating at least partial acceptance of biodiesel as an acceptable diesel fuel additive. In 2007, DaimlerChrysler indicated intention to increase warranty coverage to 20 per cent biodiesel blends if biofuel quality in the United States can be standardized.

Railway Usage

The British businessman Richard Branson's Virgin Voyager train, number 220007 *Thames Voyager*, billed as the world's first "biodiesel train", was converted to run on 80 per cent petrodiesel and only 20 per cent biodiesel, and it is claimed it will save 14 per cent on direct emissions.

The Royal Train on 15 September 2007 completed its first ever journey run on 100 per cent biodiesel fuel supplied by Green Fuels Ltd. His Royal Highness, The Prince of Wales, and Green Fuels managing director, James Hygate, were the first passengers on a train fueled entirely by biodiesel fuel. Since 2007 the Royal Train has operated successfully on B100 (100 per cent biodiesel).

Similarly, a state-owned short-line railroad in Eastern Washington ran a test of a 25 per cent biodiesel/75 per cent petrodiesel blend during the summer of 2008, purchasing fuel from a biodiesel producer seated along the railroad tracks. The train will be powered by biodiesel made in part from canola grown in agricultural regions through which the short line runs.

Also in 2007 Disneyland began running the park trains on B98 biodiesel blends (98 per cent biodiesel). The Programme was discontinued in 2008 due to storage issues, but in January 2009 it was announced that the park would then be running all trains on biodiesel manufactured from its own used cooking oils. This is a change from running the trains on soy-based biodiesel.

Aircraft Use

A test flight has been performed by a Czech jet aircraft completely powered on biodiesel. Other recent jet flights using biofuel, however, have been using other types of renewable fuels.

As A Heating Oil

Biodiesel can also be used as a heating fuel in domestic and commercial boilers, a mix of heating oil and biofuel which is standardized and taxed slightly differently than diesel fuel used for transportation. It is sometimes known as "bioheat" (which is a registered trademark of the National Biodiesel Board [NBB] and the National Oilheat Research Alliance [NORA] in the U.S., and Columbia Fuels in Canada). Heating biodiesel is available in various blends; up to 20 per cent biofuel is considered acceptable for use in existing furnaces without modification.

Older furnaces may contain rubber parts that would be affected by biodiesel's solvent properties, but can otherwise burn biodiesel without any conversion required. Care must be taken at first, however, given that varnishes left behind by petrodiesel will be released and can clog pipes-fuel filtering and prompt filter replacement is required. Another approach is to start using biodiesel as blend, and decreasing the petroleum proportion over time can allow the varnishes to come off more gradually and be less likely to clog. Thanks to its strong solvent properties, however, the furnace is cleaned out and generally becomes more efficient. A technical research paper describes laboratory research and field trials project using pure biodiesel and biodiesel blends as a heating fuel in oil fired boilers. During the Biodiesel Expo 2006 in the UK, Andrew J. Robertson presented his biodiesel heating oil research from his technical paper and suggested that B20 biodiesel could reduce UK household CO_2 emissions by 1.5 million tons per year.

A law passed under Massachusetts Governor Deval Patrick requires all home heating diesel in that state to be 2 per cent biofuel by July 1, 2010, and 5 per cent biofuel by 2013.

Historical Background

Transesterification of a vegetable oil was conducted as early as 1853 by scientists E. Duffy and J. Patrick, many years before the first diesel engine became functional. Rudolf Diesel's prime model, a single 10 ft (3 m) iron cylinder with a flywheel at its base, ran on its own power for the first time in Augsburg, Germany, on August 10, 1893. In remembrance of this event, August 10 has been declared "International Biodiesel Day".

The French Otto Company (at the request of the French government) demonstrated a Diesel engine running on peanut oil at the World Fair in Paris, France in 1900, where it received the *Grand Prix* (highest prize).

This engine stood as an example of Diesel's vision because it was powered by peanut oil—a biofuel, though not *biodiesel*, since it was not transesterified. He believed that the utilization of biomass fuel was the real future of his engine. In a 1912 speech Diesel said, "the use of vegetable oils for engine fuels may seem insignificant today but such oils may become, in the course of time, as important as petroleum and the coal-tar products of the present time."

During the 1920s, diesel engine manufacturers altered their engines to utilize the lower viscosity of petrodiesel (a fossil fuel), rather than vegetable oil (a biomass fuel). The petroleum industries were able to make inroads in fuel markets because their fuel was much cheaper to produce than the biomass alternatives. The result, for many years, was a near elimination of the biomass fuel production infrastructure. Only recently have environmental impact concerns and a decreasing price differential made biomass fuels such as biodiesel a growing alternative.

Despite the widespread use of fossil petroleum-derived diesel fuels, interest in vegetable oils as fuels for internal combustion engines was reported in several countries during the 1920s and 1930s and later during World War II. Belgium, France, Italy, the United Kingdom, Portugal, Germany, Brazil, Argentina, Japan and China were reported to have tested and used vegetable oils as diesel fuels during this time. Some operational problems were reported due to the high viscosity of vegetable oils compared to petroleum diesel fuel, which results in poor atomization of the fuel in the fuel spray and often leads to deposits and coking of the injectors, combustion chamber and valves. Attempts to overcome these problems included heating of the vegetable oil, blending it with petroleum-derived diesel fuel or ethanol, pyrolysis and cracking of the oils.

On August 31, 1937, G. Chavanne of the University of Brussels (Belgium) was granted a patent for a "Procedure for the transformation of vegetable oils for their uses as fuels" (fr. "*Procédé de Transformation d'Huiles Végétales en Vue de Leur Utilisation comme Carburants*") Belgian Patent 422,877. This patent described the alcoholysis (often referred to as transesterification) of vegetable oils using ethanol (and mentions methanol) in order to separate the fatty acids from the glycerol by replacing the glycerol with short linear alcohols. This appears to be the first account of the production of what is known as "biodiesel" today.

More recently, in 1977, Brazilian scientist Expedito Parente invented and submitted for patent, the first industrial process for the production of biodiesel. This process is classified as biodiesel by international norms, conferring a "standardized identity and quality. No other proposed biofuel has been validated by the motor industry." Currently, Parente's company Tecbio is working with Boeing and NASA to certify bioquerosene (bio-kerosene), another product produced and patented by the Brazilian scientist.

Research into the use of transesterified sunflower oil, and refining it to diesel fuel standards, was initiated in South Africa in 1979. By 1983, the process for producing fuel-quality, engine-tested biodiesel was completed and published internationally. An Austrian company, Gaskoks, obtained the technology from the South African Agricultural Engineers; the company erected the first biodiesel pilot plant in November 1987, and the first industrial-scale plant in April 1989 (with a capacity of 30,000 tons of rapeseed per annum).

Throughout the 1990s, plants were opened in many European countries, including the Czech Republic, Germany and Sweden. France launched local production of biodiesel fuel (referred to as *diester*) from rapeseed oil, which is mixed into regular diesel fuel at a level of 5 per cent, and into the diesel fuel used by some captive fleets (e.g. public transportation) at a level of 30 per cent. Renault, Peugeot and other manufacturers have certified truck engines for use with up to that level of partial biodiesel; experiments with 50 per cent biodiesel are underway. During the same period, nations in other parts of the world also saw local production of biodiesel starting up: by 1998, the Austrian Biofuels Institute had identified 21 countries with commercial biodiesel projects. 100 per cent Biodiesel is now available at many normal service stations across Europe.

In September 2005 Minnesota became the first U.S. state to mandate that all diesel fuel sold in the state contain part biodiesel, requiring a content of at least 2 per cent biodiesel.

In 2008, ASTM published new Biodiesel Blend Specifications Standards.

Properties

Biodiesel has better lubricating properties and much higher cetane ratings than today's lower sulfur diesel fuels. Biodiesel addition reduces fuel system wear, and in low levels in high pressure systems increases the life of the fuel injection equipment that relies on the fuel for its lubrication. Depending on the engine, this might include high pressure injection pumps, pump injectors (also called *unit injectors*) and fuel injectors.

The calorific value of biodiesel is about 37.27 MJ/L. This is 9 per cent lower than regular Number 2 petrodiesel. Variations in biodiesel energy density is more dependent on the feedstock used than the production process. Still these variations are less than for petrodiesel. It has been claimed biodiesel gives better lubricity and more complete combustion thus increasing the engine energy output and partially compensating for the higher energy density of petrodiesel.

Biodiesel is a liquid which varies in colour—between golden and dark brown—depending on the production feedstock. It is immiscible with water, has a high boiling point and low vapour pressure. *The flash point of biodiesel (>130 °C, >266 °F) is significantly higher than that of petroleum diesel (64 °C, 147 °F) or gasoline (-45 °C, -52 °F). Biodiesel has a density of ~ 0.88 g/ cm³, less than that of water.

Biodiesel has virtually no sulfur content, and it is often used as an additive to Ultra-Low Sulfur Diesel (ULSD) fuel.

Material Compatibility

- *Plastics*: High density polyethylene is compatible but PVC is slowly degraded. Polystyrenes are dissolved on contact with biodiesel.
- *Metals*: Biodiesel has an effect on copper-based materials (e.g. brass), and it also affects zinc, tin, lead, and cast iron. Stainless steels (316 and 304) and aluminum are unaffected.
- *Rubber*: Biodiesel also affects types of natural rubbers found in some older engine components. Studies have also found that fluorinated elastomers (FKM) cured with peroxide and base-metal oxides can be degraded when biodiesel loses its stability caused by oxidation. However testing with FKM-GBL-S and FKM-GF-S were found to be the toughest elastomer to handle biodiesel in all conditions.

Technical Standards

Biodiesel has a number of standards for its quality including European standard EN 14214, ASTM International D6751, and others.

Gelling

The cloud point, or temperature at which pure (B100) biodiesel starts to gel, varies significantly and depends upon the mix of esters and therefore the feedstock oil used to produce the biodiesel. For example, biodiesel produced from low erucic acid varieties of canola seed (RME) starts to gel at approximately - 10 °C (14 °F). Biodiesel produced from tallow tends to gel at around +16 °C (61 °F). As of 2006, there are a very limited number of products that will significantly lower the gel point of straight biodiesel. A number of studies have shown that winter operation is possible with biodiesel blended with other fuel oils including 2 low sulfur diesel fuel and 1 diesel/kerosene. The exact blend depends on the operating environment: successful operations have run using a 65 per cent LS 2, 30 per cent K 1, and 5 per cent bio blend. Other areas have run a 70 per cent Low Sulfur 2, 20 per cent Kerosene 1, and 10 per cent bio blend or an 80 per cent K1, and 20 per cent biodiesel blend. According to the National Biodiesel Board (NBB), B20 (20 per cent biodiesel, 80 per cent petrodiesel) does not need any treatment in addition to what is already taken with petrodiesel.

To permit the use of biodiesel without mixing and without the possibility of gelling at low temperatures, some people modify their vehicles with a second fuel tank for biodiesel in addition to the standard fuel tank. Alternately, a vehicle with two tanks is chosen. The second fuel tank is insulated and a heating coil using engine coolant is run through the tank. When a temperature sensor indicates that the fuel is warm enough to burn, the driver switches from the petrodiesel tank to the biodiesel tank. This is similar to the method used for running straight vegetable oil.

Contamination by water

Biodiesel may contain small but problematic quantities of water. Although it is not miscible with water, it is, like ethanol, hygroscopic (absorbs water from atmospheric moisture). One of the reasons biodiesel can absorb water is the persistence of mono and diglycerides left over from an incomplete reaction. These molecules can act as an emulsifier, allowing water to mix with the biodiesel. In addition, there may be water that is residual to processing or resulting from storage tank condensation. The presence of water is a problem because:

- Water reduces the heat of combustion of the bulk fuel. This means more smoke, harder starting, less power.
- Water causes corrosion of vital fuel system components: fuel pumps, injector pumps, fuel lines, etc.
- Water and microbes cause the paper element filters in

the system to fail (rot), which in turn results in premature failure of the fuel pump due to ingestion of large particles.

- Water freezes to form ice crystals near 0 °C (32 °F). These crystals provide sites for nucleation and accelerate the gelling of the residual fuel.
- Water accelerates the growth of microbe colonies, which can plug up a fuel system. Biodiesel users who have heated fuel tanks therefore face a year-round microbe problem.
- Additionally, water can cause pitting in the pistons on a diesel engine.

Previously, the amount of water contaminating biodiesel has been difficult to measure by taking samples, since water and oil separate. However, it is now possible to measure the water content using water-in-oil sensors.

Water contamination is also a potential problem when using certain chemical catalysts involved in the production process, substantially reducing catalytic efficiency of base (high pH) catalysts such as potassium hydroxide. However, the super-critical methanol production methodology, whereby the transesterification process of oil feedstock and methanol is effectuated under high temperature and pressure, has been shown to be largely unaffected by the presence of water contamination during the production phase.

Availability and Prices

In some countries biodiesel is less expensive than conventional diesel.

Global biodiesel production reached 3.8 million tons in 2005. Approximately 85 per cent of biodiesel production came from the European Union.

In 2007, in the United States, average retail (at the pump) prices, including federal and state fuel taxes, of B2/B5 were lower than petroleum diesel by about 12 cents, and B20 blends were the same as petrodiesel. However, as part as a dramatic shift in diesel pricing over the last year, by July 2009, the US DOE was reporting average costs of B20 15 cents per gallon higher than petroleum diesel ($2.69/gal vs. $2.54/gal). B99 and B100 generally cost more than petrodiesel except where local governments provide a subsidy.

Production

Biodiesel is commonly produced by the transesterification of the vegetable oil or animal fat feedstock. There are several methods for carrying out this transesterification reaction including the common batch process, supercritical processes, ultrasonic methods, and even microwave methods.

Chemically, transesterified biodiesel comprises a mix of mono-alkyl esters of long chain fatty acids. The most common form uses methanol (converted to sodium methoxide) to produce methyl esters (commonly referred to as Fatty Acid Methyl Ester—FAME) as it is the cheapest alcohol available, though ethanol can be used to produce an ethyl ester (commonly referred to as Fatty Acid Ethyl Ester—FAEE) biodiesel and higher alcohols such as isopropanol and butanol have also been used. Using alcohols of higher molecular weights improves the cold flow properties of the resulting ester, at the cost of a less efficient transesterification reaction. A lipid transesterification production process is used to convert the base oil to the desired esters. Any free fatty acids (FFAs) in the base oil are either converted to soap and removed from the process, or they are esterified (yielding more biodiesel) using an acidic catalyst. After this processing, unlike straight vegetable oil, biodiesel has combustion properties very similar to those of petroleum diesel, and can replace it in most current uses.

A by-product of the transesterification process is the production of glycerol. For every 1 tonne of biodiesel that is manufactured, 100 kg of glycerol are produced. Originally, there was a valuable market for the glycerol, which assisted the economics of the process as a whole. However, with the increase in global biodiesel production, the market price for this crude glycerol (containing 20 per cent water and catalyst residues) has crashed. Research is being conducted globally to use this glycerol as a chemical building block. One initiative in the UK is The Glycerol Challenge.

Usually this crude glycerol has to be purified, typically by performing vacuum distillation. This is rather energy intensive. The refined glycerol (98 per cent+ purity) can then be utilised directly, or converted into other products. The following announcements were made in 2007: A joint venture of Ashland Inc. and Cargill announced plans to make propylene glycol in Europe from glycerol and Dow Chemical announced similar plans for North America. Dow also plans to build a plant in China to make epichlorhydrin from glycerol. Epichlorhydrin is a raw material for epoxy resins.

Production Levels

In 2007, biodiesel production capacity was growing rapidly, with an average annual growth rate from 2002-06 of over 40 per cent. For the year 2006, the latest for which actual production figures could be obtained, total world biodiesel production was about 5-6 million tonnes, with 4.9 million tonnes processed in Europe (of which 2.7 million tonnes was from Germany) and most of the rest from the USA. In July of 2009, a duty was added to American imported biodiesel in the European Union in order to balance the competition from European, especially German producers. In 2007 production in Europe alone had risen to 5.7 million tonnes. The capacity for 2008 in Europe totalled 16 million tonnes. This compares with a total demand for diesel in the US and Europe of approximately 490 million tonnes (147 billion gallons). Total world production of vegetable oil for all purposes in 2005/06 was about 110 million tonnes, with about 34 million tonnes each of palm oil and soybean oil.

Biodiesel Feedstocks

A variety of oils can be used to produce biodiesel. These include:

- Virgin oil feedstock; rapeseed and soybean oils are most commonly used, soybean oil alone accounting for about ninety percent of all fuel stocks in the US. It also can be obtained from field pennycress and jatropha and other crops such as mustard, flax, sunflower, palm oil, coconut, hemp;

- Waste vegetable oil (WVO);
- Animal fats including tallow, lard, yellow grease, chicken fat, and the by-products of the production of Omega-3 fatty acids from fish oil;
- Algae, which can be grown using waste materials such as sewage and without displacing land currently used for food production;
- Oil from halophytes such as *salicornia bigelovii*, which can be grown using saltwater in coastal areas where conventional crops cannot be grown, with yields equal to the yields of soybeans and other oilseeds grown using freshwater irrigation.

Many advocates suggest that waste vegetable oil is the best source of oil to produce biodiesel, but since the available supply is drastically less than the amount of petroleum-based fuel that is burned for transportation and home heating in the world, this local solution does not scale well.

Animal fats are a by-product of meat production. Although it would not be efficient to raise animals (or catch fish) simply for their fat, use of the by-product adds value to the livestock industry (hogs, cattle, poultry). However, producing biodiesel with animal fat that would have otherwise been discarded could replace a small percentage of petroleum diesel usage. Today, multi-feedstock biodiesel facilities are producing high quality animal-fat based biodiesel. Currently, a 5-million dollar plant is being built in the USA, with the intent of producing 11.4 million litres (3 million gallons) biodiesel from some of the estimated 1 billion kg (2.3 billion pounds) of chicken fat produced annually the local Tyson poultry plant. Similarly, some small-scale biodiesel factories use waste fish oil as feedstock.

Quantity of Feedstocks Required

Current worldwide production of vegetable oil and animal fat is not sufficient to replace liquid fossil fuel use. Furthermore, some object to the vast amount of farming and the resulting fertilization, pesticide use, and land use conversion that would be needed to produce the additional vegetable oil. The estimated transportation diesel fuel and home heating oil used in the United States is about 160 million tonnes (350 billion pounds) according to the Energy Information Administration, US Department of Energy. In the United States, estimated production of vegetable oil for all uses is about 11 million tonnes (24 billion pounds) and estimated production of animal fat is 5.3 million tonnes (12 billion pounds).

If the entire arable land area of the USA (470 million acres, or 1.9 million square kilometers) were devoted to biodiesel production from soy, this would just about provide the 160 million tonnes required (assuming an optimistic 98 US gal/acre of biodiesel). This land area could in principle be reduced significantly using algae, if the obstacles can be overcome. The US DOE estimates that if algae fuel replaced all the petroleum fuel in the United States, it would require 15,000 square miles (38,849 square kilometers), which is a few thousand square miles larger than Maryland, or 1.3 Belgiums, assuming a yield of 140 tonnes/hectare (15,000 US gal/acre). Given a more realistic yield of 36 tonnes/hectare (3834 US gal/acre) the area required is about 152,000 square kilometers, or roughly equal to that of the state of Georgia or England and Wales. The advantages of algae are that it can be grown on non-arable land such as deserts or in marine environments, and the potential oil yields are much higher than from plants.

Yield

Feedstock yield efficiency per unit area affects the feasibility of ramping up production to the huge industrial levels required to power a significant percentage vehicles.

Algae fuel yields have not yet been accurately determined, but DOE is reported as saying that algae yield 30 times more energy per acre than land crops such as soybeans. Yields of 36 tonnes/hectare are considered practical by Ami Ben-Amotz of the Institute of Oceanography in Haifa, who has been farming Algae commercially for over 20 years.

Some Typical Yields

Crop	Yield	
	L/ha	US gal/acre
Algae1	~3,000	~300
Chinese tallow2,3	772	97
Palm oil4	780-1490	508
Coconut	2150	230
Rapeseed4	954	102
Soy (Indiana)	76-161	8-17
Peanut4	138	90
Sunflower4	126	82
Hemp	242	26

est. see soy figures and DOE quote below

Klass, Donald, "Biomass for Renewable Energy, Fuels, and Chemicals", page 341. Academic Press, 1998.

Kitani, Osamu, "Volume V: Energy and Biomass Engineering, IGR Handbook of Agricultural Engineering", Amer Society of Agricultural, 1999.

Biofuels: some numbers

The jatropha plant has been cited as a high-yield source of biodiesel but yields are highly dependent on climatic and soil conditions. The estimates at the low end put the yield at about 200 US gal/acre (1.5-2 tonnes per hectare) per crop; in more Favourable climates two or more crops per year have been achieved. It is grown in the Philippines, Mali and India, is drought-resistant, and can share space with other cash crops such as coffee, sugar, fruits and vegetables. It is well-suited to semi-arid lands and can contribute to slow down desertification, according to its advocates.

Efficiency and Economic Arguments

According to a study by Drs. Van Dyne and Raymer for the Tennessee Valley Authority, the average US farm consumes fuel at the rate of 82 litres per hectare (8.75 US gal/acre) of land to produce one crop. However, average crops of rapeseed produce oil at an average rate of 1,029 L/ha (110 US gal/acre), and high-yield rapeseed fields produce about 1,356 L/ha (145 US gal/acre). The ratio of input to output in these cases is roughly 1:12.5 and

1:16.5. Photosynthesis is known to have an efficiency rate of about 3-6 per cent of total solar radiation and if the entire mass of a crop is utilized for energy production, the overall efficiency of this chain is currently about 1 per cent While this may compare unFavourably to solar cells combined with an electric drive train, biodiesel is less costly to deploy (solar cells cost approximately US$1,000 per square meter) and transport (electric vehicles require batteries which currently have a much lower energy density than liquid fuels).

However, these statistics by themselves are not enough to show whether such a change makes economic sense. Additional factors must be taken into account, such as: the fuel equivalent of the energy required for processing, the yield of fuel from raw oil, the return on cultivating food, the effect biodiesel will have on food prices and the relative cost of biodiesel versus petrodiesel.

The debate over the energy balance of biodiesel is ongoing. Transitioning fully to biofuels could require immense tracts of land if traditional food crops are used (although non food crops can be utilized). The problem would be especially severe for nations with large economies, since energy consumption scales with economic output.

If using only traditional food plants, most such nations do not have sufficient arable land to produce biofuel for the nation's vehicles. Nations with smaller economies (hence less energy consumption) and more arable land may be in better situations, although many regions cannot afford to divert land away from food production.

For third world countries, biodiesel sources that use marginal land could make more sense; e.g., honge oil nuts grown along roads or jatropha grown along rail lines.

In tropical regions, such as Malaysia and Indonesia, oil palm is being planted at a rapid pace to supply growing biodiesel demand in Europe and other markets. It has been estimated in Germany that palm oil biodiesel has less than one third of the production costs of rapeseed biodiesel. The direct source of the energy content of biodiesel is solar energy captured by plants during photosynthesis. Regarding the positive energy balance of biodiesel:

When straw was left in the field, biodiesel production was

strongly energy positive, yielding 1 GJ biodiesel for every 0.561 GJ of energy input (a yield/cost ratio of 1.78).

When 'straw was burned as fuel and oilseed rapemeal was used as a fertilizer, the yield/cost ratio for biodiesel production was even better (3.71). In other words, for every unit of energy input to produce biodiesel, the output was 3.71 units (the difference of 2.71 units would be from solar energy).

Energy Security

One of the main drivers for adoption of biodiesel is energy security. This means that a nation's dependence on oil is reduced, and substituted with use of locally available sources, such as coal, gas, or renewable sources. Thus a country can benefit from adoption of biofuels, without a reduction in greenhouse gas emissions. While the total energy balance is debated, it is clear that the dependence on oil is reduced. One example is the energy used to manufacture fertilizers, which could come from a variety of sources other than petroleum. The US National Renewable Energy Laboratory (NREL) states that energy security is the number one driving force behind the US biofuels Programme, and a White House "Energy Security for the 21st Century" paper makes it clear that energy security is a major reason for promoting biodiesel. The EU commission president, Jose Manuel Barroso, speaking at a recent EU biofuels conference, stressed that properly managed biofuels have the potential to reinforce the EU's security of supply through diversification of energy sources.

Environmental Effects

The surge of interest in biodiesels has highlighted a number of environmental effects associated with its use. These potentially include reductions in greenhouse gas emissions, deforestation, pollution and the rate of biodegradation.

Food, Land and Water vs. Fuel

In some poor countries the rising price of vegetable oil is causing problems. Some propose that fuel only be made from non-edible vegetable oils such as camelina, jatropha or seashore mallow which can thrive on marginal agricultural land where many trees and crops will not grow, or would produce only low yields.

Others argue that the problem is more fundamental. Farmers may switch from producing food crops to producing biofuel crops to make more money, even if the new crops are not edible. The law of supply and demand predicts that if fewer farmers are producing food the price of food will rise. It may take some time, as farmers can take some time to change which things they are growing, but increasing demand for first generation biofuels is likely to result in price increases for many kinds of food. Some have pointed out that there are poor farmers and poor countries who are making more money because of the higher price of vegetable oil.

Biodiesel from sea algae would not necessarily displace terrestrial land currently used for food production and new algaculture jobs could be created.

Current Research

There is ongoing research into finding more suitable crops and improving oil yield. Using the current yields, vast amounts of land and fresh water would be needed to produce enough oil to completely replace fossil fuel usage. It would require twice the land area of the US to be devoted to soybean production, or two-thirds to be devoted to rapeseed production, to meet current US heating and transportation needs.

Specially bred mustard varieties can produce reasonably high oil yields and are very useful in crop rotation with cereals, and have the added benefit that the meal leftover after the oil has been pressed out can act as an effective and biodegradable pesticide.

The NFESC, with Santa Barbara-based Biodiesel Industries, Inc, is working to develop biodiesel technologies for the US navy and military, one of the largest diesel fuel users in the world.

A group of Spanish developers working for a company called Ecofasa announced a new biofuel made from trash. The fuel is created from general urban waste which is treated by bacteria to produce fatty acids, which can be used to make biodiesel.

Algae Biodiesel

From 1978 to 1996, the U.S. NREL experimented with using algae as a biodiesel source in the "Aquatic Species Programme". A self-published article by Michael Briggs, at the UNH Biodiesel Group, offers estimates for the realistic replacement of all vehicular fuel with biodiesel by utilizing algae that have a natural oil content greater than 50 per cent, which Briggs suggests can be grown on algae ponds at wastewater treatment plants. This oil-rich algae can then be extracted from the system and processed into biodiesel, with the dried remainder further reprocessed to create ethanol.

The production of algae to harvest oil for biodiesel has not yet been undertaken on a commercial scale, but feasibility studies have been conducted to arrive at the above yield estimate. In addition to its projected high yield, algaculture—unlike crop-based biofuels—does not entail a decrease in food production, since it requires neither farmland nor fresh water. Many companies are pursuing algae bio-reactors for various purposes, including scaling up biodiesel production to commercial levels.

Fungus

A group at the Russian academy of Sciences in Moscow published a paper in September 2008, stating that they had isolated large amounts of lipids from single-celled fungi and turned it into biodiesel in an economically efficient manner. More research on this fungal species; C. japonica, and others, is likely to appear in the near future.

The recent discovery of a variant of the fungus Gliocladium roseum points toward the production of so-called myco-diesel from cellulose. This organism was recently discovered in the rainforests of northern Patagonia and has the unique capability of converting cellulose into medium length hydrocarbons typically found in diesel fuel.

Biodiesel from Used Coffee Grounds

Researchers at the University of Nevada, Reno, have successfully produced biodiesel from oil derived from used coffee grounds. Their analysis of the used grounds showed a 10 per cent to 15 per cent oil content (by weight). Once the oil was extracted, it underwent conventional processing into biodiesel. It is estimated that finished biodiesel could be produced for about one US dollar per gallon. Further, it was reported that "the technique is not difficult" and that "there is so much coffee around that several hundred million gallons of biodiesel could potentially be made annually."

MYTH OF BIODIESEL BUSTED

Biodiesel is a clean burning alternative fuel, produced from domestic, renewable resources such as plant oils, animal fats, used cooking oil and even new sources such as algae. Biodiesel contains no petroleum, but it can be blended at any level with petroleum diesel to create a biodiesel blend. Biodiesel blends can be used in most compression-ignition (diesel) engines with little or no modifications. Biodiesel is simple to use, biodegradable, non-toxic, and essentially free of sulfur and aromatics.

Biodiesel is not raw vegetable oil. Fuel-grade biodiesel must be produced to strict industry specifications (ASTM D6751) in order to ensure proper performance. Biodiesel that meets ASTM D6751 and is legally registered with the Environmental Protection Agency is a legal motor fuel for sale and distribution. Raw vegetable oil cannot meet biodiesel fuel specifications, it is not registered with the EPA, and it is not a legal motor fuel.

Biodiesel is also not the same as ethanol. Ethanol is a renewable biofuel made primarily from corn and intended for use in gasolinepowered engines, while biodiesel is a renewable biofuel made from a variety of materials and designed for use in diesel engines, with different properties and benefits.

Biodiesel is one of the most thoroughly tested alternative fuels on the market. A number of independent studies—

performed by the U.S. Department of Energy, the U.S. Department of Agriculture, Stanadyne Corp. (the largest diesel fuel injection equipment manufacturer in the U.S.), Lovelace Respiratory Research Institute, and Southwest Research Institute—have shown that biodiesel performs similar to petroleum diesel with greater benefits to the environment and human health.

The biodiesel industry has been active in setting quality standards for biodiesel for more than 15 years. ASTM specifications exist for diesel fuel and biodiesel fuel blends from 6 to 20 percent (B6—B20 (D7467-09)), biodiesel blends up to B5 to be used for onand off-road diesel applications (D975-08a), and home heating and boiler applications (D396-08b). ASTM approved the original specification for pure B100 (D6751) in December 2001. These ASTM specifications apply regardless of the fat or plant oil used to make the fuel. Copies of specifications are available from ASTM at www.astm.org.

One of the major advantages of biodiesel is the fact that it can be used in most existing engines and fuel injection equipment in blends up to 20 percent with little impact to operating performance. Biodiesel has a higher cetane number than U.S. diesel fuel. In more than 50 million miles of in-field demonstrations, B20 showed similar fuel consumption, horsepower, torque, and haulage rates as conventional diesel fuel. Biodiesel also has superior lubricity, and it has the highest BTU content of any alternative fuel (falling in the range between 1 and 2 diesel fuel).

All major U.S. automakers and engine manufacturers accept the use of up to at least B5, and many major engine companies have stated formally that the use of high quality biodiesel blends up to B20 will not void their parts and workmanship warranties. For a listing of specific statements from the engine companies, please visit the National Biodiesel Board Web site at www.biodiesel.org/resources/oems.

A study released in 2008 by the National Renewable Energy Laboratory (NREL) shows the biodiesel industry has substantially met national fuel quality standards. The study demonstrated that plants certified under BQ-9000 consistently hit the mark. BQ-

9000 is a voluntary fuel quality assurance Programme that couples the foundations of universally accepted quality management systems with the product specification (ASTM D6751). The Programme covers storage, sampling, testing, blending, shipping, distribution and fuel management practices. Biodiesel production facilities certified as producers under the Programme cover nearly 80 percent of the U.S. biodiesel market volume.

The current industry recommendation is that biodiesel be used within six months, or reanalyzed after six months to ensure the fuel meets ASTM specifications. Most fuel today is used up long before six months, and many petroleum companies do not recommend storing petroleum diesel for more than six months. A longer shelf life is possible depending on the fuel composition and the use of storage-enhancing additives.

Properly managed, high quality biodiesel blends are used successfully in the coldest of climates. Biodiesel will gel in very cold temperatures, just as common 2 diesel does. Although pure biodiesel has a higher cloud point than 2 diesel fuel, typical blends of 20 percent biodiesel are managed with similar management techniques as 2 diesel. Blends of 5 percent biodiesel and less have virtually no impact on cold weather operability.

U.S. biodiesel is a green sustainable part of the solution. It reduces lifecycle carbon dioxide, a greenhouse gas, by 78 percent compared to petroleum diesel. A 2008 USDA/University of Idaho study shows for every unit of fossil energy needed to create biodiesel, 4.5 units of energy are returned. New cropland is not needed to grow materials for biodiesel, because there is already a surplus of soybean oil on the market. Advances in technology enable us to grow more using the same acres of land. The National Biodiesel Board and its members support sustainable production of biodiesel. There is no scientific basis for assigning any significant responsibility for rainforest destruction to U.S. biodiesel, and the vast majority of U.S. biodiesel is made from homegrown resources.

Produced from a wide variety of renewable resources, including plant oils, fats and even recycled restaurant grease, biodiesel is the most diversified fuel on the planet. And soybean-based biodiesel has a positive impact on the world's food supply.

Processing soybeans for biodiesel uses only the oil, leaving 80 percent of the bean for protein-rich soybean meal. Put more simply, when the demand is increased for soybean oil for use in biodiesel, the price of soybean meal actually decreases what it otherwise might cost.

TRANSESTERIFICATION PROCESS

Strengths

1. Quite simple chemistry
2. Small volume bio-diesel machines are commercially available

Weaknesses

1. Relatively high labour input
2. Higher capital investment
3. Methanol involved in process (toxic and derived from fossil fuel)
4. Glycerol is a by-product
5. Relatively high direct energy input/costs

Opportunities

1. Re-use glycerol (e.g. for heat process)
2. Methanol recovery (e.g. using waste heat)
3. Potential to develop process using bio-ethanol (rather than methanol)
4. Funding is currently available for bio-fuel production and research
5. Strengthening bio-fuel network

Threats

1. Potential for increase in cost of methanol
2. Potential for increase in cost of sodium hydroxide (catalyst)
3. Large companies introducing 5 per cent blend

Opportunity of Biodiesel

1. Ever increasing Crude oil price.
2. Employment generation capacity in rural areas.
3. Better Utilization of fallow cultivable waste land.
4. Low gestation period comparative to other non-edible oil sources.
5. Having carbon credit value (Kyoto protocol).
6. Required in large quantity to sustain huge demand.
7. With use of Biotechnology encouraging primary result.

Threats towards Biodiesel

1. Over publicity.
2. Abundance of misleading information.
3. Mall practice in input materials.
4. Costly input materials.
5. Low (no) support price for seed.
6. No sustainable Procurement Mechanism available in the Market.
7. Requirement of seed in large quantity even to fulfill demand of 5 per cent blending with diesel.
8. Government strategies to wards Bio diesel project are not implemented properly.

Biodiesel is a safe alternative fuel to replace traditional petroleum diesel. It has high-lubricity, is a clean-burning fuel and can be a fuel component for use in existing, unmodified diesel engines. This means that no retrofits are necessary when using biodiesel fuel in any diesel powered combustion engine. It is the only alternative fuel that offers such convenience. Biodiesel acts like petroleum diesel, but produces less air pollution, comes from renewable sources, is biodegradable and is safer for the environment. Producing biodiesel fuels can help create local economic revitalization and local environmental benefits. Many groups interested in promoting the use of biodiesel already exist at the local, state and national level.

Biodiesel is designed for complete compatibility with

petroleum diesel and can be blended in any ratio, from additive levels to 100 percent biodiesel. In the United States today, biodiesel is typically produced from soybean or rapeseed oil or can be reprocessed from waste cooking oils or animal fats such as waste fish oil. Because it is made of these easily obtainable plant-based materials, it is a completely renewable fuel source.

Benefits Of Biodiesel

Biodiesel can be considered a new technology, taking into account all the years consumers have had to settle for traditional diesel.

1. Biodiesel is not harmful to the environment. A vehicle tends topollute the environment and emits harmful gasses, if injected with HSD whereas if the engine is using biodiesel it emits no harmful gasses rather keeps the environment pollution free.
2. Biodiesel may not require an engine modification. Biodiesel can be blended with diesel so as to improve the efficiency of the engine without any hassles.
3. Biodiesel is cheap. You can even make biodiesel in your backyard. If your engine can work with biodiesel fuel alone, then you really need not go to the gas station to buy fuel. You can just manufacture some for your own personal use.
4. Any Vehicle using Biodiesel has very low idle stating noise. It is noted that biodiesel has a Cetane number of over 100. Cetane number is used to measure the quality of the fuel's ignition. If your fuel has a high Cetane number, you can be sure that what you get is a very easy cold starting coupled with a low idle noise.
5. Biodiesel is cost effective because it is produced locally.

Biodiesel as a fuel not only helps reducing the pollution, reduces health hazards and gives our society A CLEANER AND GREENER TOMORROW.

1. Biodiesel can be used in existing engines, vehicles and

infrastructure with practically no changes. Biodiesel can be pumped, stored and burned just like petroleum diesel fuel, and can be used pure, or in blends with petroleum diesel fuel in any proportion. Power and fuel economy using biodiesel is practically identical to petroleum diesel fuel, and year round operation can be achieved by blending with diesel fuel.

2. The degree to which fuel provides proper lubrication is its lubricity. Low lubricity petroleum diesel fuel can cause premature failure of injection system components and decreased performance. Biodiesel provides excellent lubricity to the fuel injection system.

3. Biodiesel provides significantly reduced emissions of carbon monoxide, particulate matter, unburned hydrocarbons, and sulphates compared to petroleum diesel fuel. Additionally, biodiesel reduces emissions of carcinogenic compounds by as much as 85 per cent compared with petrodiesel. When blended with petroleum diesel fuel, these emissions reductions are generally directly proportional to the amount of biodiesel in the blend.

Greenhouse Gases: Carbon dioxide produced from biodiesel combustion does not contribute to new emissions of CO_2 as it is part of the carbon cycle. Closed carbon cycle—80 per cent reduction in CO_2. Example: growing oil feedstock consumes four to six times more CO_2 than biodiesel exhaust. **Smog-Forming Pollutants (Ozone precursors):** Compared to diesel, biodiesel effects on engine exhaust are in the table that follows: B100 refers to neat biodiesel; B20 refers to 20 per cent biodiesel blended with 2 Diesel:

HC	36.73% reduction with B100 (2)	7.35% reduction with B20
CO	46.23% reduction with B100	9.25% reduction with B20
PM	68.07% reduction with B100	13.61% reduction with B20
Sox (3)	100% reduction with B100	20% reduction with B20

The above data on HC, CO, PM, and SOx is based on data reported by Dr. Groboski in 1998 paper "Combustion of Fat and Vegetable Oil Derived Fuels in Diesel Engines." OAE-BIO3 formulation with low-NOx additive will bring NOx to diesel baseline. Data on effect of biodiesel on NOx is scattered. OAE testing of in-use diesel engines showed NOx reductions of 10-18 per cent compared to diesel baseline. Biodiesel use in in-use diesel engines results in cleaning of injectors which restores the original spray patter (also improves fuel economy), resulting in lowering of NOx. Testing of heavy-duty laboratory engines by research scientist has not shown the reduction levels achieved with the in-use engines. If low-NOx is desired, OAE also offers several other proprietary engine technologies to reduce NOx by 30-40 per cent. Contact OAE representative for low-NOx options. The above data is with laboratory engines (also called golden engines). In use engines tested by OAE showed NOx reductions of 18 per cent with the use of biodiesel due primarily to the fact that biodiesel cleaned injectors and lines resulting in better spray pattern. Ozone forming potential of biodiesel HC is 50 per cent less for B100, 10 per cent less for B20. Calculated by OAE due to no sulphur in biodiesel.

Biodiesel's Toxic Emissions

Polycyclic Aromatic Hydrocarbons (PAH) reduction w/B20	80% reduction w/B100	13%
NPAH (nitrated PAH) reduction w/B20	90% reduction w/B100	50%

Grown, Produced and Distributed Locally: Worldwide, energy security is becoming a hot topic in government and society. Nearly every country in the world depends on imports of various forms of fossil fuel energy, including oil, coal and natural gas. Without a steady supply of affordable energy a country's economy grinds to a halt, with no fuel for transportation, energy to run power plants and factories, or heat homes. Biodiesel can improve energy security wherever it is produced in several ways:

Domestic Energy Crops: The dependence on foreign oil supplies is reduced to a large extent because of the presence of feedstock and other sources that are used to produce biodiesel.

Increased Refining Capacity: Today biodiesel is produced in refineries which are locally made, reducing the dependence on other countries for importing finished expensive products.

Toxicity, Biodegradability, Safety and Recycling: Occasional spills of the fuel do occur in and around the establishment of the Processing Plant. The impact of it on plants, animals and human beings should be considered. Biodiesel is less toxic than diesel and is biodegradable. All these attributes make biodiesel less harmful to the environment.

Non-Toxic: Biodiesel is considered to be less toxic as it originates from vegetable oil.

Biodegradability: In both soil and water, biodiesel degraded at a rate 4 times faster than regular diesel fuel, with nearly 80 per cent of the carbon in the fuel being readily converted by soil.

Safe and Stable Fuel: Due to the existence of low volatility nature of biodiesel, it is easier and safe to handle than petroleum. The danger of accidental ignition increases when the fuel is being stored, transported, or transferred because of high energy content in all liquid fuels. The possibility of having an accidental ignition is related to the temperature at which the fuel will create enough vapours to ignite, known as the flash point temperature. The lower the flash point of a fuel is, the lower the temperature at which the fuel can form a combustible mixture. Biodiesel has a flash point of over 2660 F, meaning it cannot form a combustible mixture until it is heated well above the boiling point of water.

Economic Development: The resources that are used to produce Biodiesel are locally available. The in-house production of Biodiesel provides host of economic benefits for the local communities. Creation of more employment, Jobs created for feedstock farming and/or collection. Skilled jobs created for biodiesel production and distribution.(Engineers, Technicians etc.)

Tax Benefits: Generation of Income for the local feedstock producers and refiners.

Viability of Biodiesel Production: Biodiesel Processing Plant Operators who have a Multi Feedstock Plant function with a competitive edge over others because of the fact that their plant can run on variety of feedstock easily available in the local market. Multi Feedstock Plant operators are less sensitive to Price Fluctuations because they can choose from a range of easily available feedstock that are used to produce Biodiesel.

WIN-WIN Situation: A win-win situation is one in which all the participants can profit from it in one way or the other. For instance here the Biodiesel Producer and the Farmers can both earn profit and continue to operate in a successful market with the flexibility of feedstock. Hence it leads to A WIN WIN SITUATION for all the participants in the industry.

UNDERSTANDING BIODIESEL AS MODERN ENGINE FUEL

Developed in the 1890s by inventor Rudolph Diesel, the diesel engine has become the engine of choice for power, reliability, and high fuel economy, worldwide. Early experimenters on vegetable oil fuels included the French government and Dr. Diesel himself, who envisioned that pure vegetable oils could power early diesel engines for agriculture in remote areas of the world, where petroleum was not available at the time. Modern biodiesel fuel, which is made by converting vegetable oils into compounds called fatty acid methyl esters, has its roots in research conducted in the 1930s in Belgium, but today's biodiesel industry was not established in Europe until the late 1980s.

The diesel engine was developed out of a desire to improve upon inefficient, cumbersome and sometimes dangerous steam engines of the late 1800s. The diesel engine works on the principal of compression ignition, in which fuel is injected into the engine's cylinder after air has been compressed to a high pressure and temperature. As the fuel enters the cylinder it self-ignites and burns rapidly, forcing the piston back down and converting the chemical energy in the fuel into mechanical energy. Dr. Rudolph Diesel, after whom the engine is named, holds the first patent for the compression ignition engine, issued in 1893. Diesel became known worldwide for his innovative engine which could use a variety of fuels.

The concept of bio fuel dates back to 1885 when Dr. Rudolf Diesel built the first diesel engine with the full intention of running it on vegetative source. In 1912 he observed, "the use of vegetable oils for engine fuels may seem insignificant today. But such oils may in the course of time become as important as petroleum and the coal tar products of present time."

In 1970, scientists discovered that the viscosity of vegetable oils could be reduced by a simple chemical process and that it could perform as diesel fuel in modern engine. Since then the technical developments have come a long way and the plant oil today has been highly established as bio fuel, equivalent to diesel.

Recent environmental (e.g. Kyoto Protocol) and economic concerns have prompted resurgence in the use of biodiesel throughout the world. In 1991, the European Community proposed a 90 per cent tax reduction for the use of bio fuels, including biodiesel. Today 21 countries worldwide produce biodiesel.

India is one of the largest petroleum consuming and importing countries. India imports about 70 per cent of its petroleum demands. The current yearly consumption of diesel oil in India is approximately 40 million tones constituting about 40 per cent of the total petro-product consumption.

Biodiesel, derived from the oils and fats of plants like sunflower, rape seeds, Canola or Jatropha (Bhagveranda) can be used as a substitute or an additive to diesel. As an alternative fuel biodiesel can provide power similar to conventional diesel fuel and thus can be used in diesel engines. Biodiesel is a renewable liquid fuel that can be produced locally thus helping reduce the country's dependence on imported crude petroleum diesel.

Early Work

The early diesel engines had complex injection systems and were designed to run on many different fuels, from kerosene to coal dust. It was only a matter of time before someone recognized that, because of their high energy content, vegetable oils would make excellent fuel. The first public demonstration of vegetable oil based diesel fuel was at the 1900 World's Fair, when the French government commissioned the Otto Company to build a diesel engine to run on peanut oil. The French government was interested in vegetable oils as a domestic fuel for their African colonies. Rudolph Diesel later did extensive work on vegetable oil fuels and became a leading proponent of such a concept, believing that farmers could benefit from providing their own fuel. However, it would take almost a century before such an idea became a widespread reality. Shortly after Dr. Diesel's death in 1913 petroleum became widely available in a variety of forms, including the class of fuel we know today as "diesel fuel". With petroleum being available and cheap, the diesel engine design was changed to match the properties of petroleum diesel fuel. The result was an engine which was fuel efficient and very powerful. For the next 80 years diesel engines would become the industry standard where power, economy and reliability are required.

Modern Engine: Modern Fuel

Due to the widespread availability and low cost of petroleum diesel fuel, vegetable oil-based fuels gained little attention, except in times of high oil prices and shortages. World War II and the oil crises of the 1970's saw brief interest in using vegetable oils to fuel diesel engines. Unfortunately, the newer diesel engine designs could not run on traditional vegetable oils, due to the much higher viscosity of vegetable oil compared to petroleum diesel fuel. A way was needed to lower the viscosity of vegetable oils to a point where they could be burned properly in the diesel engine. Many methods have been proposed to perform this task, including pyrolysis, blending with solvents, and even emulsifying the fuel with water or alcohols, none of which have provided a suitable solution. It was a Belgian inventor in 1937 who first proposed using Transesterfication to convert vegetable oils into fatty acid alkyl esters and use them as a diesel fuel replacement. The process of Transesterfication converts vegetable oil into three smaller molecules which are much less viscous and easy to burn in a diesel engine. The Transesterfication reaction is the basis for the production of modern biodiesel, which is the trade name for fatty acid methyl esters. In the early 1980s concerns over the environment, energy security, and agricultural overproduction once again brought the use of vegetable oils to the forefront, this time with Transesterfication as the preferred method of producing such fuel replacements.

BIODIESEL BY REGION

Australia

The Fuel Standard (Biodiesel) Determination 2003 was signed by the Minister for the Environment and Heritage on 18 September 2006. The determination sets out the physical and chemical parameters of the Biodiesel standard. It also sets out the associated test methods that the Government will use to determine compliance.

Biodiesel subsidies are to be phased out by 2011, after the passing of the Fuel Tax Bill 2006.

All of the metropolitan trains and most of the metropolitan buses in Adelaide (capital of South Australia) operate on a B5 blend. The South Australian Government has stated that it will soon move to B20 or possibly higher blends.

Several councils (local Governments) across Australia are using B20 (including Townsville City Council, Adelaide City Council, Sydney City Council and Newcastle City Council).

Retailers

- SAFF—South Australian Farmers Fuel (SAFF) began retailing B100 in South Australia in 2001 and now also sells B20 (marketed as "Premium Diesel") at some 52 service stations across 4 states. SAFF currently sells B100 at 14 of these service stations.
- Gull—a Western Australian based company, introduced B20 Biodiesel to several Gull service stations on April 3, 2006 which has since expanded to a total of 21 sites of purchase. In addition, pure Biodiesel (B100) along with other blends can be purchased in bulk. Gull was the second rollout, after SAFF, of Biodiesel by a service station network. Gull is also involved with the Western Australian Government to provide B5 Biodiesel for use in Transperth buses. Eventually the fleet will be provided with B10 or B20 blends. Currently seven percent of Transperth's bus fleet is running Biodiesel.

- reeFUEL—a retailer in Townsville, Queensland. reeFUEL sells only B100 and as of September 2006, was selling 50,000 litres per week into a community of .about 160,000. This is believed to be the highest penetration of biodiesel per capita in Australia.
- Conservo—a small biodiesel retail outlet in Melbourne's inner suburbs, looking to expand to other locations within Melbourne's suburbs.

In February 2005 the first retail outlet for Biodiesel opened in the Sydney suburb of Marrickville. It offers B20 and B50 blends to the general public, and caters to qualified fleets wishing to utilize B100.

Brazil

Brazil opened a commercial biodiesel refinery in March 2005. It is capable of producing 12,000 m³ (3.2 million US gallons) per year of biodiesel fuel. Feedstocks can be a variety of sunflower oil, soybean oil, or castor bean oil. The finished product will be currently a blend of gas oil with 2 per cent biodiesel and, after 2011, 5 per cent biodiesel, both usable in unmodified diesel engines. As of 2005, there were 3 refineries and 7 that are planned to open. These three factories were capable of producing 45.6 million of litres per year.

Petrobras (the Brazilian national petroleum company) launched an innovative system, making biodiesel (called H-Bio) from the petroleum refinery. In Brazil, castor bean is the best option to make biodiesel, because it's easier to plant and costs less than soybean, sunflower or other seeds.

On December 27, 2006, Brazil's government announced they will advance the 5 per cent biodiesel blend mandate to 2010 instead of 2013.

Belgium

In Belgium, there are refineries in Ertvelde (belonging to the company Oleon) and at Feluy.

Cambodia

In Cambodia, the biodiesel industry was founded with the establishment of Biodiesel Cambodia. Biodiesel Cambodia worked on different grass-roots projects, developing models that produced biodiesel from both waste cooking oil and from the oil found within the seed of the local, wild growing, Jatropha Curcas plant. Biodiesel Cambodia has been actively promoting biodiesel use and the cultivation of Jatropha as a biodiesel feedstock crop. The company established the first Jatropha-biodiesel refinery in Phnom Penh, where it produced B100 biodiesel fuel which was used to run the British Embassy's fleet of diesel vehicles with excellent results.

Elsewhere in Cambodia, there is a refinery in Sihanoukville belonging to the charitable organization Planet Biodiesel Outreach Cambodia. They are producing 1,000 liters per month and selling it in Phnom Penh and Sihanoukville to help raise funds for their charity. They run a school for impoverished Cambodian children and provide food, clothing, education, school supplies and transportation free of charge. They run their school bus on 100 per cent biodiesel that is sustainably produced from waste vegetable oil.

Canada

The Government of Canada exempted biodiesel from the federal excise tax on diesel in the March 2003 budget.

Quebec: Rothsay of Ville Ste Catherine, Quebec, produces 35,000 m³ of biodiesel per year. The shuttle bus connecting students between the two campuses of Concordia University are run solely on Biodiesel.

Nova Scotia: The Provincial Government of Nova Scotia uses biodiesel in some public buildings for heating as well as (in more isolated cases) for public transportation. Halifax Regional Municipality has converted its bus fleet to biodiesel, with a future demand of 7,500 m³ of B20 (20 per cent biodiesel fuel mixture) to B50—reducing biodiesel content in low temperatures to avoid gelling issues—and 3,000 m³ split between B20 and B100 for building heat. The municipality forecasts a greenhouse gas reduction of over 9,000 tonnes CO_2 equivalents (4,250 tonnes from fleet use and 5,000 tonnes from building heating) if fully implemented. Private sector uptake is slower—but not unheard of—possibly due to a lack of price differential with petroleum fuel and a lack of federal and provincial tax rebating. Ocean Nutrition Canada produces 6 million gallons (23,000 m³) of fatty acid ethyl esters annually as a byproduct of its Omega-3 fatty acid processing.

Ontario: Biox Corporation of Oakville is building a biodiesel processing plant in the Hamilton harbor industrial lands, due for completion in the first half of 2006. There are also a few retail filling stations selling biodiesel to motorists in Toronto and Unionville.

Manitoba: A rush of building of biodiesel plants in 2005 and 2006 started in June 2005 with Bifrost Bio-Diesel in Arborg. In addition, biodiesel is made by individuals and farmers for personal use.

British Columbia: the cooperative association proves a successful structure for micro-economy-of-scale biodiesel production reaching the end-user. Vancouver Biodiesel Co-op (located at 360 Industrial Ave, Vancouver, BC), Nelson Biodiesel Co-op, WISE Energy and Island Biodiesel Co-op are notable examples.

The Canadian government has stated a goal of producing 500 million liters of biodiesel by 2010. "Welcome to Canada Clean Fuels". http://www.canadacleanfuels.com/biodiesel.html. Retrieved 2008-05-12.

China

At least two publicly traded companies, China Clean Energy, Inc. and Gushan, manufacture and sell significant amounts of biodiesel in China.

A biodiesel plant is proposed to be built in Tseung Kwan O Industrial Estate by the ASD Biodiesel (Hong Kong) Limited.

Costa Rica

Costa Rica is a large producer of crude palm oil and this has spurred interest in biodiesel. Currently several small biodiesel production projects are starting in the country. There are also biodiesel reactor manufacturers in Costa Rica which provide equipment to the Central American and Caribbean region.

Czech Republic

Czech production of biodiesel was already above 60,000 m³ per year by the early 1990s and is now even larger. Many of the plants are very large, including one in Olomouc which produces almost 40,000 m³ per year. From the summer of 2004, Czech producers of biodiesel for blend receive a subsidy of roughly CEK 9.50/kg. All Škoda diesels built since 1996 are warrantied for DIN EN 590 biodiesel use.

Estonia

Biodiesel is available at Favoura fuel stations.

European Union

According EU Strategy for Biofuels at year 2010 target is 5.75 per cent market share for biofuels. At year 2030 the target is 10 per cent market share.

European standard DIN EN 14214 and DIN EN 590 describes the physical properties of diesel fuels.

Finland

Neste Oil is producer of NExBTL renewable diesel oil. NExBTL is first renewable fuel suitable for all diesel engines in the world. Neste Oil claims EN-590 outperforms both regular diesel as well as other biodiesels on the market.

Germany

In Germany biodiesel is, for the most part, produced from

rapeseed. Sales in Germany stood at two billion litres (about 600 million US gallons) in 2006. This amount was sufficient to meet the average yearly consumption of well over 2,000,000 automobiles. Diesel engines have become increasingly popular in Germany and almost half of all newly manufactured cars are diesel powered. This is in part due to the greater efficiency of diesel engines, the desire by consumers to use environmentally friendlier technologies and lower taxes on diesel fuel that make it cheaper than gasoline.

Verbio is a biodiesel and bioethanol producer in Germany. They currently produce 400,000 tons of biodiesel from rape seed oil, soybean oil and fatty acids, selling the product to markets in the E.U.

With 1,900 sales points, equal to one in every ten public gas stations, biodiesel is the first alternative fuel to be available nationwide. The industry is expecting a surge in demand since the authorisation at the beginning of 2004, through European Union legislation, of a maximum 5 per cent biodiesel addition to conventional diesel fuel. In Germany biodiesel is also sold at a lower price than fossil diesel fuel.

India

Biodiesel is now being produced locally in India for use in three-wheeler motor rickshaws. These engines actually run on regular diesel fuel or CNG, but in the past kerosene was used because it was far cheaper, and worked just as well. However, kerosene was dirty and wasn't as clean-burning. Biodiesel is rapidly replacing both kerosene and diesel as a more efficient, cheap, and clean alternative. Today plans are being chalked out to cultivate Jatropha plants on barren land to use its oil for biodiesel production. Now it is used for Railway engines and the plantations are recommended to plant these plants everywhere in unused areas through government sectors. Biodiesel is being used experimentally to run state transport corporation buses in Karnataka. University of Agriculture Sciences at Bangalore has identified many elite lines of Jatropha Curcas and Pongamia pinnata. Large scale activities have been initiated quite recently. For example, large-scale plantations have been initiated in North-East India and Jharkhand by D1 Williamson Magor Bio Fuel Limited, a joint venture between D1 Oils of U.K. and Williamson Magor of India. The hilly areas of the North-East are ideal for growing this hardy, low-maintenance plant.

Indian Oil Corporation has tied up with Indian Railways to introduce biodiesel crops over 1 million square kilometers. Also, Jharkhand and Madhya Pradesh have tied up with Indian Oil to cultivate large tracts of land with jatropha, the choice of crop for Indian biodiesel plans. In order to organize the industry, BioDiesel Society of India has been formed to encourage energy plantations for increasing feedstock supplies.

Indonesia

Since September 2005, Eterindo Group has been producing Biodiesel using Palm Oil derivatives as its raw materials. Currently the production capacity of Eterindo Group has reached 120,000 tonnes of Biodiesel annually. Meeting the standard requirements of ASTM D-6751 and EN 14214, in 2006 the group begin to export its Biodiesel to United States, Germany and Japan. It is now exploring another export destinations, i.e.: Asia Pacific countries, etc.

Israel

Biodiesel is not yet sold on the market, things start to change and biodiesel is being produced in two small-scale experiments. The amounts produced in these experiments are up to 10,000 liters a month. The lack of production of biodiesel in Israel is in contrary with the Research and Development abilities of the country, for Israel is a center of development for agriculture technologies. The Israel North Recycle Group (INRG) is forecasting much progress in the next year, including consumption agreements with municipal bodies, as part of the wider view of the municipalities on the subject.

Malaysia

Biodiesel called the Envo Diesel was launched by the Prime Minister Datuk Seri Abdullah Ahmad Badawi on Tuesday 22 March 2006. Malaysia currently produces 500,000 tonnes of biofuel annually and the government hopes to increase this number this year. Envo diesel blends 5 per cent processed palm oil (vegetable oil) with 95 per cent petrodiesel. In contrast, EU's B5 blends 5 per cent methyl ester with 95 per cent petrodiesel. Diesel engine manufacturers prefer the use of palm oil methyl ester blends as diesel engines are designed to handle 5 per cent methyl esters meeting the EN14214 biodiesel standard, which palm oil cannot meet.

Projects requiring Malaysian and Indonesian palm oil as feedstocks have been criticized by some environmental advocates. Friends of the Earth has published a report asserting that clearance of forests for oil-palm plantations is threatening some of the last habitat of the orangutan. Also, in a column for The Guardian, writer George Monbiot claimed that land clearance by cutting and burning large forest trees frees large amounts of carbon dioxide that is never reabsorbed by the smaller oil palms. If true, then biodiesel production from plantation-grown palm oil may be a net source of atmospheric carbon dioxide. How these issues are resolved may determine whether Malaysia eventually becomes a major producer of biodiesel.

The palm oil industry has recognized this concerned and in conjunction with the WWF has formed the Roundtable of Sustainable Palm Oil (RSPO) which endeavours to ensure development of palm oil in a sustainable way.

With the increase in awareness and importance attached to environmental issues such as global warming, more environment-friendly fuels are being developed as alternatives to fossil fuel. One such fuel, which has been gaining prominence in recent years, is biofuel. Clean and renewable, biofuel has been touted as the answer to the issue of the diminishing of energy reserves.

Led by Y. Bhg. Tan Sri Datuk Dr. Yusof Basiron, former Director General of MPOB, MPOB has been the pioneer and is at the forefront in researching into palm biodiesel project. Since the 1980s, MPOB in collaboration with the local oil giant, PETRONAS, has begun to develop a patented technology to transform crude palm oil into a viable diesel substitute. This process involves the transesterification of crude palm oil into palm oil methyl esters or palm biodiesel. It has also been successfully demonstrated in a 3000 tonnes per year pilot plant located in the MPOB headquarters.

Palm biodiesel has been systematically and exhaustively evaluated as diesel fuel substitute from 1983 to 1994. These included laboratory evaluation, stationary engine testing and field trials on a large number of vehicles including taxis, trucks, passenger cars and buses. Exhaustive field trials with 30 Mercedes Benz of Germany mounted onto passenger buses have been successfully completed with each bus covered 300,000 km, the expected life of the engines.

The advantages of palm biodiesel, drawn from the field trials are no modification of the engines is required, good engine performance, cleaner exhaust emission and comparable fuel consumption in comparison with the petroleum diesel. The palm biodiesel can be used neat or blended with petroleum diesel in any proportions. Recently, to overcome the long standing pour point problem of palm biodiesel (pour point = 15°C), MPOB has developed a process to produce low pour point palm biodiesel (-21°C to 0°C) which is suitable for temperate countries.

Lithuania

There are two biodysel producing plants in Lithuania. One in close to Mažeikiai (Samogitia) and other in Klaipëda free economic zone. Most of biodysel produced in Lithuania is consumed at the local market and a fraction exported.

Norway

Biodiesel has been launched across all of Norway. Starting from 2008, the B5 diesel has been the standard fuel at almost all stations, and the transformation from normal diesel will be completed within 2008

Papua New Guinea—Bouganville

Biodiesel is produced from Copra oil (oil extracted from coconuts) in a processing plant at Buka on the island of Bouganville, vehicles which run on this fuel have a sticker on the doors which says "powered by Coconuts". This fuel is cheaper and more readily available than imported PetroDiesel. The oil-company StatoilHydro has recently started to distribute biodiesel too.

Spain

It is possible to buy biodiesel, mixed with diesel fuel, in more than 480 petrol stations around the country.

Singapore

Two biodiesel plants will be built on Jurong Island, Singapore's petrochemicals hub. The first plant, by Peter Cremer (S) GMBH, will have a capacity of 200,000 tons/year and it is expected to be ready by early 2007, while the second is a joint venture between Wilmar Holdings and Archer Daniels Midland Company, to be operational by end 2006 with an initial capacity of 150,000 tons/year.

Natural Fuel Pte Ltd, has mechanically completed a 600,000 tons/year biodiesel plant in early 2008—making it one of the world's largest biodiesel located in a single site.

Singapore was selected for the companies' first biodiesel plant in Asia because of its excellent connectivity. There is easy access to abundant palm oil feedstock from the neighboring countries of Malaysia and Indonesia. Also, Singapore has terminal facilities which allow the biodiesel to be shipped to markets around the world.

Neste Oil is building NExBTL renewable diesel oil plant, production 800,000 tons/year. According plan it will be ready 2010.

Taiwan

In 2004, several companies started making biodiesel, and has produced more than 5,000 kilotons in a year since then. In 2006, the Bureau of Energy launched the first biodiesel buses on Earth Day.

Thailand

Thailand was the first country to launch biodiesel as a national Programme on July 10 2001. It was reported that the work was initiated by the Royal Chitralada Project, a royal-sponsored project to help rural farmers. International co-operation among ASEAN country was also starting by the Renewable Energy Institute of Thailand (Dr. Samai Jai-In) and Asia-Pacific Roundtable for Sustainable consumption and Production (Dr. Olivia Castillo,. The primary aims of the project in Thailand are:

- an alternative output for excess agricultural produce;
- substituting diesel imports.

In 2007, several biodiesel plants are operating in Thailand using the excess palm oil/palm stearin and in some cases, waste vegetable oil as raw materials. The production capacity is about 1 million litre/day and should reach 2 million litre by early 2008. About 400 petrol stations are now distributing B5 (5 per cent biodiesel with 95 per cent diesel) in Chiangmai and Bangkok. The national biodiesel standard has been developed based on the European standard. The target of the Government is to mandate B2 by 2 April 2008 and to increase to B5 by 2011 which will require almost 4 Million litres/day of biodiesel.

The raw material will most likely come from palm oil, coconut oil, Jatropha Curcas Linn, and tallow. Several pilot plants are now operating such as the Royal Chitralada Projects, Rajabiodiesel in Surattani, Department of Alternative Energy Development and Efficiency, Royal Naval Dockyard, MTEC, and Tistr [www.tistr.or.th].

United Kingdom

Biodiesel is sold by a small but growing number of filling stations in B5 and B100 blend, including a significant fraction of supermarket filling stations. Some farmers have also been using small plants to create their own biodiesel for farm machinery since the 1990s. Several Co-ops and small scale production facilities have recently begun production, typically selling fuel several pence per litre less than petrodiesel. The first large scale plant, capable of producing 50 million litres (13 million US gallons) a year, opened in Scotland in 2005, soon followed by a large plant co-owned by Tesco and Greenergy (Tesco sell 5 per cent biodiesel at many of their petrol stations. The Fuel conforms to standard EN590 which allows up to 5 per cent biodiesel inclusion. Biodiesel is treated like any other vehicle fuel in the UK and the paperwork required to register as a producer is a major limiting factor to growth in the market. Although since July 2007 home users may produce 2500 litres per year for personal use without registering or paying duty.

United States

Biodiesel is commercially available in most oilseed-producing states in the United States. As of 2005, it is somewhat more expensive than fossil diesel, though it is still commonly produced in relatively small quantities (in comparison to petroleum products and ethanol). Many farmers who raise oilseeds use a biodiesel blend in tractors and equipment as a matter of policy, to foster production of biodiesel and raise public awareness. It is sometimes easier to find biodiesel in rural areas than in cities. Similarly, some agribusinesses and others with ties to oilseed farming use biodiesel for public relations reasons. As of 2003 some tax credits were available in the U.S. for using biodiesel. In 2004 almost 30 million US gallons (110,000 m^3) of commercially produced biodiesel were sold in the U.S., up from less than 0.1 million US gallons (380 m^3) in 1998. Projections for 2005 were 75 million gallons produced from 45 factories and 150 million gallons (570 million liters). Due to increasing pollution control requirements and tax relief, the U.S. market is expected to grow to 1 or 2 billion US gallons (4,000,000 to 8,000,000 m^3) by 2010.

Uruguay

Uruguayan law 18.195 stipulates a minimum of 2 per cent of biodiesel in diesel since January 2009 and 5 per cent from January 2012. As of March 2009, state fuel monopoly ANCAP pretends to start blending biodiesel for automotive use in late May or early June 2009.

EVALUATING BIODIESEL IN THE UNITED STATES

Biodiesel is commercially available in most oilseed-producing states in the United States. As of 2005, it is more expensive than petroleum-diesel, though it is still commonly produced in relatively small quantities (in comparison to petroleum products and ethanol fuel).

Due to increasing pollution control, climate change requirements, and tax incentives, the U.S. market is expected to grow to upwards of 2 billion US gallons (7.6×10^6 m^3) by 2010.

The total U.S. production capacity for biodiesel reached 2.24 billion US gallons per year (8.5×10^6 m^3/a) in 2007, although poor market conditions held 2007 production to about 450 million US gallons (1.7×10^6 m^3), according to the National Biodiesel Board (NBB).

In 2004, almost 30 million US gallons (110×10^3 m^3) of commercially produced biodiesel were sold in the U.S., up from less than 100 thousand US gallons (380 m^3) in 1998.

Price

The price of biodiesel in the United States has come down from an average $3.50 per US gallon ($0.92/l) in 1997 to $1.85 per US gallon ($0.49/l) in 2002. This appears economically viable with current petrodiesel prices, which as of 19 September 2005 varied from $2.648 to $3.06 per US gallon.

Feedstock Sourcing

A pilot project in Unalaska/Dutch Harbor, Alaska, is producing fish oil biodiesel from the local fish processing industry in conjunction with the University of Alaska Fairbanks. It is rarely economic to ship the fish oil elsewhere and Alaskan communities are heavily dependent on diesel power generation. The Alaskan Energy Authority factories project 8 million US gallons (30×10^3 m³) of fish oil annually.

Many farmers who raise oilseeds use a biodiesel blend in tractors and equipment. Similarly, some agribusinesses and others with ties to oilseed farming use biodiesel.

Soybeans are not a very efficient crop solely for the production of biodiesel, but their common use in the United States for food products has led to soybean biodiesel becoming the primary source for biodiesel in that country. Soybean producers have lobbied to increase awareness of soybean biodiesel, expanding the market for their product.

Production

Imperium Renewables in Washington has the largest biodiesel production facility in the US, capable of making 100 million US gallons per year (380×10^3 m³/a).

In 2006, Fuel Bio Opened the largest biodiesel manufacturing plant on the east coast of the United States in Elizabeth, New Jersey. Fuel Bio's operation is capable of producing a name plate capacity of 50 million US gallons per year (190×10^3 m³/a) of biodiesel.

In 2008, ASTM published new Biodiesel Blend Specifications.

Commercialization

In 2005, U.S. entertainer Willie Nelson was selling B20 Biodiesel in four states under the name BioWillie. By late 2005 it was available at 13 gas stations and truck stops (mainly in Texas). Most purchasers were truck drivers. It was also used to fuel the buses and trucks for Mr. Nelson's tours as well as his personal automobiles.

On October 16, 2006, the city of Kalamazoo, Michigan announced an agreement with local Western Michigan University's biodiesel R and D Programme to use the biodiesel research to build a 100 thousand US gallons per year (380 m³/a) per year production system at the city wastewater treatment plant, and convert the city bus system to run entirely off of the fuel. Its use of "trap grease" from the waste tanks of restaurants around the city may be the first of its kind in the US.

Mcgyan Process Announced

The Mcgyan Process flows super critical alcohol and feedstock through a tube reactor packed with sulfated metal oxide microspheres to produce biodiesel in seconds with virtually no waste stream. The unreacted alcohol and any residual fatty acids can be recycled through the reactor making the process entirely continuous and able to achieve 100 per cent conversion. The process was invented by SarTec Co. and Augsburg College and the discovery was announced on Friday March 7, 2008. Plans to build a prototype commercial production facility that will employ this novel process have been announced by Ever Cat in Isanti, MN.

Incentives

Tax credits

As of 2003, some tax credits were available in the U.S. for using biodiesel.

By State

Biodiesel retailers can be found in all states but Alaska, though all may not offer high percentage blends or B100.

Minnesota

Minnesota Governor Tim Pawlenty signed a bill on May 12, 2008, that will require all diesel fuel sold in the state for use in internal compression engines to contain at least 20 per cent biodiesel by May 1, 2015.

In March 2002, the Minnesota State Legislature passed a bill which mandated that all diesel sold in the state must contain at least 2 per cent biodiesel. The requirement took effect on June 30, 2005, and was the first biodiesel mandate in the US.

Washington State

In March 2006, Washington became the second state to pass a 2 per cent biodiesel mandate, with a start-date set for December 1, 2008.

CASE STUDY: GRAYS HARBOR BIODIESEL PLANT

Grays Harbor Biodiesel Plant is the largest biodiesel production facility in the United States, with an annual capacity of 100 million gallons per year. The facility is sited on a 12-acre (49,000 m^2) parcel of land at the Port of Grays Harbor, Washington.

The site includes eight main tanks, which can hold 2,000,000 gallons each, and two reserves that can each hold 500,000 gallons. The facility is "feedstock agnostic," meaning it can create biodiesel from numerous different feedstocks, even simultaneously. The majority of their oil comes from canola oil and soybean oil grown in Canada and Washington.

In 2006, the National Biodiesel Board estimates that more than 250,000,000 gallons of biodiesel were consumed in the U.S., up from 75,000,000 in 2005.

Like other biodiesel production facilities around the country, the Imperium plant has been hit hard by the economic downturn and the drastic changes in the cost of petroleum fuels and biodiesel feedstocks. In early 2008 Imperium cut staff from the high of 107; again in March 2009 a further reduction brought the staff to 24. At the current time Imperium will not release any production statistics.

EVALUATING BIODIESEL IN INDIA

BioDiesel in India is virtually a non-starter. There are many reasons for that. The Main Reasons are non-availability of vegetable oil and government's policies.

Non Availability of Oil

- In India Edible oils are in short supply, and country

has to import up to 40 per cent of its requirements (It is now partly offset by Bumper Crop of Soy). Hence prices of edible oils are higher than that of Petroleum Diesel. Due to this, these are not viable and use of non-edible oils was suggested for BioDiesel manufacture.

- Even though the consumption of Edible oils in India is high, the availability of used cooking oil is very small as it is used till the end.

- Indian Culture uses vegetable oil lamps for lighting in homes and in temples (like candles in other cultures). When prices of edible oil shot up, some people turned to a bit cheaper non-edible oils. The requirement of this sector is more than 15 million tons (BioKerosine). Since seeds can be collected and crushed in a small scale in far flung villages, the use of non-edible oils for lamps is picking up very fast. This is the best way of use for millions of Rural Indians. This is depriving BioDiesel industry its supply of oil.

- All over the world Edible oils are used for manufacture of BioDiesel. These are Rape seed in Europe, Soy in Americas and Palm in South East Asia. Rape seed and soy are used for its deoiled meal as cattle feed and oil is not that important. Hence these were in excess, and had to be disposed off at lower prices. Hence initially it was a viable raw material for BioDiesel manufacture and a lot of manufacturing units came up based on these oils. Now excess oil is committed, and fresh sources need to be developed.

- Collection of non-edible oil seeds is a manual operation, and for large BioDiesel plant it is a logistical nightmare. In a day, a person can collect up to 80 kilograms of seeds, which can produce 20 to 23 liters of oil. The collection is done for 3 months, once or twice a year. For a 100 tons per day (8 million gallons per year) plant, you need 15,000 people to collect it. Collecting and organizing such a large manpower is a challenge.

- The price of Seeds of Jatropha is currently very high because most of it is used for plantation purposes. At this price, the manufacturing cost of BioDiesel is 3 times the pump price of Petroleum Diesel.

- . Most of the edible oils used currently are Stable (do not get rancid). These do not decompose much on storage. Hence these are preferred for Trans-Esterification Process. Non-Edible oils are not that stable, and need a lot of pre-treatment adding to the cost of manufacture of BioDiesel. If these are used as lamp oil, even oils with 50 per cent free fatty acids can be used.

- The use of lamp oil is increasing rapidly in India, as there is no electrical power supply for 10 to 14 hours a day in rural areas. Soon people will face shortage of these oils for lighting purposes.

- Cottage soap industry can use vegetable oils with high free fatty acid contents. Since prices of edible oils have doubled, many soap manufacturers in unorganized sector are using these oils as these are a bit cheaper.

- There are billions of other trees (Karanj, Mahua, Neem), all over India, with oil bearing seeds. Traditionally Karanj (Pongamia Pinatta) is planted along the highways. Petrol Pump owners along the highways, should be encouraged to collect the Karanj seeds. Neem (Azadirachta Indica) is planted everywhere for purification of air. Mahua (Madhuca Indica) and Sal (Shorea robusta) grows wildly in Forests. Collection and processing mechanism for these seeds is not yet developed. Hence a most of these seeds lie on the ground (and ultimately get converted into BioFertilizer).

Government's Policies

- Government of India started BioFuel mission in 2003, but it announced BioFuel Policy on 11th September 2008. The Union Cabinet in its meeting gave its approval for the National Policy on Biofuel prepared by the Ministry of New and Renewable Energy, and also approved for setting up of an empowered National

Biofuel Coordination Committee, headed by Prime Minister of India and a Biofuel Steering Committee headed by Cabinet Secretary.

- Ministry of New and Renewable Energy has been given the responsibility for the National Policy on Biofuels and overall co-ordination by Prime Minister under the Allocation of Business Rules. A proposal on "National Policy on Biofuels and its Implementation" was prepared after wide scale consultations and inter-Ministerial deliberations. The draft Policy was considered by a Group of Ministers (GoM) under the Chairmanship of Shri Sharad Pawar, Union Minister of Agriculture, Food and Public Distribution. After considering the suggestions of Planning Commission and other Members, the Group of Ministers recommended the National Biofuel Policy to the Cabinet.

Salient Features of the National Biofuel Policy are as Under:

- An indicative target of 20 per cent by 2017 for the blending of biofuels—bioethanol and bio-diesel has been proposed.
- Bio-diesel production will be taken up from non-edible oil seeds in waste/degraded/marginal lands.
- The focus would be on indigenous production of bio-diesel feedstock and import of Free Fatty Acid (FFA) based such as oil, palm etc. would not be permitted.
- Bio-diesel plantations on community/Government/forest waste lands would be encouraged while plantation in fertile irrigated lands would not be encouraged.
- Minimum Support Price (MSP) with the provision of periodic revision for bio-diesel oil seeds would be announced to provide fair price to the growers. The details about the MSP mechanism, enshrined in the National Biofuel Policy, would be worked out carefully subsequently and considered by the Bio-fuel Steering Committee.

- Minimum Purchase Price (MPP) for the purchase of bio-ethanol by the Oil Marketing Companies (OMCs) would be based on the actual cost of production and import price of bio-ethanol. In case of biodiesel, the MPP should be linked to the prevailing retail diesel price.
- The National Biofuel Policy envisages that bio-fuels, namely, biodiesel and bio-ethanol may be brought under the ambit of "Declared Goods" by the Government to ensure unrestricted movement of biofuels within and outside the States. It is also stated in the Policy that no taxes and duties should be levied on bio-diesel.
- Oil companies have declared their own BioDiesel Purchase Policy. These companies offer a price of Rs. 26.50 per liter of BioDiesel, which is less than half the current manufacturing cost of BioDiesel.
- Large patches of land are required for plantation of Jatropha. Government holds large tracts of land as Forest Lands and Revenue lands. In some states, like Chattisgarh, these are leased to State owned Oil company like IOC.
- UP Jatropha mission of Uttar Pradesh is a Joint venture of BPCL, Nandan Biomatrics and Shapoorji Pallonji and is supported by UP Government. Presently, they are going to tie-up with every Panchayat (Local Body) of selecetd Districts to plant Jatropha on Panchayat lands. Jatropha seeds will be bought back by BPCL in long run. Operation of this mission may be very slow due to Government's and Panchayat's involvement and because it is implemented through National Rural Employment Guarantee Scheme. IOC is also planning to enter in UP as well as some others. Individual farmers are ready to plant Jatropha in their useless and waste lands and want buy-back agreement like contract farming with good price (But price offered is generally very low). Bundelkhand has lots of waste land (600,000—700,000 hectares) which are best suited for

Jatropha and on the other hand Districts of Eastern
UP have saline and waste lands which are also suitable
for plantation.

If 2008 was bad year for Natural Calamities, 2009 is going
to be worse, and 2010 will be worst. To reduce such calamities,
only way to reduce Carbon-di-oxide from atmosphere is to mop
it up by fast growing Bushes like Jatropha and Castor on barren
and desert lands, which can fix up to 500 kgs of Carbon-di-
oxide per acre per year.

CASE STUDY: BIODIESEL TECHNOLOGIES, CALCUTTA, INDIA

Headquartered in Kolkata India, Biodiesel Technologies was
conceived in 2002 in response to the serious environmental and
health hazards arising out of the various polluting emissions
casing our environment. Biodiesel Technologies was conceived
by a group of Technologists with a comprehensive professional
experience with multinational companies under the able
stewardship of Mr. Amitabha Sinha who is the M.D. and Chief
of Technology of the company. Mr. Amitabha Sinha proposed a
journey for the present society at large from the present polluted
and harmful environment back to the green and pure
environment as it was ages before. He thought of developing an
alternative source of energy that can reduce pollution levels in
our country, organic in nature. Soon his thought were
materialised into action when the idea/concept of manufacturing
Biodiesel Processing Plants crystallized. The feedstock used was
organic in character which produced Biodiesel as per the ASTM,
EN and BIS Standards. This marked the beginning of our
organization. Since the operation of the first Biodiesel processing
Plant in Hyderabad, Biodiesel Technologies has built a strong
reputation as a leading pioneer in the manufacturing, fabricating
and assembling Biodiesel Processing Plants.

CASE STUDY: BDT SERIES BIODIESEL PROCESSORS

Biodiesel processing units of Biodiesel Technologies, Kolkata,

India produce ASTM D6751-compliant biodiesel from any vegetable oil or animal fats with an Acid Value of 5 per cent or less. Production rates range from 5000 Lts. To 5,00,000 Lts. per day. This production rate includes down time for regular maintenance and upkeeps well as personnel shift changes throughout the day.

BDT-Series Biodiesel Processing Unit Highlights

Guaranteed to produce ASTM D6751-compliant biodiesel. Continuous Flow/Modular Batch Transesterifiers, available. Waterless wash-zero water in—zero waste water out. Skid mounted, versatile and expandable. Includes complete Installation, start up, and transitional assistance. All units include a 5-day training curriculum and proficiency testing with a certified technician for up to FIVE of your employees. Turn-key custom installation with complete build-out available

UNDERSTANDING IMPORTANCE OF BIO DIESEL

Bio-diesel is the most valuable form of renewable energy that can be used directly in any existing, unmodified diesel engine.

I. **Energy Independence:** Considering that oil priced at $ 60 per barrel has had a disproportionate impact on the poorest countries, 38 of which are net importers and 25 of Which import all of their oil; the question of trying to achieve greater energy independence one day through the development of bio-fuels has become one of 'when' rather than 'if,' and, now on a near daily basis, a bio-fuels programme is being launched somewhere in the developing world.

II. **Smaller Trade Deficit:** Rather than importing other countries' ancient natural resources, we could be using our own living resources to power our development and enhance our economies. Instead of looking to the Mid-east for oil, the world could look to the tropics for bio-fuels. producing more bio-fuels will save foreign exchange and reduce energy expenditures and allow developing countries to put more of their resources into health, education and other services for their neediest citizens.

III.   **Economic Growth:** Bio-fuels create new markets for agricultural products and stimulate rural development because bio-fuels are generated from crops; they hold enormous potential for farmers. In the near future— especially for the two-thirds of the people in the developing world who derive their incomes from agriculture. Today, many of these farmers are too small to compete in the global market, especially with the playing field tilted against them through trade distorting agricultural subsidies. They are mostly subsistence farmers who, in a good year, produce enough to feed their families, and in a bad year, grow even poorer or starve. But bio-fuels have enormous potential to change this situation for the better. At the community level, farmers that produce dedicated energy crops can grow their incomes and grow their own supply of affordable and reliable energy. At the national level, producing more bio-fuels will generate new industries, new technologies, new jobs and new markets.

IV.   **Cleaner Air:** Bio-fuels burn more cleanly than gasoline and diesel. Using biofuels means producing fewer emissions of carbon monoxide, particulates, and toxic chemicals that cause smog, aggravate respiratory and heart disease, and contribute to thousands of premature deaths each year.

V.   **Less Global Warming:** Bio-fuels contain carbon that was taken out of the atmosphere by plants and trees as they grew. The Fossil fuels are adding huge amounts of stored carbon dioxide (CO_2) to the atmosphere, where it traps the Earth's heat like a heavy blanket and causes the world to warm. Studies show that bio-diesel reduces CO_2 emissions to a considerable extent and in some cases all most nearly to zero.

Biodiesel in Environment Concern

Biodiesel fuel burns up to 75 per cent cleaner than diesel fuel made from fossil fuels. Bio diesel substantially reduces unburned hydrocarbons, carbon monoxide and particulate matter in exhaust fumes. Sulphur dioxide emissions are 100 per cent eliminated (bio diesel contains no sulphur). This alternative fuel is plant-based and adds absolutely no CO_2 to the atmosphere.

Biofuel exhaust is not offensive and doesn't cause eye irritation. Vehicles do not spew out vile black fumes/particulates. In fact if you make your fuel from used cooking oil it may even smell of chips. Biodiesel is environmentally friendly: it is renewable, "more biodegradable than sugar and less toxic than table salt" (US National Biodiesel Board).

Biodiesel was the first renewable fuel to successfully complete the EPA-required Health Effects esting under the Clean Air Act. Mutagenicity studies show that biofuel dramatically reduces potential risks of cancer and birth defects.

Biodiesel helps preserve natural resources. For every unit of energy needed to produce biodiesel, 3.24 units of energy are gained—nearly four times more than diesel. Graphics will be uploaded soon!

In Nut-Shell

BioDiesel is environmental friendly and ideal for heavily polluted cities. Bio Diesel is as biodegradable as salt Bio Diesel produces 80 per cent less carbon dioxide and 100 per cent less sulfur dioxide emissions. It provides a 90 per cent reduction in cancer risks. Bio Diesel can be used alone or mixed in any ratio with mineral oil diesel fuel. The preferred ratio if mixture ranges between 5 and 20 per cent (B5-B20). Bio Diesel extends the live of diesel engines. Bio Diesel is cheaper then mineral oil diesel. Bio Diesel is conserving natural resources.

Features of Biodiesel

- Biodiesel is a clean burning fuel.

- Biodiesel does not have any toxic emissions like mineral diesel.
- Biodiesel is made from any vegetable oil such as Soya, Rice bran, Canola, Palm, Coconut, mustard or peanut or from any animal fat like Lard or tallow.
- Biodiesel is a complete substitute of Mineral diesel (HSD).
- Biodiesel is made through a chemical process which converts oils and fats of natural origin into fatty acid methyl esters (FAME).
- Biodiesel is intended to be used as a replacement for petroleum diesel fuel, or can be blended with petroleum diesel fuel in any proportion.
- Biodiesel does not require modifications to a diesel engine to be used.
- Biodiesel has reduced exhaust emissions compared to petroleum diesel fuel.
- Biodiesel has lower toxicity compared to petroleum diesel fuel.
- Biodiesel is safer to handle compared to petroleum diesel fuel.
- Biodiesel quality is governed by ASTM D 6751 quality parameters.
- Biodiesel is biodegradable.

7

Biogas, Biohydrogen and Allied Fuels: Production Process and Technology

BIOGAS AS BIOENERGY

Biogas typically refers to a gas produced by the biological breakdown of organic matter in the absence of oxygen. Biogas originates from biogenic material and is a type of biofuel.

One type of biogas is produced by anaerobic digestion or fermentation of biodegradable materials such as biomass, manure or sewage, municipal waste, green waste and energy crops. This type of biogas comprises primarily methane and carbon dioxide. The other principal type of biogas is wood gas which is created by gasification of wood or other biomass. This type of biogas is comprised primarily of nitrogen, hydrogen, and carbon monoxide, with trace amounts of methane.

The gases methane, hydrogen and carbon monoxide can be combusted or oxidized with oxygen. Air contains 21 percent oxygen. This energy release allows biogas to be used as a fuel. Biogas can be used as a low-cost fuel in any country for any heating purpose, such as cooking. It can also be used in modern waste management facilities where it can be used to run any type of heat engine, to generate either mechanical or electrical power. Biogas can be compressed, much like natural gas, and used to power motor vehicles and in the UK for example is estimated to have the potential to replace around 17 percent of vehicle fuel. Biogas is a renewable fuel, so it qualifies for renewable energy subsidies in some parts of the world.

Production

Biogas is practically produced as landfill gas (LFG) or digester gas.

A biogas plant is the name often given to an anaerobic digester that treats farm wastes or energy crops.

Biogas can be produced utilizing anaerobic digesters. These plants can be fed with energy crops such as maize silage or biodegradable wastes including sewage sludge and food waste. There are two key processes: Mesophilic and Thermophilic digestion.

Landfill gas is produced by wet organic waste decomposing under anaerobic conditions in a landfill. The waste is covered and mechanically compressed by the weight of the material that is deposited from above. This material prevents oxygen exposure thus allowing anaerobic microbes thrive. This gas builds up and is slowly released into the atmosphere if the landfill site has not been engineered to capture the gas. Landfill gas is hazardous for three key reasons. Landfill gas becomes explosive when it escapes from the landfill and mixes with oxygen. The lower explosive limit is 5 percent methane and the upper explosive limit is 15 percent methane. The methane contained within biogas is 20 times more potent as a greenhouse gas than carbon dioxide. Therefore uncontained landfill gas which escapes into the atmosphere may significantly contribute to the effects of global warming. In addition landfill gas' impact in global warming, volatile organic compounds (VOCs) contained within landfill gas contribute to the formation of photochemical smog.

Sweden produces biogas from confiscated alcoholic beverages.

Composition

The composition of biogas varies depending upon the origin of the anaerobic digestion process. Landfill gas typically has methane concentrations around 50 per cent. Advanced waste treatment technologies can produce biogas with 55-75 per centCH$_4$ or higher using in situ purification techniques As-produced, biogas also contains water vapour, with the fractional water vapour volume a function of biogas temperature; correction of measured volume for water vapour content and thermal expansion is easily done via algorithm.

In some cases biogas contains siloxanes. These siloxanes are formed from the anaerobic decomposition of materials commonly found in soaps and detergents. During combustion of biogas containing siloxanes, silicon is released and can combine with free oxygen or various other elements in the combustion gas. Deposits are formed containing mostly silica (SiO_2) or silicates (Si_xO_y) and can also contain calcium, sulfur, zinc, phosphorus. Such white mineral deposits accumulate to a surface thickness of several millimeters and must be removed by chemical or mechanical means.

Practical and cost-effective technologies to remove siloxanes and other biogas contaminants are currently available.

Applications

Biogas can be utilized for electricity production on sewage works, in a CHP gas engine, where the waste heat from the engine is conveniently used for heating the digester; cooking; space heating; water heating; and process heating. If compressed, it can replace compressed natural gas for use in vehicles, where it can fuel an internal combustion engine or fuel cells and is a much more effective displacer of carbon dioxide than the normal use in on-site CHP plants.

Methane within biogas can be concentrated via a biogas upgrader to the same standards as fossil natural gas, and becomes **biomethane**. If the local gas network allows for this, the producer of the biogas may utilize the local gas distribution networks. Gas must be very clean to reach pipeline quality, and must be of the correct composition for the local distribution network to accept. Carbon dioxide, water, hydrogen sulphide and particulates must be removed if present. If concentrated and compressed it can also be used in vehicle transportation. Compressed biogas is becoming widely used in Sweden,

Switzerland, and Germany. A biogas-powered train has been in service in Sweden since 2005.

Biogas has also powered automobiles. In 1974, a British documentary film entitled *Sweet as a Nut* detailed the biogas, production process from pig manure, and how the biogas fueled a custom-adapted combustion engine.

Scope and Potential Quantities

In the UK, sewage gas electricity production is tiny compared to overall power consumption—a mere 80 MW of generation, compared to 70 GW on the grid. Estimates vary but could be a considerable fraction from digestion of.

In Developing Nations

In India biogas produced from the anaerobic digestion of manure in small-scale digestion facilities is called gober gas; it is estimated that such facilities exist in over 2 million households. The digester is an airtight circular pit made of concrete with a pipe connection. The manure is directed to the pit, usually directly from the cattle shed. The pit is then filled with a required quantity of wastewater. The gas pipe is connected to the kitchen fire place through control valves. The combustion of this biogas has very little odour or smoke. Owing to simplicity in implementation and use of cheap raw materials in villages, it is one of the most environmentally sound energy sources for rural needs. Some designs use vermiculture to further enhance the slurry produced by the biogas plant for use as compost.

Biogas is also extensively used or being aggressively developed in rural China, Nepal, Vietnam, rural Costa Rica, Colombia, Rwanda, and other regions of the world where waste management and industry closely interface.

Deenabandhu Model (India)

The Deenabandhu Model is a new biogas-production model popular in India. (*Deenabandhu* means "helpful for the poor.") The unit usually has a capacity of 2 to 3 cubic metres. It is constructed using bricks or by a ferrocement mixture. The brick model costs approximately 18,000 rupees and the ferrocment model 14,000 rupees, however India's Ministry of Non-conventional Energy Sources offers a subsidy of up to 3,500 rupees per model constructed.

Arti Biogas Plant

The Appropriate Rural Technology Institute developed a compact biogas plant which uses food waste, rather than dung or manure, to create biogas. The plant is sufficiently compact to be used by urban households, and about 2,000 are currently in use—both in urban and rural households in Maharashtra, India. Few ARTI biogas plants have been installed in other parts of India or the world. The design and development of the ARTI biogas plant won the 2006 Ashden Award for Sustainable Energy 2006 in the Food Security category.

Legislation

The European Union presently has some of the strictest legislation regarding waste management and landfill sites called the Landfill Directive. The United States legislates against landfill gas as it contains these VOCs. The United States Clean Air Act and Title 40 of the Code of Federal Regulations (CFR) requires landfill owners to estimate the quantity of non-methane organic compounds (NMOCs) emitted. If the estimated NMOC emissions exceeds 50 tonnes per year the landfill owner is required to collect the landfill gas and treat it to remove the entrained NMOCs. Treatment of the landfill gas is usually by combustion. Because of the remoteness of landfill sites it is sometimes not economically feasible to produce electricity from the gas.

Gober gas

The airtight circular cylinder of a gober gas plant, which moves up and down depending upon the collection of gas
Gober gas (also spelled **gobar gas**, from the Hindi word *gober* for cow dung) is biogas generated from cow dung. A gober gas

plant is an airtight circular pit made of concrete with a pipe connection. First, manure is dumped in the pit. Then, water or wastewater is added to the manure and the concoction is sealed in the airtight concrete pit with a gas pipe leading to stove unit in the kitchen serving as the only egress for gas. When the control valve on the gas pipe is opened the biogas is combusted for cooking in a largely odourless and smokeless [manner]. After the anaerobic microbial process has been exhausted, the residue left in the concrete pit is often used as fertiliser. Owing to the process's simplicity in implementation and use of cheap raw materials, it is often regarded as one of the most environmentally sound energy sources for rural needs.

In India, gober gas is generated using countless household micro plants (an estimated more than 2 million). The concept is also rapidly growing in Pakistan. The Government of Pakistan subsidises the construction of movable gas chamber biogas plants by up to 50 percent.

UNDERSTANDING ANAEROBIC DIGESTION

Anaerobic digestion is a series of processes in which microorganisms break down biodegradable material in the absence of oxygen and is widely used to treat wastewater. As part of an integrated waste management system, anaerobic digestion reduces the emission of landfill gas into the atmosphere. Anaerobic digestion is widely used as a renewable energy source because the process produces a methane and carbon dioxide rich biogas suitable for energy production helping replace fossil fuels. Also, the nutrient-rich digestate can be used as fertiliser.

The digestion process begins with bacterial hydrolysis of the input materials in order to break down insoluble organic polymers such as carbohydrates and make them available for other bacteria. Acidogenic bacteria then convert the sugars and amino acids into carbon dioxide, hydrogen, ammonia, and organic acids. Acetogenic bacteria then convert these resulting organic acids into acetic acid, along with additional ammonia, hydrogen, and carbon dioxide. Methanogens, finally are able to convert these products to methane and carbon dioxide.

Previously, the technical expertise required to maintain anaerobic digesters coupled with high capital costs and low process efficiencies had limited the level of its industrial application as a waste treatment technology. Anaerobic digestion facilities have, however, been recognised by the United Nations Development Programme as one of the most useful decentralised sources of energy supply, as they are less capital intensive than large power plants.

History

Scientific interest in the gases produced by the natural decomposition of organic matter, was first reported in the seventeenth century by Robert Boyle and Stephen Hale, who noted that flammable gas was released by disturbing the sediment of streams and lakes. In 1808, Sir Humphry Davy determined that methane was present in the gasses produced by cattle manure. The first anaerobic digester was built by a leper colony in Bombay, India in 1859. In 1895 the technology was developed in Exeter, England, where a septic tank was used to generate gas for street lighting. Also in England, in 1904, the first dual purpose tank for both sedimentation and sludge treatment was installed in Hampton. In 1907, in Germany, a patent was issued for the Imhoff tank, an early form of digester.

Through scientific research anaerobic digestion gained academic recognition in the 1930s. This research led to the discovery of anaerobic bacteria, the microorganisms that facilitate the process. Further research was carried out to investigate the conditions under which methanogenic bacteria were able to grow and reproduce. This work was developed during World War II where in both Germany and France there was an increase in the application of anaerobic digestion for the treatment of manure.

Applications

Anaerobic digestion is particularly suited to wet organic material and is commonly used for effluent and sewage treatment.

Anaerobic digestion is a simple process that can greatly reduce the amount of organic matter which might otherwise be destined to be landfilled or burnt in an incinerator.

Almost any organic material can be processed with anaerobic digestion. This includes biodegradable waste materials such as waste paper, grass clippings, leftover food, sewage and animal waste. The exception to this is woody wastes that are largely unaffected by digestion as most anaerobes are unable to degrade lignin. The exception being xylophalgeous anaerobes (lignin consumers), as used in the process for organic breakdown of cellulosic material by a cellulosic ethanol start-up company in the U.S. Anaerobic digesters can also be fed with specially grown energy crops such as silage for dedicated biogas production. In Germany and continental Europe these facilities are referred to as *biogas plants*. A *co-digestion* or *co-fermentation* plant is typically an agricultural anaerobic digester that accepts two or more input materials for simultaneous digestion.

In developing countries simple home and farm-based anaerobic digestion systems offer the potential for cheap, low-cost energy for cooking and lighting. Anaerobic digestion facilities have been recognized by the United Nations Development Programme as one of the most useful decentralized sources of energy supply. From 1975, China and India have both had large government-backed schemes for adaptation of small biogas plants for use in the household for cooking and lighting. Presently, projects for anaerobic digestion in the developing world can gain financial support through the United Nations Clean Development Mechanism if they are able to show they provide reduced carbon emissions.

Pressure from environmentally-related legislation on solid waste disposal methods in developed countries has increased the application of anaerobic digestion as a process for reducing waste volumes and generating useful by-products. Anaerobic digestion may either be used to process the source separated fraction of municipal waste, or alternatively combined with mechanical sorting systems, to process residual mixed municipal waste. These facilities are called mechanical biological treatment plants.

Utilising anaerobic digestion technologies can help to reduce the emission of greenhouse gasses in a number of key ways:

- Replacement of fossil fuels
- Reducing methane emission from landfills
- Displacing industrially-produced chemical fertilizers
- Reducing vehicle movements
- Reducing electrical grid transportation losses

Methane and power produced in anaerobic digestion facilities can be utilized to replace energy derived from fossil fuels, and hence reduce emissions of greenhouse gasses. This is due to the fact that the carbon in biodegradable material is part of a carbon cycle. The carbon released into the atmosphere from the combustion of biogas has been removed by plants in order for them to grow in the recent past. This can have occurred within the last decade, but more typically within the last growing season. If the plants are re-grown, taking the carbon out of the atmosphere once more, the system will be carbon neutral. This contrasts to carbon in fossil fuels that has been sequestered in the earth for many millions of years, the combustion of which increases the overall levels of carbon dioxide in the atmosphere.

If the putrescible waste processed in anaerobic digesters was disposed of in a landfill, it would break down naturally and often anaerobically. In this case the gas will eventually escape into the atmosphere. As methane is about twenty times more potent as a greenhouse gas as carbon dioxide this has significant negative environmental effects.

Digestate liquor can be used as a fertilizer supplying vital nutrients to soils. The solid, fibrous component of digestate can be used as a soil conditioner. The liquor can be used as a substitute for chemical fertilizers which require large amounts of energy to produce and transport. The use of manufactured fertilizers is therefore more carbon intensive than the use of anaerobic digestate fertiliser. This solid digestate can be used to boost the organic content of soils. There are some countries, such as Spain where there are many organically depleted soils, and here the markets for the digestate can be just as important as the biogas.

In countries that collect household waste, the utilization of local anaerobic digestion facilities can help to reduce the amount of waste that requires transportation to centralized landfill sites or incineration facilities. This reduced burden on transportation has and will reduce carbon emissions from the collection vehicles. If localized anaerobic digestion facilities are embedded within an electrical distribution network, they can help reduce the electrical losses that are associated with transporting electricity over a national grid.

In Oakland, California at the East Bay Municipal Utility District's (EBMUD) Main Wastewater Treatment Plant (MWWTP), food waste is currently co-digested with primary and secondary municipal wastewater solids and other high-strength wastes. Compared to municipal wastewater solids digestion, food waste digestion has many benefits. Anaerobic digestion of food waste pulp from the EBMUD food waste process provides a higher normalized energy benefit, compared to municipal wastewater solids:

- 730 to 1,300 kWh per dry ton of food waste applied.
- 560 to 940 kWh per dry ton of municipal wastewater solids applied.

POWER GENERATION

Biogas is usually used to run a gas engine to produce electrical power; some or all of which can be used to run the sewage works. The waste heat from the engine is then used to heat the digester. It turns out that the waste heat is generally enough to heat the digester to the required temperatures. The power potential from sewage works is trivial—in the UK there are about 80MW total of such generation, and this could be conceivably increased to 150 MW, which is insignificant compared to the average power demand in the UK of about 35,000 MW. The scope for biogas generation from non-sewage waste biological matter—energy crops, food waste, abattoir waste etc is much higher, estimated to be capable of about 3,000MW. Farm biogas plants using animal waste and energy crops are expected to contribute to reducing CO2 emissions and strengthen the grid while providing UK farmers with additional revenues.

UNDER THE PROCESS

There are a number of microorganisms that are involved in the process of anaerobic digestion including acetic acid-forming bacteria (acetogens) and methane-forming archaea (methanogens). These organisms feed upon the initial feedstock, which undergoes a number of different processes converting it to intermediate molecules including sugars, hydrogen and acetic acid before finally being converted to biogas.

Different species of bacteria are able to survive at different temperature ranges. Ones living optimally at temperatures between 35-40°C are called mesophiles or mesophilic bacteria. Some of the bacteria can survive at the hotter and more hostile conditions of 55-60°C, these are called thermophiles or thermophilic bacteria. Methanogens come from the primitive group of archaea. This family includes species that can grow in the hostile conditions of hydrothermal vents. These species are more resistant to heat and can therefore operate at thermophilic temperatures, a property that is unique to bacterial families.

As with aerobic systems the bacteria in anaerobic systems the growing and reproducing microorganisms within them require a source of elemental oxygen to survive.

In an anaerobic system there is an absence of gaseous oxygen. Gaseous oxygen is prevented from entering the system through physical containment in sealed tanks. Anaerobes access oxygen from sources other than the surrounding air. The oxygen source for these microorganisms can be the organic material itself or alternatively may be supplied by inorganic oxides from within the input material. When the oxygen source in an anaerobic system is derived from the organic material itself, then the 'intermediate' end products are primarily alcohols, aldehydes, and organic acids plus carbon dioxide. In the presence of specialised methanogens, the intermediates are converted to the 'final' end products of methane, carbon dioxide with trace levels of hydrogen sulfide. In an anaerobic system the majority of the chemical energy contained within the starting material is released by methanogenic bacteria as methane.

Populations of anaerobic microorganisms typically take a significant period of time to establish themselves to be fully effective. It is therefore common practice to introduce anaerobic microorganisms from materials with existing populations. This process is called 'seeding' the digesters and typically takes place with the addition of sewage sludge or cattle slurry.

Stages

The key process stages of anaerobic digestion

There are four key biological and chemical stages of anaerobic digestion: Hydrolysis

1. Acidogenesis
2. Acetogenesis
3. Methanogenesis

In most cases biomass is made up of large organic polymers. In order for the bacteria in anaerobic digesters to access the energy potential of the material, these chains must first be broken down into their smaller constituent parts. These constituent parts or monomers such as sugars are readily available by other bacteria. The process of breaking these chains and dissolving the smaller molecules into solution is called hydrolysis. Therefore hydrolysis of these high molecular weight polymeric components is the necessary first step in anaerobic digestion. Through hydrolysis the complex organic molecules are broken down into simple sugars, amino acids, and fatty acids.

Acetate and hydrogen produced in the first stages can be used directly by methanogens. Other molecules such as volatile fatty acids (VFA's) with a chain length that is greater than acetate must first be catabolised into compounds that can be directly utilised by methanogens.

The biological process of acidogenesis is where there is further breakdown of the remaining components by acidogenic (fermentative) bacteria. Here VFAs are created along with ammonia, carbon dioxide and hydrogen sulphide as well as other by-products. The process of acidogenesis is similar to the way that milk sours.

The third stage anaerobic digestion is acetogenesis. Here simple molecules created through the acidogenesis phase are further digested by acetogens to produce largely acetic acid as well as carbon dioxide and hydrogen.

The terminal stage of anaerobic digestion is the biological process of methanogenesis. Here methanogens utilise the intermediate products of the preceding stages and convert them into methane, carbon dioxide and water. It is these components that makes up the majority of the biogas emitted from the system. Methanogenesis is sensitive to both high and low pHs and occurs between pH 6.5 and pH 8. The remaining, non-digestable material which the microbes cannot feed upon, along with any dead bacterial remains constitutes the digestate.

A simplified generic chemical equation for the overall processes outlined above is as follows:

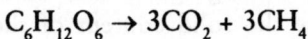

$$C_6H_{12}O_6 \rightarrow 3CO_2 + 3CH_4$$

Feedstock

Anaerobic lagoon and generators at the Cal Poly Dairy, United States 2003

The most important initial issue when considering the application of anaerobic digestion systems is the feedstock to the process. Digesters typically can accept any biodegradable material, however if biogas production is the aim, the level of putrescibility is the key factor in its successful application. The more putrescible the material the higher the gas yields possible from the system.

Substrate composition is a major factor in determining the methane yield and methane production rates from the digestion of biomass. Techniques are available to determine the compositional characteristics of the feedstock, whilst parameters such as solids, elemental and organic analyses are important for digester design and operation.

Anaerobes can breakdown material to varying degrees of success from readily in the case of short chain hydrocarbons such as sugars, to over longer periods of time in the case of cellulose and hemicellulose. Anaerobic microorganisms are unable to break down long chain woody molecules such as lignin. Anaerobic digesters were originally designed for operation using sewage sludge and manures. Sewage and manure are not, however, the material with the most potential for anaerobic digestion as the biodegradable material has already had the energy content taken out by the animal that produced it.

A second consideration related to the feedstock will be moisture content. The wetter the material the more suitable it will be to handling with standard pumps instead of energy intensive concrete pumps and physical means of movement. Also the wetter the material, the more volume and area it takes up relative to the levels of gas that are produced. The moisture content of the target feedstock will also affect what type of system is applied to its treatment. In order to use a high solids anaerobic digester for dilute feedstocks, bulking agents such as compost should be applied to increase the solid content of the input material. Another key consideration is the carbon : nitrogen ratio of the input material. This ratio is the balance of food a microbe requires in order to grow. The optimal C : N ratio for the 'food' a microbe is 20—30:1. Excess N can lead to ammonia inhibition of digestion.

The level of contamination of the feedstock material is a key consideration. If the feedstock to the digesters has significant levels of physical contaminants such as plastic, glass or metals then pre-processing will be required in order for the material to be used. If it is not removed then the digesters can be blocked and will not function efficiently. It is with this logic in mind that mechanical biological treatment plants are designed. The higher the level of pre-treatment a feedstock requires, the more processing machinery will be required and hence the project will have higher capital costs.

After sorting or screening to remove any physical contaminants, such as metals and plastics, from the feedstock the material is often shredded, minced and mechanically or hydraulically pulped to increase the surface area available to microbes in the digesters and hence increase the speed of digestion. The feedstock material is then fed into the airtight digester where anaerobic treatment takes place.

Configuration

Farm-based maize silage digester located near Neumünster in Germany, 2007. Green inflatable biogas holder is shown on top of the digester

Anaerobic digesters can be designed and engineered to operate using a number of different process configurations:

- Batch or continuous
- Temperature: Mesophilic or thermophilic
- Solids content: High solids or low solids
- Complexity: Single stage or multistage

Batch or Continuous

A batch system is the simplest form of digestion. Biomass is added to the reactor at the start of the process in a batch and is sealed for the duration of the process. Batch reactors suffer from odour issues that can be a severe problem when they are emptied. Typically biogas production will be formed with a normal distribution pattern over time. The operator can use this fact to determine when they believe the process of digestion of the organic matter has completed. As the batch digestion is simple and requires less equipment and lower levels of design work it is typically a cheaper form of digestion.

In continuous digestion processes organic matter is constantly or added in stages to the reactor. Here the end products are constantly or periodically removed, resulting in constant production of biogas. Examples of this form of anaerobic digestion include, continuous stirred-tank reactors (CSTRs), Upflow anaerobic sludge blanket (UASB), Expanded granular sludge bed (EGSB) and Internal circulation reactors (IC).

Temperature

There are two conventional operational temperature levels for anaerobic digesters, which are determined by the species of methanogens in the digesters:

- *Mesophilic* which takes place optimally around 37°-41°C or at ambient temperatures between 20°-45°C where mesophiles are the primary microorganism present
- *Thermophilic* which takes place optimally around 50°-52° at elevated temperatures up to 70°C where thermophiles are the primary microorganisms present

A limit case has been reach in Bolivia, with anaerobic digestion in temperature working conditions less than 10°C. The anaerobic process is very slow, taking more than three times the mesophilicnormal time process.

There are a greater number of species of mesophiles than thermophiles. These bacteria are also more tolerant to changes in environmental conditions than thermophiles. Mesophilic systems are therefore considered to be more stable than thermophilic digestion systems.

As mentioned above, thermophilic digestion systems are considered to be less stable, however the increased temperatures facilitate faster reaction rates and hence faster gas yields. Operation at higher temperatures facilitates greater sterilisation of the end digestate. In countries where legislation, such as the Animal By-Products Regulations in the European Union, requires end products to meet certain levels of reduction in the amount of bacteria in the output material, this may be a benefit.

A drawback of operating at thermophilic temperatures is that more heat energy input is required to achieve the correct operational temperatures. This increase in energy may not be outweighed by the increase in the outputs of biogas from the systems. It is therefore important to consider an energy balance for these systems.

Solids

Typically there are two different operational parameters associated with the solids content of the feedstock to the digesters:

- High-solids
- Low-solids

Digesters can either be designed to operate in a high solids content, with a total suspended solids (TSS) concentration greater than ~20 per cent, or a low solids concentration less than ~15 per cent.

High-solids digesters process a thick slurry that requires more energy input to move and process the feedstock. The thickness of the material may also lead to associated problems with abrasion. High-solids digesters will typically have a lower land requirement due to the lower volumes associated with the moisture.

Low-solids digesters can transport material through the system using standard pumps that require significantly lower energy input. Low-solids digesters require a larger amount of land than high-solids due to the increase volumes associated with the increased liquid-to-feedstock ratio of the digesters. There are benefits associated with operation in a liquid environment as it enables more thorough circulation of materials and contact between the bacteria and their food. This enables the bacteria to more readily access the substances they are feeding off and increases the speed of gas yields.

Number of Stages

Two-stage, low-solids, UASB digestion component of a mechanical biological treatment system near Tel Aviv, process water is seen in balance tank and sequencing batch reactor, 2005

Digestion systems can be configured with different levels of complexity:

- One-stage or single-stage
- Two-stage or multistage

A single-stage digestion system is one in which all of the biological reactions occur within a single sealed reactor or holding tank. Utilising a single stage reduces construction costs, however facilitates less control of the reactions occurring within the system. Acidogenic bacteria, through the production of acids, reduce the pH of the tank. Methanogenic bacteria, as outlined earlier, operate in a strictly defined pH range. Therefore the biological reactions of the different species in a single stage reactor can be in direct competition with each other. Another one-stage reaction system is an anaerobic lagoon. These lagoons are pond-like earthen basins used for the treatment and long-term storage of manures. Here the anaerobic reactions are contained within the natural anaerobic sludge contained in the pool.

In a two-stage or multi-stage digestion system different digestion vessels are optimised to bring maximum control over the bacterial communities living within the digesters. Acidogenic bacteria produce organic acids and more quickly grow and reproduce than methanogenic bacteria. Methanogenic bacteria require stable pH and temperature in order to optimise their performance.

Typically hydrolysis, acetogenesis and acidogenesis occur within the first reaction vessel. The organic material is then heated to the required operational temperature (either mesophilic or thermophilic) prior to being pumped into a methanogenic reactor. The initial hydrolysis or acidogenesis tanks prior to the methanogenic reactor can provide a buffer to the rate at which feedstock is added. Some European countries require a degree of elevated heat treatment in order to kill harmful bacteria in the input waste. In this instance there may be a pasteurisation or sterilisation stage prior to digestion or between the two digestion tanks. It should be noted that it is not possible to completely isolate the different reaction phases and often there is some biogas that is produced in the hydrolysis or acidogenesis tanks.

Residence

The residence time in a digester varies with the amount and type of feed material, the configuration of the digestion system and whether it be one-stage or two-stage.

In the case of single-stage thermophilic digestion residence times may be in the region of 14 days, which comparatively to mesophilic digestion is relatively fast. The plug-flow nature of some of these systems will mean that the full degradation of the material may not have been realised in this timescale. In this event digestate exiting the system will be darker in colour and will typically have more odour.

In two-stage mesophilic digestion, residence time may vary between 15 and 40 days.

In the case of mesophilic UASB digestion hydraulic residence times can be (1hour-1day) and solid retention times can be up to 90 days. In this manner the UASB system is able to separate solid an hydraulic retention times with the utilisation of a sludge blanket.

Continuous digesters have mechanical or hydraulic devices, depending on the level of solids in the material, to mix the contents enabling the bacteria and the food to be in contact. They also allow excess material to be continuously extracted to maintain a reasonably constant volume within the digestion tanks.

Products

There are three principal products of anaerobic digestion: biogas, digestate and water.

Biogas

Biogas holder with lightning protection rods and back-up gas flare

Typical Composition of Biogas

Matter	%
Methane, CH_4	50-75
Carbon dioxide, CO_2	25-50
Nitrogen, N_2	0-10
Hydrogen, H_2	0-1
Hydrogen sulfide, H_2S	0-3
Oxygen, O_2	0-2

ANAEROBIC DIGESTER TYPES

The following is a partial list of different **types of anaerobic digesters**. These processes and systems harness anaerobic digestion for purposes such as sewage treatment and biogas generation.

Examples include:

- Anaerobic activated sludge process
- Anaerobic clarigester
- Anaerobic contact process
- Anaerobic expanded-bed reactor
- Anaerobic filter
- Anaerobic fluidised bed
- Anaerobic lagoon
- Anaerobic migrating blanket reactor
- Batch system anaerobic digester
- Continuous stirred-tank reactor (CSTR)
- Expanded granular sludge bed digestion (EGSB)
- Hybrid reactor
- Imhoff tank
- Internal circulation reactor (IC)
- One-stage anaerobic digester
- Submerged media anaerobic reactor
- Two-stage anaerobic digester
- Upflow anaerobic sludge blanket digestion (UASB)
- Upflow and down-flow anaerobic attached growth

ROLE OF THERMOPHILIC DIGESTER

A **thermophilic digester** or **thermophilic biodigester** is a kind of biodigester that operates in temperatures above 50°C producing biogas. It has some advantages: it does not needs agitation and is faster in fermentation than a mesophilic digester. In fact, it can be as much as six to ten times faster than a normal biodigester. The problem is that for use in this biodigester, the source must enter at high temperature. Vinasse is produced at more than

70°C and can be used in this kind of biodigester. For one gallon of ethanol, about eight gallons of vinasse is produced. In Brazil, this kind of biodigester is used to process vinasse as a cheap source of methane.

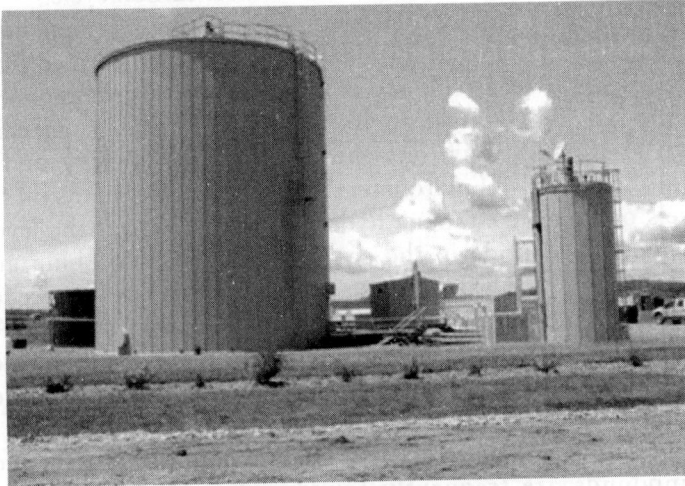

BIOGAS CARRYING PIPES

Biogas is the ultimate waste product of the bacteria feeding off the input biodegradable feedstock, and is mostly methane and carbon dioxide, with a small amount hydrogen and trace hydrogen sulfide. (As-produced, biogas also contains water vapour, with the fractional water vapour volume a function of biogas temperature.) Most of the biogas is produced during the middle of the digestion, after the bacterial population has grown, and tapers off as the putrescible material is exhausted. The gas is normally stored on top of the digester in an inflatable gas bubble or extracted and stored next to the facility in a gas holder.

The methane in biogas can be burned to produce both heat and electricity, usually with a reciprocating engine or microturbine often in a cogeneration arrangement where the electricity and waste heat generated are used to warm the digesters or to heat buildings. Excess electricity can be sold to suppliers or put into the local grid. Electricity produced by anaerobic digesters is considered to be renewable energy and may attract subsidies. Biogas does not contribute to increasing atmospheric carbon dioxide concentrations because the gas is not released directly into the atmosphere and the carbon dioxide comes from an organic source with a short carbon cycle.

Biogas may require treatment or 'scrubbing' to refine it for use as a fuel. Hydrogen sulfide is a toxic product formed from sulfates in the feedstock and is released as a trace component of the biogas. National environmental enforcement agencies such as the U.S. Environmental Protection Agencyý or the English and Welsh Environment Agency put strict limits on the levels of gasses containing hydrogen sulfide, and if the levels of hydrogen sulfide in the gas are high, gas scrubbing and cleaning equipment (such as amine gas treating) will be needed to process the biogas to within regionally accepted levels. An alternative method to this is by the addition of ferrous chloride $FeCl_2$ to the digestion tanks in order to inhibit hydrogen sulfide production.

Volatile siloxanes can also contaminate the biogas; such compounds are frequently found in household waste and wastewater. In digestion facilities accepting these materials as a component of the feedstock, low molecular weight siloxanes volatilise into biogas. When this gas is combusted in a gas engine, turbine or boiler, siloxanes are converted into silicon dioxide (SiO_2) which deposits internally in the machine, increasing wear and tear. Practical and cost-effective technologies to remove siloxanes and other biogas contaminants are available at the present time. In certain applications, *in situ* treatment can be used to increase the methane purity by reducing the carbon dioxide content.

In countries such as Switzerland, Germany and Sweden the methane in the biogas may be concentrated in order for it to be used as a vehicle transportation fuel or alternatively input directly into the gas mains. In countries where the driver for the utilisation of anaerobic digestion are renewable electricity subsidies, this route of treatment is less likely as energy is required in this processing stage and reduces the over all levels available to sell.

Digestate

Digestate is the solid remnants of the original input material to the digesters that the microbes cannot use. It also consists of the mineralised remains of the dead bacteria from within the digesters. Digestate can come in three forms; fibrous, liquor or a sludge-based combination of the two fractions. In two-stage systems the different forms of digestate come from different digestion tanks. In single stage digestion systems the two fractions will be combined and if desired separated by further processing.

The second by-product (acidogenic digestate) is a stable organic material comprised largely of lignin and cellulose, but also of a variety of mineral components in a matrix of dead bacterial cells; some plastic may be present. The material resembles domestic compost and can be used as compost or to make low grade building products such as fibreboard.

The third by-product is a liquid (methanogenic digestate) that is rich in nutrients and can be used as a fertiliser dependent on the quality of the material being digested. Levels of potentially toxic elements (PTEs) should be chemically assessed. This will be dependent upon the quality of the original feedstock. In the case of most clean and source-separated biodegradable waste streams the levels of PTEs will be low. In the case of wastes originating from industry the levels of PTEs may be higher and will need to be taken into consideration when determining a suitable end use for the material.

Digestate typically contains elements such as lignin that cannot be broken down by the anaerobic microorganisms. Also the digestate may contain ammonia that is phytotoxic and will hamper the growth of plants if it is used as a soil improving material. For these two reasons a maturation or composting stage may be employed after digestion. Lignin and other materials are available for degradation by aerobic microorganisms such as fungi helping reduce the overall volume of the material for transport. During this maturation the ammonia will be broken down into nitrates, improving the fertility of the material and making it more suitable as a soil improver. Large composting stages are typically used by dry anaerobic digestion technologies.

Wastewater

The final output from anaerobic digestion systems is water. This water originates both from the moisture content of the original waste that was treated but also includes water produced during the microbial reactions in the digestion systems. This water may be released from the dewatering of the digestate or may be implicitly separate from the digestate.

The wastewater exiting the anaerobic digestion facility will typically have elevated levels of biochemical oxygen demand (BOD) and chemical oxygen demand (COD), these are measures of the reactivity of the effluent and show an ability to pollute. Some of this material is termed 'hard COD' meaning it cannot be accessed by the anaerobic bacteria for conversion into biogas. If this effluent was put directly into watercourses it would negatively affect them by causing eutrophication. As such further treatment of the wastewater is often required. This treatment will typically be an oxidation stage where air is passed through the water in a sequencing batch reactors or reverse osmosis unit.

SILOXANE AS FUEL

A **siloxane** is any chemical compound composed of units of the form R_2SiO, where R is a hydrogen atom or a hydrocarbon group. They belong to the wider class of organosilicon compounds.

Siloxanes can have branched or unbranched backbones consisting of alternating silicon and oxygen atoms -Si-O-Si-O-, with side chains R attached to the silicon atoms. More complicated structures are also known, for example eight silicon atoms at the corners of a cube, connected by 12 oxygen atoms as the cube edges.

The word *siloxane* is derived from the words silicon, oxygen, and alkane.

Polymerized siloxanes with organic side chains (R ? H) are commonly known as silicones or as *polysiloxanes*. Representative examples are $[SiO (CH_3)_2]_n$ (polydimethylsiloxane) and $[SiO (C_6H_5)_2]_n$ (polydiphenylsiloxane). These compounds can be viewed as a hybrid of both organic and inorganic compounds.

The organic side chains confer hydrophobic properties while the -Si-O-Si-O- backbone is purely inorganic.

Cyclic Siloxanes	Linear Siloxanes
D3: hexamethylcyclotrisiloxane	MM: hexamethyldisiloxane
D4: octamethylcyclotetrasiloxane	MDM: octamethyltrisiloxane
D5: decamethylcyclopentasiloxane	MD2M: decamethyltetrasiloxane
D6: dodecamethylcyclohexasiloxane	MDnM: polydimethylsiloxane

Applications

Siloxanes can be found in products such as cosmetics, deodorant, defoamers, water repelling windshield coating, food additives and some soaps. They occur in landfill gas and are being evaluated as alternatives to perchloroethylene for drycleaning. Perchloroethylene is widely considered environmentally undesirable.

Siloxanes in Biogas

In internal combustion engines, siloxane deposits on pistons and cylinder heads are extremely abrasive and cause damage to the internal components of the engine. Engines can require a complete overhaul at 5,000 h or less of operation. Deposits on the turbine of the turbocharger will eventually reduce the components efficiency.

Stirling engines are more resistant against siloxanes, though deposits on the tubes of the heat exchanger will reduce the efficiency.

UNDERSTANDING BIOHYDROGEN

Biohydrogen is hydrogen produced via biological processes. Biohydrogen is not the same as biological hydrogen produced by algae.

Fermentation

Biohydrogen gas can be extracted from biomass through dark fermentation and/or photofermentation by bacteria.

Fermentation Plants

Fermentative hydrogen production plants are proposed industrial plants for the production of hydrogen. They would typically involve processes such as thermophilic fermentation, dark fermentation and/or photofermentation and gas cleaning. Biohydrogen production can also involve an element of anaerobic digestion where the methane from biogas is converted through steam reforming into hydrogen. Biohydrogen produced from· organic waste materials is a promising alternatives for sustainable energy sources.

Hydrogen can be produced by bacterial species such as *Rhodobacter sphaeroides* and *Enterobacter cloacae*. Biohydrogen is always and in no other way produced through dark fermentation either by mix culture of hydrogen producing sludge or pure culture of anaerobic bacteria such as *Clostridium butyricum*.

Proton exchange membrane fuel cells (PEMFC) are the essential technology that made possible the conversion of the energy content in biohydrogen to electricity.

PROCESS OF BIOLOGICAL HYDROGEN PRODUCTION

Biological hydrogen production is done in a closed photobioreactor based on the production of hydrogen by algae. Algae produce hydrogen under certain conditions. In 2000 it was discovered that if *C. reinhardtii* algae are deprived of sulfur they will switch from the production of oxygen, as in normal photosynthesis, to the production of hydrogen.

Bioreactor Design Issues

- Restriction of photosynthetic hydrogen production by accumulation of a proton gradient.
- Competitive inhibition of photosynthetic hydrogen production by carbon dioxide.
- Requirement for bicarbonate binding at photosystem II (PSII) for efficient photosynthetic activity.
- Competitive drainage of electrons by oxygen in algal hydrogen production.

- Economics must reach competitive price to other sources of energy and the economics are dependent on several parameters.
- A major technical obstacle is the efficiency in converting solar energy into chemical energy stored in molecular hydrogen.

Attempts are in progress to solve these problems via bioengineering.

Milestones

- 1939 German researcher by the name of Hans Gaffron discovered while working at the University of Chicago, that algae can switch between producing oxygen and hydrogen.
- 1997 Professor Anastasios Melis discovered, after following Hans Gaffron's work, that the deprivation of sulfur will cause the algae to switch from producing oxygen to producing hydrogen. The enzyme, hydrogenase, he found was responsible for the reaction.
- 2006—Researchers from the University of Bielefeld and the University of Queensland have genetically changed the single-cell green alga *Chlamydomonas reinhardtii* in such a way that it produces an especially large amount of hydrogen. The Stm6 can, in the long run, produce five times the volume made by the wild form of alga and up to 1.6-2.0 percent energy efficiency.
- 2007—It was discovered that if copper is added to block oxygen generation algae will switch from the production of oxygen to hydrogen
- 2007—Anastasios Melis studying solar-to-chemical energy conversion efficiency in *tlaX* mutants of *Chlamydomonas reinhardtii*, achieved 15 per cent efficiency, demonstrating that truncated Chl antenna size would minimize wasteful dissipation of sunlight by individual cells This solar-to-chemical energy conversion process could be coupled to the production of a variety of bio-fuels including hydrogen.

- 2008—Anastasios Melis studying solar-to-chemical energy conversion efficiency in *tlaR* mutants of *Chlamydomonas reinhardtii*, achieved 25 per cent efficiency out of a theoretical maximum of 30 per cent.

Research

- 2006—At the University of Karlsruhe, a prototype of a bio-reactor containing 500-1,000 litres of algae cultures is being developed. The reactor is to be used to prove the economic feasibility of the system in the next five years.
- A joint venture between El Paso's Valcent Products and the Canadian Alternative Energy firm, Global Green Solutions has built a $3 Million dollar laboratory to further develop a system that will allow for low cost, mass production of algae in just about any location across the globe. The algae is grown in a "closed loop" and produces more hydrogen than that of naturally occurring algae. While algae grow well in an "open pond", the Vertigro system uses a greenhouse filled with tall, clear plastic bags, suspended end to end in rows, to breed algae. In September 2008, Global Green Solutions sold its 50 per cent interest in Vertigro to partner Valcent. Valcent continues development of the system, while Global Green Solutions is concentrating on a biomass to steam energy they call "green steam". Similar vertical cultivation stacking systems are under development elsewhere, notably at Greenfuel Technologies who are using thin poly sheets joined to form serpentine patterns resembling the children's game "snakes and ladders" in which algae and water flow downward as carbon dioxide is bubbled upward from the bottom.

Economics

It would take an algae farm the size of the state of Texas to produce enough hydrogen to supply the energy needs of the whole world. It would take about 25,000 square kilometres to be sufficient to displace gasoline use in the US; this is less than a tenth ơf the area devoted to growing soya in the US but would equal the size of the state of Vermont, or three times the size of the everglades swamp in Florida, all dedicated to raising this form of algae.

The US Department of Energy has targeted a selling price of $2.60/kg as a goal for making renewable hydrogen econqmically viable. 1 kg is approximately the energy equivalent to a gallon of gasoline. To achieve this, the efficiency of light-to-hydrogen conversion must reach 10 per cent while current efficiency is only 1 per cent and selling price is estimated at $13.53/kg.

According to the DOE cost estimate, for a refueling station to supply 100 cars per day, it would need 300 kg. With current technology, a 300 kg per day stand-alone system will require 110,000 m^2 of pond area, 0.2 g/L cell concentration, a truncated antennae mutant and 10 cm pond depth.

Areas of research to increase efficiency include developing oxygen-tolerant hydrogenases and increased hydrogen production rates through improved electron transfer.

History

In 1939 a German researcher named Hans Gaffron, while working at the University of Chicago, observed that the algae he was studying, *Chlamydomonas reinhardtii* (a green-algae), would sometimes switch from the production of oxygen to the production of hydrogen. Gaffron never discovered the cause for this change and for many years other scientists failed in their attempts at its discovery. In the late 1990s professor Anastasios Melis a researcher at the University of California at Berkeley discovered that if the algae culture medium is deprived of sulfur it will switch from the production of oxygen (normal photosynthesis), to the production of hydrogen. He found that the enzyme responsible for this reaction is hydrogenase, but that the hydrogenase lost this function in the presence of oxygen. Melis found that depleting the amount of sulfur available to the algae interrupted its internal oxygen flow, allowing the hydrogenase an environment in which it can react, causing the algae to produce hydrogen. *Chlamydomonas moewusii* is also a good strain for the production of hydrogen. Scientists at the U.S. Department of Energy's Argonne National Laboratory are currently trying to find a way to take the part of the hydrogenase enzyme that creates the hydrogen gas and introduce it into the photosynthesis process. The result would be a large amount of hydrogen gas, possibly on par with the amount of oxygen created.

UNDERSTANDING PHOTOHYDROGEN

Photohydrogen is hydrogen produced with the help of artificial or natural light This is how the leaf of a tree splits water molecules into protons (hydrogen ions), electrons (to make carbohydrates) and oxygen (released into the air as a waste product). Photohydrogen may also be produced by the photodissociation of water by ultraviolet light.

Photohydrogen is sometimes discussed in the context of obtaining renewable energy from sunlight, by using microscopic organisms such as bacteria or algae. These organisms create hydrogen with the help of hydrogenase enzymes which convert protons derived from the water splitting reaction into hydrogen gas which can then be collected and used as a biofuel.

UNDERSTANDING BIOGASOLINE

Biogasoline is gasoline produced from biomass such as algae. Like traditionally produced gasoline, it contains between 6 (hexane) and 12 (dodecane) carbon atoms per molecule and can be used in internal-combustion engines. Biogasoline is chemically different from biobutanol and bioethanol, as these are alcohols, not hydrocarbons.

BG100, or 100 per cent biogasoline, can immediately be used as a drop-in substitute for petroleum gasoline in any conventional gasoline engine, and can be distributed in the same fueling infrastructure, as the properties match traditional gasoline from petroleum. Dodecane requires a small percentage of octane booster to match gasoline. Ethanol fuel (E85) requires a special engine and has lower combustion energy and corresponding fuel economy.

Companies such as Diversified Energy Corporation are developing approaches to take triglyceride inputs and through a process of deoxygenation and reforming (cracking, isomerizing, aromatizing, and producing cyclic molecules) producing biogasoline. This biogasoline is intended to match the chemical, kinetic, and combustion characteristics of its petroleum counterpart, but with much higher octane levels. Others are pursuing similar approaches based on hydrotreating. And lastly still others are focused on the use of woody biomass for conversion to biogasoline using enzymatic processes.

Properties of Common Fuels

Fuel	Energy Density MJ/L	Air-fuel ratio	Specific Energy MJ/kg	Heat of Vaporization MJ/kg	RON	MON
Gasoline	34.6	14.6	46.9	0.36	91-99	81-89
Butanol fuel	29.2	11.2	36.6	0.43	96	78
Ethanol fuel	24.0	9.0	30.0	0.92	129	102
Methanol fuel	19.7	6.5	15.6	1.2	136	104

UNDERSTANDING BUTANOL FUEL

Butanol may be used as a fuel in an internal combustion engine. Because its longer hydrocarbon chain causes it to be fairly non-polar, it is more similar to gasoline than it is to ethanol. Butanol has been demonstrated to work in some vehicles designed for use with gasoline without any modification. It can be produced from biomass (as "biobutanol") as well as fossil fuels (as "petrobutanol"); both biobutanol and petrobutanol have the same chemical properties.

Production of Biobutanol

Butanol from biomass is called biobutanol. It can be used in unmodified gasoline engines.

Technologies

Biobutanol can be produced by fermentation of biomass by the A.B.E. process. The process uses the bacterium *Clostridium acetobutylicum*, also known as the *Weizmann organism*. It was Chaim Weizmann who first used this bacteria for the production of acetone from starch (with the main use of acetone being the making of Cordite) in 1916. The butanol was a by-product of this fermentation (twice as much butanol was produced). The process also creates a recoverable amount of H_2 and a number of other by-products: acetic, lactic and propionic acids, acetone, isopropanol and ethanol.

The difference from ethanol production is primarily in the fermentation of the feedstock and minor changes in distillation. The feedstocks are the same as for ethanol: energy crops such as sugar beets, sugar cane, corn grain, wheat and cassava as well as agricultural byproducts such as straw and corn stalks. According to DuPont, existing bioethanol plants can cost-effectively be retrofitted to biobutanol production.

Algae Butanol

Biobutanol can be made entirely with solar energy, from algae (called Solalgal Fuel) or diatoms.

Producers

ButylFuel, LLC used a U.S. Department of Energy Small Business Technology Transfer grant to develop a process aimed at making biobutanol production economically competitive with petrochemical production processes.

DuPont and BP plan to make biobutanol the first product of their joint effort to develop, produce, and market next-generation biofuels. In Europe the Swiss company Butalco is developing genetically modified yeasts for the production of biobutanol from cellulosic materials.

Distribution

Butanol better tolerates water contamination and is less corrosive than ethanol and more suitable for distribution through existing pipelines for gasoline. In blends with diesel or gasoline, butanol is less likely to separate from this fuel than ethanol if the fuel is contaminated with water. There is also a vapour pressure co-blend synergy with butanol and gasoline containing ethanol, which facilitates ethanol blending. This facilitates storage and distribution of blended fuels.

Energy Content and Effects on Fuel Economy

Switching a gasoline engine over to butanol would in theory result in a fuel consumption penalty of about 10 per cent but butanol's effect on mileage is yet to be determined by a scientific study. While the energy density for any mixture of gasoline and butanol can be calculated, tests with other alcohol fuels have demonstrated that the effect on fuel economy is not proportional to the change in energy density.

Octane Rating

The octane rating of n-butanol is similar to that of gasoline but lower than that of ethanol and methanol. n-Butanol has a RON (Research Octane number) of 96 and a MON (Motor octane number) of 78 while t-butanol has octane ratings of 105 RON and 89 MON. t-Butanol is used as an additive in gasoline but cannot be used as a fuel in its pure form because its relatively high melting point of 25.5 °C causes it to gel and freeze near room temperature.

A fuel with a higher octane rating is less prone to knocking (extremely rapid and spontaneous combustion by compression) and the control system of any modern car engine can take advantage of this by adjusting the ignition timing. This will improve energy efficiency, leading to a better fuel economy than the comparisons of energy content different fuels indicate. By increasing the compression ratio, further gains in fuel economy, power and torque can be achieved. Conversely, a fuel with lower octane rating is more prone to knocking and will lower efficiency. Knocking can also cause engine damage.

Air-Fuel Ratio

Alcohol fuels, including butanol and ethanol, are partially

oxidized aHd t<h3>herefore need to run at richer mixtures than gasoline. Standard gasoline engines in cars can adjust the air-fuel ratio to accommodate variations in the fuel, but only within certain limits depending on model. If the limit is exceeded by running the engine on pure butanol or a gasoline blend with a high percentage of butanol, the engine will run lean, something which can critically damage components. Compared to ethanol, butanol can be mixed in higher ratios with gasoline for use in existing cars without the need for retrofit as the air-fuel ratio and energy content are closer to that of gasoline.

Specific Energy

Alcohol fuels have less energy per unit weight and unit volume than gasoline. To make it possible to compare the net energy released per cycle a measure called the fuels specific energy is sometimes used. It is defined as the energy released per air fuel ratio. The net energy released per cycle is higher for butanol than ethanol or methanol and about 10 per cent higher than for gasoline.

Substance	Kinematic Viscosity at 20°C
Butanol	3.64 cSt
Ethanol	1.52 cSt
Methanol	0.64 cSt
Gasoline	0.4-0.8 cSt
Diesel	>3 cSt
Water	1.0 cSt

The viscosity of alcohols increase with longer carbon chains. For this reason, butanol is used as an alternative to shorter alcohols when a more viscous solvent is desired. The kinematic viscosity of butanol is several times higher than that of gasoline and about as viscous as high quality diesel fuel.

Heat of Vaporization

The fuel in an engine has to be vaporized before it will burn. Insufficient vaporization is a known problem with alcohol fuels during cold starts in cold weather. As the heat of vaporization of butanol is less than half of that of ethanol, an engine running on butanol should be easier to start in cold weather than one running on ethanol or methanol.

Potential Problems with the Use of Butanol Fuel

The potential problems with the use of butanol are similar to those of ethanol:

- To match the combustion characteristics of gasoline, the utilization of butanol fuel as a substitute for gasoline requires fuel-flow increases (though butanol has only slightly less energy than gasoline, so the fuel-flow increase required is only minimal, maybe 10 per cent, compared to 40 per cent for ethanol.)
- Alcohol-based fuels are not compatible with some fuel system components.
- Alcohol fuels may cause erroneous gas gauge readings in vehicles with capacitance fuel level gauging.
- While ethanol and methanol have lower energy densities than butanol, their higher octane number allows for greater compression ratio and efficiency. Higher combustion engine efficiency allows for lesser greenhouse gas emissions per unit motive energy extracted.
- Methanol is toxic. Ethanol is considered non-toxic for the most part. Butanol is less toxic than methanol, but more toxic than ethanol, and is somewhat soluble in water. The possibility of butanol finding its way to water supplies should be considered.
- Butanol use in internal combustion engines may also create more toxic exhaust. Its effect on environment, human health etc. need also be considered.

As an advantage, butanol production from biomass could be more efficient (i.e. unit engine motive power delivered per unit solar energy consumed) than ethanol or methanol routes. Also, some bacteria that produce butanol are able to digest cellulose, not just starch and sugars.

Possible Butanol Fuel Mixtures

Standards for the blending of ethanol and methanol in gasoline exist in many countries, including the EU, the US and Brazil. Approximate equivalent butanol blends can be calculated from the relations between the stochiometric fuel-air ratio of butanol, ethanol and gasoline. Common ethanol fuel mixtures for fuel sold as gasoline currently range from 5 per cent to 10 per cent. The share of butanol can be 60 per cent greater than the equivalent ethanol share, which gives a range from 8 per cent to 16 per cent. "Equivalent" in this case refers only to the vehicle's ability to adjust to the fuel. Other properties such as energy density, viscosity and heat of vaporisation will vary and may further limit the percentage of butanol that can be blended with gasoline. Consumer acceptance may be limited due to the offensive smell of butanol.

Current Use of Butanol in Vehicles

Currently no production vehicle is known to be approved by the manufacturer for use with 100 per cent butanol. As of early 2009, only few vehicles are approved for even using E85 fuel (i.e. 85 per cent ethanol + 15 per cent gasoline). BP and Dupont, engaged in a joint venture to produce and promote butanol fuel, claim that "biobutanol can be blended up to 10 per cent v/v in European gasoline and 11.5 per cent v/v in US gasoline".

David Ramey drove from Blacklick, Ohio to San Diego, California using butanol in an unmodified 1992 Buick Park Avenue.

Research

The Swiss company Butalco GmbH uses a special technology to modify yeasts in order to produce butanol instead of ethanol. Yeasts as production organisms for butanol have decisive advantages compared to bacteria. The company Gevo in Pasadena, Ca, is developing a biotechnology process to mass-produce iso-butanol from renewable resources.

UNDERSTANDING METHANOL FUEL

Methanol has been proposed as a *fuel* for internal combustion and other engines, mainly in combination with gasoline. Methanol fuel has received less attention than ethanol fuel as an alternative to petroleum based fuels. However, in 2005 Nobel prize winner George A. Olah advocated an entire methanol economy based on energy storage in synthetically produced methanol in an essay and in 2006 he and two co-authors published a book around this theme.

History and Production

Historically, methanol was first produced from pyrolysis of wood, resulting in its common English name of wood alcohol.

Presently, methanol is usually produced using methane (the chief constituent of natural gas) as a raw material.

It may also be produced by pyrolysis of many organic materials or by Fischer Tropsch from synthetic gas, so be called biomethanol. Production of methanol from synthesis gas using Biomass-To-Liquid can offer methanol production from biomass at efficiencies up to 75 per cent. Widespread production by this route has a postulated potential to offer methanol fuel at a low cost and with benefits to the environment. These production methods, however, are not suitable for small scale production.

Use as Internal Combustion Engine Fuel

Both methanol and ethanol burn at lower temperatures than gasoline, and both are more volatile, making engine starting in cold weather more difficult. Using methanol as a fuel in spark ignition engines can offer an increased thermal efficiency and increased power output (as compared to gasoline) due to its high octane rating (114) and high heat of vaporisation. However, its low energy content of 19.7 MJ/kg and stoichiometric air fuel ratio of 6.42:1 mean that fuel consumption (on volume or mass basis) will be higher than hydrocarbon fuels. The extra water produced also makes the charge rather wet (similar to hydrogen/

oxygen combustion engines) and combined with the formation of acidic products during combustion, the wearing of valves, valveseats and cylinder might be higher than with hydrocarbon burning. Certain additives may be added to motor oil in order to neutralize these acids.

Methanol, just like ethanol, contains soluble and insoluble contaminents. These soluble contaminants, halide ions such as cloride ions, have a large effect on the corrosivity of alcohol fuels. Halide ions increase corrosion in two ways; they chemically attack passivating oxide films on several metals causing pitting corrosion, and they increase the conductivity of the fuel. Increased electrical conductivity promotes electric, galvanic, and ordinary corrosion in the fuel system. Soluble contaminants, such as aluminium hydroxide, itself a product of corrosion by halide ions, clog the fuel system over time.

Methanol is hygroscopic, meaning it will absorb water vapor directly from the atmosphere. Because absorbed water dilutes the fuel value of the methanol (although, it suppresses engine knock), and may cause phase separation of methanol-gasoline blends, containers of methanol fuels must be kept tightly sealed.

Toxicity

Methanol is poisonous; ingestion of only 10ml can cause blindness and 60ml can be fatal, and it doesn't have to be swallowed to be dangerous since the liquid can be absorbed through the skin, and the vapours through the lungs. US maximum allowed exposure in air (40 h/week) is 1900 mg/m³ for ethanol, 900 mg/m³ for gasoline, and 1260 mg/m³ for methanol. However, it is less volatile than gasoline, and therefore decreases evaporative emissions. Use of methanol, like ethanol, significantly reduces the emissions of certain hydrocarbon-related toxins such as benzene and 1,3 butadiene. But as gasoline and ethanol are already quite toxic, safety protocol is the same.

Safety

Since methanol vapour is heavier than air, it will linger close to the ground or in a pit unless there is good ventilation, and if the concentration of methanol is above 6.7 per cent in air it can be lit by a spark, and will explode above 54 F/12 C. Once ablaze, the flames give out very little light making it very hard to see the fire or even estimate its size, especially in bright daylight. If you are unlucky enough to be exposed to the poisonous substance through your breathing system, its pungent odor should give you some warning of its presence. However, it is difficult to smell methanol in the air at less than 2,000 ppm (0.2%).

Use in Racing

Beginning in 1965, pure methanol was used widespread in USAC Indy car competition, which at the time included the Indianapolis 500.

A seven-car crash on the second lap of the 1964 Indianapolis 500 resulted in USAC's decision to encourage, and later mandate, the use of methanol. Eddie Sachs and Dave MacDonald died in the crash when their gasoline-fueled cars exploded. The gasoline-triggered fire created a dangerous cloud of thick black smoke, which completely blocked the view of the track for oncoming cars. Johnny Rutherford, one of the other drivers involved, drove a methanol-fueled car which also leaked following the crash. While this car burned from the impact of the first fireball, it formed a much lesser inferno than the gasoline cars, and one that burned invisibly. That testimony, and pressure from **Indianapolis Star** writer George Moore, led to the switch to alcohol fuel in 1965.

Methanol was used by the CART circuit during its entire campaign (1979-2007). It is also used by and many short track organizations, especially midget, sprint cars and speedway bikes. Pure methanol was used by the IRL from 1996-2006.

In 2006, in partnership with the ethanol industry, the IRL used a mixture of 10 per cent ethanol and 90 per cent methanol as its fuel. Starting in 2007, the IRL switched to "pure" ethanol, E100.

Methanol fuel is also used extensively in drag racing, primarily in the Top Alcohol category.

Formula One racing continues to use gasoline as its fuel, but in pre war grand prix racing methanol was often used in the fuel.

Methanol Fuel by Country

United States

The State of California ran an experimental Programme from 1980 to 1990 which allowed anyone to convert a gasoline vehicle to 85 per cent methanol with 15 per cent additives of choice. Over 500 vehicles were converted to high compression and dedicated use of the 85/15 methanol and ethanol, with great results. Detroit was not willing to produce any methanol or ethanol vehicles without government subsidy.

In 1982 the big three were each given $5,000,000 for design and contracts for 5,000 vehicles to be bought by the State. That was the beginning of the low compression flexible-fuel vehicles which we can still buy today.

In 2005, California's Governor, Arnold Schwarzenegger, terminated the use of methanol after 25 years and 200,000,000 miles of success, to join the expanding use of ethanol driven by producers of corn. In spite of this, he was optimistic about the future of the Programme, claiming "it will be back." Ethanol is currently (as of 2007) priced at 3 to 4 dollars per gallon, while methanol made from natural gas remains at 47 cents per gallon.

Presently there are over 60 operating gas stations in California supplying methanol in their pumps.

Brazil

A drive to add a significant percentage of methanol to gasoline got very close to implementation in Brazil, following a pilot test set up by a group of scientists involving blending gasoline with methanol between 1989 and 1992. The larger-scale pilot experiment that was to be conducted in São Paulo was vetoed at the last minute by the city's mayor, out of concern for the health of gas station workers (who are mostly illiterate and could not be expected to follow safety precautions). As of 2006, the idea has not resurfaced.

UNDERSTANDING ALCOHOL FUEL

Although fossil fuels have become the dominant energy resource for the modern world, alcohol has been used as a fuel throughout history. The first four aliphatic alcohols (methanol, ethanol, propanol, and butanol) are of interest as fuels because they can be synthesized biologically, and they have characteristics which allow them to be used in current engines. One advantage shared by all four alcohols is octane rating. Biobutanol has the advantage that its energy density is closer to gasoline than the other alcohols (while still retaining over 25 per cent higher octane rating)— however, these advantages are outweighed by disadvantages (compared to ethanol and methanol) concerning production, for instance. Generally speaking, the chemical formula for alcohol fuel is $C_nH_{2n+1}OH$. The larger n is, the higher the energy density.

Alcohol fuels are usually of biological rather than petroleum sources. When obtained from biological sources, they are known as **bioalcohols** (e.g. **bioethanol**). There is no chemical difference between biologically produced alcohols and those obtained from other sources. However, ethanol that is derived from petroleum should not be considered safe for consumption as this alcohol contains about 5 per cent methanol and may cause blindness or death. This mixture may also not be purified by simple distillation, as it forms an azeotropic mixture.

Methanol and Ethanol

Methanol and ethanol can both be derived from fossil fuels or from biomass. Ethanol is produced through fermentation of sugars and methanol from synthesis gas.

As a fuel methanol and ethanol both have advantages and disadvantages over fuels such as petrol and diesel. In spark ignition engines both alcohols can run at a much higher EGR rates and with higher compression ratios. Both alcohols have a high octane rating, with ethanol at 109 RON, 90 MON, (which equates to 99.5 AKI) and methanol at 109 RON, 89 MON (which equates to 99 AKI). Ordinary European petrol is typically 95 RON, 85 MON, equal to 90 AKI. Note that AKI refers to

'Anti-Knock Index' which averages the RON and MON ratings (RON+MON)/2, and is used on U.S. gas station pumps. As a compression ignition engine fuel, both alcohols create very little particulates, but their low cetane number means that an ignition improver like glycol must be mixed into the fuel with approx. 5 per cent.

With SI engines alcohols have the potential to reduce NOx, CO, HC and particulates. A test with E85 fueled Chevrolet Luminas showed that NMHC went down by 20-22 per cent, NOx by 25-32 per cent and CO by 12-24 per cent compared to reformulated gasoline. Toxic emissions of benzene and 1,3 Butadiene also decreased while aldehyde emissions increased (acetaldehyde in particular).

Tailpipe emissions of CO_2 also decrease due to the lower carbon-to-hydrogen ratio of these alcohols, and the improved engine efficiency.

Methanol and ethanol contain soluble and insoluble contaminants. Halide ions, which are soluble contaminants, such as chloride ions, have a large effect on the corrosivity of alcohol fuels. Halide ions increase corrosion in two ways: they chemically attack passivating oxide films on several metals causing pitting corrosion, and they increase the conductivity of the fuel. Increased electrical conductivity promotes electrical, galvanic and ordinary corrosion in the fuel system. Soluble contaminants such as aluminium hydroxide, itself a product of corrosion by halide ions, clogs the fuel system over time. To prevent corrosion the fuel system must be made of suitable materials, electrical wires must be properly insulated and the fuel level sensor must be of pulse and hold type (or similar). In addition, high quality alcohol should have a low concentration of contaminants and have a suitable corrosion inhibitor added.

Methanol and ethanol are also incompatible with some polymers. The alcohol is solved by the polymers causing swelling, and over time the oxygen breaks down the carbon-carbon bonds in the polymer causing a reduction in tensile strength. For the past few decades though, most cars have been designed to tolerate up to 10 per cent ethanol (E10) without problem. This include both fuel system compatibility and lambda compensation of fuel delivery with fuel injection engines featuring closed loop lambda control. In some engines ethanol may degrade some compositions of plastic or rubber fuel delivery components designed for conventional petrol, and also be unable to lambda compensate the fuel properly.

"FlexFuel" vehicles have upgraded fuel system and engine components which are designed for long life using E85 or M85, and the ECU can adapt to any fuel blend between gasoline and E85 or M85. Typical upgrades include modifications to: fuel tanks, fuel tank electrical wiring, fuel pumps, fuel filters, fuel lines, filler tubes, fuel level sensors, fuel injectors, seals, fuel rails, fuel pressure regulators, valve seats and inlet valves. The cost of this E85 upgrade to a modern engine is inexpensive and is less than $100. "Total Flex" Autos destined for the Brazilian market can use E100 (100 per cent Ethanol).

One liter of ethanol contain 21.1 MJ, a liter of methanol 15.8 MJ and a liter of gasoline approximately 32.6 MJ. In other words, for the same energy content as one liter or one gallon of gasoline, one needs 1.6 liters/gallons of ethanol and 2.1 liters/gallons of methanol. Although actual fuel consumption doesn't increase as much as energy content numbers indicate.

Methanol has been proposed as a future biofuel. Methanol has a long history as a racing fuel. Early Grand Prix Racing used blended mixtures as well as pure methanol. The use of the fuel was primarily used in North America after the war. However, methanol for racing purposes has largely been based on natural gas and therefore would not be considered as biofuel. Methanol it is possible biofuel, however compared to ethanol, its primary advantage is its much greater well-to-wheel efficiency when produced from syngas (methanol might be produced from carbon dioxide and captive hydrogen derived using nuclear power or any renewable energy source).

Ethanol is already being used extensively as a fuel additive, and the use of ethanol fuel alone or as part of a mix with gasoline is increasing. Compared to methanol its primary advantage is that the fuel is non-toxic, although the fuel will produce some toxic exhaust emissions. From 2007, the Indy Racing League will use ethanol as its exclusive fuel, after 40 years of using methanol. Since September 2007 petrol stations in NSW, Australia are mandated to supply all their petrol with 2 per cent Ethanol content

Methanol combustion is:

$$2CH_3OH + 3O_2 \rightarrow 2CO_2 + 4H_2O + heat$$

Ethanol combustion is:

$$C_2H_5OH + 3O_2 \rightarrow 2CO_2 + 3H_2O + heat$$

Butanol

Propanol and butanol are considerably less toxic and less volatile than methanol. In particular, butanol has a high flashpoint of 35 °C, which is a benefit for fire safety, but may be a difficulty for starting engines in cold weather. The concept of flash point is however not directly applicable to engines as the compression of the air in the cylinder means that the temperature is several hundred degrees Celsius before ignition takes place.

The fermentation processes to produce propanol and butanol from cellulose are fairly tricky to execute, and the Weizmann organism (Clostridium acetobutylicum) currently used to perform these conversions produces an extremely unpleasant smell, and this must be taken into consideration when designing and locating a fermentation plant. This organism also dies when the butanol content of whatever it is fermenting rises to 7 per cent. For comparison, yeast dies when the ethanol content of its feedstock hits 14 per cent. Specialized strains can tolerate even greater ethanol concentrations—so-called turbo yeast can withstand up to 16 per cent ethanol. However, if ordinary Saccharomyces yeast can be modified to improve its ethanol resistance, scientists may yet one day produce a strain of the Weizmann organism with a butanol resistance higher than the natural boundary of 7 per cent. This would be useful because butanol has a higher energy density than ethanol, and because waste fibre left over from sugar crops used to make ethanol could be made into butanol, raising the alcohol yield of fuel crops without there being a need for more crops to be planted.

Despite these drawbacks, DuPont and British Petroleum have recently announced that they are jointly to build a small scale butanol fuel demonstration plant alongside the large bioethanol plant they are jointly developing with Associated British Foods.

Energy Environment International developed a method for producing butanol from biomass, which involves the use of two separate micro-organisms in sequence to minimize production of acetone and ethanol byproducts.

The Swiss company Butalco GmbH uses a special technology to modify yeasts in order to produce butanol instead of ethanol. Yeasts as production organisms for butanol have decisive advantages compared to bacteria.

Butanol combustion is:

$$C_4H_9OH + 6O_2 \rightarrow 4CO_2 + 5H_2O + heat$$

The 3-carbon alcohol, propanol (C_3H_7OH), is not used as a direct fuel source for petrol engines that often (unlike ethanol, methanol and butanol), with most being directed into use as a solvent. However, it is used as a source of hydrogen in some types of fuel cell; it can generate a higher voltage than methanol, which is the fuel of choice for most alcohol-based fuel cells. However, since propanol is harder to produce than methanol (biologically OR from oil), methanol fuel cells are still used a lot more often than those that utilise propanol.

By Country

Alcohol in Brazil

Brazil was until recently the largest producer of alcohol fuel in the world, typically fermenting ethanol from sugarcane. The country produces a total of 18 billion liters (4.8 billion gallons) annually, of which 3.5 billion liters are exported, 2 billion of them to the U.S. Alcohol cars debuted in the Brazilian market in 1978 and became quite popular because of heavy subsidy, but in the 80's prices rose and gasoline regained the leading market share.

However, from 2003 on, alcohol is rapidly rising its market share once again because of new technologies involving flexible-

fuel engines, called "Flex" by all major car manufacturers
(Volkswagen, General Motors, Fiat, etc.). "Flex" engines work
with gasoline, alcohol or any mixture of both fuels. As of
February 2007, approx. 80 per cent of new vehicles sold in Brazil
are hybrid fuel

Because of the Brazilian leading production and technology,
many countries became very interested in importing alcohol fuel
and adopting the "Flex" vehicle concept. In March 7th of 2007,
US president George W. Bush visited the city of São Paulo to
sign agreements with Brazilian president Lula on importing
alcohol and its technology as an alternative fuel.

Alcohol in Russia

Other than Brazil, the only two other countries with an extensive
Programme for supplying alcohol fuel to supplement petroleum
use is Russia, which has reduced its dependency on oil by using
methanol made from the destructive pyrolysis of eucalyptus wood
and fibre. The other, China has reported with a 70 per cent
methanol use to conventional gasoline an independence from
crude oil.

Alcohol in the United States

The United States at the end of 2007 was producing 7 billion
gallons (26.9 billion liters) per year. E10 or Gasohol is commonly
marketed in Delaware and E85 is found in many states,
particularly in the Mid West where ethanol from corn is produced
locally. Due to government subsidies, many new vehicles are
sold each year that can use E85, although the majority are run
solely on gasoline due to the limited availability of E85.

Many states and municipalities have mandated that all
gasoline fuel be blended with 10 percent alcohol (usually ethanol)
during some or all of the year. This is to reduce pollution and
allows these areas to comply with federal pollution limits.
Because alcohol is partially oxygenated, it produces less overall
pollution, including ozone. In some areas (California in
particular) the regulations may also require other formulations
or added chemicals that reduce pollution, but add complexity
to the fuel distribution and increase the cost of the fuel.

Alcohol in the European Union

Consumption of Bioethanol (GWh)

Country	2005	2006	2007	2008
1. France	871	1,719	3,164	4,693
2. Germany	1,682	3,544	3,448	4,675
3. Sweden	1,681	1,894	2,119	2,488
4. Netherlands	0	179	1,023	1,512
5. Spain	1,314	1,332	1,512	1,454
6. Poland	329	611	837	1,382
7. United Kingdom	502	563	906	1,223
8. Finland	0	10	20	858
9. Austria	0	0	199	633
10. Hungary	28	136	314	454
11. Czech Republic	0	13	1	378
12. Ireland	0	13	59	207
13. Lithuania	10	64	134	182
14. Belgium	0	0	0	145
15. Slovakia	0	4	140	76
16. Bulgaria	–	0	0	72
17. Denmark	0	42	60	50
18. Slovenia	0	2	9	28
19. Estonia	0	0	0	17
20. Latvia	5	12	0	0
21. Luxembourg	0	0	14	11
22. Portugal	0	0	0	0
23. Italy	59	0	0	0
24. Greece	0	0	0	0
25. Romania	–	0	0	0
26. Malta	0	0	0	0
27. Cyprus	0	0	0	0
27. European Union	6,481	10,138	13,962	20,538

1 toe = 11,63 MWh, 0 = no data

Alcohol consumption does not specify the traffic fuel use

The 2008 data is not confirmed yet

8

Ethanol: Current Production Process and Future Prospects

ETHANOL AS FUEL

Ethanol fuel is ethanol (ethyl alcohol), the same type of alcohol found in alcoholic beverages. It can be used as a fuel, mainly as a biofuel alternative to gasoline, and is widely used by flex-fuel light vehicles in Brazil, and as an oxygenate to gasoline in the United States. Together, both countries were responsible for 89 percent of the world's ethanol fuel production in 2008. Because it is easy to manufacture and process and can be made from very common crops such as sugar cane, potato, manioc and corn, in several countries ethanol fuel is increasingly being blended as gasohol or used as an oxygenate in gasoline. Bioethanol, unlike petroleum, is a renewable resource that can be produced from agricultural feedstocks.

Anhydrous ethanol (ethanol with less than 1 per cent water) can be blended with gasoline in varying quantities up to pure ethanol (E100), and most modern gasoline engines will operate well with mixtures of 10 per cent ethanol (E10). Most cars on the road today in the U.S. can run on blends of up to 10 per cent ethanol, and the use of 10 per cent ethanol gasoline is mandated in some cities.

Ethanol can be mass-produced by fermentation of sugar or by hydration of ethylene (ethene $CH_2=CH_2$) from petroleum and other sources. Current interest in ethanol mainly lies in bio-ethanol, produced from the starch or sugar in a wide variety of crops, but there has been considerable debate about how useful bio-ethanol will be in replacing fossil fuels in vehicles. Concerns relate to the large amount of arable land required for crops, as well as the energy and pollution balance of the whole cycle of ethanol production. Recent developments with cellulosic ethanol production and commercialization may allay some of these concerns.

According to the International Energy Agency, cellulosic ethanol could allow ethanol fuels to play a much bigger role in the future than previously thought. Cellulosic ethanol offers promise as resistant cellulose fibers, a major and universal component in plant cells walls, can be used to generate ethanol.

Chemistry

Structure of Ethanol Molecule: All bonds are single bonds

Glucose (a simple sugar) is created in the plant by photosynthesis.

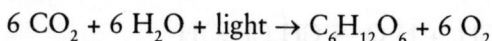

$$6 \ CO_2 + 6 \ H_2O + light \rightarrow C_6H_{12}O_6 + 6 \ O_2$$

During ethanol fermentation, glucose is decomposed into ethanol and carbon dioxide.

$$C_6H_{12}O_6 \rightarrow 2 \ C_2H_5OH + 2 \ CO_2 + heat$$

During combustion ethanol reacts with oxygen to produce carbon dioxide, water, and heat:

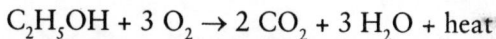

$$C_2H_5OH + 3 \ O_2 \rightarrow 2 \ CO_2 + 3 \ H_2O + heat$$

After doubling the combustion reaction because two molecules of ethanol are produced for each glucose molecule, and adding all three reactions together, there are equal numbers of each type of molecule on each side of the equation, and the net reaction for the overall production and consumption of ethanol is just:

$$light \rightarrow heat$$

The heat of the combustion of ethanol is used to drive the piston in the engine by expanding heated gases. It can be said that sunlight is used to run the engine.

Glucosé itself is not the only substance in the plant that is fermented. The simple sugar fructose also undergoes fermentation. Three other compounds in the plant can be fermented after breaking them up by hydrolysis into the glucose or fructose molecules that compose them. Starch and cellulose are molecules that are strings of glucose molecules, and sucrose (ordinary table sugar) is a molecule of glucose bonded to a molecule of fructose. The energy to create fructose in the plant ultimately comes from the metabolism of glucose created by photosynthesis, and so sunlight also provides the energy generated by the fermentation of these other molecules.

Ethanol may also be produced industrially from ethene (ethylene). Addition of water to the double bond converts ethene to ethanol:

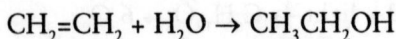

$$CH_2=CH_2 + H_2O \rightarrow CH_3CH_2OH$$

This is done in the presence of an acid which catalyzes the reaction, but is not consumed. The ethene is produced from petroleum by steam cracking.

When ethanol is burned in the atmosphere rather than in pure oxygen, other chemical reactions occur with different components of the atmosphere such as N_2. This leads to the production of nitrous oxides NO_x, a major air pollutant.

Sources

Ethanol is a renewable energy source because the energy is generated by using a resource, sunlight, which is naturally replenished. Creation of ethanol starts with photosynthesis causing a feedstock, such as sugar cane or corn, to grow. These feedstocks are processed into ethanol.

About 5 per cent of the ethanol produced in the world in 2003 was actually a petroleum product. It is made by the catalytic hydration of ethylene with sulfuric acid as the catalyst. It can also be obtained via ethylene or acetylene, from calcium carbide, coal, oil gas, and other sources. Two million tons of petroleum-derived ethanol are produced annually. The principal suppliers are plants in the United States, Europe, and South Africa. Petroleum derived ethanol (synthetic ethanol) is chemically identical to bio-ethanol and can be differentiated only by radiocarbon dating.

Bio-ethanol is usually obtained from the conversion of carbon based feedstock. Agricultural feedstocks are considered renewable because they get energy from the sun using photosynthesis, provided that all minerals required for growth (such as nitrogen and phosphorus) are returned to the land. Ethanol can be produced from a variety of feedstocks such as sugar cane, bagasse, miscanthus, sugar beet, sorghum, grain sorghum, switchgrass, barley, hemp, kenaf, potatoes, sweet potatoes, cassava, sunflower, fruit, molasses, corn, stover, grain, wheat, straw, cotton, other biomass, as well as many types of cellulose waste and harvestings, whichever has the best well-to-wheel assessment.

An alternative process to produce bio-ethanol from algae is being developed by the company Algenol. Rather than grow algae and then harvest and ferment it the algae grow in sunlight and produce ethanol directly which is removed without killing the algae. It is claimed the process can produce 6000 gallons per acre per year compared with 400 gallons for corn production.

Currently, the first generation processes for the production of ethanol from corn use only a small part of the corn plant: the corn kernels are taken from the corn plant and only the starch, which represents about 50 per cent of the dry kernel mass, is transformed into ethanol. Two types of second generation processes are under development. The first type uses enzymes and yeast to convert the plant cellulose into ethanol while the second type uses pyrolysis to convert the whole plant to either a liquid bio-oil or a syngas. Second generation processes can also be used with plants such as grasses, wood or agricultural waste material such as straw.

Production Process

The basic steps for large scale production of ethanol are:

microbial (yeast) fermentation of sugars, distillation, dehydration, and denaturing (optional). Prior to fermentation, some crops require saccharification or hydrolysis of carbohydrates such as cellulose and starch into sugars. Saccharification of cellulose is called cellulolysis. Enzymes are used to convert starch into sugar.

Fermentation

Ethanol is produced by microbial fermentation of the sugar. Microbial fermentation will currently only work directly with sugars. Two major components of plants, starch and cellulose, are both made up of sugars, and can in principle be converted to sugars for fermentation. Currently, only the sugar (e.g. sugar cane) and starch (e.g. corn) portions can be economically converted. However, there is much activity in the area of cellulosic ethanol, where the cellulose part of a plant is broken down to sugars and subsequently converted to ethanol.

Distillation

For the ethanol to be usable as a fuel, water must be removed. Most of the water is removed by distillation, but the purity is limited to 95-96 per cent due to the formation of a low-boiling water-ethanol azeotrope. The 95.6 per cent m/m (96.5 per cent v/v) ethanol, 4.4 per cent m/m (3.5 per cent v/v) water mixture may be used as a fuel alone, but unlike anhydrous ethanol, is immiscible in gasoline, so the water fraction is typically removed in further treatment in order to burn with in combination with gasoline in gasoline engines.

Dehydration

There are basically five dehydration processes to remove the water from an azeotropic ethanol/water mixture. The first process, used in many early fuel ethanol plants, is called azeotropic distillation and consists of adding benzene or cyclohexane to the mixture. When these components are added to the mixture, it forms an heterogeneous azeotropic mixture in vapor-liquid-liquid equilibrium, which when distilled produces anhydrous ethanol in the column bottom, and a vapor mixture of water and cyclohexane/benzene. When condensed, this becomes a two-phase liquid mixture. Another early method, called extractive distillation, consists of adding a ternary component which will increase ethanol relative volatility. When the ternary mixture is distilled, it will produce anhydrous ethanol on the top stream of the column.

With increasing attention being paid to saving energy, many methods have been proposed that avoid distillation all together for dehydration. Of these methods, a third method has emerged and has been adopted by the majority of modern ethanol plants. This new process uses molecular sieves to remove water from fuel ethanol. In this process, ethanol vapor under pressure passes through a bed of molecular sieve beads. The bead's pores are sized to allow absorption of water while excluding ethanol. After a period of time, the bed is regenerated under vacuum to remove the absorbed water. Two beds are used so that one is available to absorb water while the other is being regenerated. This dehydration technology can account for energy saving of 3,000 btus/gallon (840 kJ/l) compared to earlier azeotropic distillation.

Technology

Ethanol-Based Engines

Ethanol is most commonly used to power automobiles, though it may be used to power other vehicles, such as farm tractors and airplanes. Ethanol (E100) consumption in an engine is approximately 51 per cent higher than for gasoline since the energy per unit volume of ethanol is 34 per cent lower than for gasoline. However, the higher compression ratios in an ethanol-only engine allow for increased power output and better fuel economy than could be obtained with lower compression ratios. In general, ethanol-only engines are tuned to give slightly better power and torque output than gasoline-powered engines. In flexible fuel vehicles, the lower compression ratio requires tunings that give the same output when using either gasoline or hydrated ethanol. For maximum use of ethanol's benefits, a much higher compression ratio should be used, which would render that engine unsuitable for gasoline use. When ethanol fuel availability allows high-compression ethanol-only vehicles to be practical, the fuel efficiency of such engines should be equal to or greater than current gasoline engines. Current high compression ethanol-only engine designs are approximately 20-30 per cent less fuel efficient than their gasoline-only counterparts.

A 2004 study and an earlier paper published by the Society of Automotive Engineers identify a method to exploit the characteristics of fuel ethanol substantially better than mixing it with gasoline. The method presents the possibility of leveraging the use of alcohol to achieve definite improvement over the cost-effectiveness of hybrid electric. The improvement consists of using dual-fuel direct-injection of pure alcohol (or the azeotrope or E85) and gasoline, in any ratio up to 100 per cent of either, in a turbocharged, high compression-ratio, small-displacement engine having performance similar to an engine having twice the displacement. Each fuel is carried separately, with a much smaller tank for alcohol. The high-compression (which increases efficiency) engine will run on ordinary gasoline under low-power cruise conditions. Alcohol is directly injected into the cylinders (and the gasoline injection simultaneously reduced) only when necessary to suppress 'knock' such as when significantly accelerating. Direct cylinder injection raises the already high octane rating of ethanol up to an effective 130. The calculated over-all reduction of gasoline use and CO_2 emission is 30 per cent. The consumer cost payback time shows a 4:1 improvement over turbo-diesel and a 5:1 improvement over hybrid. In addition, the problems of water absorption into pre-mixed gasoline (causing phase separation), supply issues of multiple mix ratios and cold-weather starting are avoided.

Ethanol's higher octane rating allows an increase of an engine's compression ratio for increased thermal efficiency. In one study, complex engine controls and increased exhaust gas recirculation allowed a compression ratio of 19.5 with fuels ranging from neat ethanol to E50. Thermal efficiency up to approximately that for a diesel was achieved. This would result in the MPG (miles per gallon) of a dedicated ethanol vehicle to be about the same as one burning gasoline.

Since 1989 there have also been ethanol engines based on

the diesel principle operating in Sweden. They are used primarily in city buses, but also in distribution trucks and waste collectors. The engines, made by Scania, have a modified compression ratio, and the fuel (known as ED95) used is a mix of 93.6 per cent ethanol and 3.6 per cent ignition improver, and 2.8 per cent denaturants. The ignition improver makes it possible for the fuel to ignite in the diesel combustion cycle. It is then also possible to use the energy efficiency of the diesel principle with ethanol.

Engine Cold Start during the Winter

The Brazilian 2008 Honda Civic flex-fuel has outside direct access to the secondary reservoir gasoline tank in the front right side, the corresponding fuel filler door is shown by the arrow.

High ethanol blends present a problem to achieve enough vapor pressure for the fuel to evaporate and spark the ignition during cold weather (since ethanol tends to increase fuel enthalpy of vaporization). When vapor pressure is below 45 kPa starting a cold engine becomes difficult. In order to avoid this problem at temperatures below 11 ° Celsius (59 °F), and to reduce ethanol higher emissions during cold weather, both the US and the European markets adopted E85 as the maximum blend to be used in their flexible fuel vehicles, and they are optimized to run at such a blend. At places with harsh cold weather, the ethanol blend in the US has a seasonal reduction to E70 for these very cold regions, though it is still sold as E85. At places where temperatures fall below −12 °C (10 °F) during the winter, it is recommended to install an engine heater system, both for gasoline and E85 vehicles. Sweden has a similar seasonal reduction, but the ethanol content in the blend is reduced to E75 during the winter months.

Brazilian flex fuel vehicles can operate with ethanol mixtures up to E100, which is hydrous ethanol (with up to 4 per cent water), which causes vapor pressure to drop faster as compared to E85 vehicles. As a result, Brazilian flex vehicles are built with a small secondary gasoline reservoir located near the engine. During a cold start pure gasoline is injected to avoid starting problems at low temperatures. This provision is particularly necessary for users of Brazil's southern and central regions, where temperatures normally drop below 15 ° Celsius (59 °F) during the winter. An improved flex engine generation was launched in 2009 that eliminates the need for the secondary gas storage tank. In March 2009 Volkswagen do Brasil launched the Polo E-Flex, the first Brazilian flex fuel model without an auxiliary tank for cold start.

Ethanol Fuel Mixtures

Hydrated ethanol x gasoline type C price table for use in Brazil
To avoid engine stall due to "slugs" of water in the fuel lines interrupting fuel flow, the fuel must exist as a single phase. The fraction of water that an ethanol-gasoline fuel can contain without phase separation increases with the percentage of ethanol. This shows, for example, that E30 can have up to about 2 per cent water. If there is more than about 71 per cent ethanol, the remainder can be any proportion of water or gasoline and phase separation will not occur. However, the fuel mileage declines with increased water content. The increased solubility of water with higher ethanol content permits E30 and hydrated ethanol to be put in the same tank since any combination of them always results in a single phase. Somewhat less water is tolerated at lower temperatures. For E10 it is about 0.5 per cent v/v at 70 F and decreases to about 0.23 per cent v/v at −30 F.

In many countries cars are mandated to run on mixtures of ethanol. Brazil requires cars be suitable for a 25 per cent ethanol blend, and has required various mixtures between 22 per cent and 25 per cent ethanol, since of July 2007 25 per cent is required. The United States allows up to 10 per cent blends, and some states require this (or a smaller amount) in all gasoline sold. Other countries have adopted their own requirements. Beginning with the model year 1999, an increasing number of vehicles in the world are manufactured with engines which can run on any fuel from 0 per cent ethanol up to 100 per cent ethanol without modification. Many cars and light trucks (a class containing minivans, SUVs and pickup trucks) are designed to be flexible-fuel vehicles (also called *dual-fuel* vehicles). In older model years, their engine systems contained alcohol sensors in the fuel and/ or oxygen sensors in the exhaust that provide input to the engine control computer to adjust the fuel injection to achieve stochiometric (no residual fuel or free oxygen in the exhaust) air-to-fuel ratio for any fuel mix. In newer models, the alcohol sensors have been removed, with the computer using only oxygen and airflow sensor feedback to estimate alcohol content. The engine control computer can also adjust (advance) the ignition timing to achieve a higher output without pre-ignition when it predicts that higher alcohol percentages are present in the fuel being burned. This method is backed up by advanced knock sensors—used in most high performance gasoline engines regardless of whether they're designed to use ethanol or not—that detect pre-ignition and detonation.

Fuel Economy

In theory, all fuel-driven vehicles have a fuel economy (measured as miles per US gallon, or liters per 100 km) that is directly proportional to the fuel's energy content. In reality, there are many other variables that come in to play that affect the performance of a particular fuel in a particular engine. Ethanol contains approx. 34 per cent less energy per unit volume than gasoline, and therefore in theory, burning pure ethanol in a vehicle will result in a 34 per cent reduction in miles per US gallon, given the same fuel economy, compared to burning pure gasoline. Since ethanol has a higher octane rating, the engine can be made more efficient by raising its compression ratio. In fact using a variable turbocharger, the compression ratio can be optimized for the fuel being used, making fuel economy almost constant for any blend. For E10 (10 per cent ethanol and 90 per cent gasoline), the effect is small (~3 per cent) when compared to conventional gasoline, and even smaller (1-2 per cent) when compared to oxygenated and reformulated blends. However, for E85 (85 per cent ethanol), Actual performance may vary depending on the vehicle. Based on EPA tests for all 2006 E85 models, the average fuel economy for E85 vehicles resulted 25.56 per cent lower than unleaded gasoline. The EPA-rated mileage of current USA flex-fuel vehicles should be considered when making price comparisons, but it must be noted that E85 is a high performance fuel, with an octane rating of about 104, and should be compared to premium. In one estimate the US retail price for E85 ethanol is 2.62 US dollar per gallon or 3.71 dollar corrected for energy equivalency compared to a gallon of gasoline priced at 3.03 dollar. Brazilian cane ethanol (100 per cent) is priced at 3.88 dollar against 4.91 dollar for E25 (as July 2007).

Consumer Production Systems

While biodiesel production systems have been marketed to home and business users for many years, commercialized ethanol production systems designed for end-consumer use have lagged in the marketplace. In 2008, two different companies announced home-scale ethanol production systems. The AFS125 Advanced Fuel System from Allard Research and Development is capable of producing both ethanol and biodiesel in one machine, while the E-100 MicroFueler from E-Fuel Corporation is dedicated to ethanol only.

Experience by Country

The world's top ethanol fuel producers in 2008 were the United States with 9.0 billion U.S. liquid gallons (bg) and Brazil (6.47 bg), accounting for 89 per cent of world production of 17.33 billion US gallons (65.6 million liters). Strong incentives, coupled with other industry development initiatives, are giving rise to fledgling ethanol industries in countries such as Canada, China, Thailand, Colombia, India, Australia, and some Central American countries. Nevertheless, ethanol has yet to make a dent in world oil consumption of approximately 4000 million tonnes/yr (84 million barrels/day) in 2006.

Brazil

Brazil has ethanol fuel available throughout the country. A typical Petrobras filling station at São Paulo with dual fuel service, marked A for alcohol (ethanol) and G for gasoline.

Typical Brazilian "flex" models from several carmakers, that run on any blend of ethanol and gasoline, from E20-E25 gasohol to E100 ethanol fuel.

Total Annual Ethanol Production (All Grades) by Country (2004-2006)
Top 15 Countries
(Millions of U.S. Liquid Gallons)

World Rank	Country	2006	2005	2004
1	United States	4,855	4,264	3,535
2	Brazil	4,491	4,227	3,989
3	China	1,017	1,004	964
4	India	502	449	462
5	France	251	240	219
6	Germany	202	114	71
7	Russia	171	198	198
8	Canada	153	61	61
9	Spain	122	93	79
10	South Africa	102	103	110
11	Thailand	93	79	74
12	United Kingdom	74	92	106
13	Ukraine	71	65	66
14	Poland	66	58	53
15	Saudi Arabia	52	32	79
	World Total	**13,489**	**12,150**	**10,770**

Annual Fuel Ethanol Production by Country (2007-2008)
Top 15 Countries/Blocks
(Millions of U.S. Liquid Gallons)

World Rank	Country/Region	2008	2007
1	United States	9,000.0	6,498.6
2	Brazil	6,472.2	5,019.2
3	European Union	733.6	570.3
4	China	501.9	486.0
5	Canada	237.7	211.3
6	Thailand	89.8	79.2
7	Colombia	79.3	74.9
8	India	66.0	52.8
9	Central America	n/a	39.6
10	Australia	26.4	26.4
11	Turkey	n/a	15.8
12	Pakistan	n/a	9.2
13	Peru	n/a	7.9
14	Argentina	n/a	5.2
15	Paraguay	n/a	4.7
	World Total	**17,335.29**	**13,101.7**

The Honda CG 150 Titan Mix was launched in the Brazilian market in 2009 and became the first flex-fuel motorcycle sold in the world.

Brazil has the largest and most successful bio-fuel Programmes in the world, involving production of ethanol fuel from sugar cane, and it is considered to have the world's first sustainable biofuels economy. In 2006 Brazilian ethanol provided 18 per cent of the country's road transport sector fuel consumption needs, and by April 2008, more than 50 per cent of fuel consumption for the gasoline market. As a result of the increasing use of ethanol, together with the exploitation of domestic deep water oil sources, Brazil, which years ago had to import a large share of the petroleum needed for domestic consumption, in 2006 reached complete self-sufficiency in oil supply.

Together, Brazil and the United States lead the industrial world in global ethanol production, accounting together for 70 per cent of the world's production and nearly 90 per cent of ethanol used for fuel. In 2006 Brazil produced 16.3 billion liters (4.3 billion U.S. liquid gallons), which represents 33.3 per cent of the world's total ethanol production and 42 per cent of the world's ethanol used as fuel. Sugar cane plantations cover 3.6 million hectares of land for ethanol production, representing just 1 per cent of Brazil's arable land, with a productivity of 7,500 liters of ethanol per hectare, as compared with the U.S. maize ethanol productivity of 3,000 liters per hectare.

The ethanol industry in Brazil is more than 30 year-old and even though is no longer subsidized, production and use of ethanol was stimulated through:

- Low-interest loans for the construction of ethanol distilleries
- Guaranteed purchase of ethanol by the state-owned oil company at a reasonable price
- Retail pricing of neat ethanol so it is competitive if not slightly Favourable to the gasoline-ethanol blend
- Tax incentives provided during the 1980s to stimulate the purchase of neat ethanol vehicles.

Guaranteed purchase and price regulation were ended some

years ago, with relatively positive results. In addition to these other policies, ethanol producers in the state of São Paulo established a research and technology transfer center that has been effective in improving sugar cane and ethanol yields. There are no longer light vehicles in Brazil running on pure gasoline. Since 1977 the government made mandatory to blend 20 per cent of ethanol (E20) with gasoline (gasohol), requiring just a minor adjustment on regular gasoline motors. Today the mandatory blend is allowed to vary nationwide between 20 per cent to 25 per cent ethanol (E25) and it is used by all regular gasoline vehicles, plus three million cars running on 100 per cent hydrated ethanol and six million of dual or flexible-fuel vehicles. The Brazilian car manufacturing industry developed full flexible-fuel vehicles that can run on any proportion of gasoline and ethanol. Introduced in the market in 2003, these vehicles became a commercial success. On August 2008, the fleet of "flex" cars and light commercial vehicles had reached 6 million new vehicles sold, which represents around 23 per cent of Brazil's light motor vehicle fleet. The ethanol-powered and "flex" vehicles, as they are popularly known, are manufactured to tolerate hydrated ethanol, an azeotrope comprised of 95.6 per cent ethanol and 4.4 per cent water.

The latest innovation within the Brazilian flexible-fuel technology is the development of flex-fuel motorcycles. The first flex motorcycle was launched to the market by Honda in March 2009. Produced by its Brazilian subsidiary Moto Honda da Amazônia, the CG 150 Titan-Mix is sold for around US$2,700.

United States

United States Fuel Ethanol Production and Imports (2001-2008)

(Millions of U.S. Liquid Gallons)

Year	Production	Imports	Demand
2001	1,770	n/a	n/a
2002	2,130	46	2,085
2003	2,800	61	2,900
2004	3,400	161	3,530
2005	3,904	135	4,049
2006	4,855	653	5,377
2007	6,500	450	6,847
2008	9,000	556	9,637

Note: Demand figures includes stocks change and small exports in 2005.

The United States produces and consumes more ethanol fuel than any other country in the world. Ethanol use as fuel dates back to Henry Ford, who in 1896 designed his first car, the "Quadricycle" to run on pure ethanol. Then in 1908, he produced the famous Ford Model T capable of running on gasoline, ethanol or a combination of both. Ford continued to advocate for ethanol as fuel even during the prohibition.

Most cars on the road today in the U.S. can run on blends of up to 10 per cent ethanol, and motor vehicle manufacturers already produce vehicles designed to run on much higher ethanol blends. In 2007 Portland, Oregon, became the first city in the United States to require all gasoline sold within city limits to contain at least 10 per cent ethanol. As of January 2008, three states—Missouri, Minnesota, and Hawaii—require ethanol to be blended with gasoline motor fuel. Many cities also require ethanol blends due to non-attainment of federal air quality goals.

Several motor vehicle manufacturers, including Ford, Chrysler, and GM, sell flexible-fuel vehicles that can use gasoline and ethanol blends ranging from pure gasoline all the way up to 85 per cent ethanol (E85). By mid-2006, there were approximately six million E85-compatible vehicles on U.S. roads.

In the USA there are currently about 1,900 stations distributing ethanol, although most stations are in the corn belt area. One of the debated methods for distribution in the US is using existing oil pipelines, which raises concerns over corrosion. In any case, some companies proposed building a 1,700-mile pipeline to carry ethanol from the Midwest through Central Pennsylvania to New York.

The production of fuel ethanol from corn in the United States is controversial for a few reasons. Production of ethanol from corn is 5 to 6 times less efficient than producing it from sugarcane. Ethanol production from corn is highly dependent upon subsidies and it consumes a food crop to produce fuel. The subsidies paid to fuel blenders and ethanol refineries have often been cited as the reason for driving up the price of corn, and in farmers planting more corn and the conversion of considerable land to corn (maize) production which generally consumes more fertilizers and pesticides than many other land uses. This is at odds with the subsidies actually paid directly to farmers that are designed to take corn land out of production and pay farmers to plant grass and idle the land, often in conjunction with soil conservation Programmes, in an attempt to boost corn prices. Recent developments with cellulosic ethanol production and commercialization may allay some of these concerns. A theoretically much more efficient way of ethanol production has been suggested to use sugar beets which make about the same amount of ethanol as corn without using the corn food crop especially since sugar beets can grow in less tropical conditions than sugar cane.

Most of the ethanol consumed in the US is in the form of low blends with gasoline up to 10 per cent. Shown a fuel pump in Maryland selling mandatory E10.

On October 2008 the first "biofuels corridor" was officially opened along I-65, a major interstate highway in the central United States. Stretching from northern Indiana to southern Alabama, this corridor consisting of more than 200 individual fueling stations makes it possible to drive a flex-fueled vehicle from Lake Michigan to the Gulf of Mexico without being further than a quarter tank worth of fuel from an E85 pump.

On April 23, 2009, the California Air Resources Board approved the specific rules and carbon intensity reference values for the California Low-Carbon Fuel Standard (LCFS) that will go into effect in January 1, 2011. During the consultation process there was controversy regarding the inclusion and modeling of indirect land use change effects. After the CARB's ruling, among other criticisms, representatives of the US ethanol industry complained that this standard overstates the environmental effects of corn ethanol, and also criticized the inclusion of indirect effects of land-use changes as an unfair penalty to home-made corn ethanol because deforestation in the developing world is being tied to US ethanol production. The initial reference value set for 2011 for LCFS means that Mid-west corn ethanol will not meet the California standard unless current carbon intensity is reduced.

A similar controversy arose after the U.S. Environmental Protection Agency (EPA) published on May 5, 2009, its notice of proposed rulemaking for the new Renewable Fuel Standard (RFS). The draft of the regulations was released for public comment during a 60-day period. EPA's proposed regulations also included the carbon footprint from indirect land-use changes. On the same day, President Barack Obama signed a Presidential Directive with the aim to advance biofuels research and improve their commercialization. The Directive established a Biofuels Interagency Working Group comprise of three agencies, the Department of Agriculture, the Environmental Protection Agency, and the Department of Energy. This group will developed a plan to increase flexible fuel vehicle use and assist in retail marketing efforts. Also they will coordinate infrastructure policies impacting the supply, secure transport, and distribution of biofuels. The group will also come up with policy ideas for increasing investment in next-generation fuels, such as cellulosic ethanol, and for reducing the environmental footprint of growing biofuels crops, particularly corn-based ethanol.

Europe

Production of Bioethanol in the European Union (GWh)

No	Country	2005	2006
1.	Germany	978	2,554
2.	Spain	1,796	2,382
3.	France	853	1,482
4.	Sweden	907	830
5.	Italy	47	759
6.	Poland	379	711
7.	Hungary	207	201
8.	Lithuania	47	107
9.	Netherlands	47	89
10.	Czech Republic	0	89
11.	Latvia	71	71
12.	Finland	77	0
27.	Total	5,411	9,274

n.a. = not available

Consumption of Bioethanol in the European Union (GWh)

No	Country	2005	2006	2007
1.	Germany	1,682	3,544	3,408
2.	France	871	1,719	3,174
3.	Sweden	1,681	1,894	2,113
4.	Spain	1,314	1,332	1,310
5.	Poland	329	611	991
6.	United Kingdom	502	563	907
7.	Bulgaria	–	0	769
8.	Austria	0	0	254
9.	Slovakia	0	4	154
10.	Lithuania	10	64	135
11.	Hungary	28	136	107
12.	Netherlands	0	179	101
13.	Denmark	–	42	70
14.	Ireland	0	13	54
15.	Latvia	5	12	20
16.	Luxembourg	0	0	10
17.	Slovenia	0	2	9
18.	Czech Republic	0	13	2
19.	Italy	59	0	0
20.	Finland	0	10	n.a.
27	EU	6,481	10,138	13,563

The consumption of bioethanol is largest in Europe in Germany, Sweden, France and Spain. Europe produces equivalent to 90 per cent of its consumption (2006). Germany produced ca 70 per cent of its consumption, Spain 60 per cent and Sweden 50 per cent (2006). In Sweden there are 792 E85 filling stations and in France 131 E85 service stations with 550 more under construction.

On Monday, September 17, 2007 the first ethanol fuel pump was opened in Reykjavik, Iceland. This pump is the only one of its kind in Iceland. The fuel is imported by Brimborg, a Volvo dealer, as a pilot to see how ethanol fueled cars work in Iceland.

In The Netherlands regular petrol with no bio-additives is slowly being outphased, since EU-legislation has been passed that requires the fraction of nonmineral origin to become minimum 5.75 per cent of the total fuel consumption volume in 2010. This can be realised by substitutions in diesel or in petrol of any biological source; or fuel sold in the form of pure biofuel. (2007) There are only a few gas stations where E85 is sold, which is an 85 per cent ethanol, 15 per cent petrol mix. Directly neighbouring country Germany is reported to have a much better biofuel infrastructure and offers both E85 and E50. Biofuel is taxed equally as regular fuel. However, fuel tanked abroad cannot be taxed and a recent payment receipt will in most cases suffice to prevent fines if customs check tank contents. (Authorities are aware of high taxation on fuels and cross-border fuel refilling is a well-known practice.)

An example of an ethanol powered bus. This is a Scania OmniCity which has been touring the United Kingdom, which does not use the fuel widely. A larger fleet of similar buses will enter service in Stockholm in 2008.

Sweden

Sweden is the leading country in Europe regarding the use of ethanol as fuel, though it has to import most of the ethanol. All Swedish gas stations are required by an act of parliament to offer at least one alternative fuel, and every fifth car in Stockholm now drives at least partially on alternative fuels, mostly ethanol. The number of bioethanol stations in Europe is highest in Sweden, with 1,200 stations and a fleet of 116 thousand flexi-fuel vehicles as of July 2008.

Stockholm will introduce a fleet of Swedish-made electric hybrid buses in its public transport system on a trial basis in 2008. These buses will use ethanol-powered internal-combustion engines and electric motors. The vehicles' diesel engines will use ethanol.

In order to achieve a broader use of biofuels several government incentives were implemented. Ethanol, as the other biofuels, were exempted of both, the CO_2 and energy taxes until 2009, resulting in a 30 per cent price reduction at the pump of E85 fuel over gasoline. Furthermore, other demand side incentives for flexifuel vehicle owners include a USD 1,800 bonus to buyers of FFVs, exemption from the Stockholm congestion tax, up to 20 per cent discount on auto insurance, free parking spaces in most of the largest cities, lower annual registration taxes, and a 20 per cent tax reduction for flexifuel company cars. Also, a part of the Programme, the Swedish Government ruled that 25 per cent of their vehicle purchases (excluding police, fire and ambulance vehicles) must be alternative fuel vehicles. By the first months of 2008, this package of incentives resulted in sales of flexible-fuel cars representing 25 per cent of new car sales.

Bioethanol Stations European Union

Country	Stations	No/10⁶ Persons
Sweden	1,200	131.26
France	211	3.27
Germany	193	2.35
Switzerland	40	5.27
Ireland	29	6.84
United Kingdom	22	0.36

China

China is promoting ethanol-based fuel on a pilot basis in five cities in its central and northeastern region, a move designed to create a new market for its surplus grain and reduce consumption of petroleum. The cities include Zhengzhou, Luoyang and Nanyang in central China's Henan province, and Harbin and Zhaodong in Heilongjiang province, northeast China. Under the Programme, Henan will promote ethanol-based fuel across the province by the end of this year. Officials say the move is of great importance in helping to stabilize grain prices, raise farmers' income and reducing petrol-induced air pollution.

Thailand

Thailand already use 10 per cent ethanol (E10) widely on big scale on the local market. Beginning in 2008 Thailand started with the sale of E20 and by late 2008 E85 flexible fuel vehicles were introduced with only two gas stations selling E85.

Thailand is now converting some of the cassava stock hold by the government into fuel ethanol. Cassava-based ethanol productions are being ramped up to help manage the agricultural outputs of both cassava and sugar cane. With its abundant biomass resources, it is believed that the fuel ethanol Programme will be a new means of job creation in the rural areas while enhancing the balance sheet of fuel imports.

Australia

Legislation in Australia imposes a 10 per cent cap on the concentration of fuel ethanol blends. Blends of 90 per cent unleaded petrol and 10 per cent fuel ethanol are commonly referred to as E10. E10 is available through service stations operating under the BP, Caltex, Shell and United brands as well as those of a number of smaller independents. Not surprisingly, E10 is most widely available closer to the sources of production in Queensland and New South Wales where Sugar Cane is grown. E10 is most commonly blended with 91 RON "regular unleaded" fuel. There is a requirement that retailers label blends containing fuel ethanol on the dispenser.

Due to ethanol's greater stability under pressure it is used by Shell in their 100 octane fuel. Similarly IFS add 10 per cent ethanol to their 91 octane fuel, label it premium fuel and sell it more cheaply that regular unleaded. This is converse to the general practice of adding ethanol to a lesser quality fuel to bring its octane rating up to 91.

Some concern was raised over the use of ethanol blend fuels in petrol vehicles in 2003, yet manufacturers widely claimed that their vehicles were engined for such fuels. Since then there have been no reports of adverse affects to vehicles running on ethanol blended fuels.

Caribbean Basin

United States Fuel Ethanol Imports by Country (2002-2007)

(Millions of U.S. Liquid Gallons)

Country	2007	2006	2005	2004	2003	2002
Brazil	188.8	433.7	31.2	90.3	0	0
Jamaica	75.2	66.8	36.3	36.6	39.3	29.0
El Salvador	73.3	38.5	23.7	5.7	6.9	4.5
Trinidad and Tobago	42.7	24.8	10.0	0	0	0
Costa Rica	39.3	35.9	33.4	25.4	14.7	12.0

All countries in Central America, northern South America and the Caribbean are located in a tropical zone with suitable climate for growing sugar cane. In fact, most of these countries have a long tradition of growing sugar cane mainly for producing sugar and alcoholic beverages.

As a result of the guerilla movements in Central America, in 1983 the United States unilateral and temporarily approved the Caribbean Basin Initiative, allowing most countries in the region to benefit from several tariff and trade benefits. These benefits were made permanent in 1990 and more recently, these benefits were replaced by the Caribbean Basin Trade and Partnership Act, approved in 2000, and the Dominican Republic—Central America Free Trade Agreement that went to effect in 2008. All these agreements have allowed several countries in the region to export ethanol to the U.S. free of tariffs. Until 2004, the countries that benefited the most were Jamaica and Costa Rica, but as the U.S. began demanding more fuel ethanol, the two countries increased their exports and two others began exporting. In 2007, Jamaica, El Salvador, Trinidad and Tobago and Costa Rica exported together to the U.S. a total of 230.5 million gallons of ethanol, representing 54.1 per cent of U.S. fuel ethanol imports. Brasil began exporting ethanol to the U.S. in 2004 and exported 188.8 million gallons representing 44.3 per cent of U.S. ethanol imports in 2007. The remaining imports that year came from Canada and China.

In March 2007, "ethanol diplomacy" was the focus of President George W. Bush's Latin American tour, in which he and Brazil's president, Luiz Inacio Lula da Silva, were seeking to promote the production and use of sugar cane based ethanol throughout Latin America and the Caribbean. The two countries also agreed to share technology and set international standards for biofuels. The Brazilian sugar cane technology transfer would allow several Central American, Caribbean and Andean countries to take advantage of their tariff-free trade agreements to increase or become exporters to the United States in the short-term. Also, in August 2007, Brazil's President toured Mexico and several countries in Central America and the Caribbean to promote Brazilian ethanol technology. The ethanol alliance between the U.S. and Brazil generated some negative reactions from Venezuela's President Hugo Chavez, and by then Cuba's President, Fidel Castro, who wrote that *"you will see how many people among the hungry masses of our planet will no longer consume corn."* *"Or even worse,"* he continued, *"by offering financing to poor countries to produce ethanol from corn or any other kind of food, no tree will be left to defend humanity from climate change.*"' Daniel Ortega, Nicaragua's President, and one of the preferencial recipients of Brazilian technical aid also voiced critics to the Bush plan, but he vowed support for sugar cane based ethanol during Lula's visit to Nicaragua.

Colombia

Colombia's ethanol Programme began in 2002, based on a law approved in 2001 mandating a mix of 10 per cent ethanol with regular gasoline, and the plan is to gradually reach a 25 per cent blend in twenty-years. Sugar cane-based ethanol production began in 2005, when the law went into effect, and as local production was not enough to supply enough ethanol to the entire country's fleet, the Programme was implemented only on cities with more than 500,000 inhabitants, such as Cali, Pereira, and the capital city of Bogotá. All of the ethanol production comes from the Department of Valle del Cauca, Colombia's traditional sugar cane region. Cassava is the second source of ethanol, and potatoes and castor oil are also being studied.

On March 2009 the Colombian government enacted a mandate to introduce E85 flexible-fuel cars. The executive decree applies to all gasoline-powered vehicles with engines smaller than 2.0 liters manufactured, imported, and commercialized in the country beginning in 2012, mandating that 60 per cent of such vehicles must have flex-fuel engines capable of running with gasoline or E85, or any blend of both. By 2014 the mandatory quota is 80 per cent and it will reach 100 per cent by 2016. All vehicles with engines bigger than 2.0 liters must be E85 capable starting in 2013. The decree also mandates that by 2011 all gasoline stations must provide infrastructure to guarantee availability of E85 throughout the country. The mandatory introduction of E85 flex-fuels has been controversial.

Costa Rica

The government, based on the National Biofuel Programme,

established the mandatory use of all gasoline sold in Costa Rica with a blend of around 7.5 per cent ethanol, starting in October 2008. The implementation phase follows a two year trial that took place in the provinces of Guanacaste and Puntarenas. The government expects to increase the percent of ethanol mixed with gasoline to 12 per cent in the next 4 to 5 years. The Costa Rican government is pursuing this policy to lower the country's dependency of foreign oil and to reduce the amount of greenhouse gases produced. The plan also calls for an increase in ethanol producing crops and tax breaks for flex-fuel vehicles and other alternative fuel vehicles. However, the introduction of the blend of 7 per cent ethanol was postponed in September 2008 until the beginning of 2009. This delay was due to a request by the national association of fuel retailers to have more time available to adapt their fueling infrastructure. Additional delays caused another postponement, as fueling stations were not ready yet for handling ethanol fuel, and now implementation is expected for November 2009.

Despite the official postponement, during the months of February and March 2009, ethanol in different blends was sold without warning to consumers, which was cause for complains. The national distribution company, RECOPE, explained that it had already bought 50,000 barrels of ethanol stored and ready for distribution, so it decided to used as an oxygenate in substitution of MTBE. Nevertheless, retail sales of E7 continue uninterrupted in the trial regions of Guanacaste and the Central Pacific for three years now.

El Salvador

As a result of the cooperation agreement between the United States and Brazil, El Salvador was chosen in 2007 to lead a pilot experience to introduce state-of-the-art technology for growing sugar cane for production of ethanol fuel in Central America, as this technical bilateral cooperation is looking for helping Central American countries to reduce their dependence on foreign oil.

Comparison of Brazil and the U.S.

Evolution of the ethanol productivity per hectare of sugarcane planted in Brazil between 1975 and 2004. Source: Goldemberg (2008).

Brazil's sugar cane-based industry is far more efficient than the U.S. corn-based industry. Brazilian distillers are able to produce ethanol for 22 cents per liter, compared with the 30 cents per liter for corn-based ethanol. Sugarcane cultivation requires a tropical or subtropical climate, with a minimum of 600 mm (24 in) of annual rainfall. Sugarcane is one of the most efficient photosynthesizers in the plant kingdom, able to convert up to 2 per cent of incident solar energy into biomass. Ethanol is produced by yeast fermentation of the sugar extracted from sugar cane.

Sugarcane production in the United States occurs in Florida, Louisiana, Hawaii, and Texas. In prime growing regions, such as Hawaii, sugarcane can produce 20 kg for each square meter exposed to the sun. The first three plants to produce sugar cane-based ethanol are expected to go online in Louisiana by mid 2009. Sugar mill plants in Lacassine, St. James and Bunkie were converted to sugar cane-based ethanol production using Colombian technology in order to make possible a profitable ethanol production. These three plants will produce 100 million gallons of ethanol within five years.

U.S. corn-derived ethanol costs 30 per cent more because the corn starch must first be converted to sugar before being distilled into alcohol. Despite this cost differential in production, in contrast to Japan and Sweden, the U.S. does not import much of Brazilian ethanol because of U.S. trade barriers corresponding to a tariff of 54-cent per gallon—a levy designed to offset the 45-cent per gallon blender's federal tax credit that is applied to ethanol no matter its country of origin. One advantage U.S. corn-derived ethanol offers is the ability to return 1/3 of the feedstock back into the market as a replacement for the corn used in the form of Distillers Dried Grain.

Comparison of Key Characteristics between the Ethanol Industries in the United States and Brazil

Characteristic	Brazil	U.S.	Units/Comments
Feedstock	Sugar cane	Maize	Main cash crop for ethanol production, the US has less than 2% from other crops.
Total ethanol fuel production (2008)	6,472	9,000	Million U.S. liquid gallons
Total arable land	355	270	Million hectares.
Total area used for ethanol crop (2006)	3.6 (1%)	10 (3.7%)	Million hectares (% total arable)
Productivity per hectare	6,800-8,000	3,800-4,000	Liters of ethanol per hectare. Brazil is 727 to 870 gal/acre (2006), US is 321 to 424 gal/acre (2003)
Energy balance (input energy productivity)	8.3 to 10.2	1.3 to 1.6	Ratio of the energy obtained from ethanol/energy expended in its production
Estimated GHG emissions reduction	86-90%	10-30%	% GHGs avoided by using ethanol instead of gasoline, using existing crop land (No ILUC).
Full life-cycle carbon intensity	73.40	105.10	Grams of CO_2 equivalent released per MJ of energy produced, includes indirect land use changes.
Estimated payback time for GHG emissions	17 years	93 years	Brazilian cerrado for sugarcane and US grassland for corn. Land use change scenarios by Fargione
Flexible-fuel vehicle fleet	7.5 million	8.0 million	Autos and light trucks only. Brazil as of April 2009 (E100 FFVs). U.S. as of early 2009 (E85 FFVs).
Ethanol fueling stations in the country	35,017 (100%)	1,963 (1%)	As % of total gas stations in the country. Brazil by 2007-12, U.S. by 2009-03 (170,000 total.)

(Contd.)

Characteristic	Brazil	U.S.	Units/Comments
Ethanol's share in the gasoline market	50%	4%	As % of total consumption on a volumetric basis. Brazil as of April 2008. US as of December 2006.
Cost of production (USD/gallon)	0.83	1.14	2006/2007 for Brazil (22¢/liter), 2004 for U.S. (35¢/liter)
Government subsidy (in USD)	0	0.45/gallon	U.S. since 2009-01-01 as a tax credit. Brazilian ethanol production is no longer subsidized.
Import tariffs (in USD)	0	0.54/gallon	As of June 2009, Brazil does not import ethanol, the U.S. does

Notes: (1) Only contiguous U.S., excludes Alaska. (2) Assuming no land use change. (3) CARB estimate for Midwest corn ethanol. California's gasoline carbon intensity is 95.86 blended with 10 per cent ethanol. (4) Assuming direct land use change. (5) If diesel-powered vehicles are included and due to ethanol's lower energy content by volume, bioethanol represented 16.9 per cent of the road sector energy consumption in 2007. (6) Brazilian ethanol production is no longer subsidized, but gasoline is heavily taxed Favouring ethanol fuel consumption (~54 per cent tax). By the end of July 2008, when oil prices were close to its latest peak and the Brazilian Real exchange rate to the US dollar was close to its most recent minimum, the average gasoline retail price at the pump in Brazil was USD 6.00 per gallon, while the average US price was USD 3.98 per gallon. The latest gas retail price increase in Brazil occurred in late 2005, when oil price was at USD 60 per barrel.

Environment

Energy Balance

Country	Type	Energy balance
United States	Corn ethanol	1.3
Brazil	Sugarcane ethanol	8
Germany	Biodiesel	2.5
United States	Cellulosic ethanol[†]	2-36[††]

† experimental, not in commercial production.

†† depending on production method.

All biomass goes through at least some of these steps: it needs to be grown, collected, dried, fermented, and burned. All of these steps require resources and an infrastructure. The total amount of energy input into the process compared to the energy released by burning the resulting ethanol fuel is known as the **energy balance** (or "Net energy gain"). Figures compiled in a 2007 by *National Geographic Magazine* point to modest results for corn ethanol produced in the US: one unit of fossil-fuel energy is required to create 1.3 energy units from the resulting ethanol. The energy balance for sugarcane ethanol produced in Brazil is more Favourable, 1:8. Energy balance estimates are not easily produced, thus numerous such reports have been generated that are contradictory. For instance, a separate survey reports that production of ethanol from sugarcane, which requires a tropical climate to grow productively, returns from 8 to 9 units of energy for each unit expended, as compared to corn which only returns about 1.34 units of fuel energy for each unit of energy expended.

Carbon dioxide, a greenhouse gas, is emitted during fermentation and combustion. However, this is canceled out by the greater uptake of carbon dioxide by the plants as they grow to produce the biomass. When compared to gasoline, depending on the production method, ethanol releases less greenhouse gases.

Air Pollution

Compared with conventional unleaded gasoline, ethanol is a particulate-free burning fuel source that combusts with oxygen to form carbon dioxide, water and aldehydes. Gasoline produces

2.44 [[CO$_2$ equivalent]] kg/l and ethanol 1.94 (this is –21 per cent CO$_2$). The Clean Air Act requires the addition of oxygenates to reduce carbon monoxide emissions in the United States. The additive MTBE is currently being phased out due to ground water contamination, hence ethanol becomes an attractive alternative additive. Current production methods include air pollution from the manufacturer of macronutrient fertilizers such as ammonia.

A study by atmospheric scientists at Stanford University found that E85 fuel would increase the risk of air pollution deaths relative to gasoline by 9 per cent in Los Angeles, USA: a very large, urban, car-based metropolis that is a worst case scenario. Ozone levels are significantly increased, thereby increasing photochemical smog and aggravating medical problems such as asthma.

Manufacture

In 2002, monitoring the process of ethanol production from corn revealed that they released VOCs (volatile organic compounds) at a higher rate than had previously been disclosed. The Environmental Protection Agency (EPA) subsequently reached settlement with Archer Daniels Midland and Cargill, two of the largest producers of ethanol, to reduce emission of these VOCs. VOCs are produced when fermented corn mash is dried for sale as a supplement for livestock feed. Devices known as thermal oxidizers or catalytic oxidizers can be attached to the plants to burn off the hazardous gases.

Carbon Dioxide

UK government calculation of carbon intensity of corn bioethanol grown in the US and burnt in the UK.

Graph of UK figures for the carbon intensity of bioethanol and fossil fuels. This graph assumes that all bioethanols are burnt in their country of origin and that previously existing cropland is used to grow the feedstock.

The calculation of exactly how much carbon dioxide is produced in the manufacture of bioethanol is a complex and inexact process, and is highly dependent on the method by which the ethanol is produced and the assumptions made in the calculation. A calculation should include:

- The cost of growing the feedstock
- The cost of transporting the feedstock to the factory
- The cost of processing the feedstock into bioethanol

Such a calculation may or may not consider the following effects:

- The cost of the change in land use of the area where the fuel feedstock is grown.
- The cost of transportation of the bioethanol from the factory to its point of use
- The efficiency of the bioethanol compared with standard gasoline
- The amount of Carbon Dioxide produced at the tail pipe.
- The benefits due to the production of useful bi-products, such as cattle feed or electricity.

The graph on the right shows figures calculated by the UK government for the purposes of the Renewable transport fuel obligation.

The January 2006 Science article from UC Berkeley's ERG, estimated reduction from corn ethanol in GHG to be 13 per cent after reviewing a large number of studies. However, in a correction to that article released shortly after publication, they reduce the estimated value to 7.4 per cent. A National Geographic Magazine overview article (2007) puts the figures at 22 per cent less CO_2 emissions in production and use for corn ethanol compared to gasoline and a 56 per cent reduction for cane ethanol. Carmaker Ford reports a 70 per cent reduction in CO_2 emissions with bioethanol compared to petrol for one of their flexible-fuel vehicles.

An additional complication is that production requires tilling new soil which produces a one-off release of GHG that it can take decades or centuries of production reductions in GHG emissions to equalize. As an example, converting grass lands to corn production for ethanol takes about a century of annual savings to make up for the GHG released from the initial tilling.

Change in Land Use

Large-scale farming is necessary to produce agricultural alcohol and this requires substantial amounts of cultivated land. University of Minnesota researchers report that if all corn grown in the U.S. were used to make ethanol it would displace 12 per cent of current U.S. gasoline consumption. There are claims that land for ethanol production is acquired through deforestation, while others have observed that areas currently supporting forests are usually not suitable for growing crops. In any case, farming may involve a decline in soil fertility due to reduction of organic matter, a decrease in water availability and quality, an increase in the use of pesticides and fertilizers, and potential dislocation of local communities. However, new technology enables farmers and processors to increasingly produce the same output using less inputs.

Cellulosic ethanol production is a new approach which may alleviate land use and related concerns. Cellulosic ethanol can be produced from any plant material, potentially doubling yields, in an effort to minimize conflict between food needs vs. fuel needs. Instead of utilizing only the starch by-products from grinding wheat and other crops, cellulosic ethanol production maximizes the use of all plant materials, including gluten. This approach would have a smaller carbon footprint because the amount of energy-intensive fertilisers and fungicides remain the same for higher output of usable material. The technology for producing cellulosic ethanol is currently in the commercialization stage.

Many analysts suggest that, whichever ethanol fuel production strategy is used, fuel conservation efforts are also needed to make a large impact on reducing petroleum fuel use.

Efficiency of Common Crops

As ethanol yields improve or different feedstocks are introduced, ethanol production may become more economically feasible in the US. Currently, research on improving ethanol yields from each unit of corn is underway using biotechnology. Also, as long as oil prices remain high, the economical use of other feedstocks, such as cellulose, become viable. By-products such as straw or wood chips can be converted to ethanol. Fast growing species like switchgrass can be grown on land not suitable for other cash crops and yield high levels of ethanol per unit area.

Crop	Annual yield (Liters/hectare)	Annual yield (US gal/acre)	Greenho use-gas savings (% vs. petrol) (1)	Comments
Miscanthus	7300	780	37-73	Low-input perennial grass. Ethanol production depends on development of cellulosic technology.
Switchgrass	3100-7600	330-810	37-73	Low-input perennial grass. Ethanol production depends on development of cellulosic technology. Breeding efforts underway to increase yields. Higher biomass production possible with mixed species of perennial grasses.
Poplar	3700-6000	400-640	51-100	Fast-growing tree. Ethanol production depends on development of cellulosic technology. Completion of genomic sequencing project will aid breeding efforts to increase yields.
Sugar cane	6800-8000	727-870	87-96	Long-season annual grass. Used as feedstock for most bioethanol produced in Brazil. Newer processing plants burn residues not used for ethanol to generate electricity. Only grows in tropical and subtropical climates.

(Contd.)

Crop	Annual yield (Liters/hec tare)	Annual yield (US gal/acre)	Greenho use-gas savings (% vs. petrol) (1)	Comments
Sweet sorghum	2500-7000	270-750	No data	Low-input annual grass. Ethanol production possible using existing technology. Grows in tropical and tempe-rate climates, but highest ethanol yield estimates assume multiple crops per year (only possible in tropical climates). Does not store well.
Corn	3100-4000	330-424	10-20	High-input annual grass. Used as feedstock for most bio-ethanol produced in USA. Only kernels can be proce-ssed using available techno-logy; development of commer-cial cellulosic tech-nology would allow stover to be used and increase etha-nol yield by 1,100—2,000 litres/ha.

Source: (except those indicated): Nature 444 (December 7, 2006): 673-676.
(1) Savings of GHG emissions assuming no land use change (using existing crop lands).

Reduced Petroleum Imports and Costs

One rationale given for extensive ethanol production in the U.S. is its benefit to energy security, by shifting the need for some foreign-produced oil to domestically-produced energy sources. Production of ethanol requires significant energy, but current U.S. production derives most of that energy from coal, natural gas and other sources, rather than oil. Because 66 per cent of oil consumed in the U.S. is imported, compared to a net surplus of coal and just 16 per cent of natural gas (2006 figures), the displacement of oil-based fuels to ethanol produces a net shift from foreign to domestic U.S. energy sources.

According to a 2008 analysis by Iowa State University, the growth in US ethanol production has caused retail gasoline prices to be US $0.29 to US $0.40 per gallon lower than would otherwise have been the case.

Criticism and Controversy

There are various current issues with ethanol production and use, which are presently being discussed in the popular media and scientific journals. These include: the effect of moderating oil prices, the "food vs fuel" debate, carbon emissions levels, sustainable biofuel production, deforestation and soil erosion, impact on water resources, human rights issues, poverty reduction potential, ethanol prices, energy balance and efficiency, and centralised vs. decentralised production models.

Food vs. fuel is about the risk of diverting farmland or crops for ethanol production to the detriment of the food supply. The debate is internationally controversial, with good-and-valid arguments on all sides of this ongoing debate. There is disagreement about how significant this is, what is causing it, what the impact is, and what can or should be done about it.

Fuel System Problems

Several of the outstanding ethanol fuel issues are linked specifically to fuel systems. Fuels with more than 10 per cent ethanol are not compatible with non E85-ready fuel system components and may cause corrosion of iron components. Ethanol fuel can negatively affect electric fuel pumps by increasing internal wear, cause undesirable spark generation, and is not compatible with capacitance fuel level gauging indicators and may cause erroneous fuel quantity indications in vehicles that employ that system. It is also not always compatible with marine craft, especially those that use fibreglass fuel tanks. Ethanol is also not used in aircraft for these same reasons.

Using 100 per cent ethanol fuel decreases fuel-economy by 15-30 per cent over using 100 per cent gasoline; this can be avoided using certain modifications that would, however, render the engine inoperable on regular petrol without the addition of an adjustable ECU. Tough materials are needed to accommodate a higher compression ratio to make an ethanol engine as efficient as it would be on petrol; these would be similar to those used in diesel engines which typically run at a CR of 20:1, vs. about 8-12:1 for petrol engines.

In April 2008 the German environmental minister cancelled a proposed 10 per cent ethanol fuel scheme citing technical problems: too many older cars in Germany are unequipped to handle this fuel. Ethanol levels in fuel will remain at 5 per cent.

Other Non-Transport Uses

There is still extensive use of kerosene for lighting and cooking in less developed countries, and ethanol can have a role in reducing petroleum dependency in this use too. A non profit named Project Gaia seeks to spread the use of ethanol stoves to replace wood, charcoal and kerosene. There is also potential for bioethanol replacing some kerosene use in domestic lighting from feedstocks grown locally. A 50 per cent ethanol water mixture has been tested in specially designed stoves and lanterns for rural areas.

UNDERSTANDING ETHANOL FERMENTATION

Ethanol fermentation is a biological process in which sugars such as glucose, fructose, and sucrose are converted into cellular energy and thereby produce ethanol and carbon dioxide as metabolic waste products.

Because yeasts perform this process in the absence of oxygen, ethanol fermentation is classified as anaerobic.

Ethanol fermentation occurs in the production of alcoholic beverages and ethanol fuel, and in the rising of bread dough.

The Chemical Process of Fermentation of Glucose

The chemical equation below summarizes the fermentation of glucose. One glucose molecule is converted into two ethanol molecules and two carbon dioxide molecules:

$$C_6H_{12}O_6 \rightarrow 2C_2H_5OH + 2CO_2$$

The process begins with a molecule of glucose being broken down by the process of glycolysis into pyruvate:

$$C_6H_{12}O_6 \rightarrow 2\ CH_3COCOO^- + 2H^+$$

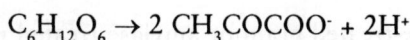

This reaction is accompanied by the size difference of two molecules of NAD^+ to NADH and a net of two ADP molecules converted to two ATP plus the two water molecules. Pyruvate is then converted to acetaldehyde and carbon dioxide by an enzyme called pyruvate decarboxylase and requiring thiamine diphosphate as cofactor. The acetaldehyde is subsequently reduced to ethanol by the NADH from the previous glycolysis, which is returned to NAD^+:

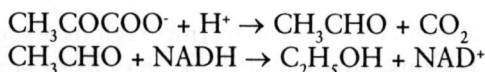

$$CH_3COCOO^- + H^+ \rightarrow CH_3CHO + CO_2$$
$$CH_3CHO + NADH \rightarrow C_2H_5OH + NAD^+$$

Many species of Yeast (K. lactis, K lipolytica) will oxidize pyruvate completely to carbon dioxide and water (respiration) if oxygen is present in the environment and will ferment only in an anaerobic environment. However, the commonly used bakers' yeast *S. cerevisiae* as well the yeast *S. pombe*, both prefer fermentation to respiration even in the presence of oxygen and will yield ethanol even under aerobic conditions given the right sources of nutrition.

Procedure of making alcohol from sugar:

Uses

Ethanol fermentation is responsible for the rising of bread dough. Yeast organisms consume sugars in the dough and produce ethanol and carbon dioxide as waste products. The carbon dioxide forms bubbles in the dough, expanding it into something of a foam. Nearly all the ethanol evaporates from the dough when the bread is baked.

The production of all alcoholic beverages, except those produced by carbonic maceration, employs ethanol fermentation by yeast. Wines and brandies are produced by fermentation of the natural sugars present in fruits, especially grapes. Beers, ales, and whiskeys employ fermentation of grain starches that have been converted to sugar by the application of the enzyme, amylase, which is present in grain kernels that have been germinated. Amylase-treated grain or amylase-treated potatoes are fermented for the production of vodka. Fermentation of cane sugar is the first step in producing rum. In all cases, the fermentation must take place in a vessel that is arranged to allow carbon dioxide to escape, but that prevents outside air from coming in, as exposure to oxygen would prevent the formation of ethanol.

Similar yeast fermentation of various carbohydrate products is used to produce much of the ethanol used for fuel.

Feedstocks for Fuel Production

The dominant ethanol feedstock in warmer regions is sugarcane. In temperate regions, this accessibility has been somewhat replicated by selective breeding of the sugar beet.

In the United States, the main feedstock for the production of ethanol is currently corn. Approximately 2.8 gallons of ethanol are produced from one bushel of corn (0.42 liter per kilogram). While much of the corn turns into ethanol, some of the corn also yields by-products such as DDGS (distillers dried grains with solubles) that can be used to fulfill a portion of the diet of livestock. A bushel of corn produces about 18 pounds of DDGS. Although most of the fermentation plants have been built in corn-producing regions, sorghum is also an important feedstock for ethanol production in the Plains states. Pearl millet is showing promise as an ethanol feedstock for the southeastern U.S. and duckweed potential is being studied.

In some parts of Europe, particularly France and Italy, wine is used as a feedstock due to a massive oversupply termed *wine lake*. Japan is hoping to use rice wine (sake) as an ethanol source.

Ethanol Market Forecast

The main players will be Brazil, USA, EU, and tropical developing countries. The EU can currently (2007) produce ethanol in large quantities with a mineral-oil based chemical process for US$0.57 per liter. The USA produces ethanol for circa US$0.32 per liter, mainly from corn starch. Brazil produces ethanol for circa US$0.27 per liter, from sugarcane. Tropical developing countries do not produce very large amounts of ethanol yet.

Brazil is the largest producer, but it will not be able to meet the EU's needs for many years to come, assuming that it will expand ethanol production at maximum possible rate. The USA is expected to become self-supplying (to avoid high oil prices), but is not expected to become a major exporter. The EU also wants to avoid high oil prices, and is starting to require a minimum ethanol percentage in automobile fuels, so it wants to import ethanol. Ethanol can be made from mineral oil or from sugars or starches, cheapest of which are starches, and starchy crop with highest energy content per acre is cassava, which grows in tropical countries.

Thailand already had a large cassava industry in the 1990s, for use as cattle food and as cheap admixture to wheat flour; Nigeria and Ghana are already establishing cassava-to-ethanol plants; Brazil is doing that too (sugarcane and cassava grow on very different types of soil); and so are many other countries.

EU expects that combined effect of increasing ethanol production will be able to meet its needs in 2012. Therefore it is expect that in 2012 price of ethanol will drop from maybe US$0.42 to maybe US$0.30 (FOB Africa). Production of ethanol from cassava is currently economically feasible when crude oil prices are above US$120 per barrel.

New varieties of cassava are being developed, so future situation remains uncertain. Currently, cassava can yield more than 40 tons per hectare (with irrigation and fertilizer), and from a ton of cassava roots, circa 200 liter of ethanol can be produced (assuming cassava with 22 per cent starch content), and a liter of ethanol contains circa 10.7 MJ of energy. Overall energy efficiency of cassava-root to ethanol conversion is circa 32 per cent.

Cassava plants can grow in poor soils, are drought resistant, and need a minimum temperature of 17 °C. They can use solar radiation up to 300 W/m² (equivalent to lightly clouded tropical sky), and optimum water use is 100 to 150 cm (slightly less than rainfall in rain forest). For compensating for nutrients taken up, Cassava's fertilizer demand is (in kilograms of nutrient per ton cassava): N:21, P:10, K:42, Ca:7, Mg:4, so if fertilizer prices go up, so does ethanol price.

Starch price (food-quality starch from Thailand) is circa 0.22 US$/kg, and from 1 kg starch, 0.9 liter of ethanol can be produced, so, producer price would be 0.24 US$/liter plus cost of conversion from starch to ethanol. A US$10 million conversion plant can convert circa 80 million liters per year, so total cost of ethanol from cassava currently is near USA's production price. Due to improvements being made in this relatively new industry, producer price would become lower, probably near that of Brazil, and maybe even lower than that.

Yeast used for processing cassava is *Endomycopsis fibuligera*, sometimes used together with bacterium *Zymomonas mobilis*.

Most of this information can be found on FAO's website.

Microbes Used in Ethanol Fermentation

- Yeast
- Zymomonas Mobilis

UNDERSTANDING CORN ETHANOL

Corn ethanol is ethanol produced from corn as a biomass through industrial fermentation, chemical processing and distillation. It is primarily used in the United States as an alternative to gasoline and petroleum (first-generation biofuel). Corn ethanol is the most common type of ethanol in the United States, but is considered less efficient than other types of ethanol (sugar cane, etc.) because only the grain is used and many petroleum-based products (fertilizer, pesticides, etc) are used in its production.

Ethanol production may occur through two corn processing methods: dry and wet corn milling; the main difference between the two is the initial treatment of grain. In dry milling operations, liquefied corn starch is produced by heating corn meal with water and enzymes. A second enzyme converts the liquefied starch to sugars, which are fermented by yeast into ethanol and carbon dioxide; released CO_2 during fermentation can be captured and sold for use in carbonating beverages and in the manufacture of dry ice, but it is often released to the atmosphere because carbon capture for food use requires specialized and expensive equipment. Wet milling operations separate the pericarp, germ (oil), and protein from the starch before it is fermented into ethanol.

Environmental Efficiency

Despite CO_2 (a greenhouse gas) being released during ethanol production and combustion, it is recaptured as a nutrient to the crops that are used in its production. Approximately 75 per cent of ethanol production is performed via the dry corn milling process, since dry corn mills are less expensive to construct. The problem with traditional grain based ethanol is that it utilizes fossil fuels to produce heat during the conversion process, generating substantial greenhouse gas emissions.

There is also controversy over the production of corn in many equatorial regions with rain forests, and South Africa. Farmers and agriculture businesses are burning down rain forests (which are natural CO2 sequesters already). While CO2 studies assume no co2 sequestration if the corn is not planted. Scientists say that this will mean that ethanol will end up contributing to global warming more than if we used oil, where rainforests are destroyed to produce it. This is also an issue in other biofuels such as biodiesel.

Problems Associated with Corn-Derived Ethanol

The cost of building 100 million gallon ethanol plants is $140 million; the cost of natural gas to operate these plants is estimated at $15-$25 million per year, while the amount of water required in the production of ethanol is roughly 2 million gallons per day.(Approximately 1700 gallons of water for every gallon of ethanol) And since corn is one of the most water-intensive crops to grow, the volume of water involved in production makes it imperative that the water used be treated sewage water, rather than the Ogallala aquifer beneath the great plains, which is being drawn down at rates exceeding 100 times replacement rate.

According to a University of Minnesota study, corn ethanol may be even worse for air quality than gasoline. The study concluded that the total environmental and health costs of making a gallon of gasoline was about 71 cents, compared with a range of 72 cents to $1.45 for corn-based ethanol, and 19 to 32 cents for cellulosic ethanol, depending upon the technology and type of plants used.

Others argue that the benefits versus rewards, from an environmental perspective, are substantially less. Whether or not ethanol production from corn is efficient is debatable. Proponents of corn-derived ethanol point to studies emphasizing an overall net positive energy gain, whereas others claim that when the complete production costs of farming, seed, fertilizer, pesticides, fuel, ethanol distillation, etc., are taken into consideration, ethanol requires 30 per cent more energy to produce than it creates. Ethanol proponents say that ethanol simply puts back the same carbon dioxide that the plants from which it is made absorbed while growing, and hence does not add to the world's balance of greenhouse gases. A 2006 University of Minnesota study shows a positive energy balance for ethanol of around 25 per cent, but also highlights many environmental and economic limitations affecting the viability of corn ethanol.

There are also ethical challenges in deciding the best use of natural resources. The "fuel or food debate" rages over the loss of dedicating more land to ethanol crops would squeeze the supply of land for food production. Demand for biofuels is raising the price of crops which may adversely affect food supplies. In June 2008, corn was selling at an all time high of well over $7.36/ bushel (compared with the recent norm of around $2/bushel). Poultry and Pork producers are feeling the pinch of rising corn prices, which may soon be felt as higher bills at the grocery checkout counter. Moreover, government-sponsored efforts to cull forest lands for "biomass" fuel stocks could deplete habitats. For example, dedicating more land to ethanol crops would squeeze the supply of land for food production. There is currently 349 million acres (1,410,000 km²) of available farmland, and an additional 388 million acres (1,570,000 km²) of idle land. The amount of land required to plant enough corn to replace imported fuels is 238 million acres (963,000 km²). Realistically, peak corn-ethanol would likely top out at between 11 and 15 billion US gallons (42 and 57 million m³) per year—only a fraction of the 140 billion US gallons (530 million m³) of gasoline consumed each year in the United States.

Producing corn is very energy intensive, and uses fossil fuels in virtually every step of the crop cycle: transporting and planting the seeds; operating farm equipment; making and applying fertilizer; and transporting the corn to market. Fertilizer, herbicide, and insecticide production consume the most fossil fuels. Fossil-fuel based fertilizers also contaminate the soil and groundwater, but they can not be replaced by natural fertilizer: there are not enough animals to provide the fertilizer to grow the corn necessary to produce all the grain-based ethanol needed to run American cars. And the herbicides and pesticides necessary to grow corn at an industrial scale leach into the groundwater, too.

There is an ongoing debate concerning the amount of energy it takes to produce ethanol from corn. For example, it takes energy equivalent to about one gallon of gasoline to make four pounds of nitrogen, the main ingredient in most fertilizer, and every one of the more than 15 million acres (61,000 km²) planted in corn is dusted with about 58 pounds of nitrogen. Given the variety of factors that go into growing corn, estimates vary widely about the amount of energy used: one estimate contends that it would require 1.5 gallons of ethanol to provide the same amount of energy as a gallon of gasoline. Others challenge these conclusions, asserting that this analysis is based on obsolete data and miscalculated key energy values and does not account for the useful by-products, such as animal feed, of making ethanol; taking all that into account, ethanol could provide up to 40 per cent more energy than is consumed in making it.

The question of sustainability arises when we consider that ethanol from corn can't possibly be grown forever because growing corn depletes the soil even if sustainable farming methods such as crop rotation are used. Some researchers argue that ethanol production from corn could wear out the soil within 30 years. Although farmers in the midwest have been growing corn on the same land for the past 200 years and are getting a larger crop today then they did 30 years ago.

Ethanol, even in gasoline blends, cannot be shipped through the country's existing gasoline pipeline system because it is easily contaminated by water and corrodes the pipes; there is no ethanol pipeline anywhere in the world, although Brazil's Petrobras claims to have one in the planning stages. Ethanol is currently shipped by truck or rail car to fuel distributors, who then mix it with gasoline before delivering it to filling stations in more trucks. This adds to the cost of ethanol and to its overall CO_2 emissions. In order to use ethanol on any large scale, transport vehicles will either have to be retrofitted for ethanol, or the government be forced to build or subsidize pipelines.

For ethanol to be currently economically viable requires massive Federal subsidies and price supports. Even the biggest of proposed ethanol supports—an increase in mandated ethanol consumption from 7.5 billion gallons a year to 15 billion gallons a year, as called for in the energy bill Congress is currently (10/07) debating—would barely dent America's oil consumption, which is approximately 150 billion gallons annually.

Another issue concerning ethanol as a gasoline substitute includes the fact that only around 5 million automobiles currently in America are "flexible-fuel vehicles"—cars that are equipped to run on a blend of 85 percent ethanol and 15 percent gasoline (known as E85). That's out of 135 million registered passenger cars in the United States. Moreover, as the Dallas Morning News reported last year, the owners of almost all of these flex-fuel vehicles tend to fill them up with regular gas, owing to a scarcity of gas stations that sell E85.

There is no shortage of disagreement on the pros and cons of using corn-ethanol as a petroleum substitute. What does emerge from the discussion is that while corn-derived ethanol may be a short term solution to America's energy problem, it casts doubt on corn-based ethanol's long-term viability, and begs the need for a long term solution: enter cellulosic ethanol.

There is actually a debated possibility that (indirectly) corn and other ethanols cause rainforests to be slashed and burnt. This happens because the fields that were formally used for corn are now used for ethanol. Thus the corn grown for eating is grown on what were grazing lands for farmers and ranchers. These people start cutting rainforest for grazing ground and also use the fertile grasslands.

The Effect of Rapid Growth in the Renewable Fuels Industry on the Water Supply

The production of corn ethanol uses water in two ways—irrigation and processing. There are two types of ethanol processing, wet milling and dry milling, and the central difference between the two processes is how they initially treat the grain. In wet milling, the corn grain is seeped in water and then separated for processing in the first step. Dry milling, which is more common requires a different process. According to a report by the National Renewable Energy Laboratory, "Over 80 per cent of U.S. ethanol is produced from corn by the dry grind process." The dry grind process proceeds as follows:

"Corn grain is milled, then slurried with water to create 'mash.' Enzymes are added to the mash and this mixture is then cooked to hydrolyze the starch into glucose sugars. Yeast ferment these sugars into ethanol and carbon dioxide and the ethanol is purified through a combination of distillation and molecular sieve dehydration to create fuel ethanol. The byproduct of this process is known as distiller's dried grains and solubles (DDGS) and is used wet or dry as animal feed."

Most ethanol plants recycle most of their water. As a result, most of the water used in the process is for energy—in the cooling towers and boiling system. Water use of ethanol production varies with location. According to a recent University of Minnesota study, ethanol production in the Midwest uses less water than that in the Western U.S. The main reason for this water usage discrepancy is that in the West, more water is needed for irrigating cornfields. The study states, "As a general trend, the [embodied water in corn ethanol] increases from the East to the West and from the Midwest to the Southwest regions of the U.S." Therefore, looking at average water usage per liter of ethanol is not useful or reflective of reality.

Using the numbers of regional ethanol from the University of Minnesota Study in 2007, the corn ethanol industry used approximately 3,260,000,000,000 L, or 861,000,000,000 gallons, of water. If one uses the average embodied water in ethanol (EWe) found in the U of M study, to produce the 57 billion liters of ethanol in 2015 required by the Energy and Independence Security Act, the U.S. would use approximately 8,094,000,000,000,000,000 L, or 2,140,000,000,000,000,000 gallons, of water. However, as stated earlier, the average EWe does not convey useful information because it does not account for regional differences in water needs for irrigation. Thus, this predicted amount of water used for corn ethanol production may be completely misleading.

In terms of local water impact, the study found that "a considerable volume of groundwater was withdrawn for bioethanol in the regions with vulnerable fossil aquifers." The U.S. Drought Monitor states that at any given time, around 10 states face drought-like conditions, and even in the Midwest there have been growing conflicts over water usage in ethanol related farming. (PDF and Monitor CITATION). On average, irrigated corn farms use 785 gallons of water for every gallon of ethanol produced, demonstrating the high demands for water in not only production, but also farming. (PDF cite).

"In 2006 alone, almost 5 billion gallons of ethanol were produced, an increase of a billion gallons over the previous year. At least 73 corn ethanol pants are currently under construction, with eight more undergoing expansion, which will add another 6 billion gallons of new capacity by 2009" (PDF citation). This expansion is part of the Bush Administrations "Twenty in Ten" project, which plans to displace 20 per cent of gasoline in 10 years, resulting in 35 billion gallons of renewable fuel by 2017 (PDF citation). This kind of expansion will likely result in huge quantities of water tied in ethanol production in the Midwest and Southwest regions, undoubtedly causing a strain on water resources available to the community and industry.

Chiu, Yi-Wen; Suh, Sangwon; and Walseth, Brian. "Water Embodied in Bioethanol." Environ. Sci. Technol., 2009, 43 (8), pp 2688—2692. http://pubs.acs.org/doi/abs/10.1021/es8031067?prevSearch=university+of+minnesota+ethanol+water andsearchHistoryKey

Aden, Andy. September/October, 2007. "Water Usage for Current and Future Ethanol Production." National Renewable Energy Laboratory. www.swhydro.arizona.edu/archive/V6_N5/feature4.pdf

UNDERSTANDING CELLULOSIC ETHANOL

Cellulosic Ethanol is a biofuel produced from wood, grasses, or the non-edible parts of plants.

It is a type of biofuel produced from lignocellulose, a structural material that comprises much of the mass of plants. Lignocellulose is composed mainly of cellulose, hemicellulose and lignin. Corn stover, switchgrass, miscanthus, woodchips and the byproducts of lawn and tree maintenance are some of the more popular cellulosic materials for ethanol production.

Production of ethanol from lignocellulose has the advantage of abundant and diverse raw material compared to sources like corn and cane sugars, but requires a greater amount of processing to make the sugar monomers available to the microorganisms that are typically used to produce ethanol by fermentation.

Switchgrass and Miscanthus are the major biomass materials being studied today, due to their high productivity per acre. Cellulose, however, is contained in nearly every natural, free-growing plant, tree, and bush, in meadows, forests, and fields all over the world without agricultural effort or cost needed to make it grow.

According to U.S. Department of Energy studies conducted by Argonne National Laboratory of the University of Chicago, one of the benefits of cellulosic ethanol is that it reduces greenhouse gas emissions (GHG) by 85 per cent over reformulated gasoline. By contrast, starch ethanol (e.g., from corn), which most frequently uses natural gas to provide energy for the process, may not reduce GHG emissions at all depending on how the starch-based feedstock is produced. A study by Nobel Prize winner Paul Crutzen found ethanol produced from corn, and sugarcane had a "net climate warming" effect when compared to oil.

History

The first attempt at commercializing a process for ethanol from wood was done in Germany in 1898. It involved the use of dilute acid to hydrolyze the cellulose to glucose, and was able to produce 7.6 liters of ethanol per 100 kg of wood waste (18 gal per ton). The Germans soon developed an industrial process optimized for yields of around 50 gallons per ton of biomass. This process soon found its way to the United States, culminating in two commercial plants operating in the southeast during World War I. These plants used what was called "the American Process"—a one-stage dilute sulfuric acid hydrolysis. Though the yields were half that of the original German process (25 gallons of ethanol per ton versus 50), the throughput of the American process was much higher. A drop in lumber production forced the plants to close shortly after the end of World War I. In the meantime, a small, but steady amount of research on dilute acid hydrolysis continued at the USDA Forest Products Laboratory. During World War II, the US again turned to cellulosic ethanol, this time for conversion to butanediol to produce synthetic rubber. The Vulcan Copper and Supply Company was contracted to construct and operate a plant to convert sawdust into ethanol. The plant was based on modifications to the original German Scholler process as developed by the Forest Products Laboratory. This plant achieved an ethanol yield of 50 gal/dry ton but was still not profitable and was closed after the war.

With the rapid development of enzyme technologies in the last 2 decades, the acid hydrolysis process has gradually been replaced by emzymatic hydrolysis. However, chemical pre-treatment of the feedstock is required to prehydrolyze (separate) hemicellulose for robust enzymatic saccharification of cellulose substrate. The dilute acid pretreatment is developed based on the early work on acid hydrolysis of wood at the USDA Forest Products Laboratory. Recently, the Forest Products Laboratory together with the University of Wisconsin at Madison developed the Sulphite Pre-treatment to overcome Recalcitrance of Lignocellulose (SPORL) for robust enzymatic hydrolysis of wood cellulose.

United States President Bush, in his State of the Union address delivered January 31, 2006, proposed to expand the use of cellulosic ethanol. In his State of the Union Address on January 23, 2007, President Bush announced a proposed mandate for

35 billion gallons of ethanol by 2017. It is widely recognized that the maximum production of ethanol from corn starch is 15 billion gallons per year, implying a proposed mandate for production of some 20 billion gallons per year of cellulosic ethanol by 2017. Bush's proposed plan includes $2 billion funding (from 2007-2017?) for cellulosic ethanol plants, with an additional $1.6 billion (from 2007-2017?) announced by the USDA on January 27, 2007.

In March 2007, the US government awarded $385 million in grants aimed at jump-starting ethanol production from non-traditional sources like wood chips, switchgrass and citrus peels. Half of the six projects chosen will use thermo-chemical methods and half will use cellulosic ethanol methods.

The American company Range Fuels announced in July 2007 that it was awarded a construction permit from the state of Georgia to build the first commercial-scale 100-million-gallon-per-year cellulosic ethanol plant in the United States. Construction began in November, 2007.

The U.S. could potentially produce 1.3 billion dry tons of cellulosic biomass per year, which has the energy content of four billion barrels of crude oil. This translates to 65 per cent of American oil consumption.

Production Methods

There are two ways of producing ethanol from cellulose:

- Cellulolysis processes which consist of hydrolysis on pre-treated lignocellulosic materials, using enzymes to break complex cellulose into simple sugars such as glucose and followed by fermentation and distillation.
- Gasification that transforms the lignocellulosic raw material into gaseous carbon monoxide and hydrogen. These gases can be converted to ethanol by fermentation or chemical catalysis.

They both include distillation as the final step to isolate the pure ethanol.

Cellulolysis (Biological Approach)

There are four or five stages to produce ethanol using a biological approach:

1. A "pre-treatment" phase, to make the lignocellulosic material such as wood or straw amenable to hydrolysis,
2. Cellulose hydrolysis (cellulolysis), to break down the molecules into sugars;
3. Separation of the sugar solution from the residual materials, notably lignin;
4. Microbial fermentation of the sugar solution;
5. Distillation to produce roughly 95 per cent pure alcohol.
6. Dehydration by molecular sieves to bring the ethanol concentration to over 99.5 per cent

Pre-treatment

Although lignocellulose is the most abundant plant material resource, its susceptibility has been curtailed by its rigid structure. As the result, an effective pre-treatment is needed to liberate the cellulose from the lignin seal and its crystalline structure so as to render it accessible for a subsequent hydrolysis step. By far, most pre-treatment are done through physical or chemical means. In order to achieve higher efficiency, both physical and chemical pre-treatment are required. Physical pre-treatment is often called size reduction to reduce biomass physical size. Chemical pre-treatment is to remove chemical barriers so that the enzymes can access to cellulose for microbial destruction.

To date, the available pre-treatment techniques include acid hydrolysis, steam explosion, ammonia fiber expansion, organosolve, sulfite pre-treatment to overcome recalcitrance of lignocellulsoe (SPORL), alkaline wet oxidation and ozone pre-treatment. Besides effective cellulose liberation, an ideal pre-treatment has to minimize the formation of degradation products because of their inhibitory effects on subsequent hydrolysis and fermentation processes. The presence of inhibitors will not only further complicate the ethanol production but also increase the cost of production due to entailed detoxification steps. Even though pre-treatment by acid hydrolysis is probably the oldest and most studied pre-treatment technique, it produces several potent inhibitors including furfural and hydroxymethyl furfural (HMF) which are by far regarded as the most toxic inhibitors present in lignocellulosic hydrolysate. Ammonia Fiber Expansion (AFEX) is a promising pre-treatment with no inhibitory effect in resulting. hydrolysate.

Most pre-treatment processes are not effective when applied to feedstocks with high lignin content, such as forest biomass. Organosolve and SPORL are the only two processes that can achieve over 90 per cent cellulsoe conversion for forest biomass, especially those of softwood species. SPORL is the most energy efficient (sugar production per unit energy consumption in pre-treatment) and robust process for pre-treatment of forest biomass with very low production of fermentation inhibitors.

Cellulolytic Processes

The cellulose molecules are composed of long chains of sugar molecules. In the hydrolysis process, these chains are broken down to free the sugar, before it is fermented for alcohol production.

There are two major cellulose hydrolysis (cellulolysis) processes: a chemical reaction using acids, or an enzymatic reaction.

Chemical Hydrolysis: In the traditional methods developed in the 19th century and at the beginning of the 20th century, hydrolysis is performed by attacking the cellulose with an acid. Dilute acid may be used under high heat and high pressure, or more concentrated acid can be used at lower temperatures and atmospheric pressure. A decrystalized cellulosic mixture of acid and sugars reacts in the presence of water to complete individual sugar molecules (hydrolysis). The product from this hydrolysis is then neutralized and yeast fermentation is used to produce ethanol. As mentioned, a significant obstacle to the dilute acid process is that the hydrolysis is so harsh that toxic degradation products are produced that can interfere with fermentation. Concentrated acid must be separated from the sugar stream for recycle (simulated moving bed (SMB) chromatographic separation for example) to be commercially attractive.

Enzymatic Hydrolysis: Cellulose chains can be broken into glucose molecules by cellulase enzymes.

This reaction occurs at body temperature in the stomach of ruminants such as cows and sheep, where the enzymes are produced by bacteria. This process uses several enzymes at various stages of this conversion. Using a similar enzymatic system, lignocellulosic materials can be enzymatically hydrolyzed at a relatively mild condition (50°C and pH5), thus enabling effective cellulose breakdown without the formation of byproducts that would otherwise inhibit enzyme activity. All major pre-treatment methods, including dilute acid pre-treatment, require an enzymatic hydrolysis step to achieve high sugar yield for ethanol fermentation. Currently, most pre-treatment studies have been laboratory based, but companies are rapidly exploring means to transition from the laboratory to pilot, or production scale.

Various enzyme companies have also contributed significant technological breakthroughs in cellulosic ethanol through the mass production of enzymes for hydrolysis at competitive prices.

The fungus Trichoderma reesei is used by Iogen Corporation, to secrete "specially engineered enzymes" for an enzymatic hydrolysis process. The raw material (wood or straw) has to be pre-treated to make it amenable to hydrolysis.

Another Canadian company, SunOpta markets a patented technology known as "Steam Explosion" to pre-treat cellulosic biomass, overcoming its "recalcitrance" to make cellulose and hemicellulose accessible to enzymes for conversion into fermentable sugars. SunOpta designs and engineers cellulosic ethanol biorefineries and its process technologies and equipment are in use in the first 3 commercial demonstration plants in the world: Verenium (formerly Celunol Corporation)'s facility in Jennings, Louisiana, Abengoa's facility in Salamanca, Spain, and a facility in China owned by China Resources Alcohol Corporation (CRAC). The CRAC facility is currently producing cellulosic ethanol from local corn stover on a 24-hour a day basis using SunOpta's process and technology.

Genencor and Novozymes are two other companies that have received United States Department of Energy funding for research into reducing the cost of cellulases, key enzymes in the production of cellulosic ethanol by enzymatic hydrolysis.

Other enzyme companies, such as Dyadic International, are developing genetically engineered fungi which would produce large volumes of cellulase, xylanase and hemicellulase enzymes which can be used to convert agricultural residues such as corn stover, distiller grains, wheat straw and sugar cane bagasse and energy crops such as switch grass into fermentable sugars which may be used to produce cellulosic ethanol.

Verenium, formed by the merger of Diversa and Celunol, operates a pilot cellulosic ethanol plant in Jennings, Louisiana and is building a 1.4 million gallon per year demonstration plant on adjacent land to be completed by the end of 2007 and begin operation in early 2008. Verenium is the first publicly traded company with integrated, end-to-end capabilities to make cellulosic biofuels.

KL Energy Corporation, formerly KL Process Design Group, began commercial operation of a 1.5 million gallon per year cellulosic ethanol facility in Upton, WY in the last quarter of 2007. The Western Biomass Energy facility is currently achieving yields of 40-45 gallons per bone dry ton. It is the first operating commercial cellulosic ethanol facility in the nation. The KL Energy process uses a thermo-mechanical breakdown and enzymatic conversion process. The primary feedstock is soft wood, however, lab tests have already proven the KL Energy process on wine pomace, sugarcane bagasse, municipal solid waste, and switch grass.

Microbial Fermentation

Traditionally, baker's yeast (*Saccharomyces cerevisiae*), has long been used in brewery industry to produce ethanol from hexoses (6-carbon sugar). Due to the complex nature of the carbohydrates present in lignocellulosic biomass, a significant amount of xylose and arabinose (5-carbon sugars derived from the hemicellulose portion of the lignocellulose) is also present in the hydrolysate. For example, in the hydrolysate of corn stover, approximately 30 per cent of the total fermentable sugars is xylose. As a result, the ability of the fermenting microorganisms to use the whole range of sugars available from the hydrolysate is vital to increase the economic competitiveness of cellulosic ethanol and potentially bio-based chemicals.

In recent years, metabolic engineering for microorganisms used in fuel ethanol production has shown significant progress. Besides *Saccharomyces cerevisiae*, microorganisms such as *Zymomonas mobilis* and *Escherichia coli* have been targeted through metabolic engineering for cellulosic ethanol production.

Recently, engineered yeasts have been described efficiently fermenting xylose , and arabinose, and even both together. Yeast cells are especially attractive for cellulosic ethanol processes as they have been used in biotechnology for hundreds of years, as they are tolerant to high ethanol and inhibitor concentrations and as they can grow at low pH values which avoids bacterial contaminations.

Combined Hydrolysis and Fermentation

Some species of bacteria have been found capable of direct conversion of a cellulose substrate into ethanol. One example is *Clostridium thermocellum*, which uses a complex cellulosome to break down cellulose and synthesize ethanol. However, *C. thermocellum* also produces other products during cellulose metabolism, including acetate and lactate, in addition to ethanol, lowering the efficiency of the process. Some research efforts are directed to optimizing ethanol production by genetically engineering bacteria that focus on the ethanol-producing pathway.

Gasification Process (Thermochemical Approach)

The gasification process does not rely on chemical decomposition of the cellulose chain (cellulolysis). Instead of breaking the cellulose into sugar molecules, the carbon in the raw material is converted into synthesis gas, using what amounts to partial combustion. The carbon monoxide, carbon dioxide and hydrogen may then be fed into a special kind of fermenter. Instead of sugar fermentation with yeast, this process uses a microorganism named *Clostridium ljungdahlii*. This microorganism will ingest (eat) carbon monoxide, carbon dioxide and hydrogen and produce ethanol and water. The process can thus be broken into three steps:

1. *Gasification*—Complex carbon based molecules are broken apart to access the carbon as carbon monoxide, carbon dioxide and hydrogen are produced
2. *Fermentation*—Convert the carbon monoxide, carbon dioxide and hydrogen into ethanol using the *Clostridium ljungdahlii* organism
3. *Distillation*—Ethanol is separated from water

A recent study has found another *Clostridium* bacterium that seems to be twice as efficient in making ethanol from carbon monoxide as the one mentioned above.

Alternatively, the synthesis gas from gasification may be fed to a catalytic reactor where the synthesis gas is used to produce ethanol and other higher alcohols through a thermochemical process. This process can also generate other types of liquid fuels, an alternative concept under investigation by at least one biofuels company.

Economics

Cellulosic ethanol has the potential to become a competitive energy resource, but currently requires extra financial support to develop the infrastructure necessary to the technology.

Construction of pilot scale lignocellulosic ethanol plants requires considerable financial support through grants and subsidies. On 28 February 2007, the U.S. Dept. of Energy announced $385 million in grant funding to six cellulosic ethanol plants. This grant funding accounts for 40 per cent of the investment costs. The remaining 60 per cent comes from the promoters of those facilities. Hence, a total of $1 billion will be invested for approximately 140 million gallon capacity. This translates into $7/annual gallon production capacity in capital investment costs for pilot plants (this would work out to $.35/ gal over the 20-year life of a facility); future capital costs are expected to be lower. Corn to ethanol plants cost roughly $1—

3/annual gallon capacity, though the cost of the corn itself is considerably greater than for switchgrass or waste biomass. As of 2007, ethanol is produced mostly from sugars or starches, obtained from fruits and grains. In contrast, cellulosic ethanol is obtained from cellulose, the main component of wood, straw and much of the structure of plants. Since cellulose cannot be digested by humans, the production of cellulose does not compete with the production of food, other than conversion of land from food production to cellulose production (which has recently started to become an issue, due to rising wheat prices.) The price per ton of the raw material is thus much cheaper than grains or fruits. Moreover, since cellulose is the main component of plants, the whole plant can be harvested. This results in much better yields—up to 10 short tons per acre (22 t/ha), instead of 4 or 5 short tons per acre (9—11 t/ha) for the best crops of grain.

The raw material is plentiful. Cellulose is present in every plant, in the form of straw, grass, and wood. It is estimated that 323 million tons of cellulose containing raw materials that could be used to create ethanol are thrown away each year in US alone. This includes 36.8 million dry tons of urban wood wastes, 90.5 million dry tons of primary mill residues, 45 million dry tons of forest residues, and 150.7 million dry tons of corn stover and wheat straw. Transforming them into ethanol using efficient and cost effective hemi (cellulase) enzymes or other processes might provide as much as 30 per cent of the current fuel consumption in the United States. Moreover, even land marginal for agriculture could be planted with cellulose-producing crops like switchgrass, resulting in enough production to substitute for all the current oil imports into the United States.

Paper, cardboard, and packaging comprise a substantial part of the solid waste sent to landfills in the United States each day, 41.26 per cent of all organic municipal solid waste (MSW) according to California Integrated Waste Management Board's city profiles. These city profiles account for accumulation of 612.3 short tons (555.5 t) daily per landfill where an average population density of 2,413 per square mile persists. Organic waste consists of 0.4 per cent manure, 1.6 per cent gypsum

Board, 4.2 per cent Glossy Paper, 4.2 per cent Paper Ledger, 9.2 per cent Wood, 10.5 per cent Envelopes, 11.9 per cent newsprint, 12.3 per cent grass and leaves, 30.0 per cent food scrap, 34.0 per cent office paper, 35.2 per cent corrugated cardboard, and 46.4 per cent agricultural composites, makes up 71.51 per cent of land fill. All these except Gypsum Board contain cellulose which is transformable into cellulosic ethanol because they are the leading cause of methane plumes. Methane, a greenhouse gas, is 21 times more potent than carbon-dioxide.

Reduction of the disposal of solid waste through cellulosic ethanol conversion would reduce solid waste disposal costs by local and state governments. It is estimated that each person in the US throws away 4.4 lb (2.0 kg) of trash each day, of which 37 per cent contains waste paper which is largely cellulose. That computes to 244 thousand tons per day of discarded waste paper that contains cellulose. The raw material to produce cellulosic ethanol is not only free, it has a negative cost—i.e., ethanol producers can get paid to take it away.

In June 2006, a U.S. Senate hearing was told that the current cost of producing cellulosic ethanol is US $2.25 per US gallon (US $0.59/litre). This is primarily due to the current poor conversion efficiency. At that price it would cost about $120 to substitute a barrel of oil (42 gallons), taking into account the lower energy content of ethanol. However, the Department of Energy is optimistic and has requested a doubling of research funding. The same Senate hearing was told that the research target was to reduce the cost of production to US $1.07 per US gallon (US $0.28/litre) by 2012. "The production of cellulosic ethanol represents not only a step toward true energy diversity for the country, but a very cost-effective alternative to fossil fuels. It is advanced weaponry in the war on oil," said Vinod Khosla, managing partner of Khosla Ventures, who recently told a Reuters Global Biofuels Summit that he could see cellulosic fuel prices sinking to $1 per gallon within ten years.

University of Massachusetts at Amherst researchers have developed a streamlined technique which uses "catalytic fast pyrolysis" (heating to 400—600 °C followed by rapid cooling) and zeolite as a catalyst to produce cellulosic ethanol in about

60 seconds. They estimate improvements in the process should be able to generate ethanol at the equivalent of $1—$1.70/gal of gasoline. As of April 2008, the process has only been developed to work at laboratory scales.

Environmental Effects: Corn-based vs. Grass-based

In 2008, there was only a small amount of switchgrass dedicated for ethanol production. In order for it to be grown on a large-scale production it must compete with existing uses of agricultural land, mainly for the production of crop commodities. Of the United States' 2.26 billion acres (9.1 million km²) of unsubmerged land, 33 per cent are forestland, 26 per cent pastureland and grassland, and 20 per cent crop land. A study done by the U.S. Departments of Energy and Agriculture in 2005 determined whether there were enough available land resources to sustain production of over 1 billion dry tons of biomass annually to replace 30 per cent or more of the nation's current use of liquid transportation fuels. The study found that there could be 1.3 billion dry tons of biomass available for ethanol use, by making little changes in agricultural and forestry practices and meeting the demands for forestry products, food, and fiber. A recent study done by the University of Tennessee reported that as many as 100 million acres (400,000 km², or 154,000 sq mi) of cropland and pasture will need to be allocated to switchgrass production in order to offset petroleum use by 25 percent.

Currently, corn is easier and less expensive to process into ethanol in comparison to cellulosic ethanol. The Department of Energy estimates that it costs about $2.20 per gallon to produce cellulosic ethanol, which is twice as much as ethanol from corn. Enzymes that destroy plant cell wall tissue cost 30 to 50 cents per gallon of ethanol compared to 3 cents per gallon for corn. The Department of Energy hopes to reduce this cost to $1.07 per gallon by 2012 to be effective. However, cellulosic biomass is cheaper to produce than corn, because it requires fewer inputs, such as energy, fertilizer, herbicide, and is accompanied by less soil erosion and improved soil fertility. Additionally, nonfermentable and unconverted solids left after making ethanol can be burned to provide the fuel needed to operate the conversion plant and produce electricity. Energy used to run corn-based ethanol plants is derived from coal and natural gas. The Institute for Local Self-Reliance estimates the cost of cellulosic ethanol from the first generation of commercial plants will be in the $1.90—$2.25 per gallon range, excluding incentives. This compares to the current cost of $1.20—$1.50 per gallon for ethanol from corn and the current retail price of over $4.00 per gallon for regular gasoline (which is subsidized and taxed).

One of the major reasons for increasing the use of biofuels is to reduce greenhouse gas emissions. In comparison to gasoline, ethanol burns cleaner with a greater efficiency, thus putting less carbon dioxide and overall pollution in the air. Additionally, only low levels of smog are produced from combustion. According to the U.S. Department of Energy, ethanol from cellulose reduces green house gas emission by 90 percent, when compared to gasoline and in comparison to corn-based ethanol which decreases emissions by 10 to 20 percent. Carbon dioxide gas emissions are shown to be 85 per cent lower than those from gasoline. Cellulosic ethanol contributes little to the greenhouse effect and has a five times better net energy balance than corn-based ethanol. When used as a fuel, cellulosic ethanol releases less sulfur, carbon monoxide, particulates, and greenhouse gases. Cellulosic ethanol should earn producers carbon reduction credits, higher than those given to producers who grow corn for ethanol, which is about 3 to 20 cents per gallon.

It takes 0.76 J of energy from fossil fuels to produce 1 J worth of ethanol from corn. This total includes the use of fossil fuels used for fertilizer, tractor fuel, ethanol plant operation, etc. Research has shown that 1 gallon of fossil fuel can produce over 5 gallons of ethanol from prairie grasses, according to Terry Riley, President of Policy at the Theodore Roosevelt Conservation Partnership. The United States Department of Energy concludes that corn-based ethanol provides 26 percent more energy than it requires for production, while cellulosic ethanol provides 80 percent more energy. Cellulosic ethanol yields 80 percent more energy than is required to grow and convert it. The process of turning corn into ethanol requires about 1,700 gallons of water for every 1 gallon of ethanol produced. Additionally, each gallon of ethanol leaves behind 12 gallons of waste that must be disposed. Grain ethanol uses only the edible portion of the plant. Expansion of corn acres for the production of ethanol poses threats to biodiversity. Corn lacks a strong root system, therefore, when produced, it causes soil erosion. This has a direct effect on soil particles, along with excess fertilizers and other chemicals, washing into local waterways, damaging water quality and harming aquatic life. Planting riparian areas can serve as a buffer to waterways, and decrease runoff.

Cellulose is not used for food and can be grown in all parts of the world. The entire plant can be used when producing cellulosic ethanol. Switchgrass yields twice as much ethanol per acre than corn. Therefore, less land is needed for production and thus less habitat fragmentation. Biomass materials require fewer inputs, such as fertilizer, herbicides, and other chemicals that can pose risks to wildlife. Their extensive roots improve soil quality, reduce erosion, and increase nutrient capture. Herbaceous energy crops reduce soil erosion by greater than 90 per cent, when compared to conventional commodity crop production. This can translate into improved water quality for rural communities. Additionally, herbaceous energy crops add organic material to depleted soils and can increase soil carbon, which can have a direct effect on climate change, as soil carbon can absorb carbon dioxide in the air. As compared to commodity crop production, biomass reduces surface runoff and nitrogen transport. Switchgrass provides an environment for diverse wildlife habitation, mainly insects and ground birds. Conservation Resource Programme (CRP) land is composed of perennial grasses, which are used for cellulosic ethanol, and may be available for use.

For years American farmers have practiced row cropping, with crops such as sorghum and corn. Because of this, much is known about the effect of these practices on wildlife. The most significant effect of increased corn ethanol would be the additional land that would have to be converted to agricultural use and the increased erosion and fertilizer use that goes along with agricultural production. Increasing our ethanol production through the use of corn could produce negative effects on wildlife, the magnitude of which will depend on the scale of production and whether the land used for this increased production was formerly idle, in a natural state, or planted with other row crops. Another consideration is whether to plant a switchgrass monoculture or use a variety of grasses and other vegetation. While a mixture of vegetation types likely would provide better wildlife habitat, the technology has not yet developed to allow the processing of a mixture of different grass species or vegetation types into bioethanol. Of course, cellulosic ethanol production is still in its infancy, and the possibility of using diverse vegetation stands instead of monocultures deserves further exploration as research continues.

Feedstocks

In general there are two types of feedstocks: **forest (woody) Biomass** and **agricultural biomass**. In the US, about 1.4 billion dry tons of biomass can be sustainabily produced annually. About 370 million tons or 30 per cent are forest biomass . Forest biomass has higher cellulose and lignin content and lower hemicellulose and ash content than agricultural biomass. Because of the difficulties and low ethanol yield in fermenting pretreatment hydrolysate, especially those with very high 5 carbon hemicellulsoe sugars such as xylose, forest biomass has significant advantages over agricultural biomass. Forest biomass also has high density which significantly reduces transportation cost. It can be harvested year around which eliminates long term storage. The close to zero ash content of forest biomass significantly reduces dead load in transportation and processing. To meet the needs for biodiversity, forest biomass will be an important biomass feestock supply mix in the future biobased economy. However, forest biomass is much more recalcitrant than agricultural biomass. Recently, the USDA Forest Products Laboratory together with the University of Wisconsin at Madison developed efficient technologies   that can overcome the strong recalcitrance of forest (woody) biomass including those of softwood species that have low xylan content. Short-rotation intensive culture or tree farming can offer an almost unlimited opportunity for forest biomass production.

wood chips from slashes and tree tops and saw dust from saw mills, and waste paper pulp are common forest biomass feedstocks for cellulosic ethanol production .

The following are a few examples of agricultural biomass:

Switchgrass (*Panicum virgatum*) is a native tallgrass prairie grass. Known for its hardiness and rapid growth, this perennial grows during the warm months to heights of 2—6 feet. Switchgrass can be grown in most parts of the United States, including swamplands, plains, streams, and along the shores and *interstate highways*. It is *self-seeding* (no tractor for sowing, only for mowing), resistant to many diseases and pests, and can produce high yields with low applications of fertilizer and other chemicals. It is also tolerant to poor soils, flooding, and drought; improves soil quality and prevents erosion due its type of root system.

Switchgrass is an approved cover crop for land protected under the federal Conservation Reserve Programme (CRP). CRP is a government Programme that pays producers a fee for not growing crops on land on which crops recently grew. This Programme reduces soil erosion, enhances water quality, and increases wildlife habitat. CRP land serves as a habitat for upland game, such as pheasants and ducks, and a number of insects. Switchgrass for biofuel production has been considered for use on Conservation Reserve Programme (CRP) land, which could increase ecological sustainability and lower the cost of the CRP Programme. However, CRP rules would have to be modified to allow this economic use of the CRP land.

Miscanthus x giganteus is another viable feedstock for cellulosic ethanol production. This species of grass is native to Asia and is the sterile triploid hybrid of *miscanthus sinensis* and *miscanthus sacchariflorus*. It can grow up to 12 feet (3.7 m) tall with little water or fertilizer input. Miscanthus is similar to switchgrass with respect to cold and drought tolerance and water use efficiency. Miscanthus is commercially grown in the European Union as a combustible energy source.

Corn cobs especially **corn stovers** are the most popular agricultural biomass.

Cellulosic Ethanol Commercialization

Cellulosic ethanol commercialization is the process of building an industry out of methods of turning cellulose-containing organic matter into fuel. Companies such as Iogen, Broin, and Abengoa are building refineries that can process biomass and turn it into ethanol, while companies such as Genencor, Diversa, Novozymes, and Dyadic are producing enzymes which could enable a cellulosic ethanol future. The shift from food crop feedstocks to waste residues and native grasses offers significant opportunities for a range of players, from farmers to biotechnology firms, and from project developers to investors.

The cellulosic ethanol industry developed some new commercial-scale plants in 2008. In the United States, plants totalling 12 million liters (3.17 million gal) per year were operational, and an additional 80 million liters (21.13 million gal.) per year of capacity—in 26 new plants—was under construction. In Canada, capacity of 6 million liters per year was operational. In Europe, several plants were operational in Germany, Spain, and Sweden, and capacity of 10 million liters per year was under construction.

Commercial Cellulosic Ethanol Plants in the U.S.
(Operational or under construction)

Company	Location	Feedstock
Abengoa Bioenergy	Hugoton, KS	Wheat straw
BlueFire Ethanol	Irvine, CA	Multiple sources
Colusa Biomass Energy Corporation	Sacramento, CA	Waste rice straw
DuPont Danisco	Vonore, TN	Corn cobs
Fulcrum BioEnergy	Reno, NV	Municipal solid waste
Gulf Coast Energy	Mossy Head, FL	Wood waste
KL Energy Corp.	Upton, WY	Wood
Mascoma	Lansing, MI	Wood
POET LLC	Emmetsburg, IA	Corn cobs
Range Fuels	Treutlen County, GA	Wood waste
SunOpta	Little Falls, MN	Wood chips
US Envirofuels	Highlands County, FL	Sweet sorghum
Xethanol	Auburndale, FL	Citrus peels

CASE STUDY: VERENIUM CORPORATION

Verenium Corporation, (NASDAQ: VRNM) is a Cambridge, Massachusetts-based biotechnology company that specialized in metagenomics and the development of high performance specialty enzymes for applications in biofuels and other industrial applications. Formerly separate firms Diversa and Celunol, Verenium Corporation was created by the former two firms' merger in 2007.

According to the *Economist* article listed in the "References" section, Diversa prospects in hot springs, ocean beds, soda lakes, and on the Arctic tundra for genes potentially useful in industry.

In February 2007, Diversa announced a merger with Celunol, a leading corporation in the emerging cellulosic ethanol industry. Celunol had received substantial backing from Silicon Valley venture capitalist Vinod Khosla. The buyout was financed by approximately 15 million shares of Diversa stock plus up to $20 million in debt financing to fund Celunol operations. After the merger, Khosla became one of the largest single shareholders of the new company.

Cellulosic Ethanol Production

Verenium's demonstration scale biorefinery, built to produce 1.4 million gallons of ethanol a year from cellulosic biomass opened in Jennings, LA on the 29th of May 2008. The plant makes ethanol from agricultural waste left over from processing sugarcane.

CASE STUDY: COSKATA, INC.

Coskata, Inc. is a Warrenville, Illinois based energy company researching the production of cellulosic ethanol from woodchips. Costaka's process combines both biological (i.e., microbes) and thermochemical (heat and chemicals) processing. The estimated cost of production via this technology is under $1 per gallon, as opposed to corn-based ethanol costing approximately $1.40 per gallon.

Coskata announced in April 2008, that the company would begin producing ethanol on a small scale at a plant being built near Pittsburgh, PA. With a capacity of about 40,000 gallons annually, the fuel will be used by General Motors to be tested in their vehicles. The pilot plant is being constructed in a modular design by Zeton Inc. in Burlington, Ontario, Canada. A full scale production plant capable of producing 50 to 100 million gallons of cellulosic ethanol is expected to go online in 2011.

Coskata recently signed a deal with US Sugar Corporation to build a cane-waste biofuels conversion facility in Florida.

CASE STUDY: DUPONT DANISCO

DuPont Danisco Cellulosic Ethanol LLC

Type	LLC Joint venture
Founded	July 21, 2008
Headquarters	Delaware, USA
Key people	Joseph R. Skurla, President
Parent	DuPont, Danisco
Website	http://www.ddce.com/

DuPont Danisco Cellulosic Ethanol LLC is a 50/50 joint venture between DuPont and Genencor, a subsidiary of Danisco. The $140 million venture seeks to commercialize technology for production of ethanol from non-edible parts of plants and other biomass, otherwise known as cellulosic ethanol, and to eventually license it to ethanol producers.

The company in partnership with the University of Tennessee Research Foundation plans to construct a pilot-scale bio-refinery and research and development facility for cellulosic ethanol in Vonore, Tennessee. The plant is expected to be in operation by the end of 2009.

OPTION OF BIOETHANOL FOR SUSTAINABLE TRANSPORT

BEST Programme ED95 trial bus operating in São Paulo, Brazil.

BioEthanol for Sustainable Transport (BEST) is a project financially supported by the European Union for promoting the introduction and market penetration of bioethanol as a vehicle fuel, and the introduction and wider use of flexible-fuel vehicles and ethanol cars on the world market. The project began in 2006 and will continue until the end of 2009, and has nine participating regions or cities in Europe, Brazil, and China.

Goals

President Luiz Inácio Lula da Silva and King Carl XVI Gustaf of Sweden inspecting one of the 400 buses running on ED95 on Stockholm.

The BEST project targets include the introduction of more than 10,000 flex-fuel or ethanol cars and 160 ethanol buses; promote the opening of 135 E85 and 13 ED95 public fuel stations; and to promote the development and testing of low ethanol blends with gasoline and diesel.

Participants

There are nine participating cities or regions and several commercial partners. Stockholm (Sweden) is the coordinating city, and other participants are Basque Country and Madrid (Spain), the Biofuel Region in Sweden, Brandenburg (Germany), La Spezia (Italy), Nanyang (China), Rotterdam (Netherlands), São Paulo (Brazil), and Somerset (UK). The commercial partners are Ford Europe, Saab Automobile and several bioethanol suppliers.

Implemented Projects

Under the auspices of the BEST project, the first ED95 bus began operations in São Paulo city on December 2007 as a trial project. The bus is a Scania with a modified diesel engine capable of running with 95 per cent hydrous ethanol with 5 per cent ignition improver. Scania adjusted the compression ratio from 18:1 to 28:1, added larger fuel injection nozzles, and altered the injection timing.

CASE STUDY: BLUE FLINT ETHANOL

Blue Flint Ethanol is an bioethanol producing company and a production plant with a same name, located in Underwood,

North Dakota approximately 50 miles (80 km) north of Bismarck. The plant is unique in the fact that rather than burning fuel such as coal or natural gas to drive the production process, waste heat from electrical generation at the Coal Creek Station is used. The US$100 million dollar plant is capable of processing 18 million bushels of corn to produce 50 million gallons of ethanol each year, and is estimated to have a net annual economic impact of $160 million on the North Dakota economy, as well as the creation of approximately 40 new jobs to run the plant. In addition to producing ethanol the plant will also produce dry distillers grains, a byproduct of the distillation process which is used as animal feed. Of the 18 million bushels of corn used each year for feedstock, the majority will be grown in southeast North Dakota and brought in via rail, with the remaining one third being produced locally. In 2006 the Blue Flint Ethanol project was awarded the Project of the Year Award by Governor John Hoeven.

History

The Coal Creek power station which provides the thermal input to the process began construction in 1974, becoming fully operational for electricity production in 1980. The power plant had already been using heat from some of the steam to dry the lignite coal it burns as fuel, improving its efficiency, however 60 per cent of the steam's heat energy was still being lost. In 2005 Headwaters Incorporated, in conjunction with Great River Energy, announced their plans for the construction of an ethanol production facility collocated with the Coal Creek Power plant. The plant will be majority owned by Headwaters who will manage the construction and plant operations, with GRE owning a minority share of the plant. Construction of the facility took place during 2006 with ethanol production beginning in February 2007.

Similar Facilities

Although the Blue Flint Ethanol plant is the first in the US to utilize waste heat for ethanol production, other plants are starting to pursue similar strategies. Great River Energy is also building a 62 megawatt baseload (37 megawatt peaking capacity) coal fired boiler in Spiritwood, North Dakota which will produce both electricity, malt, and 100 million gallons of ethanol yearly. Additionally a 20 megawatt plant in Goodland, Kansas will produce electricity, ethanol, and biodiesel.

EVALUATING ETHANOL FUEL IN THE UNITED STATES

The United States is the world's largest producer of ethanol fuel since 2005. The U.S. produced 9.0 billion U.S. liquid gallons of ethanol fuel in 2008, and together with Brazil, both countries accounted for 89 per cent of the world's production in that year. Ethanol fuel is mainly used in the U.S. as an oxygenate to gasoline in the form of low-level blends, and to a lesser extent, as fuel for E85 flex-fuel vehicles. Most ethanol fuel in the U.S. is produced using corn as feedstock.

As of 2007, ethanol market share was about 3 per cent of the U.S. gasoline-vehicle fuel consumption, but domestic production capacity has increased tenfold since 1990, from 900 million of gallons back then, through 1,630 million in 2000, reaching 9,000 million gallons in 2008. Most cars on the road today in the U.S. can run on blends of up to 10 per cent ethanol, and motor vehicle manufacturers already produce vehicles designed to run on much higher ethanol blends. Portland, Oregon, recently became the first city in the United States to require all gasoline sold within city limits to contain at least 10 per cent ethanol Flexible-fuel cars, trucks, and minivans can use gasoline and ethanol blends ranging from pure gasoline up to 85 per cent ethanol (E85). By early 2009, there were approximately eight million E85-compatible vehicles on U.S. roads.

The Renewable Fuels Association reports 170 U.S. ethanol distilleries in operation and another 24 under construction as of January 2009. Ethanol production is likely to soar over the next several years, since the Energy Policy Act of 2005 set a renewable fuels standard mandating 7.5 billion gallons of annual domestic renewable-fuel production by 2012. Former president George W. Bush sought to generate a western-hemisphere dominated industry that can produce as much as 35 billion gallons (130 billion liters) a year, equal to the entire world's production as of 2007. Expanding ethanol industries provide jobs in plant construction, operations, and maintenance, mostly in rural communities. However, by early 2009, the industry is under financial stress due to the effects of the economic crisis of 2008 as motorists are driving less, gasoline prices have dropped sharply, there is excess production capacity, and less financing available.

Since most U.S. ethanol is produced from corn and the required electricity from many distilleries comes mainly from coal plants, there has been considerable debate about how sustainable corn-based bio-ethanol could be in replacing fossil fuels in vehicles. Controversy and concerns relate to the large amount of arable land required for crops and its impact on grain supply, direct and indirect land use change effects, as well as issues regarding its energy balance and carbon intensity considering the full life cycle of ethanol production. Recent developments with cellulosic ethanol production and commercialization may allay some of these concerns.

History of Ethanol in the US

In 1826, Samuel Morey, experimented with a prototypical internal combustion engine that used ethanol (combined with turpentine and ambient air then vaporized) as fuel. At the time, his discovery was overlooked mostly due to the success of steam power. And while ethanol was known of for decades, it received little attention as a fuel until 1860 when Nicholas Otto began experimenting with internal combustion engines. In 1859, oil was found in Pennsylvania which provided a new supply of fuel for the United States. A popular fuel in the U.S. before petroleum was a blend of alcohol and turpentine called "camphene", also known as "burning fluid." With the discovery of a ready supply of oil, kerosene's popularity grew.

in 1896, Henry Ford designed his first car, the "Quadricycle" to run on pure ethanol. Then in 1908, he produced the famous

Ford Model T capable of running on gasoline, ethanol or a combination of both Ford continued to advocate for ethanol as fuel even during the prohibition, but cheaper oil caused gasoline to prevail.

Gasoline containing up to 10 per cent ethanol has been in increasing use in the United States since the late 1970s. The demand for ethanol fuel produced from field corn was spured by the discovery that methyl tertiary butyl ether (MTBE) was contaminating groundwater. MTBE use as a oxygenate additive was widespread due to mandates of the Clean Air Act amendments of 1992 to reduce carbon monoxide emissions. As a result, MTBE use in gasoline was banned in almost 20 states by 2006. There was also concern that widespread and costly litigation might be taken against the U.S. gasoline suppliers, and a 2005 decision refusing legal protection for MTBE, opened a new market for ethanol fuel, the primary substitute for MTBE. At a time when corn prices were around US$2 a bushel, corn growers recognized the potential of this new market and delivered accordingly. This demand shift took place at a time when oil prices were already significantly rising. By 2006, about 50 percent of the gasoline used in the U.S. and more than 85 percent of Hawaii's gasoline contains ethanol at different proportions.

Current Trends

The world's top ethanol fuel producers in 2008 were the United States with 9,000 billion U.S. liquid gallons (bg) and Brazil (6.47 bg), accounting for 89 percent of world production of 17.33 billion US gallons (65.6 million liters). By early 2009, the U.S. ethanol production industry consisted of 170 plants operating in 26 states. The 2008 production from these plants was 38 percent higher over the previous year, and growth has been so steep that the U.S. surpassed Brazil as the worlds largest ethanol producer since 2005. Dozens more plants are under development and due to come on line in 2008 and 2009, increasing U.S. production capacity to nearly 12 billion gallons.

However, since late 2008 and early 2009, the industry has been under financial stress due to the effects of the economic crisis of 2008 as motorists are driving less, gasoline prices have dropped sharply, and corn prices have been fluctuating and remaining relatively high. There is also excess production capacity and less bank financing available. As a result, some plants are working below capacity, several firms have closed some of their plants while others have laid off staff, some firms filed for bankruptcy protection, including former industry leader VeraSun Energy, some new plants construction have been suspended, and several oil companies have been buying ethanol plant at discounted prices. The Energy Information Administration has raised concerns that the ethanol industry will not meet the targets set by Congress in 2007.

Most cars on the road today in the U.S. can run on blends of up to 10 per cent ethanol, and motor vehicle manufacturers already produce vehicles designed to run on much higher ethanol blends. In 2007, Portland, Oregon, recently became the first city in the United States to require all gasoline sold within city limits to contain at least 10 per cent ethanol. As of January 2008, three states—Missouri, Minnesota, and Hawaii—require ethanol to be blended with gasoline motor fuel. Florida made mandatory such blends by the end of 2010. Many cities are also required to use an ethanol blend due to non-attainment of federal air quality goals.

Ford, Chrysler, and GM are among the automobile companies that sell flexible-fuel vehicles that can run gasoline and ethanol blends ranging from pure gasoline up to 85 per cent ethanol (E85), and by 2008 almost any type of automobile and light duty vehicles is available in the market with the flex-fuel option, including sedans, vans, SUVs and pick-up trucks. By mid-2008, there were more than seven million E85-compatible vehicles on U.S. roads, though actual used of E85 fuel is limited, not only because the ethanol fueling infrastructures is limited, but also, as found by a 2005 survey, 68 per cent of American flex-fuel car owners were not aware they owned an E85 flex. This is due to the fact that the exterior of flex and non-flex vehicles look exactly the same; there is no sale price difference between them; the lack of consumer's awareness about E85s; and also the decision of American automakers of not putting any kind of exterior labeling, so buyers can be aware they are getting an E85 vehicle. In contrast, all Brazilian automakers clearly mark FFVs with badging or a high quality sticker in the exterior body, with a text that is some variant of the word Flex. As of 2007, many new FFV models in the US now feature a yellow cap to close the fueling line to remind drivers of the E85 capabilities, and GM is also using badging with the text "Flexfuel/ E85 Ethanol" to clearly mark the car as an E85 FFV.

E-85 flex-fuel vehicles are becoming increasingly common in the Midwest, where corn is a major crop and is the primary feedstock for ethanol fuel production. A major restriction hampering sales of E85 flex vehicles or fueling with E85, is the limited infrastructure available to sell E85 to the public, as by July 2008 there were only 1,706 gasoline filling stations selling E85 to the public in the entire US, with a great concentration of E85 stations in the Corn Belt states, lead by Minnesota with 353 stations, more than any other state, followed by Illinois with 181, and Wisconsin with 114. About another 200 stations that dispense ethanol are restricted to local city, state, and federal government vehicles without access to the general public.

The Energy Information Administration (EIA) predicts in its Annual Energy Outlook 2007 that ethanol consumption will reach 11.2 billion gallons by 2012, outstripping the 7.5 billion gallons required in the Renewable Fuel Standard that was enacted as part of the Energy Policy Act of 2005.

Expanding ethanol (and biodiesel) industries provide jobs in plant construction, operations, and maintenance, mostly in rural communities. According to the Renewable Fuels Association, the ethanol industry created almost 154,000 U.S. jobs in 2005 alone, boosting household income by $5.7 billion. It also contributed about $3.5 billion in tax revenues at the local, state, and federal levels.

Ethanol has less energy than an equivalent volume of gasoline, but can be produced from domestic renewable sources such as corn, sugar beets and sugar cane. There are also industrial processes which use ethanol as an intermediate or final product, mouthwash for example. Companies that serve industrial users of ethanol have been observing the green movement and have found ways to collect and reprocess ethanol. For instance, Veolia, an environmental services company, has been a pioneer in the field of gathering waste ethanol and producing it from other resources. This is a new alternative to the low yielding corn production. Waste ethanol is cleaned up through distillation, which increases the alcohol content to match ethanol produced from corn.

Ethanol is also produced from off specification alcoholic beverages and through the fermentation process of other off specification beverages and products such as out-of-date cola syrup. The resulting liquids are processed through the fermentation and distillation equipment. All the packaging, plastic, aluminium, steel, cardboard and wood pallets, is recycled. The final ethanol product is sold back in to the ethanol market where much of it is used as fuel. In the United States, the newest ethanol recovery plant opened in Medina, Ohio, in July 2008, by Veolia Environmental Services, confirming the green movement in alternative fuel production

Reduced Petroleum Imports and Costs

One rationale given for extensive ethanol production in the U.S. is its benefit to energy security, by shifting the need for some foreign-produced oil to domestically-produced energy sources. Production of ethanol requires significant energy, but current U.S. production derives most of that energy from coal, natural gas and other sources, rather than oil. Because 66 per cent of oil consumed in the U.S. is imported, compared to a net surplus of coal and just 16 per cent of natural gas (2006 figures), the displacement of oil-based fuels to ethanol produces a net shift from foreign to domestic U.S. energy sources.

According to a 2008 analysis by Iowa State University, the growth in US ethanol production has caused retail gasoline prices to be US $0.29 to US $0.40 per gallon lower than would otherwise have been the case.

Cellulosic Ethanol

In his State of the Union Address on January 31, 2006, President

George W. Bush stated, "We'll also fund additional research in cutting-edge methods of producing ethanol, not just from corn, but from wood chips and stalks or switchgrass. Our goal is to make this new kind of ethanol practical and competitive within six years." The U.S. Department of Energy released a report on July 7, 2006 with an ambitious new research agenda for the development of cellulosic ethanol as an alternative to gasoline. The 200-page scientific roadmap cites recent advances in biotechnology that have made cost-effective production of ethanol from cellulose, or inedible plant fiber, an attainable goal, with federal loan guarantees for new cellulosic biorefineries. The report outlines a detailed research plan for developing new technologies to transform cellulosic ethanol— a renewable, cleaner-burning, and carbon-neutral alternative to gasoline—into an economically viable transportation fuel. The Department of Energy has invested in research on enzymatic, thermochemical, acid hydrolysis, hybrid hydrolysis/enzymatic, and a variety of other approaches toward achieving success in discovering an efficient and low cost method of converting cellulose to ethanol.

President Bush's 2007 budget earmarked $150 million for the research effort—more than double the 2006 budget in Favour of the cellulosic lobby. Taxpayers and consumers are already shouldering part of the cost: each gallon of ethanol sold is subsidized by a 51-cent/gallon federal tax credit paid to U.S. producers. These subsidies, along with state incentive Programmes, cost the nation over $2 billion a year, leading legislators to pledge to invest in cellulosic ethanol. Another dampening factor is the short term loss of income to American refiners of crude oil. The U.S. market is especially lucrative, sometimes earning its refiners $30 or more on every barrel of crude oil they refine. Exxon Mobil Corp. earned $1.3 billion in its refining arm in the second quarter, up 11 per cent from a year before. The expectation, over the long run, is that the U.S. economy would more than earn its share back if our primary source of energy were manufactured and processed in the United States, but individual companies could be adversely affected.

Sugar-Based Ethanol

Technologically, the process of producing ethanol from sugar is simpler than converting corn into ethanol. Converting corn into ethanol requires additional cooking and the application of enzymes, whereas the conversion of sugar requires only a yeast fermentation process. The energy requirement for converting sugar into ethanol is about half that for corn. With sugar cane there is more than enough energy to do the conversion with energy left over. A 2006 USDA report found that at the current market prices for ethanol, converting sugarcane, sugar beets and molasses to ethanol would be profitable. Research is taking place to improve the productivity of sugarcane ethanol production by breeding new varieties adapted to the US soil and weather conditions, as well as to take advantage of cellulosic ethanol technologies to also convert waste sugarcane bagasse into ethanol.

Sugarcane production in the United States occurs in Florida, Louisiana, Hawaii, and Texas. The first three plants to produce sugar cane-based ethanol are expected to go online in Louisiana by mid 2009. Sugar mill plants in Lacassine, St. James and Bunkie were converted to sugar cane-based ethanol production using Colombian technology in-order to make possible a profitable ethanol production. These three plants will produce 100 million gallons of ethanol per year within five years.

By 2009 two other sugarcane ethanol production projects are being developed in Kauai, Hawaii and Imperial Valley, California. The Hawaiian plant will have a capacity to produce between 12 to 15 million gallons of ethanol a year and will supply the local market only, as sugar prices do not make it competitive for Hawaiian sugarcane ethanol to be sold in the continental US. This production plant is expected to go on line by 2010. The Californian plant will have a capacity to produce 60 million gallons a year and it is expected to go online in 2011.

Presidents George W. Bush and Luiz Inácio Lula da Silva during Bush's visit to Brazil, March 2007.

In March 2007, "ethanol diplomacy" was the focus of President George W. Bush's Latin American tour, in which he and Brazil's president, Luiz Inacio Lula da Silva, were seeking to promote the production and use of sugar cane based ethanol throughout the Caribbean Basin. The two countries also agreed to share technology and set international standards for biofuels. The Brazilian sugar cane technology transfer will permit various Central American, such as Honduras, El Salvador, Nicaragua, Costa Rica and Panama, several Caribbean countries, and various Andean Countries tariff-free trade with the U.S. thanks to existing concessionary trade agreements. The expectation is that using Brazilian technology for refining sugar cane based ethanol, such countries could become exporters to the United States in the short-term.

In 2007, Jamaica, El Salvador, Trinidad and Tobago and Costa Rica exported together to the U.S. a total of 230.5 million gallons of sugar cane-based ethanol, representing 54.1 per cent of U.S. fuel ethanol imports. Brasil began exporting ethanol to the U.S. in 2004 and exported 188.8 million gallons representing 44.3 per cent of U.S. ethanol imports in 2007. The remaining imports that year came from Canada and China.

Comparison with Brazilian Ethanol

Brazil's sugar cane-based ethanol industry is more efficient than the U.S. corn-based industry. Sugar cane ethanol has an energy balance 7 times greater than ethanol produced from corn. Brazilian distillers are able to produce ethanol for 22 cents per liter, compared with the 30 cents per liter for corn-based ethanol. U.S. corn-derived ethanol costs 30 per cent more because the corn starch must first be converted to sugar before being distilled into alcohol. Despite this cost differential in production, the U.S. does not import more Brazilian ethanol because of U.S. trade barriers corresponding to a tariff of 54-cent per gallon—a levy designed to offset the 45-cent per gallon blender's federal tax credit that is applied to ethanol no matter its country of origin. One advantage U.S. corn-derived ethanol offers is the ability to return 1/3 of the feedstock back into the market as a replacement for the corn used in the form of Distillers Dried Grain.

Comparison of Key Characteristics between the Ethanol Industries in the United States and Brazil

Characteristic	Brazil	U.S.	Units/Comments
Main feedstock	Sugar cane	Maize	Main cash crop for ethanol production, the US has less than 2% from other crops.
Total ethanol fuel production (2008)	6,472	9,000	Million U.S. liquid gallons
Total arable land	355	270[1]	Million hectares.
Total area used for ethanol crop (2006)	3.6 (1%)	10 (3.7%)	Million hectares (% total arable)
Productivity per hectare	6,800-8,000	3,800-4,000	Liters of ethanol per hectare. Brazil is 727 to 870 gal/acre (2006), US is 321 to 424 gal/acre (2003-05)
Energy balance (input energy productivity)	8.3 to 10.2	1.3 to 1.6	Ratio of the energy obtained from ethanol/energy expended in its production
Estimated greenhouse gas emission reduction	86-90%[2]	10-30%[2]	% GHGs avoided by using ethanol instead of gasoline, using existing crop land, without ILUC effects.
Full life-cycle carbon intensity	73.40	105.10[3]	Grams of CO_2 equivalent released per MJ of energy produced, includes indirect land use changes.
Estimated payback time for greenhouse gas emission	17 years[4]	93 years [4]	Brazilian cerrado for sugar cane and US grassland for com. Land use change scenarios by Fargione et al.
Flexible-fuel vehicle fleet (autos and light trucks)	7.5 million	8 million	Brazil as of April 2009 (FFVs use any blend up to E100). U.S. as of early 2009 (FFVs use E85).

(Contd.)

Characteristic	Brazil	U.S.	Units/Comments
Ethanol fueling stations in the country	35,017 (100%)	1,963 (1%)	As % of total gas stations in the country. Brazil by 2007-12, U.S. by 2009-03 (170,000 total.)
Ethanol's share within the gasoline market	50%[5]	4%	As % of total consumption on a volumetric basis. Brazil as of April 2008. US as of December 2006.
Cost of production (USD/gallon)	0.83	1.14	2006/2007 for Brazil (22¢/liter), 2004 for U.S. (35¢/liter)
Government subsidy (in USD)	0[6]	0.45/gallon	U.S. since 2009-01-01 as a tax credit. Brazilian ethanol production is no longer subsidized.[6]
Import tariffs (in USD)	0	0.54/gallon	As of June 2009, Brazil does not import any ethanol, the U.S. does.

Notes: (1) Only contiguous U.S., excludes Alaska. (2) Assuming no land use change. (3) CARB estimate for Midwest corn ethanol. California's gasoline carbon intensity is 95.86 blended with 10 per cent ethanol. (4) Assuming direct land use change. (5) If diesel-powered vehicles are included and due to ethanol's lower energy content by volume, bioethanol represented 16.9 per cent of the road sector energy consumption in 2007. (6) Brazilian ethanol production is no longer subsidized, but gasoline is heavily taxed Favouring ethanol fuel consumption (~54 per cent tax). By the end of July 2008, the average gasoline retail price at the pump in Brazil was USD 6.00 per gallon, while the average U.S. price was USD 3.98 per gallon. The latest gas retail price increase in Brazil occurred in late 2005, when oil price was at USD 60 per barrel.

Latest Developments

On April 23, 2009, the California Air Resources Board approved the specific rules and carbon intensity reference values for the California Low-Carbon Fuel Standard (LCFS) that will go into effect in January 1, 2011. During the consultation process there was controversy regarding the inclusion and modeling of indirect land use change effects. After the CARB's ruling, among other criticisms, representatives of the US ethanol industry complained that this standard overstates the environmental effects of corn ethanol, and also criticized the inclusion of indirect effects of land-use changes as an unfair penalty to home-made corn ethanol because deforestation in the developing world is being tied to US ethanol production. The initial reference value set for 2011 for LCFS means that Mid-west corn ethanol will not meet the California standard unless current carbon intensity is reduced.

A similar controversy arose after the U.S. Environmental Protection Agency (EPA) published on May 5, 2009, its notice of proposed rulemaking for the new Renewable Fuel Standard (RFS). The draft of the regulations was released for public comment during a 60-day period. EPA's proposed regulations also included the carbon footprint from indirect land-use changes. On the same day, President Barack Obama signed a Presidential Directive with the aim to advance biofuels research and improve their commercialization. The Directive established a Biofuels Interagency Working Group comprise of three agencies, the Department of Agriculture, the Environmental Protection Agency, and the Department of Energy. This group will developed a plan to increase flexible fuel vehicle use and assist in retail marketing efforts. Also they will coordinate infrastructure policies impacting the supply, secure transport, and distribution of biofuels. The group will also come up with policy ideas for increasing investment in next-generation fuels, such as cellulosic ethanol, and for reducing the environmental footprint of growing biofuels crops, particularly corn-based ethanol.

E85 IN THE UNITED STATES

For each state the total number of facilities is given. Numbers in parentheses are the number of publicly accessible facilities, when that number is less than the total in a state.

State	Stations	State	Stations	State	Stations	State	Stations
Alabama	14 (2)	Alaska	0	Arizona	31 (30)	Arkansas	10
California	32 (28)	Colorado	83	Connecticut	4	Delaware	1
District of Columbia	3 (1)	Florida	27 (5)	Georgia	34	Hawaii	0
Idaho	5(4)	Illinois	223	Indiana	136	Iowa	132 (91)
Kansas	54	Kentucky	13	Louisiana	9	Maine	0
Maryland	17	Massachusetts	2	Michigan	112 (61)	Minnesota	378
Mississippi	5	Missouri	116	Montana	3	Nebraska	62
Nevada	23	New Hampshire	0	New Jersey	2	New Mexico	7
New York	29	North Carolina	19	North Dakota	41	Ohio	63
Oklahoma	9	Oregon	12	Pennsylvania	27	Rhode Island	0
South Carolina	59 (58)	South Dakota	82	Tennessee	26	Texas	47
Utah	5	Vermont	0	Virginia	4 (3)	Washington	12
West Virginia	3 (2)	Wisconsin	133	Wyoming	8		

Notes:

- Not all stations are publicly accessible.
- Some facilities that are publicly accessible only accept fleet purchasing cards.
- Data is outdated. Please check the links before you use this data.
- Data sourced from U.S. Department of Energy and NGO National Ethanol Vehicle Coalition and NGO USA Energy Independence Publications may differ
- Station count usually the greater of two numbers when sources differ.

There are approximately 1900 E85 Filling Stations in the U.S. as of January, 2009. Minnesota has the largest number of E85 fuel locations of any U.S. state with close to 350 stations, while Illinois has the second-greatest number of E85 pumps with about 160. Although Minnesota has the most E85 pumps they only represent a tiny fraction of the total fuel outlets. According to Oil Price Information Service (OPIS) there are approximately 140,000 publicly accessible retail gasoline stations in the United States. (All filling stations in Minnesota are however required to sell E10, a mixture of 10 per cent ethanol and 90 per cent gasoline.)

Constraints

Concerns about rising gasoline prices and energy dependence have led to a resurgence of interest in E85 fuel; for example, Nebraska mandated the use of E85 in state vehicles whenever possible in May 2005. Similarly, whereas selling any fuel containing more than 10 per cent ethanol is still currently illegal in some states, even this is rapidly changing. For example, Florida proposed changing state law to permit the sale of alternative fuels such as E85 at an October 7, 2005 meeting, and held public hearings on October 24. The expected outcome of having held this hearing is the changing of Florida state law to permit the selling of alternative fuels such as E85 by the end of 2005 to the general public. (At that time, only county, state, and Federal fleet vehicles could purchase E85 in Florida, from only 3 pumps in the state.) Several other states have similar laws still on their books that prevent the sale of E85 to the general public. The expected general outcome, though, is the rapidly widening acceptance of E85 sales to the general public in all of the United States by the end of 2006.

Federal Use

US Federal fleet flexible-fuel vehicles (FFVs) are required to operate on alternative fuels 100 per cent of the time upon the signing of the Energy Policy Act of 2005 into law by President

Bush on August 8, 2005. Formerly, such FFVs were required to be operated by the end of 2005 on alternative fuels only 51 per cent of the time (i.e., the majority of the time) by Executive Order 13149. This means that the US Government's use of E85 is effectively doubled as of August 8, 2005 with the signing into law of the Energy Policy Act of 2005. This jump in consumption had the effect of limiting public availability of E85 coincident with shortages of gasoline due to impacts of hurricanes in the Gulf of Mexico during the 2005 hurricane season. Although the price of corn had not changed greatly, the usage of E85 nonetheless jumped, thereby creating a shortage of E85, and causing E85 prices to rise coincident with gasoline prices during the 2005 Hurricane Season.

Price

As of 2005, E85 is frequently sold for up to 36 per cent lower price per quantity than gasoline. Much of this discount can be attributed to various government subsidies, and, at least in the United States, the elimination of state taxes that typically apply to gasoline and can amount to 47 cents, or more, per gallon of fuel. The US federal tax exemption that keeps ethanol economically competitive with petroleum fuel products is due to expire in 2007, but this exemption may be extended through legislative action. In the aftermath of Hurricane Katrina in 2005, the price of E85 rose to nearly on par with the cost of 87 octane gasoline in many states in the United States, and was for a short time the only fuel available when gasoline was sold out, but within four weeks of Katrina, the price of E85 had fallen once more to a 20 per cent to 35 per cent lower cost than 87 octane gasoline.

Fuel Economy

Because ethanol contains less energy than gasoline, fuel economy is reduced for most 2002 and earlier American FFVs (flexible-fuel vehicles) that are currently on the road by about 30 per cent (most after 2003 lose only 15-17 per cent, or less) when operated on pure E85 (summer blend). Some of the newest American vehicles can lessen this reduction to only 5-15 per cent, but as recently as 2007 the Environmental Protection Agency stated on its website that several of the most current American FFVs were still losing 25-30 per cent fuel efficiency when running on E85. Some Swedish engineered cars with engine management systems provide much better fuel economy on E85 than on gasoline; for example, the Saab Aero-X turbocharged concept car produces higher fuel economy and higher power on 100 per cent ethanol (E100) than gasoline through using a higher compression ratio engine with advanced SAAB engine control computers. Another car that has higher power on ethanol is the Koenigsegg CCXR, which on ethanol is the second most powerful production car with 1020 hp. This according to the manufacturer is due to the cooling properties of ethanol. Still, for almost all American-made FFVs, more E85 is typically needed to do the same work as can be achieved with a lesser volume of gasoline. This difference is sometimes offset by the lower cost of the E85 fuel, depending on E85's current price discount relative to the current price of gasoline. As described earlier, the best thing for drivers to do is to record fuel usage with both fuels and calculate cost/distance for them. Only by doing that, can the end-user economy of the two fuels be compared.

For example, an existing pre-2003 model year American-made FFV vehicle that normally achieves, say, 30 MPG on pure gasoline will typically achieve about 20 MPG, or slightly better, on E85 (summer blend.) When operated on E85 winter blend, which is actually E70 (70 per cent ethanol, 30 per cent gasoline), fuel economy will be higher than when operating on the summer blend. To achieve any short-term operational fuel cost savings, the price of E85 should therefore be 30 per cent or more below the price of gasoline to equalize short term fuel costs for most older pre-2003 FFVs for both winter and summer blends of E85. Life-cycle costs over the life of the FFV engine are theoretically lower for E85, as ethanol is a cooler and cleaner burning fuel than gasoline. Provided that one takes a longterm life-cycle operating cost view, a continuous price discount of only 20 per cent to 25 per cent below the cost of gasoline is probably about the break-even point in terms of vehicle life-cycle operating costs for operating most FFVs on E85 exclusively (for summer, spring/fall, and winter blends).

Fuel economy in fuel-injected non-FFVs operating on a mix of E85 and gasoline varies greatly depending on the engine and fuel mix. For a 60:40 blend of gasoline to E85 (summer blend), a typical fuel economy reduction of around 23.7 per cent resulted in one person's carefully executed experiment with a 1998 Chevrolet S10 pickup with a 2.2L 4-cylinder engine, relative to the fuel economy achieved on pure gasoline. Similarly, for a 50:50 blend of gasoline to E85 (summer blend), a typical fuel economy reduction of around 25 per cent resulted for the same vehicle. (Fuel economy performance numbers were measured on a fixed commute of approximately 110 miles roundtrip per day, on a predominantly freeway commute, running at a fixed speed (62 mph), with cruise control activated, air conditioning ON, at sea level, with flat terrain, traveling to/from Kennedy Space Center, FL.). It is important to note, however, that if the engine had been specifically tuned for consumption of ethanol (higher compression, different fuel-air mixture, etc.) the mileage would have been much better than the results above. The aforementioned fact leads some to believe that the "FFV" engine is more of an infant technology rather than fully mature.

The amount of reduction in mileage, therefore, is highly dependent upon the particulars of the vehicle design, exact composition of the ethanol-gasoline blend and state of engine tune (fuel air mixture and compression ratio primarily).

Vehicles

As of 2006, approximately one in forty vehicles on the road in the U.S. can run on E85.

In 2006, GM Chief Executive Rick Wagoner stated that GM would continue to produce FFVs due to the positive image that they create for the company. He went on to say that GM would focus heavily on increasing the distribution system for E85, and that the fueling infrastructure had to grow "very, very rapidly."

As of 2006, GM had built 1.7 million E85 capable vehicles, with plans for 400,000 more before years end.

E85 and the 2006 Minnesota Governor's Race

In the 2006 election for Minnesota Governor, the issue of E85 may have helped re-elect Republican Tim Pawlenty. In the final week of the campaign, DFL Lieutenant Governor candidate Judi Dutcher was stumping for her running mate Mike Hatch. A reporter asked Dutcher what Hatch's stance on E85 was and Dutcher appeared to be unfamiliar with E85. As Hatch tried to cover up Dutcher's gaffe, he became angry with reporters which many people think led to a lack of support and Pawlenty's re-election in an election year when Republicans did not fare well as a group.

EVALUATING ETHANOL FUEL IN HAWAII

State law makes all gasoline sold for vehicles to have 10 per cent ethanol blended (E-10). No E-85 or 85 per cent ethanol blend is yet sold for flex fuel use. Presently, over 40 million gallons of ethanol are imported from the Caribbean or continental US. Shipping costs, which amount to 20 to 30 cents per gallon, are passed on to the consumer. Several companies have registered to make ethanol over 3 years ago in Hawaii, but none have started construction as of 2008/9. The two sugar plantations are expected to make ethanol, but only Gay and Robinson on Kauai has said it will stop sugar operations to make ethanol using two processes. The first technology is fermentation of a faster growing cane and the secondary process will be with Clear Fuels cellulosic process from the waste (bagasse) of the first process. Enough ethanol is expected to be made for the island of Kauai's use, but it must be shipped to Oahu for blending and back for sale at the pump.

New thermophilic algae by La Wahie to make ethanol is being researched in Kona and the University of Hawaii. Sun makes algae grow faster in the tropics and there may animal feed to be had from cane left overs. CO_2 is absorbed by the catalyst. Several other ideas have come forward with agricultural basis on the big island of Hawaii where cane fields are fallow since sugar stopped production. One company wants to use the wood grown in old cane areas that were planted with eucalyptus. Cost are high for these kinds of operations as wood products are sold to a world market. Another newer technology is being proposed for Oahu. The island has a big landfill problem and the City of Honolulu operates an incineration boiler to electricity operation (h-power) which has been in operation since 1992. It presently handles about half of the garbage and green waste on the island. A new boiler is projected to be online soon at a cost of $65,000,000. The 2 old boilers are down often for repair.

Another proposal by Diamond Head Renewable Resources LLC (DHRR), is to make alcohol fuels including ethanol from garbage, biomass and green waste. The new technologies being piloted around the world for this cellulosic based feedstock will find a ready supply of cheap material. The projected conversion of waste to fuel is needed to reduce green house gases and even use methane from landfill as a source for making ethanol. The plant makes a syngas (synthesis gas) that is converted by catalyst to fuels and syngas can also be made into electricity and the plant will do just that making over 10 megawatts of power some for its own use and the rest sold to the grid as renewable electrical sourced energy. Net saving of approx. 74,000 tons of CO_2 by displacing fossil fuel with biofuel fuels at just 10 per cent. More with electricity and using the methane from landfill.

This process and technology is similar to the Clear Fuels proposed plant on Kauai, without the first part of a sugar to ethanol operation. First generation or food based technology has had difficulty with financing their projects since the incentives from the Farm act of 2008 are for new technology from cellulosic biomass to ethanol. Incentives include a dollar a gallon tax credit, loan guarantees, and grants or funds available for small beginning operations. State incentive are also available with tax incentives and a department to assist business (DBEDT). Second generation ethanol is a priority for permitting by health officials.

An investor 100 per cent state tax incentive for high tech and new generation ethanol may be reduced or completely abolished. Other locations looking to make ethanol may view Hawai'i as a leader in garbage to ethanol. Small plants expanded in phases reduce risk and market fluctuations. Once this plant is operational on Oahu they will use 350 dry tons a day to make product and electricity. Expected first stage of DHRR output is 12 million gallons of alcohols a year growing to 40. A mix of ethanol, methanol and other alcohols.

EVALUATING ETHANOL FUEL IN BRAZIL

Brazil is the world's second largest producer of ethanol fuel and the world's largest exporter. Together, Brazil and the United States lead the industrial production of ethanol fuel, accounting together for 89 per cent of the world's production in 2008. In 2008 Brazil produced 24.5 billion litres (6.47 billion U.S. liquid gallons), which represents 37.3 per cent of the world's total ethanol used as fuel.

Brazil is considered to have the world's first sustainable biofuels economy and the biofuel industry leader, a policy model for other countries; and its sugarcane ethanol *"the most successful alternative fuel to date."* However, some authors consider that the successful Brazilian ethanol model is sustainable only in Brazil due to its advanced agri-industrial technology and its enormous amount of arable land available; while for other authors it is a solution only for some countries in the tropical zone of Latin America, the Caribbean, and Africa. Brazil's 30-year-old ethanol fuel Programme is based on the most efficient agricultural technology for sugarcane cultivation in the world, uses modern equipment and cheap sugar cane as feedstock, the residual cane-waste (bagasse) is used to process heat and power, which results in a very competitive price and also in a high energy balance (output energy/input energy), which varies from 8.3 for average conditions to 10.2 for best practice production.

Brazil has ethanol fuel available throughout the country. Shown here a typical Petrobras gas station at São Paulo with dual fuel service, marked A for alcohol (ethanol) and G for gasoline.

There are no longer any light vehicles in Brazil running on

pure gasoline. Since 1976 the government made it mandatory to blend anhydrous ethanol with gasoline, fluctuating between 10 per cent to 22 per cent. and requiring just a minor adjustment on regular gasoline motors. In 1993 the mandatory blend was fixed by law at 22 per cent anhydrous ethanol (E22) by volume in the entire country, but with leeway to the Executive to set different percentages of ethanol within pre-established boundaries. In 2003 these limits were set at a minimum of 20 per cent and a maximum of 25 per cent. Since July 1, 2007 the mandatory blend is 25 per cent of anhydrous ethanol and 75 per cent gasoline or E25 blend.

The Brazilian car manufacturing industry developed flexible-fuel vehicles that can run on any proportion of gasoline (E20-E25 blend) and hydrous ethanol (E100). Introduced in the market in 2003, flex vehicles became a commercial success, reaching a record 94 per cent share of all new cars sales in August 2009. As of July 2009, the fleet of flex-fuel cars and light commercial vehicles had reached 8,25 million vehicles, representing 23.4 per cent of Brazil's registered light motor vehicle fleet. The success of "flex" vehicles, together with the mandatory E25 blend throughout the country, have allowed ethanol fuel consumption in the country to achieve a 50 per cent market share of the gasoline-powered fleet by February 2008. Considering diesel-powered vehicles, sugarcane ethanol represented 16.7 per cent of the country's total energy consumption by the automotive sector in 2007.

History

Sugarcane has been cultivated in Brazil since 1532 as sugar was one of the first commodities exported to Europe by the Portuguese settlers. The first use of sugarcane ethanol as a fuel in Brazil dates back to the late twenties and early thirties of the twentieth century, with the introduction of the automobile in the country. Ethanol fuel production peaked during World War II and, as German submarine attacks threatened oil supplies, the mandatory blend became as high as 50 per cent in 1943.

After the end of the war cheap oil caused gasoline to prevail, and ethanol blends were only used sporadically, mostly to take advantage of sugar surpluses, until the seventies, when the first oil crisis resulted in gasoline shortages and awareness of the dangers of oil dependence. As a response to this crisis, the Brazilian government began promoting bioethanol as a fuel. The National Alcohol Programme -*Pró-Álcool*- (Portuguese: 'Programme a Nacional do Álcool'), launched in 1975, was a nation-wide Programme financed by the government to phase out automobile fuels derived from fossil fuels, such as gasoline, in Favour of ethanol produced from sugar cane.

The 1979 Brazilian Fiat 147 was the first modern automobile launched to the market capable of running only on hydrous ethanol fuel (E100).

The first phase of the Programme concentrated on production of anhydrous ethanol for blending with gasoline. The Brazilian government made mandatory the blending of ethanol fuel with gasoline, fluctuating from 1976 until 1992 between 10 per cent to 22 per cent. Due to this mandatory minimum gasoline blend, pure gasoline (E0) is no longer sold in the country. A federal law was passed in October 1993 establishing a mandatory blend of 22 per cent anhydrous ethanol (E22) in the entire country. This law also authorized the Executive to set different percentages of ethanol within pre-established boundaries; and since 2003 these limits were fixed at a maximum of 25 per cent (E25) and a minimum of 20 per cent (E20) by volume. Since then, the government has set the percentage of the ethanol blend according to the results of the sugarcane harvest and the levels of ethanol production from sugarcane, resulting in blend variations even within the same year. Since July 2007 the mandatory blend is 25 per cent of anhydrous ethanol and 75 per cent gasoline or E25 blend.

Historical trend of Brazilian total production of light vehicles, neat ethanol (alcohol), flex fuel, and gasoline vehicles from 1979 to 2008.

After testing in government fleets with several prototypes developed by the local carmakers, and compelled by the second oil crisis, the Fiat 147, the first modern commercial ethanol-only powered car (E100 only) was launched to the market in

July 1979. The Brazilian government provided three important initial drivers for the ethanol industry: guaranteed purchases by the state-owned oil company Petrobras, low-interest loans for agro-industrial ethanol firms, and fixed gasoline and ethanol prices where hydrous ethanol sold for 59 per cent of the government-set gasoline price at the pump. These incentives made ethanol production competitive. After reaching more than 4 million cars and light trucks running on pure ethanol by the late 1980s, representing one third of the country's motor vehicle fleet, ethanol production and sales of ethanol-only cars tumbled due to several factors. First, gasoline prices fell sharply as a result of lower gasoline prices, but mainly because of a shortage of ethanol fuel supply in the local market left thousands of vehicles in line at gas stations or out of fuel in their garages by mid 1989. As supply could not keep pace with the increasing demand required by the now significant ethanol-only fleet, the Brazilian government began importing ethanol in 1991.

The 2003 Brazilian VW Gol 1.6 Total Flex was the first flexible-fuel car capable of running on any blend of gasoline and ethanol.

Confidence on ethanol-powered vehicles was restored only with the introduction in the Brazilian market of flexible-fuel vehicles. In March 2003 Volkswagen launched in the Brazilian market the Gol 1.6 Total Flex, the first commercial flexible fuel vehicle capable of running on any blend of gasoline and ethanol. By 2009, popular manufacturers that build flexible fuel vehicles are Chevrolet, Fiat, Ford, Peugeot, Renault, Volkswagen, Honda, Mitsubishi, Toyota, Citröen, and Nissan. Flexible fuel cars were 22 per cent of the car sales in 2004, 73 per cent in 2005, 87.6 per cent in July 2008, and reached a record 94 per cent in August 2009. The rapid adoption and commercial success of "flex" vehicles, as they are popularly known, together with the mandatory blend of alcohol with gasoline as E25 fuel, have increased ethanol consumption up to the point that by February 2008 a landmark in ethanol consumption was achieved when ethanol retail sales surpassed the 50 per cent market share of the gasoline-powered fleet. This level of ethanol fuel consumption had not been reached since the end of the 1980s, at the peak of the *Pró-Álcool* Programme. Also, from 1979 until July 2009, Brazil has successfully reduced by 14 million the number of vehicles running just on gasoline (5.7 million neat ethanol and 8.2 million flex-fuel light vehicles, plus 102 thousand flex-fuel motorcycles), thereby reducing the country's dependence on oil imports. The number of neat ethanol vehicles still in use is estimated between 2 to 3 million vehicles.

The 2009 Honda CG 150 Titan Mix was launched in the Brazilian market and became the first flex-fuel motorcycle sold in the world.

Under the auspices of the BioEthanol for Sustainable Transport (BEST) project, the first ethanol-powered (ED95) bus began operations in São Paulo city on December 2007 as a one-year trial project. During the trial period performance and emissions will be monitored as significant reductions are expected in carbon monoxide and particulate matter emissions, and as previous tests have shown a reduction in fuel economy of around 60 per cent when ED95 is compared to regular diesel.

The latest innovation within the Brazilian flexible-fuel technology is the development of flex-fuel motorcycles. The first flex motorcycle was launched by Honda in March 2009. Produced by its Brazilian subsidiary Moto Honda da Amazônia, the CG 150 Titan Mix is sold for around US$2,700. In order to avoid cold start problems, the fuel tank must have at least 20 per cent of gasoline at temperatures below 15°C (59°F). During the first six months after its market launch the CG 150 Titan Mix sold 102,782 motorcycles, capturing a 10.1 per cent market share, and ranking fourth in sales of new motorcycles in the Brazilian market in 2009.

Production

Ethanol production in Brazil uses sugarcane as feedstock and relies on first-generation technologies based on the use of the sucrose content of sugarcane. Ethanol yield has grown 3.77 per cent per year since 1975 and productivity gains been based on improvements in the agricultural and industrial phases of the production process. Further improvements on best practices are expected to allow in the short to mid-term an average ethanol productivity of 9,000 liters per hectare.

There were 378 ethanol plants operating in Brazil by July 2008, 126 dedicated to ethanol production and 252 producing both sugar and ethanol. There are 15 additional plants dedicated exclusively to sugar production. These plants have an installed capacity of crushing 538 million metric tons of sugarcane per year, and there are 25 plants under construction expected to be on line by 2009 that will add an additional capacity of crushing 50 million tons of sugarcane per year. The typical plant cost approximately USD 150 million and requires a nearby sugarcane plantation of 30,000 hectares.

Ethanol production is concentrated in the Central and Southeast regions of the country, led by São Paulo state, with around 60 per cent of the country's total ethanol production, followed by Paraná (8 per cent), Minas Gerais (8 per cent) and Goiás (5 per cent). These two regions have been responsible for 90 per cent of Brazil's ethanol production since 2005 and the harvest season goes from April to November. The Northeast Region is responsible for the remaining 10 per cent of ethanol production, lead by Alagoas with 2 per cent of total production. The harvest season in the North-Northeast region goes from September to March, and the average productivity in this region is lower than the South-Central region. Due to the difference in the two main harvest seasons, Brazilian statistics for sugar and ethanol production are commonly reported on a harvest two-year basis rather than on a calendar year.

For the 2008/09 harvest it is expected that about 44 per cent of the sugarcane will be used for sugar, 1 per cent for alcoholic beverages, and 55 per cent for ethanol production. An estimate of between 24.9 billion litres (6.58 billion U.S. liquid gallons) to 27.1 billion litres (7.16 billion gallons) of ethanol are expected to be produced in 2008/09 harvest year, with most of the production being destined for the internal market, and only 4.2 billion liters (1.1 billion gallons) for exports, with an estimated 2.5 billion liters (660 million gallons) destined for the US market. Sugarcane cultivated area grew from 7 million to 7.8 million hectares of land from 2007 to 2008, mainly using abandoned pasture lands. In 2008 Brazil has 276 million hectares of arable land, 72 per cent use for pasture, 16.9 per cent for grain crops, and 2.8 per cent for sugarcane, meaning that ethanol is just requiring approximately 1.5 per cent of all arable land available in the country.

As sugar and ethanol share the same feedstock and their industrial processing is fully integrated, formal employment statistics are usually presented together. In 2000 there were 642,848 workers employed by these industries, and as ethanol production expanded, by 2005 there were 982,604 workers employed in the sugarcane cultivation and industrialization, including 414,668 workers in the sugarcane fields, 439,573 workers in the sugar mills, and 128,363 workers in the ethanol distilleries. While employment in the ethanol distilleries grew 88.4 per cent from 2000 to 2005, employment in the sugar fields just grew 16.2 per cent as a direct result of expansion of mechanical harvest instead manual harvesting, which avoids burning the sugarcane fields before manual cutting and also increases productivity. The states with the most employment in 2005 were São Paulo (39.2 per cent), Pernambuco (15 per cent), Alagoas (14.1 per cent), Paraná (7 per cent), and Minas Gerais (5.6 per cent).

Agricultural Technology

Sugarcane (*Saccharum officinarum*) plantation ready for harvest, Ituverava, São Paulo State.

Evolution of the ethanol productivity per hectare of sugarcane planted in Brazil between 1975 and 2004. Source: Goldemberg (2008).

Typical ethanol distillery and dehydration facility, Piracicaba, São Paulo State.

Variation of ethanol prices to producers in 2007 reflecting the harvest season supply. Yellow is for anhydrous ethanol and green is for hydrated ethanol (R$ per liter).

A key aspect for the development of the ethanol industry in Brazil was the investment in agricultural research and development by both the public and private sector. The work of

EMBRAPA, the state-owned company in charge for applied research on agriculture, together with research developed by state institutes and universities, especially in the State of São Paulo, have allowed Brazil to became a major innovator in the fields of biotechnology and agronomic practices, resulting in the most efficient agricultural technology for sugarcane cultivation in the world. Efforts have been concentrated in increasing the efficiency of inputs and processes to optimize output per hectare of feedstock, and the result has been a threefold increase of sugarcane yields in 29 years, as Brazilian average ethanol yields went from 2,024 liters per ha in 1975 to 5,917 liters per ha in 2004; allowing the efficiency of ethanol production to grow at a rate of 3.77 per cent per year. Brazilian biotechnologies include the development of sugarcane varieties that have a larger sugar or energy content, one of the main drivers for high yields of ethanol per unit of planted area. The increase of the index total recoverable sugar (TRS) from sugarcane has been very significant, 1.5 per cent per year in the period 1977 to 2004, resulting in an increase from 95 to 140 kg/ha. Innovations in the industrial process have allowed an increase in sugar extraction in the period 1977 to 2003. The average annual improvement was 0.3 per cent; some mills have already reached extraction efficiencies of 98 per cent.

Biotechnology research and genetic improvement have led to the development of strains which are more resistant to disease, bacteria, and pests, and also have the capacity to respond to different environments, thus allowing the expansion of sugarcane cultivation to areas previously considered inaqueate for such cultures. By 2008 more than 500 sugarcane varieties are cultivated in Brazil, and 51 of them were released just during the last ten years. Four research Programmes, two private and two public, are devoted to further genetic improvement. Since the mid nineties, Brazilian biotechnology laboratories have developed transgenic varieties, still non commerciallized. Identification of 40,000 cane genes was completed in 2003 and there are a couple dozen research groups working on the functional genome, still on the experimental phase, but commercial results are expected within five years.

Also, there is ongoing research regarding sugarcane biological nitrogen fixation, with the most promising plant varieties showing yields three times the national average in soils of very low fertility, thus avoiding nitrogenous fertilization. There is also research for the development of second-generation or cellulosic ethanol. In São Paulo state an increase of 12 per cent in sugar cane yield and 6.4 per cent in sugar content is expected over the next decade. This advance combined with an expected 6.2 per cent improvement in fermentation efficiency and 2 per cent in sugar extraction, may increase ethanol yields by 29 per cent, raising average ethanol productivity to 9,000 liters/ha. Approximately US$50 million has recently been allocated for research and projects focused on advancing the abstention of ethanol from sugarcane in São Paulo state.

Production Process

Sucrose extracted from sugarcane accounts for little more than 30 per cent of the chemical energy stored in the mature plant; 35 per cent is in the leaves and stem tips, which are left in the fields during harvest, and 35 per cent are in the fibrous material (bagasse) left over from pressing. Most of the industrial processing of sugarcane in Brazil is done through a very integrated production chain, allowing sugar production, industrial ethanol processing, and electricity generation from byproducts. The typical steps for large scale production of sugar and ethanol include milling, electricity generation, fermentation, distillation of ethanol, and dehydration.

Milling and Refining

Once harvested, sugarcane is usually transported to the plant by semi-trailer trucks. After quality control sugarcane is washed, chopped, and shredded by revolving knives. The feedstock is fed to and extracted by a set of mill combinations to collect a juice, called garapa in Brazil, that contain 10—15 per cent sucrose, and bagasse, the fiber residue. The main objective of the milling process is to extract the largest possible amount of sucrose from the cane, and a secondary but important objective is the production of bagasse with a low moisture content as boiler fuel, as bagasse is burned for electricity generation, allowing the plant to be self-sufficient in energy and to generate electricity for the local power grid. The cane juice or garapa is then filtered and treated by chemicals and pasteurized. Before evaporation, the juice is filtered once again, producing vinasse, a fluid rich in organic compounds. The syrup resulting from evaporation is then precipitated by crystallization producing a mixture of clear crystals surrounded by molasses. A centrifuge is used to separate the sugar from molasses, and the crystals are washed by addition of steam, after which the crystals are dried by an airflow. Upon cooling, sugar crystallizes out of the syrup. From this point, the sugar refining process continues to produced different types of sugar, and the molasses continue a separate process to produce ethanol.

Fermentation, Distillation and Dehydration

The resulting molasses are treated to become a sterilized molasse free of impurities, ready to be fermented. In the fermentation process sugars are transformed into ethanol by addition of yeast. Fermentation time varies from four to twelve hours resulting in an alcohol content of 7-10 per cent by total volume (°GL), called fermented wine. The yeast is recovered from this wine through a centrifuge. Making use of the different boiling points the alcohol in the fermented wine is separated from the main resting solid components. The remaining product is hydrated ethanol with a concentration of 96°GL, the highest concentration of ethanol that can be achieved via azeotropicdistillation, and by national specification can contain up to 4.9 per cent of water by volume. This hydrous ethanol is the fuel used by ethanol-only and flex vehicles in the country. Further dehydration is normally done by addition of chemicals, up to the specified 99.7°GL in order to produce anhydrous ethanol, which is used for blending with pure gasoline to obtain the country's E25 mandatory blend. The additional processing required to convert hydrated into anhydrous ethanol increases the cost of the fuel, as in 2007 the average producer price difference between the two was around 14 per cent for São Paulo State. This production price difference, though small, contributes to the competitiveness of the hydrated ethanol (E100) used in Brazil, not only with regard to local gasoline prices but also as compared to other countries such as the US and Sweden, that only use anhydrous ethanol for their flex fuel fleet.

Electricity Generation from Bagasse

Since the early days bagasse was burnt in the plant to provide the energy required for the industrial part of the process. Today, the Brazilian best practice uses high-pressure boilers that increases energy recovery, allowing most sugar-ethanol plants to be energetically self-sufficient and even sell surplus electricity to utilities. By 2000, the total amount of sugarcane bagasse produced per year was 50 million tons/dry basis out of more than 300 million tons of harvested sugarcane. Several authors estimated a potential power generation from the use of sugarcane bagasse ranging from 1,000 to 9,000 MW, depending on the technology used and the use of harvest trash. One utility in São Paulo is buying more than 1 per cent of its electricity from sugar mills, with a production capacity of 600 MW for self-use and 100 MW for sale. According to analysis from Frost and Sullivan, Brazil's sugarcane bagasse used for power generation has reached 3.0 GW in 2007, and it is expected to reach 12.2 GW in 2014. The analysis also found tha sugarcane bagasse cogeneration accounts for 3 per cent of the total Brazilian energy matrix. The energy is especially valuable to utilities because it is produced mainly in the dry season when hydroelectric dams are running low.

According to a study commissioned by the Dutch government in 2006 to evaluate the sustainability of Brazilian bioethanol *"...there are also substantial gains possible in the efficiency of electricity use and generation: The electricity used for distillery operations has been estimated at 12.9 kWh/tonne cane, with a best available technology rate of 9.6 kWh/tonne cane. For electricity generation the efficiency could be increased from 18 kWh/tonne cane presently, to 29.1 kWh/tonne cane maximum. The production of surplus electricity could in theory be increased from 5.3 kWh/tonne cane to 19 kWh/tonne cane".*

Overall Energy Use

Energy-use associated with the production of sugarcane ethanol derives from three primary sources: the agricultural sector, the industrial sector, and the distribution sector. In the agricultural sector, 35.98 GJ of energy are used to plant, maintain, and harvest one hectare (10,000 m2) of sugarcane for usable biofuel. This includes energy from numerous inputs, including nitrogen, phosphate, potassium oxide, lime, seed, herbicides, insecticides, labour and diesel fuel. The industrial sector, which includes the milling and refining sugarcane and the production of ethanol fuel, uses 3.63 GJ of energy and generates 155.57 GJ of energy per hectare of sugarcane plantation. Scientists estimate that the potential power generated from the cogeneration of bagasse could range from 1,000 to 9,000 MW, depending on harvest and technology factors. In Brazil, this is about 3 per cent of the total energy needed. The burning of bagasse can generate 18 kilowatt-hours, or 64.7 MJ per Mg of sugarcane. Distillery facilities require about 45 MJ to operate, leaving a surplus energy supply of 19.3 MJ, or 5.4 kWh. In terms of distribution, researchers calculates sugarcane ethanol's transport energy requirement to be .44 GJ per cubic-meter, thus one hectare of land would require 2.82 GJ of energy for successful transport and distribution. After taking all three sectors into account, the EROEI (Energy Return over Energy Invested) for sugarcane ethanol is about 8.

There are several improvements to the industrial processes, such as adopting a hydrolysis process to produce ethanol instead of surplus electricity, or the use of advanced boiler and turbine technology to increase the electricity yield, or a higher use of excess bagasse and harvest trash currently left behind in the fields, that together with various other efficiency improvements in sugarcane farming and the distribution chain have the potential to allow further efficiency increases, translating into higher yields, lower production costs, and also further improvements in the energy balance and the reduction of greenhouse gas emissions.

Exports

Brazil is the world's largest exporter of ethanol. In 2007 it exported 933.4 million gallons (3,532.7 million liters), representing almost 20 per cent of its production, and accounting for almost 50 per cent of the global exports. Since 2004 Brazilian exporters have as their main customers the United States, Netherlands, Japan, Sweden, Jamaica, El Salvador, Costa Rica, Trinidad and Tobago, Nigeria, Mexico, India, and South Korea.

The countries in the Caribbean Basin import relative high quantities of Brazilian ethanol, but not much is destined for domestic consumption. These countries reprocess the product, usually converting Brazilian hydrated ethanol into anhydrous ethanol, and then re-export it to the United States, gaining value-added and avoiding the 2.5 per cent duty and the USD 0,54 per gallon tariff, thanks to the trade agreements and benefits granted by Caribbean Basin Initiative (CBI). This process is limited by a quota, set at 7 per cent of U.S. ethanol consumption. Although direct U.S. exports fell in 2007, imports from four CBI countries almost doubled, increasing from 15.5 per cent in 2006 to 25.8 per cent in 2007, reflecting increasing re-exports to the U.S., thus partially compensating the loss of Brazilian direct exports to the U.S. This situation has caused some concerns in the United States, as it and Brazil are trying to build a partnership to increase ethanol production in Latin American and the Caribbean. As the U.S. is encouraging "new ethanol production in other countries, production that could directly compete with U.S.-produced ethanol".

The U.S., potentially the largest market for the Brazilian ethanol, currently imposes trade restrictions on Brazilian ethanol of $USD 0.54 per gallon, in order to encourage domestic ethanol production, most of which has so far been based on processing corn instead of sugar cane or soybeans, which is only 25-30 per cent as efficient. There is concern that allowing the Brazilian ethanol to enter the U.S. market without taxation will undercut the budding ethanol industry in the United States. One of the arguments for that is that Brazil currently subsidises its ethanol production, which is false, as the subsidies Programme finished in the 1990s. Others argue that rather than impose trade restrictions on the import of the Brazilian product, that the U.S. should make subsidies of its own available to support its fledgling domestic producers. Exports of Brazilian ethanol to the U.S. reached a total of US$ 1 billion in 2006, an increase of 1,020 per cent over 2005 (US$ 98 millions), but fell significantly in 2007 due to sharp increases in American ethanol production from maize.

As shown in the table, together, the United States, the European Union, the CBI countries with Mexico, and Japan, were the destination of 91 per cent of Brazilian ethanol exports, both in 2007 and 2006. As of 2007, the European Union region, led by the Netherlands, is the main importer of Brazilian ethanol, with 265.3 million gallons (1,004.2 million liters). However, and despite of reduced direct imports, the United States continues to be the single one country where Brazilian ethanol is exported, reaching 228.96 million gallons (866.6 million liters) to the continental U.S., 13.78 million gallons (52.1 million liters) shipped to the U.S. Virgin Islands, and 3.68 million gallons (14.0 million litters) shipped to Puerto Rico, for a total export for the U.S. in 2007 of 246.4 million gallons (932.75 million liters), down from 469.6 million gallons (1.77 billion liters) in 2006.

Prices and Effect on Oil Consumption

Alcohol and gasoline prices per liter at Rio de Janeiro (left) and São Paulo (right), corresponding to a price ratio of E100 ethanol to E25 gasoline of 0.64 and 0.56.

Historical variation of ethanol production by region from 1990/91 to 2006/07 (harvest year). Light green is the production for the State of São Paulo.

Most automobiles in Brazil run either on hydrous alcohol (E100) or on gasohol (E25 blend), as the mixture of 25 per cent anhydrous ethanol with gasoline is mandatory in the entire country. Since 2003, dual-fuel ethanol flex vehicles that run on any proportion of hydrous ethanol and gasoline have been gaining popularity, surpassing 6 million new cars and light commercial vehicles sold by August 2008, and as the same month, 75 per cent of light vehicle manufacturing production is flexible fuel without additional cost for buyers. Customers have 49 models available to chose from. Brazilian full flex-fuel vehicles have electronic sensors that automatically detect the type of fuel and the blend mix, and accordingly adjust the engine combustion. Users have the freedom to choose depending on the free market prices of each fuel.

Due to the lower energy content of ethanol fuel, full flex-fuel vehicles get fewer miles per gallon. Ethanol price has to be between 25-30 per cent cheaper per gallon to reach the break even point. As a rule of thumb, Brazilian consumers are frequently advised by the media to use more alcohol than gasoline in their mix only when ethanol prices are 30 per cent lower or more than gasoline, as ethanol price fluctuates heavily depending on the harvest yields and seasonal fluctuation of sugarcane harvest. Since 2005, ethanol prices have been very competitive without any subsidies, even with gasoline prices kept constant in local currency since mid-2005, at a time when oil was just approaching USD 60 a barrel. The price ratio between gasoline and ethanol fuel has been well above 30 per cent during this period for most states, except during low sugar cane supply between harvests and for states located far away from the ethanol production centers. According to Brazilian producers, ethanol can remain competitive if the price of oil does not fall below USD 30 a barrel.

By 2008 consumption of ethanol fuel by the Brazilian fleet of light vehicles, as pure ethanol and in gasohol, is replacing gasoline at the rate of about 27,000 cubic metres per day, and by February 2008 the combined consumption of anhydrous and hydrated ethanol fuel surpassed 50 per cent of the fuel that would be needed to run the light vehicle fleet on pure gasoline alone. Consumption of anhydrous ethanol for the mandatory E25 blend, together with hydrous ethanol used by flex vehicles, reached 1.432 billion liters, while pure gasoline consumption was 1.411 billion liters.

However, the effect on the country's overall petroleum consumption was smaller than that, as domestic oil consumption still far outweighs ethanol consumption. In 2005, Brazil consumed 2 million barrels (320,000 m³) of oil per day, versus 280,000 barrels (45,000 m³) of ethanol. Although Brazil is a major oil producer and now exports gasoline (19,000 m³/day), it still must import oil because of internal demand for other oil byproducts, chiefly diesel fuel, which cannot be easily replaced by ethanol. When trucks and other diesel-powered vehicles are considered ethanol represented 16.9 per cent of total energy consumption by the road transport sector in terms of energy equivalent to crude oil, and 14.9 per cent of the entire transport sector.

Comparison with the United States

Brazil's sugar cane-based industry is more efficient than the U.S. corn-based industry. Sugar cane ethanol has an energy balance seven times greater than ethanol produced from corn. Brazilian distillers are able to produce ethanol for 22 cents per liter, compared with the 30 cents per liter for corn-based ethanol. U.S. corn-derived ethanol costs 30 per cent more because the corn starch must first be converted to sugar before being distilled into alcohol. Despite this cost differential in production, the U.S. does not import more Brazilian ethanol because of U.S. trade barriers corresponding to a tariff of 54-cent per gallon, first imposed in 1980, but kept to offset the 45-cent per gallon blender's federal tax credit that is applied to ethanol no matter its country of origin.

Sugarcane cultivation requires a tropical or subtropical climate, with a minimum of 600 mm (24 in) of annual rainfall. Sugarcane is one of the most efficient photosynthesizers in the plant kingdom, able to convert up to 2 per cent of incident solar energy into biomass. Sugarcane production in the United States occurs in Florida, Louisiana, Hawaii, and Texas. The first three plants to produce sugarcane-based ethanol are expected to go online in Louisiana by mid 2009. Sugar mill plants in Lacassine,

Consumer Price Spread between E25 Gasoline And E100 By State. Red and Orange Show States with Average Prices below the Break Even Range. Ethanol Price should be between 25 to 30% Cheaper than Gasoline to Compensate its Lower Fuel Economy

State	Average Retail Price (R$/liter)		Price Spread E25-E100 (%)	State	Average retail price (R$/liter)		Price spread E25-E100 (%)	State	Average retail price (R$/liter)		Price spread E25-E100 (%)
	E100	E25	(%)		E100	E25	(%)		E100	E25	(%)
Acre (AC)	2.080	2.943	29.32	Maranhão (MA)	1.709	2.628	34.97	Rio de Janeiro (RJ)	1.676	2.531	33.78
Alagoas (AL)	1.844	2.766	33.33	Mato Grosso (MT)	1.452	2.677	45.76	Rio Grande do Norte (RN)	1.940	2.669	27.31
Amapá (AP)	2.246	2.686	16.38	Mato Grosso do Sul (MS)	1.683	2.676	37.11	Rio Grande do Sul (RS)	1.779	2.574	30.89
Amazonas (AM)	2.773	2.452	27.69	Minas Gerais (MG)	1.610	2.377	32.27	Rondônia (RR)	1.839	2.669	31.10
Bahia (BA)	1.630	2.522	35.37	Pará (PA)	2.120	2.772	23.52	Roraima (RO)	2.154	2.710	20.52
Brasília (DF)	1.884	2.586	27.15	Paraíba (PB)	1.883	2.553	26.24	Santa Catarina (SC)	1.697	2.556	33.61
Ceará (CE)	1.768	2.510	29.56	Paraná (PR)	1.445	2.429	40.51	São Paulo (SP)	1.306	2.398	45.54
Espírito Santo (ES)	1.795	2.662	32.57	Pernambuco (PE)	1.700	2.573	33.93	Sergipe (SE)	1.888	2.518	25.02
Goiás (GO)	1.581	2.565	38.36	Piauí (PI)	1.927	2.655	27.42	Tocantins (TO)	1.708	2.748	37.85
Country average	1.513	2.511	39.75								

Source: Agência Nacional do Petróleo (ANP). Average retail prices for week of 26/10/2008 to 01/11/2008.

Note: Data is presented in local currency because the exchange rate for the Brazilian real has been fluctuating heavily due to the ongoing global financial crisis. Exchange rate for 2008-10-31 was USD 1 = R$ 2.16.

St. James and Bunkie were converted to sugar cane-based ethanol production using Colombian technology in order to make possible a profitable ethanol production. These three plants will produce 100 million gallons of ethanol within five years. By 2009 two other sugarcane ethanol production projects are being developed in Kauai, Hawaii and Imperial Valley, California.

Comparison of Key Characteristics between the Ethanol Industries in the United States and Brazil

Characteristic	Brazil	U.S.	Units/Comments
Feedstock	Sugar cane	Maize	Main cash crop for ethanol production, the US has less than 2% from other crops.
Total ethanol fuel production (2008)	6,472	9,000	Million U.S. liquid gallons
Total arable land	355	270	Million hectares.
Total area used for ethanol crop (2006)	3.6 (1%)	10 (3.7%)	Million hectares (% total arable)
Productivity per hectare	6,800-8,000	3,800-4,000	Liters of ethanol per hectare. Brazil is 727 to 870 gal/acre (2006), US is 321 to 424 gal/acre (2003)
Energy balance (input energy productivity)	8.3 to 10.2	1.3 to 1.6	Ratio of the energy obtained from ethanol/energy expended in its production
Estimated GHG emissions reduction	86-90%	10-30%	% GHGs avoided by using ethanol instead of gasoline, using existing crop land (No ILUC).
Full life-cycle carbon intensity	73.40	105.10	Grams of CO_2 equivalent released per MJ of energy produced, includes indirect land use changes.
Estimated payback time for GHG emissions	17 years	93 years	Brazilian cerrado for sugarcane and US grassland for corn. Land use change scenarios by Fargione

(Contd.)

Characteristic	Brazil	U.S.	Units/Comments
Flexible-fuel vehicle fleet	8.2 million	8.0 million	Autos and light trucks only. Brazil as of July 2009 (E100 FFVs). U.S. as of early 2009 (E85 FFVs).
Ethanol fueling stations in the country	35,017 (100%)	1,963 (1%)	As % of total gas stations in the country. Brazil by 2007-12, U.S. by 2009-03 (170,000 total.)
Ethanol's share in the gasoline market	50%	4%	As % of total consumption on a volumetric basis. Brazil as of April 2008. US as of December 2006.
Cost of production (USD/gallon)	0.83	1.14	2006/2007 for Brazil (22¢/liter), 2004 for U.S. (35¢/liter)
Government subsidy (in USD)	0	0.45/gallon	U.S. since 2009-01-01 as a tax credit. Brazilian ethanol production is no longer subsidized.
Import tariffs (in USD)	0	0.54/gallon	As of June 2009, Brazil does not import ethanol, the U.S. does

Notes: (1) Only contiguous U.S., excludes Alaska. (2) Assuming no land use change. (3) CARB estimate for Midwest corn ethanol. California's gasoline carbon intensity is 95.86 blended with 10 per cent ethanol. (4) Assuming direct land use change. (5) If diesel-powered vehicles are included and due to ethanol's lower energy content by volume, bioethanol represented 16.9 per cent of the road sector energy consumption in 2007. (6) Brazilian ethanol production is no longer subsidized, but gasoline is heavily taxed Favouring ethanol fuel consumption (~54 per cent tax). By the end of July 2008, when oil prices were close to its latest peak and the Brazilian Real exchange rate to the US dollar was close to its most recent minimum, the average gasoline retail price at the pump in Brazil was USD 6.00 per gallon, while the average US price was USD 3.98 per gallon. The latest gas retail price increase in Brazil occurred in late 2005, when oil price was at USD 60 per barrel.

Ethanol Diplomacy

President Luiz Inácio Lula da Silva and King Carl XVI Gustaf of

Sweden inspecting one of the 400 buses running on ED95 on Stockholm.

In March 2007, "ethanol diplomacy" was the focus of President George W. Bush's Latin American tour, in which he and Brazil's president, Luiz Inácio Lula da Silva, were seeking to promote the production and use of sugar cane based ethanol throughout Latin America and the Caribbean. The two countries also agreed to share technology and set international standards for biofuels. The Brazilian sugar cane technology transfer will permit various Central American countries, such as Honduras, Nicaragua, Costa Rica and Panama, several Caribbean countries, and various Andean Countries tariff-free trade with the U.S. thanks to existing concessionary trade agreements.

Even though the U.S. has imposed a USD 0.54 tariff on every gallon of imported ethanol since 1980, the Caribbean nations and Central American countries are exempt from such duties based on the benefits granted by the Caribbean Basin Initiative (CBI). CBI provisions allow tariff-free access to the US market from ethanol produced from foreign feedstock (outside CBI countries) up to 7 per cent of the previous year US consumption. Also additional quotas are allowed if the beneficiary countries produce at least 30 per cent of the ethanol from local feedstocks up to an additional 35 million gallons. Thus, several countries have been importing hydrated ethanol from Brazil, processing it at local distilleries to dehydrate it, and then re-exporting it as anhydrous ethanol. American farmers have complained about this loophole to legally bypass the tariff. The 2005 Dominican Republic—Central America Free Trade Agreement (CAFTA) maintained the benefits granted by the CBI, and CAFTA provisions established country-specific shares for Costa Rica and El Salvador within the overall quota. An initial annual allowance was established for each country, with gradually-increasing annual levels of access to the US market. The expectation is that using Brazilian technology for refining sugar cane based ethanol, such countries could become net exporters to the United States in the short-term. In August 2007, Brazil's President toured Mexico and several countries in Central America and the Caribbean to promote Brazilian ethanol technology.

The Memorandum of Understanding (MOU) that the American and Brazilian presidents signed in March 2007 has been described as a success for foreign policy, bringing Brazil and the United States closer especially on energy policy. Nevertheless, policy makers have cited a lack of "substantive progress" implementing the three pillars found in that agreement and have called for an expansion of international engagement beyond the executive branches.

EMBRAPA's African Regional Office in Ghana.

Brazil has also extended its technical expertise to several African countries, including Ghana, Mozambique, Angola, and Kenya. This effort is led by EMBRAPA, the state-owned company in charge for applied research on agriculture, and responsible for most of the achievements in increasing sugarcane productivity during the last thirty years. Another 15 African countries have shown interest in receiving Brazilian technical aid to improve sugarcane productivity and to produced ethanol efficiently. Brazil also has bilateral cooperation agreements with several other countries in Europe and Asia.

As President Lula wrote for The Economist regarding Brazil's global agenda: *"Brazil's ethanol and biodiesel programmes are a benchmark for alternative and renewable fuel sources. Partnerships are being established with developing countries seeking to follow Brazil's achievements—a 675m-tonne reduction of greenhouse-gas emissions, a million new jobs and a drastic reduction in dependence on imported fossil fuels coming from a dangerously small number of producer countries. All of this has been accomplished without compromising food security, which, on the contrary, has benefited from rising agricultural output. We are setting up offices in developing countries interested in benefiting from Brazilian know-how in this field."*.

Environmental and Social Impacts

Environmental Effects

Benefits

Ethanol produced from sugarcane provides energy that is

renewable and less carbon intensive than oil. Bioethanol reduces air pollution thanks to its cleaner emissions, and also contributes to mitigate climate change by reducing greenhouse gas emissions.

Energy Balance

One of the main concerns about bioethanol production is the energy balance, the total amount of energy input into the process compared to the energy released by burning the resulting ethanol fuel. This balance considers the full cycle of producing the fuel, as cultivation, transportation and production require energy, including the use of oil and fertilizers. A comprehensive life cycle assessment commissioned by the State of São Paulo found that Brazilian sugarcane based ethanol has a Favourable energy balance, varying from 8.3 for average conditions to 10.2 for best practice production. This means that for average conditions one unit of fossil-fuel energy is required to create 8.3 energy units from the resulting ethanol. These findings have been confirmed by other studies.

UK estimates for the carbon intensity of bioethanol and fossil fuels. As shown, Brazilian ethanol from sugarcane is the most efficient biofuel currently under commercial production in terms of GHG emission reduction.

Greenhouse Gas Emissions

Another benefit of bioethanol is the reduction of greenhouse gas emissions as compared to gasoline, because as much carbon dioxide is taken up by the growing plants as is produced when the bioethanol is burnt, with a zero theoretical net contribution. Several studies have shown that sugarcane based ethanol reduces greenhouse gases by 86 to 90 per cent if there is no significant land use change, and ethanol from sugarcane is regarded the most efficient biofuel currently under commercial production in terms of GHG emission reduction. However, two studies published in 2008 are critical of previous assessments of greenhouse gas emissions reduction, as the authors considered that previous studies did not take into account the effect of land use changes.

Air Pollution

The widespread use of ethanol brought several environmental benefits to urban centers regarding air pollution. Lead additives to gasoline were reduced through the 1980s as the amount of ethanol blended in the fuel was increased, and these additives were completely eliminated by 1991. The addition of ethanol blends instead of lead to gasoline lowered the total carbon monoxide (CO), hydrocarbons, sulfur emissions, and particulate matter significantly. The use of ethanol-only vehicles has also reduced CO emissions drastically. Before the *Pró-Álcool* Programme started, when gasoline was the only fuel in use, CO emissions were higher than 50 g/km driven; they had been reduced to less than 5.8 g/km in 1995. Several studies have also shown that São Paulo has benefit with significantly less air pollution thanks to ethanol's cleaner emissions. Furthermore, Brazilian flex-fuel engines are being designed with higher compression ratios, taking advantage of the higher ethanol blends and maximizing the benefits of the higher oxygen content of ethanol, resulting in lower emissions and improving fuel efficiency.

Even though all automotive fossil fuels emit aldehydes, one of the drawbacks of the use of hydrated ethanol in ethanol-only engines is the increase in aldehyde emissions as compared with gasoline or gasohol. However, the present ambient concentrations of aldehyde, in São Paulo city are below the reference levels recommended as adequate to human health found in the literature. Other concern is that because formaldehyde and acetaldehyde emissions are significantly higher, and although both aldehydes occur naturally and are frequently found in the open environment, additional emissions may be important because of their role in smog formation. However, more research is required to establish the extent and direct consequences, if any, on health.

Issues

Sugar cane harvest loading operation for transport to the sugar/ethanol processing plant, without previous burning of the plantation, São Paulo state.

Mechanized sugarcane harvesting operation. Use of harvesting machines avoids the need for burning the plantation, São Paulo state.

Typical vehicle used for harvest transport to the sugar/ethanol processing plant at São Paulo state.

Water Use and Fertilizers

Ethanol production has also raised concerns regarding water overuse and pollution, soil erosion and possible contamination by excessive use of fertilizers. A study commissioned by the Dutch government in 2006 to evaluate the sustainability of Brazilian bioethanol concluded that there is sufficient water to supply all foreseeable long-term water requirements for sugarcane and ethanol production. Also, and as a result of legislation and technological progress, the amount of water collected for ethanol production has decreased considerably during the previous years. The overuse of water resources seems a limited problem in general in São Paulo, particularly because of the relatively high rainfall, yet, some local problems may occur. Regarding water pollution due to sugarcane production, Embrapa classifies the industry as level 1, which means "no impact" on water quality.

This evaluation also found that consumption of agrochemicals for sugar cane production is lower than in citric, corn, coffee and soybean cropping. Disease and pest control, including the use of agrochemicals, is a crucial element in all cane production. The study found that development of resistant sugar cane varieties is a crucial aspect of disease and pest control and is one of the primary objectives of Brazil's cane genetic improvement Programmes. Disease control is one of the main reasons for the replacement of a commercial variety of sugar cane.

Field Burning

Advancements in fertilizers and natural pesticides have all but eliminated the need to burn fields. Sugarcane fields are traditionally burned just before harvest to avoid harm to the workers, by removing the sharp leaves and killing snakes and other harmful animals, and also to fertilize the fields with ash. There has been less burning due to pressure from the public and health authorities, and as a result of the recent development of effective harvesting machines. A 2001 state law banned burning in sugarcane fields in São Paulo state by 2021, and machines will gradually replace human labour as the means of harvesting cane, except where the abrupt terrain does not allow for mechanical harvesting. However, 150 out of 170 of São Paulo's sugar cane processing plants signed in 2007 a voluntary agreement with the state government to comply by 2014. Independent growers signed in 2008 the voluntary agreement to comply, and the deadline was extended to 2017 for sugar cane fields located in more abrupt terrain. By the 2008 harvest season, around 47 per cent of the cane was collected with harvesting machines. Mechanization will reduce pollution from burning fields and has higher productivity than people, but also will create unemployment for these seasonal workers, many of them coming from the poorest regions of Brazil. Due to mechanization the number of temporary workers in the sugarcane plantations has already declined.

Effects of Land Use Change

Two studies published in 2008 questioned the benefits estimated in previous assessments regarding the reduction of greenhouse gas emissions from sugarcane based ethanol, as the authors consider that previous studies did not take into account the direct and indirect effect of land use changes. The authors found a "biofuel carbon debt" is created when Brazil and other developing countries convert land in undisturbed ecosystems, such as rainforests, savannas, or grasslands, to biofuel production, and to crop production when agricultural land is diverted to biofuel production. This land use change releases more CO_2 than the annual greenhouse gas (GHG) reductions that these biofuels would provide by displacing fossil fuels. Among others, the study analyzed the case of Brazilian Cerrado being converted for sugarcane ethanol production. The biofuel carbon debt on converted Cerrado is estimated to be repaid in 17 years, the least amount of time of the scenarios that were analyzed, as for example, ethanol from US corn was estimated to have a 93 year payback time. The study conclusion is that the net effect of biofuel production via clearing of carbon-rich habitats is to increase CO_2 emissions for decades or centuries relative to fossil fuel use.

Regarding this concern, previous studies conducted in Brazil have shown there are 355 million ha of arable land in Brazil, of which only 72 million ha are in use. Sugarcane is only taking 2 per cent of arable land available, of which ethanol production represented 55 per cent in 2008. Embrapa estimates that there is enough agricultural land available to increase at least 30 times the existing sugarcane plantation without endangering sensible ecosystems or taking land destined for food crops. Most future growth is expected to take place on abandoned pasture lands, as it has been the historical trend in São Paulo state. Also, productivity is expected to improve even further based on current biotechnology research, genetic improvement, and better agronomic practices, thus contributing to reduce land demand for future sugarcane cultures. This trend is demonstrated by the increases in agricultural production that took place in São Paulo state between 1990 and 2004, where coffee, orange, sugarcane and other food crops were grown in an almost constant area.

Also regarding the potential negative impacts of land use changes on carbon emissions, a study commissioned by the Dutch government concluded that "*it is very difficult to determine the indirect effects of further land use for sugar cane production (i.e. sugar cane replacing another crop like soy or citrus crops, which in turn causes additional soy plantations replacing pastures, which in turn may cause deforestation), and also not logical to attribute all these soil carbon losses to sugar cane*". Other authors have also questioned these indirect effects, as cattle pastures are displaced to the cheaper land near the Amazon. Studies rebutting this concern claim tha land devoted to free grazing cattle is shrinking, as density of cattle on pasture land increased from 1.28 heads of cattle/ha to 1.41 from 2001 to 2005, and further improvements are expected in cattle feeding practices.

Deforestation

Location of environmentally valuable areas with respect to sugarcane plantations. São Paulo, located in the Southeast Region of Brazil, concentrates two-thirds of sugarcane cultures.

Other criticism have focused on the potential for clearing rain forests and other environmentally valuable land for sugarcane production, such as the Amazonia, the Pantanal or the Cerrado. Embrapa has rebutted this concern explaining that 99.7 per cent of sugarcane plantations are located at least 2,000 km from the Amazonia, and expansion during the last 25 years took place in the Center-South region, also far away from the Amazonia, the Pantanal or the Atlantic forest. In São Paulo state growth took place in abandoned pasture lands.

The impact assessment regarding future changes in land use, forest protection and risks on biodiversity conducted as part of the study commissioned by the Dutch government concluded that *"the direct impact of cane production on biodiversity is limited, because cane production replaces mainly pastures and/ or food crop and sugar cane production takes place far from the major biomes in Brazil (Amazon Rain Forest, Cerrado, Atlantic Forest, Caatinga, Campos Sulinos and Pantanal)"*. However, *"...the indirect impacts from an increase of the area under sugar cane production are likely more severe. The most important indirect impact would be an expansion of the area agricultural land at the expense of cerrados. The cerrados are an important biodiversity reserve. These indirect impacts are difficult to quantify and there is a lack of practically applicable criteria and indicators."*

Brazil's president, Luiz Inácio Lula da Silva has also claimed this concern is not valid. According to him *"The Portuguese discovered a long time ago that the Amazon isn't a place to plant cane."* In order to guarantee a sustainable development of ethanol production, the government is working on a countrywide zoning plan to restrict sugarcane growth in or near environmentally sensitive areas, allowing only the eight existing plants to remain operating in these sensitive areas, but without further extension of their sugarcane fields. The proposed restricted area has 4.6 million square kilometers, almost half of the Brazilian territory.

Social Implications

Typical sugarcane worker during the harvest season, São Paulo state.

Sugarcane has had an important social contribution to the some of the poorest people in Brazil by providing income usually above the minimum wage, and a formal job with fringe benefits. Formal employment in Brazil accounts an average 45 per cent across all sectors, while the sugarcane sector has a share of 72.9 per cent formal jobs in 2007, up from 53.6 per cent in 1992, and in the more developed sugarcane ethanol industry in São Paulo state formal employment reached 93.8 per cent in 2005. Average wages in sugar cane and ethanol production are above the official minimum wage, but minimum wages may be insufficient to avoid poverty. The North-Northeast regions stands out for having much lower levels of education among workers and lower monthly income. The average number workers with 3 or less school years in Brazil is 58.8 per cent, while in the Southeast this percentage is 46.2 per cent, in the Northeast region is 76,4 per cent. Therefore, earnings in the Center-South are not surprisingly higher than those in the North-Northeast for comparable levels of education. In 2005 sugarcane harvesting workers in the Center-South region received an average wage 58.7 per cent higher than the average wage in the North-Northeast region. The main social problems are related to cane cutters which do most of the low-paid work related to ethanol production.

The total number of permanent employees in the sector fell by one-third between 1992 and 2003, in part due to the increasing reliance on mechanical harvesting, especially in the richest and more mature sugarcane producers of São Paulo state. During the same period, the share of temporary or seasonal workers has fluctuated, first declining and then increasing in recent years to about one-half of the total jobs in the sector, but in absolute terms the number of temporary workers has declined also. The sugarcane sector in the poorer Northeast region is more labour intensive as production in this region represents only 18.6 per cent of the country's total production but employs 44.3 per cent of worker force in the sugarcane sector.

The manual harvesting of sugarcane has been associated with hardship and poor working conditions. In this regard, the study commissioned by the Dutch government confirmed that the main problem is indeed related to manual cane harvesting. A key problem in working conditions is the high work load. As a result of mechanization the workload per worker has increased from 4 to 6 ton per day in the eighties to 8 to 10 ton per day in the nineties, up to 12 to 15 ton per day in 2007. If the quota is not fulfilled, workers can be fired. Producers say this problem will disappear with greater mechanization in the next dècade. Also, as mechanization of the harvesting is increasing and only feasible in flat terrain, more workers are being used in areas where conditions are not suitable for mechanized harvesting equipment, such as rough areas where the crops are planted irregularly, making working conditions harder and more hazardous.

Also unhealthy working conditions and even cases of slavery and deaths from overwork (cane cutting) have been reported, but these are likely worst-case examples. Even though sufficiently strict labour laws are present in Brazil, enforcement is weak. Displacement and seasonal labour also implies physical and cultural disruption of multifunctional family farms and traditional communities.

Regarding social responsibility the ethanol production sector maintains more than 600 schools, 200 nursery centers and 300 day care units, as legislation requires that 1 per cent of the net sugar cane price and 2 per cent of the net ethanol price must be devoted to medical, dental, pharmaceutical, sanitary, and educational services for sugar cane workers. In practice more than 90 per cent of the mills provide health and dental care, transportation and collective life insurance, and over 80 per cent provide meals and pharmaceutical care. However, for the temporary low wage workers in cane cutting these services may not be available.

Effect on Food Prices

Some environmentalists, such as George Monbiot, have expressed fears that the marketplace will convert crops to fuel for the rich, while the poor starve and biofuels cause environmental problems. Environmental groups have raised concerns about this trade-off for several years. The food vs fuel debate reached a global scale in 2008 as a result of the international community's concerns regarding the steep increase in food prices. On April 2008, Jean Ziegler, back then United Nations Special Rapporteur on the Right to Food, called biofuels a *"crime against humanity"*, a claim he had previously made in October 2007, when he called for a 5-year ban for the conversion of land for the production of biofuels. Also on April 2008, the World Bank's President, Robert Zoellick, stated that *"While many worry about filling their gas tanks, many others around the world are struggling to fill their stomachs. And it's getting more and more difficult every day."*

Luiz Inácio Lula da Silva gave a strong rebuttal, calling these claims *"fallacies resulting from commercial interests"*, and putting the blame instead on U.S. and European agricultural subsidies, and a problem restricted to U.S. ethanol produced from maize. The Brazilian President has also claimed on several occasions that his country's sugar cane based ethanol industry has not contributed to the food price crises.

A report released by Oxfam in June 2008 criticized biofuel policies of rich countries as neither a solution to the climate crisis nor the oil crisis, while contributing to the food price crisis. The report concluded that from all biofuels available in the market, Brazilian sugarcane ethanol is "far from perfect" but it is the most Favourable biofuel in the world in term of cost and greenhouse gas balance. The report discusses some existing problems and potential risks, and asks the Brazilian government for caution to avoid jeopardizing its environmental and social sustainability. The report also says that: *"Rich countries spent up to $15 billion last year supporting biofuels while blocking cheaper Brazilian ethanol, which is far less damaging for global food security."*

A World Bank research report published on July 2008 found that from June 2002 to June 2008 *"biofuels and the related consequences of low grain stocks, large land use shifts, speculative activity and export bans"* accounted for 70-75 per cent of total price rises. The study found that higher oil prices and a weak dollar explain 25-30 per cent of total price rise. The study said that "...*large increases in biofuels production in the United States and Europe are the main reason behind the steep rise in global food prices*" and also stated that "*Brazil's sugar-based ethanol did not push food prices appreciably higher*". The report argues that increased production of biofuels in these developed regions were supported by subsidies and tariffs on imports, and considers that without such policies, price increases worldwide would have been smaller. This research paper also concluded that Brazil's sugar cane based ethanol has not raised sugar prices significantly, and recommends removing tariffs on ethanol imports by both the U.S. and E.U., to allow more efficient producers such as Brazil and other developing countries, including many African countries, to produce ethanol profitably for export to meet the mandates in the E.U. and U.S.

An economic assessment report also published in July 2008 by the OECD agrees with the World Bank report regarding the negative effects of subsidies and trade restrictions, but found that the impact of biofuels on food prices are much smaller. The OECD study is also critical of the limited reduction of GHG emissions achieved from biofuels produced in Europe and North America, concluding that the current biofuel support policies would reduce greenhouse gas emissions from transport fuel by no more than 0.8 per cent by 2015, while Brazilian ethanol from sugar cane reduces greenhouse gas emissions by at least 80 per cent compared to fossil fuels. The assessment calls on governments for more open markets in biofuels and feedstocks in order to improve efficiency and lower costs.

A study by the Brazilian research unit of the Fundação Getúlio Vargas regarding the effects of biofuels on grain prices. concluded that the major driver behind the 2007-2008 rise in food prices was speculative activity on futures markets under conditions of increased demand in a market with low grain stocks. The study also concluded that expansion of biofuel production was not a relevant factor and also that there is no correlation between Brazilian sugarcane cultivated area and average grain prices, as on the contrary, the spread of sugarcane was accompanied by rapid growth of grain crops in the country.

Panoramic view of the Costa Pinto Production Plant located in Piracicaba, São Paulo state. This industrial plan is set up to produce sugar, ethanol fuel (both anhydrous and hydrous), industrial grade ethanol, and alcohol for beverages. The foreground shows the receiving operation of the sugarcane harvest, immediately followed by the mill process, and in the right side of the background is located the distillation facility where ethanol is produced. This plant produces the electricity it needs from baggasse residuals from sugar cane left over by the milling process, and it sells the surplus electricity to public utilities.

9

Algal and Microbial Fuel:
Culture, Production and Technology Involved

FUEL FROM ALGAE

Algae fuel, also called **algal fuel, oilgae, algaeoleum** or **third-generation biofuel**, is a biofuel from algae.

High oil prices, competing demands between foods and other biofuel sources and the world food crisis have ignited interest in algaculture (farming algae) for making vegetable oil, biodiesel, bioethanol, biogasoline, biomethanol, biobutanol and other biofuels. Among algal fuels' attractive characteristics: they do not affect fresh water resources, can be produced using ocean and wastewater, and are biodegradable and relatively harmless to the environment if spilled. Algae cost more per unit mass yet can yield over 30 times more energy per unit area than other, second-generation biofuel crops. One biofuels company has claimed that algae can produce more oil in an area the size of a two car garage than a football field of soybeans, because almost the entire algal organism can use sunlight to produce lipids, or oil. The United States Department of Energy estimates that if algae fuel replaced all the petroleum fuel in the United States, it would require 15,000 square miles (40,000 km²). This is less than 1/7th area of corn harvested in the United States in 2000.

During photosynthesis, algae and other photosynthetic organisms capture carbon dioxide and sunlight and convert it into oxygen and biomass. Up to 99 per cent of the carbon dioxide in solution can be converted, which was shown by Weissman and Tillett (1992) in large-scale open-pond systems. The production of biofuels from algae does not reduce atmospheric carbon dioxide (CO_2), because any CO_2 taken out of the atmosphere by the algae is returned when the biofuels are burned. They do however eliminate the introduction of new CO_2 by displacing fossil hydrocarbon fuels.

As of 2008, such fuels remain too expensive to replace other commercially available fuels, with the cost of various algae species typically between US$5—10 per kilogram. But several companies and government agencies are funding efforts to reduce capital and operating costs and make algae oil production commercially viable.

History

The **Aquatic Species Programme** launched in 1978. The U.S. research Programme, funded by the U.S. DoE, was tasked with investigating the use of algae for the production of energy. The Programme initially focused efforts on the production of hydrogen, however, shifted primary research to studying oil production in 1982. From 1982 through its culmination, the majority of the Programme research was focused on the production of transportation fuels, notably biodiesel, from algae. In 1995, as part of the over-all efforts to lower budget demands, the DoE decided to end the Programme. Research stopped in 1996 and staff began compiling their research for publication. In July 1998, the DoE published the report "A Look Back at the U.S. Department of Energy's Aquatic Species Programme: Biodiesel from Algae".

Factors

Dry algae factor is the percentage of algae cells in relation with the media where it is cultured, e.g. if the dry algae factor is 50 per cent, one would need 2 kg of wet algae (algae in the media) to get 1 kg of algae cells.

Lipid factor is the percentage of vegoil in relation with the algae cells needed to get it, i.e. if the algae lipid factor is 40 per cent, one would need 2.5 kg of algae cells to get 1 kg of oil.

Yield

Yields cover a vast range from 5,000 to 150,000 US gallons of oil per acre per year (4,700—140,000 m^3/km^2·a). If all aspects of the cultivation are controlled—temperature, CO_2 levels, sunlight and nutrients (including carbohydrates as a food source), then extremely high yields can be obtained. Such variation can make calculations on which to base 'fuel the world' scenarios very difficult.

For example, Glen Kertz of Valcent Products, claims that "algae can produce 100,000 gallons of oil per acre" per year (94,000 m^3/km^2·a). This relies on growing the algae in an entirely closed loop system. More recently, Valcent have claimed 150,000 US gallons may be possible; their most recent actual reported yields could reach 33,000 gallons per acre per year (21,153,000 gal/sq mi·a or 31,000 m^3/km^2·a) if they could achieve their yields with an algal species having 50 per cent lipids. In 2007, the U.S. consumed 20.680 million barrels per day (3.2879×10[6] m^3/d) of petroleum or 317 billion US gallons per year (1.20×10[9] m^3/a). Thus, with the production capabilities of Valcent, it would only require 15,000 square miles (39,000 km^2) of land to completely displace petroleum use in the U.S.

Current projections, however, do not take into account the energy losses due to converting the algae lipids into fuels.

Fuels

The vegoil algae product can then be harvested and converted into biodiesel; the algae's carbohydrate content can be fermented into bioethanol and biobutanol.

Biodiesel

Currently most research into efficient algal-oil production is being done in the private sector, but predictions from small scale production experiments bear out that using algae to produce biodiesel may be the only viable method by which to produce enough automotive fuel to replace current world diesel usage.

Microalgae have much faster growth-rates than terrestrial crops. The per unit area yield of oil from algae is estimated to be from between 5,000 to 20,000 US gallons per acre per year (4,700 to 18,000 $m^3/km^2 \cdot a$); this is 7 to 30 times greater than the next best crop, Chinese tallow (700 US gal/acre·a or 650 $m^3/km^2 \cdot a$).

Studies show that algae can produce up to 60 per cent of their biomass in the form of oil. Because the cells grow in aqueous suspension where they have more efficient access to water, CO_2 and dissolved nutrients, microalgae are capable of producing large amounts of biomass and usable oil in either high rate algal ponds or photobioreactors. This oil can then be turned into biodiesel which could be sold for use in automobiles. The more efficient this process becomes the larger the profit that is turned by the company. Regional production of microalgae and processing into biofuels will provide economic benefits to rural communities.

Biobutanol

Butanol can be made from algae or diatoms using only a solar powered biorefinery. This fuel has an energy density 10 per cent less than gasoline, and greater than that of either ethanol or methanol. In most gasoline engines, butanol can be used in place of gasoline with no modifications. In several tests, butanol consumption is similar to that of gasoline, and when blended with gasoline, provides better performance and corrosion resistance than that of ethanol or E85.

The green waste left over from the algae oil extraction can be used to produce butanol.

Biogasoline

Biogasoline is gasoline produced from biomass such as algae. Like traditionally produced gasoline, it contains between 6 (hexane) and 12 (dodecane) carbon atoms per molecule and can be used in internal-combustion engines.

Methane

Through the use of algaculture grown organisms and cultures, various polymeric materials can be broken down into methane.

SVO

The algal-oil feedstock that is used to produce biodiesel can also be used for fuel directly as "Straight Vegetable Oil", (SVO). The benefit of using the oil in this manner is that it doesn't require the additional energy needed for transesterification, (processing the oil with an alcohol and a catalyst to produce biodiesel). The drawback is that it does require modifications to a normal diesel engine. Transesterified biodiesel can be run in an unmodified modern diesel engine, provided the engine is designed to use ultra-low-sulfur diesel, which, as of 2006, is the new diesel fuel standard in the United States.

Hydrocracking to Traditional Transport Fuels

Vegetable oil can be used as feedstock for an oil refinery where methods like hydrocracking or hydrogenation can be used to transform the vegetable oil into standard fuels like gasoline and diesel.

Jet Fuel

Rising jet fuel prices are putting severe pressure on airline companies, creating an incentive for algal jet fuel research. The International Air Transport Association, for example, supports research, development and deployment of algal fuels. IATA's goal is for its members to be using 10 per cent alternative fuels by 2017.

On January 8, 2009, Continental Airlines ran the first test for the first flight of an algae-fueled jet. The test was done using a twin-engine commercial jet consuming a 50/50 blend of biofuel and normal aircraft fuel. It was the first flight by a U.S. carrier to use an alternative fuel source on this specific type of aircraft. The flight from Houston's Bush International Airport completed a circuit over the Gulf of Mexico. The pilots on-board, executed a series of tests at 38,000 feet (12,000 m), including a mid-flight engine shutdown. Larry Kellner, chief executive of Continental Airlines, said they had tested a drop-in fuel which meant that no modification to the engine was required. The fuel was praised for having a low flash point and sufficiently low freezing point, issues that have been problematic for other bio-fuels.

Algae Cultivation

Algae can produce 15-300 times more oil per acre than conventional crops, such as rapeseed, palms, soybeans, or jatropha. As Algae has a harvesting cycle of 1-10 days, it permits several harvests in a very short time frame, a differing strategy to yearly crops (Chisti 2007). Algae can also be grown on land that is not suitable for other established crops, for instance, arid land, land with excessively saline soil, and drought-stricken land. This minimizes the issue of taking away pieces of land from the cultivation of food crops (Schenk *et al.* 2008).

They can grow 20 to 30 times faster than food crops.

Not only does algae produce biofuel, it also helps with reducing CO_2 emissions. Algae, like other fuels, releases carbon dioxide when it is burned. Fortunately, Algae takes in CO_2 and replaces it with Oxygen during the process of photosynthesis. Ultimately, its net emissions are zero because the CO_2 released in burning is the same amount that was absorbed initially.

The hard part about algae production is growing the algae in a controlled way and harvesting it efficiently.

PhotoBioreactors

Most companies pursuing algae as a source of biofuels are pumping nutrient-laden water through plastic tubes (called "bioreactors") that are exposed to sunlight (and so called photobioreactors or PBR).

Running a PBR is more difficult than an open pond, and more costly.

Algae can also grow on marginal lands, such as in desert areas where the groundwater is saline, rather than utilise fresh water.

The difficulties in efficient biodiesel production from algae lie in finding an algal strain with a high lipid content and fast growth rate that isn't too difficult to harvest, and a cost-effective cultivation system (i.e., type of photobioreactor) that is best suited to that strain. There is also a need to provide concentrated CO_2 to turbocharge the production.

Closed Loop System

Another obstacle preventing widespread mass production of algae for biofuel production has been the equipment and structures needed to begin growing algae in large quantities. Maximum use of existing agriculture processes and hardware is the goal.

In a closed system (not exposed to open air) there is not the problem of contamination by other organisms blown in by the air. The problem for a closed system is finding a cheap source of sterile CO_2. Several experimenters have found the CO_2 from a smokestack works well for growing algae. To be economical, some experts think that algae farming for biofuels will have to be done next to power plants, where they can also help soak up the pollution.

Open Pond

Open-pond systems for the most part have been given up for the cultivation of algae with high-oil content. Many believe that a major flaw of the Aquatic Species Programme was the decision to focus their efforts exclusively on open-ponds; this makes the entire effort dependent upon the hardiness of the strain chosen, requiring it to be unnecessarily resilient in order to withstand wide swings in temperature and pH, and competition from invasive algae and bacteria. Open systems using a monoculture are also vulnerable to viral infection. The energy that a high-oil strain invests into the production of oil is energy that is not invested into the production of proteins or carbohydrates, usually resulting in the species being less hardy, or having a slower growth rate. Algal species with a lower oil content, not having to divert their energies away from growth, have an easier time in the harsher conditions of an open system.

Some open sewage ponds trial production has been done in Marlborough, New Zealand.

Algae Types

A feasibility study using marine microalgae in a photobioreactor is being done by The International Research Consortium on Continental Margins at the International University Bremen.

Research into algae for the mass-production of oil is mainly focused on microalgae; organisms capable of photosynthesis that are less than 0.4 mm in diameter, including the diatoms and cyanobacteria; as opposed to macroalgae, e.g. seaweed. However, some research is being done into using seaweeds for biofuels, probably due to the high availability of this resource. This preference towards microalgae is due largely to its less complex structure, fast growth rate, and high oil content (for some species). Some commercial interests into large scale algal-cultivation systems are looking to tie in to existing infrastructures, such as coal power plants or sewage treatment facilities. This approach not only provides the raw materials for the system, such as CO_2 and nutrients; but it changes those wastes into resources.

Aquaflow Bionomic Corporation of New Zealand announced that it has produced its first sample of homegrown bio-diesel fuel with algae sourced from local sewerage ponds. A small quantity of laboratory produced oil was mixed with 95 per cent regular diesel.

The Department of Environmental Science at Ateneo de Manila University in the Philippines, is working on producing biofuel from algae, using a local species of algae.

NBB's Feedstock Development Programme is addressing production of algae on the horizon to expand available material for biodiesel in a sustainable manner.

The following species listed are currently being studied for their suitability as a mass-oil producing crop, across various locations worldwide:

- Botryococcus braunii
- Chlorella
- Dunaliella tertiolecta
- Gracilaria
- Pleurochrysis carterae (also called CCMP647).
- Sargassum, with 10 times the output volume of Gracilaria.

Nutrients

Nutrients like nitrogen (N), phosphorus (P), and potassium (K), are important for plant growth and are essential parts of fertilizer.

Silica and iron, as well as several trace elements, may also be considered important marine nutrients as the lack of one can limit the growth of, or productivity in, an area.

One company, Green Star Products, announced their development of a micronutrient formula to increase the growth rate of algae. According to the company, its formula can increase the daily growth rate by 34 per cent and can double the amount of algae produced in one growth cycle.

Wastewater

A possible nutrient source is waste water from the treatment of sewage, agricultural, or flood plain run-off, all currently major pollutants and health risks. However, this waste water cannot feed algae directly and must first be processed by bacteria, through anaerobic digestion. If waste water is not processed before it reaches the algae, it will contaminate the algae in the reactor, and at the very least, kill much of the desired algae strain. In biogas facilities, organic waste is often converted to a mixture of carbon dioxide, methane, and organic fertilizer. Organic fertilizer that comes out of digester is liquid, and nearly suitable for algae growth, but it must first be cleaned and sterilized.

The utilization of wastewater and ocean water instead of freshwater is strongly advocated due to the continuing depletion of freshwater resources. However, heavy metals, trace metals, and other contaminants in wastewater can decrease the ability of cells to produce lipids biosynthetically and also impact various other workings in the machinery of cells. The same is true for ocean water, but the contaminants are found in different concentrations. Thus, agricultural-grade fertilizer is the preferred source of nutrients, but heavy metals are again a problem, especially for strains of algae that are susceptible to these metals. In open pond systems the use of strains of algae that can deal with high concentrations of heavy metals could prevent other organisms from infesting these systems (Schenk *et al.* 2008).

At the Woods Hole Oceanographic Institution and the Harbor Branch Oceanographic Institution the wastewater from domestic and industrial sources contain rich organic compounds that are being used to accelerate the growth of algae.

Also the Department of Biological and Agricultural Engineering of the University of Georgia is exploring microalgal biomass production using industrial wastewater.

Algaewheel, based in Indianapolis, Indiana, presented a proposal to build a facility in Cedar Lake, Indiana that uses algae to treat municipal wastewater and uses the sludge byproduct to produce biofuel.

Investment

There is always uncertainty about the success of new products and investors have to consider carefully the proper energy sources in which to invest. A drop in fossil fuel oil prices might make consumers and therefore investors lose interest in renewable energy. Algal fuel companies are learning that investors have different expectations about returns and length of investments. AlgaePro Systems found in its talks with investors that while one wants at least 5 times the returns on their investment, others would only be willing to invest in a profitable operation over the long term. Every investor has its own unique stipulations that are obstacles to further algae fuel development.

Universities

US Universities working on Oil from Algae:

- The University of Texas at Austin.
- Cal Poly State University, San Luis Obispo.
- Montana State University, Utah State University.
- University of Virginia.
- Arizona State University
- Ohio University
- University of Maine, Orono
- University of Kansas
- Old Dominion University.
- Brooklyn College
- Colorado State University.

- The University of Toledo
- University of California, San Diego.

UK universities working on oil from Algae:

- Brighton University

Research and Promotion

The Ukraine Cabinet plans to produce biofuel of a special type of algae.

Also the CSIC's Instituto de Bioquímica Vegetal y Fotosíntesis (Microalgae Biotechnology Group, in Sevilla, Spain is researching the algal fuels.

Organizations

Algal Biomass Organization (ABO) is formed by Boeing Commercial Airplanes, A2BE Carbon Capture Corporation, National Renewable Energy Labs, Institution of Oceanography, Benemann Associates, Mont Vista Capital and Montana State University.

Global air carriers Air New Zealand, Continental, Virgin Atlantic Airways, and biofuel technology developer UOP LLC, a Honeywell company, will be the first wave of aviation-related members, together with Boeing, to join Algal Biomass Organization.

The National Algae Association (NAA) is a non-profit organization comprised of algae researchers, algae production companies and the investment community who share the goal of commercializing algae oil as an alternative feedstock for the biofuels markets. The NAA gives its members a forum to efficiently evaluate various algae technologies for potential early stage company opportunities.

BOTRYOCOCCUS BRAUNII

Botryococcus braunii (Bb) is a green, pyramid shaped planktonic microalga of the order Chlorococcales (class

Chlorophyceae) that is of potentially great importance in the field of biotechnology. Colonies held together by a lipid biofilm matrix can be found in temperate or tropical oligotrophic lakes and estuaries, and will bloom when in the presence of elevated levels of dissolved inorganic phosphorus. The species is notable for its ability to produce high amounts of hydrocarbons, especially oils in the form of Triterpenes, that are typically around 30-40 percent of their dry weight. Compared to other green algae species it has a relatively thick cell wall that is accumulated from previous cellular divisions; making extraction of cytoplasmic components rather difficult. Fortunately, much of the useful hydrocarbon oil is outside of the cell.

Optimal Growth Environment

Botryococcus braunii has been shown to grow best at a temperature of 23°C, a light intensity of 60 W/M^2, with a light period of 12 hours per day, and a salinity of 0.15 Molar NaCl. However, this was the results of testing with one strain, others certainly vary to some degree. B. braunii is commonly grown in cultures of Chu 13 medium.

Biofuel Applications of Botryococcus Oils

The practice of farming cultivating is known as algaculture. Botryococcus braunii has great potential for algaculture because of the hydrocarbons it produces, which can be chemically converted into fuels. Up to 86 per cent of the dry weight of Botryococcus braunii can be long chain hydrocarbons. The vast majority of these hydrocarbons are botryocuccus oils: botryococcenes, alkadienes and alkatrienes. Transesterification can NOT be used to make biodiesel from botryococcus oils. This is because these oils are not 'vegetable oils' in the common meaning, in which they are fatty acid triglycerides. While botryococcus oils are oils of vegetable origin, they are inedible and chemically very different, being triterpenes, and lack the free oxygen atom needed for transesterification. Botryococcus oils can be used as feedstock for hydrocracking in an oil refinery to produce octane (gasoline, a.k.a. petrol), kerosene, and diesel. Botryococcenes are preferred over alkadienes and alkatrienes for hydrocracking as botryococcenes will likely be transformed into a fuel with a higher octane rating.

Oils

Three major races of Botryococcus braunii are known, and they are distinguished by the structure of their oils. Botryococcenes are unbranched isoprenoid triterpenes having the formula C_nH_{2n-10}. The A race produces alkadienes and alkatrienes (derivatives of fatty acids) wherein n is an odd number 23 through 31. The B race produces botryococcenes wherein n is in the range 30 through 37 biofuels of choice for hydrocracking to gasoline-type hydrocarbons. The "L" strain makes an oil not formed by other strains of Botryococcus braunii. Within this major classification, various strains of Botryococcus will differ in the precise structure and concentrations of the constituent hydrocarbons oils.

According to page 30 on Aquatic Species Programme (ASP) report, the A-strain of *Botryococcus* would not function well as a feedstock for lipid based fuel production due to its slow growth (one doubling every 72 hours). However, subsequent research by Qin showed that the doubling time could be reduced to 48 hours in its optimal growth environment. In view of findings by Frenz (5), the doubling times may not be as important as the method of hydrocarbon harvest. The ASP also found A-strain *Botryococcus* oil to be less than ideal, having most of its lipids as C_{29} to C_{34} aliphatic hydrocarbons, and less abundance of C_{18} fatty acids. This evaluation of Bb oils was done in relation to their suitability for transesterification (i.e. creating biodiesel), which was the focus of the ASP at the time Bb was evaluated. The ASP did not study Bb oils for their suitability in hydrocracking, as some subsequent studies have done on the "B" race.

Hydrocarbon Oil Constituents of Botryococcus braunii

Compound	% mass
Isobotryococcene	4%
Botryococcene	9%
$C_{34}H_{58}$	11%
$C_{36}H_{62}$	34%
$C_{36}H_{62}$	4%
$C_{37}H_{64}$	20%
Other hydrocarbons	18%

Note: The two listed $C_{36}H_{62}$ entries are not typos; they are for two different isomers.

Botryococcus Braunii Specimen Sources

The prices below are subject to change, and many organizations have restrictions and special pricing regimes. None of the prices include shipping costs.

- SERI US strain BOTRY1
- CCALA CZ strains DROOP 1950, HEGEWALD 1977, SANTOS 1975, SANTOS 1997 €13
- CCAC DE strain CCAC02129 (Vienna 1997) €30 non-profit, €100 commercial—cannot resell to other parties
- SAG DE strains DROOP 1950, HEGEWALD 1977 €12.50 subsidized rate, 50 standard rate
- MCC-NIES JP strain KAGOSHIMA 1997 ¥6000 academic, ¥10000 commercial
- CCAP UK strains DROOP 1950, JAWORSKI 1984 £30 non-profit, £60 commercial
- UTEX US strains HEGEWALD 1977, DROOP 1950 $30 US Government and Academic, $75 commercial
- ACOI PT strains SANTOS 1975, SANTOS 1997 €10 academic inside Portugal, €20 other academic, €50 commercial
- BCC US has been 'coming soon' for some time now $15 academic, $50 commercial
- Susquehanna Biotech LLC US Live, Dense culture available

Patented Strains

In 1988, UCBerkeley was granted US Plant Patent 6169 for *Botryococcus braunii variety Showa*, developed by UC Berkeley scientist Dr. Arthur Nonomura, in the Melvin Calvin Laboratory as part of the Nobel laureate's groundbreaking interdisciplinary Programme for the development of renewable transport fuels. The proprietary variety was notable, says the patent application, because of its highly reproducible botryococcenes hydrocarbon content comprising 20 per cent of the dry weight of "Showa." It is clear that *Showa* was borne out as the top source of hydrocarbons of its time. The patent expired in April 2008.

In May 2006, Dr. Nonomura filed an international patent application disclosing novel growth and harvesting processes for the Chlorophyta. A separate patent for plants is also filed on *Botryococcus braunii variety Ninsei* that exhibits the feature of extracolonial secretion of it botryococcenoids that can be processed in existing gasoline refineries to transport fuels.

UNDERSTANDING ALGACULTURE

Algaculture is a form of aquaculture involving the farming of species of algae.

The majority of algae that are intentionally cultivated fall into the category of microalgae (also referred to as phytoplankton, microphytes, or planktonic algae). Macroalgae, commonly known as seaweed, also have many commercial and industrial uses, but due to their size and the specific requirements of the environment in which they need to grow, they do not lend themselves as readily to cultivation.

Commercial and industrial algae cultivation produces food ingredients food, fertilizer, bioplastics, dyes and colorants, chemical feedstock, pharmaceuticals, pollution control, algal fuel and many others.

Growing, Harvesting, and Processing Algae

Monoculture

Most growers prefer monocultural production and go to

considerable lengths to maintain the purity of their cultures. With mixed cultures, one species comes to dominate over time and if a non-dominant species is believed to have particular value, it is necessary to obtain pure cultures in order to cultivate this species. Individual species cultures are also needed for research purposes.

A common method of obtaining pure cultures is serial dilution. Cultivators dilute a wild sample or a lab sample containing the desired algae with filtered water and introduce small aliquots into a large number of small growing containers. Dilution follows a microscopic examination of the source culture that predicts that a few of the growing containers contain a single cell of the desired species. Following a suitable period on a light table, cultivators again use the microscope to identify containers to start larger cultures.

Alternatively, mixed algae cultures can work well for larval mollusks. First, the cultivator filters the sea water to remove algae which are too large for the larvae to eat. Next, the cultivator adds nutrients and possibly aerates the result. After one or two days in a greenhouse or outdoors, the resulting thin soup of mixed algae is ready for the larvae. An advantage of this method is low maintenance.

Growing algae

Water, carbon dioxide, minerals and light are all important factors in cultivation, and different algae have different requirements. The basic reaction in water is carbon dioxide + light energy = glucose + oxygen .

Temperature

The water must be in a temperature range that will support the specific algal species being grown.

Light And Mixing

In most algal-cultivation systems, light only penetrates the top 3 to 4 inches (76-100 mm) of the water. As the algae grow and multiply, the culture becomes so dense that it blocks light from reaching deeper into the water. Direct sunlight is too strong for most algae, which need only about 1/10 the amount of light they receive from direct sunlight.

To use deeper ponds, growers agitate the water, circulating the algae so that it does not remain on the surface. Paddle wheels can stir the water and compressed air coming from the bottom lifts algae from the lower regions. Agitation also helps prevent over-exposure to the sun.

Another means of supplying light is to place the light *in* the system. Glow plates made from sheets of plastic or glass and placed within the tank offer precise control over light intensity.

Odor and Oxygen

The odor associated with bogs, swamps, indeed any stagnant waters, can be due to oxygen depletion caused by the decay of deceased algal blooms. Under anoxic conditions, the bacteria inhabiting algae cultures break down the organic material and produce hydrogen sulfide and ammonia which causes the odor. This hypoxia often results in the death of aquatic animals. In a system where algae is intentionally cultivated, maintained, and harvested, neither eutrophication nor hypoxia are likely to occur.

Some living algae also produce odorous chemicals, particularly certain blue-green algae (cyanobacteria) such as *Anabaena*. The most well-known of these odor-causing chemicals are MIB (2-methylisoborneol) and geosmin. They give a musty or earthy odor that can be quite strong. Eventual death of the cyanobacteria releases additional gas that is trapped in the cells. These chemicals are detectable at very low levels, in the parts per billion range, and are responsible for many "taste and odor" issues in drinking water treatment and distribution. Cyanobacteria can also produce chemical toxins that have been a problem in drinking water.

Nutrients

Nutrients can accelerate growth, but excess nutrients go to waste.

Pond and Bioreactor Cultivation Methods

Algae can be cultured in open-ponds (such as raceway-type ponds and lakes) and photobioreactors. Raceway ponds may be less expensive.

Open-ponds: Raceway-type ponds and lakes are open to

the elements. Open ponds are highly vulnerable to contamination by other microorganisms, such as other algal species or bacteria. Thus cultivators usually choose closed systems for monocultures. Open systems also do not offer control over temperature and lighting. The growing season is largely dependent on location and, aside from tropical areas, is limited to the warmer months.

Open pond systems are cheaper to construct, at the minimum requiring only a trench or pond. Large ponds have the largest production capacities relative to other systems of comparable cost. Also, open pond cultivation can exploit unusual conditions that suit only specific algae. For instance, *Spirulina sp.* thrives in water with a high concentration of sodium bicarbonate and *Dunaliela salina* grow in extremely salty water. Open culture can also work if there is a system of culling the desired algae and inoculating new ponds with a high starting concentration of the desired algae.

Some chain diatoms fall into this category since they can be filtered from a stream of water flowing through an outflow pipe. A "pillow case" of a fine mesh cloth is tied over the outflow pipe allowing other algae to escape. The chain diatoms are held in the bag and feed shrimp larvae (in Eastern hatcheries) and inoculate new tanks or ponds.

Enclosing a pond with a transparent or translucent barrier effectively turns it into a greenhouse. This solves many of the problems associated with an open system. It allows more species to be grown; it allows the species that are being grown to stay dominant; and it extends the growing season—and if heated the pond can produce year round.

Photobioreactors: Algae can also be grown in a photobioreactor (PBR). A PBR is a bioreactor which incorporates a light source. Virtually any translucent container could be called a PBR, however the term is more commonly used to define a closed system, as opposed to an open tank or pond.

Because PBR systems are closed, the cultivator must provide all nutrients, including CO_2.

A PBR can operate in "batch mode", which involves restocking the reactor after each harvest, but it is also possible to grow and harvest continuously. Continuous operation requires precise control of all elements to prevent immediate collapse. The grower provides sterilized water, nutrients, air, and carbon dioxide at the correct rates. This allows the reactor to operate for long periods. An advantage is that algae that grows in the "log phase" is generally of higher nutrient content than old "senescent" algae. Maximum productivity occurs when the "exchange rate" (time to exchange one volume of liquid) is equal to the "doubling time" (in mass or volume) of the algae.

Different types of PBRs include:

- tanks
- Polyethylene sleeves or bags
- Glass or plastic tubes.

Harvesting

Algae can be harvested using microscreens, by centrifugation, by flocculation. and by froth flotation.

Interrupting the carbon dioxide supply can cause algae to flocculate on its own, which is called "autoflocculation".

"Chitosan", a commercial flocculant, more commonly used for water purification, is far more expensive. The powdered shells of crustaceans are processed to acquire chitin, a polysaccharide found in the shells, from which chitosan is derived via deacetylation. Water that is more brackish, or saline requires larger amounts of flocculant. Flocculation is often too expensive for large operations.

Alum and ferric chloride are other chemical flocculants.

In froth flotation, the cultivator aerates the water into a froth, and then skims the algae from the top.

Ultrasound and other harvesting methods are currently under development.

Oil Extraction

Algae oils have a variety of commercial and industrial uses, and are extracted through a wide variety of methods. Estimates of the cost to extract oil from microalgae vary, but are likely to be around $1.80 (US$)/kg (compared to $0.50 (US$)/kg for palm oil).

Physical Extraction

In the first step of extraction, the oil must be separated from the rest of the plant. The simplest method is mechanical crushing. When algae is dried it retains its oil content, which then can be "pressed" out with an oil press. Many commercial manufacturers of vegetable oil use a combination of mechanical pressing and chemical solvents in extracting oil. Since different strains of algae vary widely in their physical attributes, various press configurations (screw, expeller, piston, etc) work better for specific algae types. Often, mechanical crushing is used in conjunction with chemical solvents, as described below.

Osmotic shock is a sudden reduction in osmotic pressure, this can cause cells in a solution to rupture. Osmotic shock is sometimes used to release cellular components, such as oil.

Ultrasonic extraction, a branch of sonochemistry, can greatly accelerate extraction processes. Using an ultrasonic reactor, ultrasonic waves are used to create cavitation bubbles in a solvent material. When these bubbles collapse near the cell walls, the resulting shock waves and liquid jets cause those cells walls to break and release their contents into a solvent. Ultrasonication can enhance basic enzymatic extraction. The combination "sonoenzymatic treatment" accelerates extraction and increases yields.

Chemical Extraction

Chemical solvents are often used in the extraction of the oils. The downside to using solvents for oil extraction are the dangers involved in working with the chemicals. Care must be taken to avoid exposure to vapors and skin contact, either of which can cause serious health damage. Chemical solvents also present an explosion hazard.

A common choice of chemical solvent is hexane, which is widely used in the food industry and is relatively inexpensive. Benzene and ether can also separate oil. Benzene is classified as a carcinogen.

Another method of chemical solvent extraction is Soxhlet extraction. In this method, oils from the algae are extracted through repeated washing, or percolation, with an organic solvent such as hexane or petroleum ether, under reflux in a special glassware. The value of this technique is that the solvent is reused for each cycle.

Enzymatic extraction uses enzymes to degrade the cell walls with water acting as the solvent. This makes fractionation of the oil much easier. The costs of this extraction process are estimated to be much greater than hexane extraction. The enzymatic extraction can be supported by ultrasonication. The combination "sonoenzymatic treatment" causes faster extraction and higher oil yields.

Supercritical CO_2 can also be used as a solvent. In this method, CO_2 is liquefied under pressure and heated to the point that it becomes supercritical (having properties of both a liquid and a gas), allowing it to act as a solvent.

Other methods are still being developed, including ones to extract specific types of oils, such as those with a high production of long-chain highly unsaturated fatty acids.

Algal Culture Collections

Specific algal strains can be acquired from algal culture collections.

Uses of Algae

Dulse is one of many edible algae.

Food

Several species of algae are raised for food.

- Purple laver (*Porphyra*) is perhaps the most widely domesticated marine algea. In Asia it is used in nori (Japan) and gim (Korea). In Wales, it is used in laverbread, a traditional food, and in Ireland it is collected and made into a jelly by stewing or boiling. Preparation also can involve frying or heating the fronds with a little water and beating with a fork to produce a pinkish jelly. Harvesting also occurs along the west coast of North America, and in Hawaii and New Zealand.

- Dulse ("Palmaria palmata") is a red species sold in Ireland and Atlantic Canada. It is eaten raw, fresh, dried, or cooked like spinach.

- Spirulina (*Arthrospira platensis*) is a blue-green microalgae with a long history as a food source in East Africa and pre-colonial Mexico. Spirulina is high in protein and other nutrients, finding use as a food supplement and for malnutrition. Spirulina thrives in open systems and commercial growers have found it well-suited to cultivation. One of the largest production sites is Lake Texcoco in central Mexico. The plants produce a variety of nutrients and high amounts of protein. Spirulina is often used commercially as a nutritional supplement.

- Chlorella, another popular microalgae, has similar nutrition to spirulina. Chlorella is very popular in Japan. It is also used as a nutritional supplement with possible effects on metabolic rate. Some allege that Chlorella can reduce mercury levels in humans (supposedly by chelation of the mercury to the cell wall of the organism).

- Irish moss (*Chondrus crispus*), often confused with *Mastocarpus stellatus*, is the source of carrageenan, which is used as a stiffening agent in instant puddings, sauces, and dairy products such as ice cream. Irish moss is also used by beer brewers as a fining agent.

- Sea lettuce (*Ulva lactuca*), is used in Scotland where it is added to soups and salads. Dabberlocks or badderlocks (*Alaria esculenta*) is eaten either fresh or cooked in Greenland, Iceland, Scotland and Ireland.

- Aphanizomenon flos-aquae is a cyanobacteria similar to spirulina, which is used as a nutitional supplement.

- Extracts and oils from algae are also used as additives in various food products. The plants also produce Omega-3 and Omega-6 fatty acids, which are commonly found in fish oils, and which have been shown to have positive health benefits.

Fertilizer and agar

For centuries seaweed has been used as fertilizer. It is also an excellent source of potassium for manufacture of potash and potassium nitrate.

Both microalgae and macroalgae are used to make agar.

Pollution Control

With concern over global warming, new methods for the thorough and efficient capture of CO_2 are being sought out. The carbon dioxide that a carbon-fuel burning plant produces can feed into open or closed algae systems, fixing the CO_2 and accelerating algae growth. Untreated sewage can supply additional nutrients, thus turning two pollutants into valuable commodities.

Algae cultivation is under study for uranium/plutonium sequestration and purifying fertilizer runoff.

Energy Production

Business, academia and governments are exploring the possibility of using algae to make gasoline, diesel and other fuels.

Other Uses

Chlorella, particularly a transgenic strain which carries an extra mercury reductase gene, has been studied as an agent for environmental remediation due to its ability to reduce Hg^{2+} to the less toxic elemental mercury.

Cultivated algae serve many other purposes, including bioplastic production, dyes and colorant production, chemical feedstock production, and pharmaceutical ingredients.

LIST OF ALGAL FUEL PRODUCERS

Israel

- In June 2008, Tel Aviv-based Seambiotic and Seattle-based Inventure Chemical announced a joint venture to use CO2 emissions-fed algae to make ethanol and biodiesel at a biofuel plant in Ashkelon, Israel.

Spain

- AlgaeLink, in Cadiz
- Bio Fuel Systems

The Netherlands

- AlgaeLink, a company from BioKing, that produces for KLM and cooperates with GE Aviation. (But that's not what KLM will confirm) (also as yet to have produced evidence of working commercial system)
- Ingrepro BV that produces algae in integrated systems using waste waters. They are the largest industrial producer of algaebiomass in Europe

Argentina

- Oil Fox, Chubut, 10 T/day.

Canada

- Algae Fuel System, Saskatoon, Canada
- International Energy, Inc. (OTCBB: IENI.OB)

USA

There are diverse companies developing biofuels from algae:

- Algae Floating Systems, Inc.
- AlgaeFuel, based in Berkeley, California.
- Algae Fuel System, Ukiah, CA.
- Algaewheel, based in Indianapolis, Indiana. AlgalOilDiesel, LLP, based in Corvallis, Oregon.
- Applied Research Associates, Inc., based in Albuquerque, New Mexico, algae biofuel research in Panama City, Florida office.
- Aquatic Energy.
- Algoil Industries, Inc.
- Aurora BioFuels

- Blue Marble Energy
- Cellana (Shell and HR BioPetroleum).
- Chevron Corporation (in collaboration with US-DOE NREL).
- Dao Energy, LLC is a Texas registered company with an office in Chengdu, Sichuan, China
- <T • Diversified Energy Corporation.
- Global Green Solutions
- GreenerBioEnergy
- GreenFuel Technologies Corporation
- Imperium Renewables, former Seattle Biodiesel, LLC.
- Inventure Chemical
- Kai BioEnergy Corp.
- Live Fuels, Inc.
- PetroSun and Algae BioFuels Inc., wholly-owned subsidiary. In Scottsdale, Arizona.
- Phycal LLC
- Sapphire Energy, financed by the former Microsoft chairman Bill Gates.
- Solazyme, Inc. that supplies algae oil to Imperium Renewables. It had entered into a biodiesel feedstock development and testing agreement with Chevron Technology Ventures.
- Solix Biofuels in Boulder, Colorado.
- Susquehanna Biotech, LLC
- Vertigro, a Valcent Products´ joint venture with Global Green Solutions, in El Paso, Texas. Uses the High Density Vertical Bioreactor.
- Virgin Green Fund
- <H2>Australia
- Biomax biodiesel produces and distributes biodiesel made from algal oil and recycled cooking oil.

New Zealand

- Aquaflow Bionomic Corporation (ABC). Boeing and Air New Zealand announced a joint project with Aquaflow Bionomic to develop algae jet fuel.

DIRECTORY OF BIODIESEL FROM ALGAE OIL

The advantages of deriving biodiesel from algae include rapid growth rates, a high per-acre yield; and algae biofuel contains no sulfur, is non-toxic, and is highly biodegradable. Some species of algae are ideally suited to biodiesel production due to their high oil content—in some species, topping out near 50 per cent.

Overview

- **Algae: 'The ultimate in renewable energy'**—Some types of algae are about 50 percent oil, suitable for biodiesel. The U.S. government has been experimenting with algae off and on for 18 years. There may be hundreds of thousands of species not yet identified. (*CNN*; March 25, 2008)
- **Oilgae.com**—Biodiesel from Algae Oil—Info, Resources, News and Links Algae range from small, single-celled organisms to multi-cellular organisms, some with fairly complex differentiated form. Algae are usually found in damp places or bodies of water and thus are common in terrestrial as well as aquatic environments. Like plants, algae require primarily three components to grow: sunlight, carbon-di-oxide and water. Photosynthesis is an important biochemical process in which plants, algae, and some bacteria convert the energy of sunlight to chemical energy. The existing large-scale natural sources are of algae are: Bogs, marshes and swamps—Salt marshes and salt lakes. Microalgae contain lipids and fatty acids as membrane components, storage products, metabolites and sources of energy. Algae contain anywhere between 2 per cent and 40 per cent of lipids/oils by weight. There are three well-known methods to extract the oil from oilseeds, and these methods should apply equally well for algae too: 1. Expeller/Press 2. Hexane solvent oil extraction 3. Supercritical Fluid extraction

Biomass Yield

These are the yields obtained in stable cultures during an entire year, as reported by the Aquatic Species Programme. All other results from that Programme show either unstable growth or yields obtained over short periods of time (often, during winter months when algae productivity drops significantly, the Aquatic Species Programme discontinued the cultures.)

Metric Tons/Hectare/Year

M. minutum (algae), 1989.....35.8

M. minutum (algae), 1989.....30.3

M. minutum (algae), 1990.....38.3

Algae (no species mentioned), 1978.....43.8

Algae (no species mentioned), 1978.....51.1

Sugarcane.....79.2 [Brazilian average, 2005]

Sorghum.....70 [Average for Andhra Pradesh, India, 2005]

Cassava.....65 [Nigeria, 1985]

Oil palm.....50 [Global average, including low yields in Africa; in Malaysia, average yields are 75 MT/ha/yr]

Arundo Donax.....50 [Grown in sub-tropics, Handbook of Energy Crops]

Oil Yield

Gallons of Oil per Acre per Year—

Corn	15	
Soybeans	48	
Safflower	83	
Sunflower	102	
Rapeseed	127	
Oil Palm	635	
Micro Algae	1850	[based on actual biomass yields]
Micro Algae	5000-15000	[theoretical laboratory yield]

EVALUATING MICROBIAL FUEL CELL

A **microbial fuel cell (MFC)** or **biological fuel cell** is a bio-electrochemical system that drives a current by mimicking bacterial interactions found in nature. Mediator-less MFCs are a much more recent development and due to this the factors that affect optimum operation, such as the bacteria used in the system, the type of ion membrane, and the system conditions such as temperature, are not particularly well understood. Bacteria in mediator-less MFCs typically have electrochemically-active redox enzymes such as cytochromes on their outer membrane that can transfer electrons to external materials (Min, et al., 2005).

Microbial Fuel Cell

A microbial fuel cell is a device that converts chemical energy to electrical energy by the catalytic reaction of microorganisms (Allen and Bennetto, 1993). A typical microbial fuel cell consists of anode and cathode compartments separated by a cation specific membrane. In the anode compartment, fuel is oxidized by microorganisms, generating electrons and protons. Electrons are transferred to the cathode compartment through an external electric circuit, and the protons are transferred to the cathode compartment through the membrane. Electrons and protons are consumed in the cathode compartment, combining with oxygen to form water. In general, there are two types of microbial fuel cells, mediator and mediator-less microbial fuel cells.

Mediator Microbial Fuel Cell

Most of the microbial cells are electrochemically inactive. The electron transfer from microbial cells to the electrode is facilitated by mediators such as thionine, methyl viologen, methyl blue, humic acid, neutral red and so on (Delaney et al., 1984; Lithgow et al., 1986). Most of the mediators available are expensive and toxic.

Mediator-Less Microbial Fuel Cell

Mediator-less microbial fuel cells have been engineered at the Korea Institute of Science and Technology, by a team led by Kim, Byung Hong. A mediator-less microbial fuel cell does not require a mediator but uses electrochemically active bacteria to transfer electrons to the electrode (electrons are carried directly from the bacterial respiratory enzyme to the electrode). Among the electrochemically active bacteria are, *Shewanella putrefaciens* (Kim *et al.*, 1999a), *Aeromonas hydrophila* (Cuong *et al.*, 2003), and others. Şome bacteria, which have pili on their external membrane, are able to transfer their electron production via these pili.

Mediator-less Microbial Fuel Cell can besides running on wastewater also derive energy directly from (certain) aquatic plants. These include reed sweetgrass, cordgrass, rice, tomatoes, lupines, algae

Microbial Electrolysis Cell

A step further than the mediator-less MFC is the Microbial electrolysis cells (MEC). This cell, that again requires no mediator and has been invented by René Rozendal.

Generating Electricity

When micro-organisms consume a substrate such as sugar in aerobic conditions they produce carbon dioxide and water. However when oxygen is not present they produce carbon dioxide, protons and electrons as described below (Bennetto, 1990):

$$C_{12}H_{22}O_{11} + 13H_2O \rightarrow 12CO_2 + 48H^+ + 48e^- \quad ...(1)$$

Microbial fuel cells use inorganic mediators to tap into the electron transport chain of cells and steal the electrons that are produced. The mediator crosses the outer cell lipid membranes and plasma wall; it then begins to liberate electrons from the electron transport chain that would normally be taken up by oxygen or other intermediates. The now-reduced mediator exits the cell laden with electrons that it shuttles to an electrode where it deposits them; this electrode becomes the electro-generic anode (negatively charged electrode). The release of the electrons means that the mediator returns to its original oxidised state ready to repeat the process. It is important to note that this can only happen under anaerobic conditions, if oxygen is present then it will collect all the electrons as it has a greater electronegativity than the mediator.

In a microbial fuel cell operation, the anode is the terminal electron acceptor recognized by bacteria in the anodic chamber. Therefore, the microbial activity is strongly dependent on the redox potential of the anode. In fact, it was recently published that a Michaelis-Menten curve was obtained between the anodic potential and the power output of an acetate driven microbial fuel cell (Cheng *et al.*, 2008). A critical anodic potential seemed to exists at which a maximum power output of a microbial fuel cell is achieved (Cheng *et al.*, 2008).

A number of mediators have been suggested for use in microbial fuel cells. These include natural red, methylene blue, thionine or resorfuin (Bennetto, *et al.*, 1983).

This is the principle behind generating a flow of electrons from most micro-organisms. In order to turn this into a usable supply of electricity this process has to be accommodated in a fuel cell.

In order to generate a useful current it is necessary to create a complete circuit, not just shuttle electrons to a single point.

The mediator and micro-organism, in this case yeast, are mixed together in a solution to which is added a suitable substrate such as glucose. This mixture is placed in a sealed chamber to stop oxygen entering, thus forcing the micro-organism to use anaerobic respiration. An electrode is placed in the solution that will act as the anode as described previously.

In the second chamber of the MFC is another solution and electrode. This electrode, called the cathode is positively charged and is the equivalent of the oxygen sink at the end of the electron transport chain, only now it is external to the biological cell. The solution is an oxidizing agent that picks up the electrons at the cathode. As with the electron chain in the yeast cell, this could be a number of molecules such as oxygen. However, this is not particularly practical as it would require large volumes of circulating gas. A more convenient option is to use a solution of a solid oxidizing agent.

Connecting the two electrodes is a wire (or other electrically conductive path which may include some electrically powered device such as a light bulb) and completing the circuit and connecting the two chambers is a salt bridge or ion-exchange membrane. This last feature allows the protons produced, as described in Eqt. 1 to pass from the anode chamber to the cathode chamber.

The reduced mediator carries electrons from the cell to the electrode. Here the mediator is oxidized as it deposits the electrons. These then flow across the wire to the second electrode, which acts as an electron sink. From here they pass to an oxidising material.

Uses

Power Generation

Microbial fuel cells have a number of potential uses. The first and most obvious is harvesting the electricity produced for a power source. Virtually any organic material could be used to 'feed' the fuel cell. MFCs could be installed to wastewater treatment plants. The bacteria would consume waste material from the water and produce supplementary power for the plant. The gains to be made from doing this are that MFCs are a very clean and efficient method of energy production. Chemical processing wastewater (Venkata Mohan, *et al.*, 2008a, b) and designed synthetic wastewater (Venkata Mohan, *et al.*, 2007, 2008c) have been used to produce biolectricty in dual and single chambered mediatorless MFCs (non-coated graphite electrodes) (Venkata Mohan, *et al.*, 2008d, e) apart from wastewater treatment. Higher power production was observed with biofilm covered anode (graphite) (Venkata Mohan, *et al.*, 2008e). A fuel cell's emissions are well below regulations (Choi, *et al.*, 2000). MFCs also use energy much more efficiently than standard combustion engines which are limited by the Carnot Cycle. In theory an MFC is capable of energy efficiency far beyond 50 per cent (Yue and Lowther, 1986). According to new research conducted by René Rozendal, using the new microbial fuel cells, conversion of the energy to hydrogen is 8x as high as conventional hydrogen production technologies.

However MFCs do not have to be used on a large scale, as

the electrodes in some cases need only be 7 ìm thick by 2 cm long (Chen, *et al.*, 2001). The advantages to using an MFC in this situation as opposed to a normal battery is that it uses a renewable form of energy and would not need to be recharged like a standard battery would. In addition to this they could operate well in mild conditions, 20°C to 40°C and also at pH of around 7 (Bullen, *et al.*, 2005). Although more powerful than metal catalysts, they are currently too unstable for long term medical applications such as in pacemakers (Biotech/Life Sciences Portal).

Besides wastewater power plants, as mentioned before, energy can also be derived directly from crops. This allows the set-up of power stations based on algae platforms or other plants incorporating a large field of aquatic plants. According to Bert Hamelers, the fields are best set-up in synergy with existing renewable plants (eg offshore windturbines). This reduces costs as the microbial fuel cell plant can then make use of the same electricity lines as the wind turbines.

Further Uses

Since the current generated from a microbial fuel cell is directly proportional to the strength of wastewater used as the fuel, an MFC can be used to measure the strength of wastewater (Kim, *et al.*, 2003). The strength of wastewater is commonly evaluated as biochemical oxygen demand (BOD) values. BOD values are determined incubating samples for 5 days with proper source of microbes, usually activate sludge collected from sewage works. When BOD values are used as a real time control parameter, 5 days' incubation is too long. An MFC-type BOD sensor can be used to measure real time BOD values. Oxygen and nitrate are preferred electron acceptors over the electrode reducing current generation from an MFC. MFC-type BOD sensors underestimate BOD values in the presence of these electron acceptors. This can be avoided by inhibiting aerobic and nitrate respirations in the MFC using terminal oxydase inhibitors such as cyanide and azide [Chang, I.S., Moon, H., Jang, J.K. and Kim, B.H. (2005) Improvement of a microbial fuel cell performance as a BOD sensor using respiratory inhibitors. Biosensors and Bioelectronics

20, 1856-1859.] This type of BOD sensor is commercially available.

Current Research Practices

Currently, most researchers in this field are biologists rather than electrochemists or engineers. This has prompted some researchers (Menicucci, 2005) to point out some undesirable practices, such as recording the maximum current obtained by the cell when connecting it to a resistance as an indication of its performance, instead of the steady-state current that is often a degree of magnitude lower. Sometimes, data about the value of the used resistance is scanty, leading to non-comparable data.

History

At the turn of the last century, the idea of using microbial cells in an attempt to produce electricity was first conceived. M.C. Potter was the first to perform work on the subject in 1911 (Potter, 1911). A professor of botany at the University of Durham Potter managed to generate electricity from *E. coli*, however the work was not to receive any major coverage. In 1931 however Barnet Cohen drew more attention to the area when he created a number of microbial half fuel cells that, when connected in series, were capable of producing over 35 volts, though only with a current of 2 milliamps (Cohen, 1931). More work on the subject came with a study by DelDuca *et al.* who used hydrogen produced by the fermentation of glucose by *Clostridium butyricum* as the reactant at the anode of a hydrogen and air fuel cell. Unfortunately, though the cell functioned it was found to be unreliable due to the unstable nature of the hydrogen production from the micro-organisms (Delduca, *et al.*, 1963). Although this issue was later resolved in work by Suzuki *et al.* in 1976 (Karube, *et al.*, 1976) the current design concept of an MFC came into existence a year later with work once again by Suzuki (Karube, *et al.*, 1977).

Even by the time of Suzuki's work in the late seventies little was understood about how these microbial fuel cells functioned, however the idea was picked up and studied later in more detail first by MJ Allen and then later by H. Peter Bennetto both from King's College London. Bennetto saw the fuel cell as a possible method for the generation of electricity for third world countries. His work, starting in the early 1980s helped build an understanding of how fuel cells operate and until his retirement was seen by many as the foremost authority on the subject.

It is now known that electricity can be produced directly from the degradation of organic matter in a microbial fuel cell, although the exact mechanisms of the process are still to be fully understood. Like a normal fuel cell an MFC has both an anode and a cathode chamber. The anaerobic anode chamber is connected internally to the cathode chamber by an ion exchange membrane, the circuit is completed by an external wire.

In May 2007, the University of Queensland, Australia, completed its prototype MFC, as a cooperative effort with Fosters Brewing Company. The prototype, a 10 liter design, converts the brewery waste water into carbon dioxide, clean water, and electricity. With the prototype proven successful, plans are in effect to produce a 660 gallon version for the brewery, which is estimated to produce 2 kilowatts of power. While it is a negligible amount of power, the production of clean water is of utmost importance to Australia, which is experiencing its worst drought in over 100 years.

Appendix 1

Bioenergy Conversion Factors

Energy contents are expressed here as Lower Heating Value (LHV) unless otherwise stated (this is closest to the actual energy yield in most cases). Higher Heating Value (HHV, including condensation of combustion products) is greater by between 5% (in the case of coal) and 10% (for natural gas), depending mainly on the hydrogen content of the fuel. For most biomass feedstocks this difference appears to be 6-7%. The appropriateness of using LHV or HHV when comparing fuels, calculating thermal efficiencies, etc. really depends upon the application. For stationary combustion where exhaust gases are cooled before discharging (e.g. power stations), HHV is more appropriate. Where no attempt is made to extract useful work from hot exhaust gases (e.g. motor vehicles), the LHV is more suitable.

ENERGY UNITS

Quantities

- 1.0 joule (J) = one Newton applied over a distance of one meter (= 1 kg m²/s²).
- 1.0 joule = 0.239 calories (cal)
- 1.0 calorie = 4.187 J
- 1.0 gigajoule (GJ) = 10^9 joules = 0.948 million Btu = 239 million calories = 278 kWh
- 1.0 British thermal unit (Btu) = 1055 joules (1.055 kJ)
- 1.0 Quad = One quadrillion Btu (10^{15} Btu) = 1.055 exajoules (EJ), or approximately 172 million barrels of oil equivalent (boe)

- 1000 Btu/lb = 2.33 gigajoules per tonne (GJ/t)
- 1000 Btu/US gallon = 0.279 megajoules per liter (MJ/l)

Power

- 1.0 watt = 1.0 joule/second = 3.413 Btu/hr
- 1.0 kilowatt (kW) = 3413 Btu/hr = 1.341 horsepower
- 1.0 kilowatt-hour (kWh) = 3.6 MJ = 3413 Btu
- 1.0 horsepower (hp) = 550 foot-pounds per second = 2545 Btu per hour = 745.7 watts = 0.746 kW

Energy Costs

- $1.00 per million Btu = $0.948/GJ
- $1.00/GJ = $1.055 per million Btu

Some Common Units of Measure

- 1.0 U.S. ton (short ton) = 2000 pounds
- 1.0 imperial ton (long ton or shipping ton) = 2240 pounds
- 1.0 metric tonne (tonne) = 1000 kilograms = 2205 pounds
- 1.0 US gallon = 3.79 liter = 0.833 Imperial gallon
- 1.0 imperial gallon = 4.55 liter = 1.20 US gallon
- 1.0 liter = 0.264 US gallon = 0.220 imperial gallon
- 1.0 US bushel = 0.0352 m^3 = 0.97 UK bushel = 56 lb, 25 kg (corn or sorghum) = 60 lb, 27 kg (wheat or soybeans) = 40 lb, 18 kg (barley)

Areas and Crop Yields

- 1.0 hectare = 10,000 m^2 (an area 100 m x 100 m, or 328 x 328 ft) = 2.47 acres
- 1.0 km^2 = 100 hectares = 247 acres
- 1.0 acre = 0.405 hectares
- 1.0 US ton/acre = 2.24 t/ha
- 1 metric tonne/hectare = 0.446 ton/acre
- 100 g/m^2 = 1.0 tonne/hectare = 892 lb/acre

Biomass Energy

- Cord: a stack of wood comprising 128 cubic feet (3.62 m³); standard dimensions are 4 × 4 × 8 feet, including air space and bark. One cord contains approx. 1.2 U.S. tons (oven-dry) = 2400 pounds = 1089 kg
  - 1.0 metric tonne wood = 1.4 cubic meters (solid wood, not stacked)
  - Energy content of wood fuel (HHV, bone dry) = 18-22 GJ/t (7,600-9,600 Btu/lb)
  - Energy content of wood fuel (air dry, 20% moisture) = about 15 GJ/t (6,400 Btu/lb)
- Energy content of agricultural residues (range due to moisture content) = 10-17 GJ/t (4,300-7,300 Btu/lb)
- Metric tonne charcoal = 30 GJ (= 12,800 Btu/lb) (but usually derived from 6-12 t air-dry wood, i.e. 90-180 GJ original energy content)
- Metric tonne ethanol = 7.94 petroleum barrels = 1262 liters
  - ethanol energy content (LHV) = 11,500 Btu/lb = 75,700 Btu/gallon = 26.7 GJ/t = 21.1 MJ/liter. HHV for ethanol = 84,000 Btu/gallon = 89 MJ/gallon = 23.4 MJ/liter
  - ethanol density (average) = 0.79 g/ml (= metric tonnes/m³)
- Metric tonne biodiesel = 37.8 GJ (33.3 - 35.7 MJ/liter)
  - biodiesel density (average) = 0.88 g/ml (= metric tonnes/m³)

Fossil Fuels

- Barrel of oil equivalent (boe) = approx. 6.1 GJ (5.8 million Btu), equivalent to 1,700 kWh. "Petroleum barrel" is a liquid measure equal to 42 U.S. gallons (35 Imperial gallons or 159 liters); about 7.2 barrels oil are equivalent to one tonne of oil (metric) = 42-45 GJ.
- Gasoline: US gallon = 115,000 Btu = 121 MJ = 32 MJ/liter (LHV). HHV = 125,000 Btu/gallon = 132 MJ/gallon = 35 MJ/liter

- Metric tonne gasoline = 8.53 barrels = 1356 liter = 43.5 GJ/t (LHV); 47.3 GJ/t (HHV)
- gasoline density (average) = 0.73 g/ml (= metric tonnes/m³)
- Petro-diesel = 130,500 Btu/gallon (36.4 MJ/liter or 42.8 GJ/t)
  - petro-diesel density (average) = 0.84 g/ml (= metric tonnes/m³)
- Note that the energy content (heating value) of petroleum products per unit mass is fairly constant, but their density differs significantly – hence the energy content of a liter, gallon, etc. varies between gasoline, diesel, kerosene.
- Metric tonne coal = 27-30 GJ (bituminous/anthracite); 15-19 GJ (lignite/sub-bituminous) (the above ranges are equivalent to 11,500-13,000 Btu/lb and 6,500-8,200 Btu/lb).
  - Note that the energy content (heating value) per unit mass varies greatly between different "ranks" of coal. "Typical" coal (rank not specified) usually means bituminous coal, the most common fuel for power plants (27 GJ/t).
- Natural gas: HHV = 1027 Btu/ft3 = 38.3 MJ/m³; LHV = 930 Btu/ft3 = 34.6 MJ/m³
  - Therm (used for natural gas, methane) = 100,000 Btu (= 105.5 MJ)

Carbon content of fossil fuels and bioenergy feedstocks

- coal (average) = 25.4 metric tonnes carbon per terajoule (TJ)
  - 1.0 metric tonne coal = 746 kg carbon
- oil (average) = 19.9 metric tonnes carbon / TJ
- 1.0 US gallon gasoline (0.833 Imperial gallon, 3.79 liter) = 2.42 kg carbon
- 1.0 US gallon diesel/fuel oil (0.833 Imperial gallon, 3.79 liter) = 2.77 kg carbon
- natural gas (methane) = 14.4 metric tonnes carbon / TJ

- 1.0 cubic meter natural gas (methane) = 0.49 kg carbon
- carbon content of bioenergy feedstocks: approx. 50% for woody crops or wood waste; approx. 45% for graminaceous (grass) crops or agricultural residues

Alcohol Fuels: Alcohol can be blended with gasoline for use as transportation fuel. It may be produced from a wide variety of organic feedstock. The common alcohol fuels are methanol and ethanol. Methanol may be produced from coal, natural gas, wood and organic waste. Ethanol is commonly made from agricultural plants, primarily corn, containing sugar.

Alternative energy: Alternative energy refers to energy sources that have no or fewer undesired consequences than fossil fuels and nuclear energy. They are renewable have lower carbon emissions, compared with conventional energy sources. Alternative energy includes biomass, wind, solar, geothermal and hydroelectric energy.

Anaerobic digestion: Decomposition of biological wastes by micro-organisms, usually under wet conditions, in the absence of air (oxygen), to produce a gas comprising mostly methane and carbon dioxide.

Animal Waste Conversion: Animal waste conversion is the process of obtaining energy from animal wastes. This is a type of biomass energy

B100 - 100% biodiesel

B20, B5, B2 - Petroleum diesel blended with 20%, 5%, and 2% biodiesel, respectively.

Bagasse: Bagasse is the fibrous residue remaining after sugarcane or sorghum stalks are crushed to extract their juice. It can be used as a feedstock for manufacturing cellulosic ethanol.

Barrel of oil equivalent: (boe) The amount of energy contained in a barrel of crude oil, i.e. approximately 6.1 GJ (5.8 million Btu), equivalent to 1,700 kWh. A "petroleum barrel" is a liquid measure equal to 42 U.S. gallons (35 Imperial gallons or 159 liters); about 7.2 barrels are equivalent to one tonne of oil (metric).

Batey: In the Dominican Republic, a batey (plural bateyes) is usually a company town on a sugar plantation where sugar workers live. Initially only for sugar cane cutters themselves, mainly Haitian laborers, many bateyes include families of these workers and have grown beyond the original purpose and size. Some have become municipalities. A batey may include only a few workers and / or their families or may have grown to several hundred families or more.

Biochemical conversion: The use of fermentation or anaerobic digestion to produce fuels and chemicals from organic sources.

Bioconversion: Bioconversion refers to processes that use plants or micro-organisms to change one form of energy into another. For example, an experimental process uses algae to convert solar energy into gas that could be used for fuel.

Biodiesel: Biodiesel is any liquid biofuel suitable as a diesel fuel substitute or diesel fuel additive or extender. Biodiesel fuels are typically made from oils such as soybeans, rapeseed, or sunflowers, or from animal tallow. Biodiesel can also be made from hydrocarbons derived from agricultural products such as rice hulls. Biodiesel is meant to be used in standard diesel engines and is thus distinct from the vegetable and waste oils used to fuel *converted* diesel engines.

Bioenergy: Bioenergy is useful, renewable energy produced from organic matter. The conversion of the complex carbohydrates in organic matter to energy. Organic matter may either be used directly as a fuel or processed into liquids and gases.

Biofuels: Biofuels are liquid fuels and blending components produced from biomass (plant) feedstocks, used primarily for transportation.

Biogas (Biomass gas): Biomass gas is a medium Btu gas containing methane and carbon dioxide, resulting from the action of microorganisms on organic materials such as a landfill

Biomass Converter: A biomass converter is a technical system that converts organic feedstock (biomass) into a technically usable energy carrier:- for example, a steam reformer.

Biomass fuel: Liquid, solid, or gaseous fuel produced by conversion of biomass. Examples include bioethanol from sugar cane or corn, charcoal or woodchips, and biogas from anaerobic decomposition of wastes.

Biomass: Biomass is organic non-fossil material of biological origin constituting a renewable energy source. Examples are wood, agricultural waste and other living-cell material that can be burned to produce heat energy. They also include algae, sewage and other organic substances that may be used to make energy through chemical processes.

Biopower: Short for biomass power.

Bone dry: Having zero percent moisture content. Wood heated in an oven at a constant temperature of 100°C (212°F) or above until its weight stabilizes is considered bone dry or oven dry.

Bottoming cycle: A cogeneration system in which steam is used first for process heat and then for electric power production.

British thermal unit: (Btu) A non-metric unit of heat, still widely used by engineers. One Btu is the heat energy needed to raise the temperature of one pound of water from 60°F to 61°F at one atmosphere pressure. 1 Btu = 1055 joules (1.055 kJ).

Cane ethanol: Cane ethanol is generally available as a by-product of sugar mills producing sugar. It can be used as a fuel, mainly as a biofuel alternative to gasoline, and is widely used in cars in Brazil. It is steadily becoming a promising alternative to gasoline throughout much of the world and thus instead of sugar may be produced as a primary product out of sugar canes processing. (Wikipedia) (article on Sugarcane)

Cap and Trade: Cap and trade is a climate change mitigation program. It is intended to deliver results with a mandatory cap on emissions while providing companies with flexibility in how the sources comply. Successful cap and trade programs reward innovation, efficiency, and early action and provide strict environmental accountability without inhibiting economic growth.

Capacity: The maximum power that a machine or system can produce or carry safely. The maximum instantaneous output of a resource under specified conditions. The capacity of generating equipment is generally expressed in kilowatts or megawatts.

Capital cost: The total investment needed to complete a project and bring it to a commercially operable status. The cost of construction of a new plant. The expenditures for the purchase or acquisition of existing facilities.

Carbon credit: Carbon credits are a key component of national and international attempts to mitigate the growth in concentrations of greenhouse gases (GHGs). One Carbon Credit is equal to one ton of Carbon. Carbon trading is an application of an emissions trading approach. Greenhouse gas emissions are capped and then markets are used to allocate the emissions among the group of regulated sources. The idea is to allow market mechanisms to drive industrial and commercial processes in the direction of low emissions or less "carbon intensive" approaches than are used when there is no cost to emitting carbon dioxide and other GHGs into the atmosphere. Since GHG mitigation projects generate credits, this approach can be used to finance carbon reduction schemes between trading partners and around the world. (Wikipedia)

Cellulose: The principal chemical constituent of cell walls of plants: a long chain of simple sugar molecules.

Cellulosic ethanol: Cellulosic ethanol is produced from cellulosic materials like grasses and wood chips. At present there are no commercial cellulosic ethanol refiners. A brochure explains the complications and steps involved toward making cellulosic ethanol commercially viable.

cfm: Cubic feet per minute (1000 cfm = 0.472 cubic meters per second, m^3/s)

Char: The remains of solid biomass that has been incompletely combusted, such as charcoal if wood is incompletely burned.

Charcoal from paper waste: Paper waste can be pounded, mashed, mixed with sawdust, squeezed into cylinders to make a dry briquette – recycled paper charcoal for use in cooking.

Chipper: A machine that produces wood chips by knife action.

Chips: Woody material cut into short, thin wafers. Chips are used as a raw material for pulping and fiberboard or as biomass fuel.

Cogeneration: Cogeneration (also combined heat and power, CHP) is the use of a heat engine or a power station to simultaneously generate both electricity and useful heat. (Wikipedia)

Combined cycle: Two or more generation processes in series or in parallel, configured to optimize the energy output of the system.

Combined-cycle power plant: The combination of a gas turbine and a steam turbine in an electric generation plant. The waste heat from the gas turbine provides the heat energy for the steam turbine.

Combustion air: The air fed to a fire to provide oxygen for combustion of fuel. It may be preheated before injection into a furnace.

Combustion efficiency: (actual heat produced by combustion) divided by (total heat potential of the fuel consumed)

Combustion: Burning. The transformation of biomass fuel into heat, chemicals, and gases through chemical combination of hydrogen and carbon in the fuel with oxygen in the air.

Commercial forest land: Forested land which is capable of producing new growth at a minimum rate of 20 cubic feet per acre/per year, excluding lands withdrawn from timber production by statute or administrative regulation.

Conifer: Tree, usually evergreen, with cones and needle-shaped or scalelike leaves, producing wood known commercially as softwood.

Coppice regeneration: The ability of certain hardwood species to regenerate by producing multiple new shoots from a stump left after harvest.

Cord: A stack of wood consisting of 128 cubic feet (3.62 cubic meters). A cord has standard dimensions of 4 x 4 x 8 feet, including air space and bark. One cord contains about 1.2 U.S. tons (oven-dry), i.e. 2400 pounds or 1089 kg.

Cross-border development (Haitian-Dominican Border): Haiti and the Dominican Republic share the island of Hispaniola in the Caribbean, with Haiti on the Western and the Dominican Republic on the Eastern side. Haiti occupies about 1/3 and the Dominican Republic about 2/3 of the island. The land border the countries share is 388 km (241 milea) long. The border area is considered potentially useful for growing jatropha and other plant matter suitable for small bioenergy production.

Diameter at breast height: (DBH) The diameter of a tree measured 4 feet 6 inches above the ground.

Digester: An airtight vessel or enclosure in which bacteria decomposes biomass in water to produce biogas.

Discount rate: A rate used to convert future costs or benefits to their present value.

Downdraft gasifier: A gasifier in which the product gases pass through a combustion zone at the bottom of the gasifier.

Dutch oven furnace: One of the earliest types of furnaces, having a large, rectangular box lined with firebrick (refractory) on the sides and top. Commonly used for burning wood. Heat is stored in the refractory and radiated to a conical fuel pile in the center of the furnace.

Effluent: Effluent is the liquid or gas discharged from a process or chemical reactor, usually containing residues from that process.

Emissions: Emissions are releases of gases to the atmosphere caused by humans. In the context of global climate change, they consist of radiatively important greenhouse gases – gases that absorb incoming solar radiation or outgoing infrared radiation, affecting the vertical temperature profile of the atmosphere

Energy Crops: Energy crops are those grown specifically for their fuel value. These include food crops such as corn and sugarcane, and nonfood crops such as poplar trees and switchgrass. Currently, two energy crops are under development: short – rotation woody crops, which are fast – growing hardwood trees harvested in five to eight years, and herbaceous energy crops, such as perennial grasses, which are harvested annually after taking two to three years to reach full productivity.

Ethanol (also known as Ethyl Alcohol or Grain Alcohol, CH3-CH2OH): Ethanol is a clear, colorless flammable oxygenated hydrocarbon with a boiling point of 173.5 degrees Fahrenheit in the anhydrous state. However, it readily forms a binary azeotrope with water, with a boiling point of 172.67 degrees Fahrenheit at a composition of 95.57 percent by weight ethanol. It is used in the United States as a gasoline octane enhancer and oxygenate (maximum 10 percent concentration). Ethanol can be used in higher concentrations (E85) in vehicles designed for its use. Ethanol is typically produced chemically from ethylene, or biologically from fermentation of various sugars from carbohydrates found in agricultural crops and cellulosic residues from crops or wood.

Externality: A cost or benefit not accounted for in the price of goods or services. Often "externality" refers to the cost of pollution and other environmental impacts.

Feedstock: Any material which is converted to another form or product.

Feller-buncher: A self-propelled machine that cuts trees with giant shears near ground level and then stacks the trees into piles to await skidding.

Fermentation: Ethanol fermentation is a biological process in which sugars such as glucose, fructose, and sucrose are converted into cellular energy and thereby produce ethanol and carbon dioxide as metabolic waste products. Because yeasts perform this process in the absence of oxygen, ethanol fermentation is classified as anaerobic.power: (firm energy) Power which is guaranteed by the supplier to be available at all times during a period covered by a commitment. That portion of a customer's energy load for which service is assured by the utility provider.

Fluidized-bed boiler: A large, refractory-lined vessel with an air distribution member or plate in the bottom, a hot gas outlet in or near the top, and some provisions for introducing fuel. The fluidized bed is formed by blowing air up through a layer of inert particles (such as sand or limestone) at a rate that causes the particles to go into suspension and continuous motion. The super-hot bed material increased combustion efficiency by its direct contact with the fuel.

Fly ash: Small ash particles carried in suspension in combustion products.

Forest health: A condition of ecosystem sustainability and attainment of management objectives for a given forest area. Usually considered to include green trees, snags, resilient stands growing at a moderate rate, and endemic levels of insects and disease. Natural processes still function or are duplicated through management intervention.

Forest residues: Material not harvested or removed from logging sites in commercial hardwood and softwood stands as well as material resulting from forest management operations such as precommercial thinnings and removal of dead and dying trees.

Fossil fuel: Solid, liquid, or gaseous fuels formed in the ground after millions of years by chemical and physical changes in plant and animal residues under high temperature and pressure. Oil, natural gas, and coal are fossil fuels.

Fuel cell: A device that converts the energy of a fuel directly to electricity and heat, without combustion.

Fuel cycle: The series of steps required to produce electricity. The fuel cycle includes mining or otherwise acquiring the raw fuel source, processing and cleaning the fuel, transport, electricity generation, waste management and plant decommissioning.

Fuel handling system: A system for unloading wood fuel from vans or trucks, transporting the fuel to a storage pile or bin, and conveying the fuel from storage to the boiler or other energy conversion equipment.

Fuelwood: Fuelwood includes wood and wood products, possibly including coppices, scrubs, branches, etc., bought or gathered, and used by direct combustion.

Furnace: An enclosed chamber or container used to burn biomass in a controlled manner to produce heat for space or process heating.

Gas turbine: (combustion turbine) A turbine that converts the energy of hot compressed gases (produced by burning fuel in compressed air) into mechanical power. Often fired by natural gas or fuel oil.

Gasification: A chemical or heat process to convert a solid fuel to a gaseous form.

Gasifier: A device for converting solid fuel into gaseous fuel. In biomass systems, the process is referred to as pyrolitic distillation.

Genetic selection: Application of science to systematic improvement of a population, e.g. through selective breeding.

Gigawatt: (GW) A measure of electrical power equal to one billion watts (1,000,000 kW). A large coal or nuclear power station typically has a capacity of about 1 GW.

Greenhouse effect: The effect of certain gases in the Earth's atmosphere in trapping heat from the sun.

Greenhouse gases: Gases that trap the heat of the sun in the Earth's atmosphere, producing the greenhouse effect. The two major greenhouse gases are water vapor and carbon dioxide. Other greenhouse gases include methane, ozone, chlorofluorocarbons, and nitrous oxide.

Grid: An electric utility company's system for distributing power.

Habitat: The area where a plant or animal lives and grows under natural conditions. Habitat includes living and non-living attributes and provides all requirements for food and shelter.

Hardwoods: Usually broad-leaved and deciduous trees.

Heat Rate: The amount of fuel energy required by a power plant to produce one kilowatt-hour of electrical output. A measure of generating station thermal efficiency, generally expressed in Btu per net kWh. It is computed by dividing the total Btu content of fuel burned for electric generation by the resulting net kWh generation.

Heat transfer efficiency: useful heat output released / actual heat produced in the firebox

Heating value: The maximum amount of energy that is available from burning a substance.

Hectare: Common metric unit of area, equal to 2.47 acres. 100 hectares = 1 square kilometer.

Herbaceous: Non-woody type of vegetation, usually lacking permanent strong stems, such as grasses, cereals and canola (rape).

Higher heating value: (HHV) The maximum potential energy in dry fuel. For wood, the range is from 7,600 to 9,600 Btu/lb (17.7 to 22.3 GJ/t).

Horsepower: (electrical horsepower; hp) A unit for measuring the rate of mechanical energy output, usually used to describe the maximum output of engines or electric motors. 1 hp = 550 foot-pounds per second = 2,545 Btu per hour = 745.7 watts = 0.746 kW

Hydrocarbon: Any chemical compound containing hydrogen, oxygen, and carbon.

Incinerator: Any device used to burn solid or liquid residues or wastes as a method of disposal. In some incinerators, provisions are made for recovering the heat produced.

Inclined grate: A type of furnace in which fuel enters at the top part of a grate in a continuous ribbon, passes over the upper drying section where moisture is removed, and descends into the lower burning section. Ash is removed at the lower part of the grate.

Incremental energy costs: The cost of producing and transporting the next·available unit of electrical energy. Short run incremental costs (SRIC) include only incremental operating costs. Long run incremental costs (LRIC) include the capital cost of new resources or capital equipment.

Independent power producer: A power production facility that is not part of a regulated utility.

Indirect liquefaction: Conversion of biomass to a liquid fuel through a synthesis gas intermediate step.

Joule: Metric unit of energy, equivalent to the work done by a force of one Newton applied over a distance of one meter (= 1 kg m2/s2). One joule (J) = 0.239 calories (1 calorie = 4.187 J).

Kilowatt hour: (kWh) A measure of energy equivalent to the expenditure of one kilowatt for one hour. For example, 1 kWh will light a 100-watt light bulb for 10 hours. 1 kWh = 3,413 Btu.

Kilowatt: (kW) A measure of electrical power equal to 1,000 watts. 1 kW = 3,413 Btu/hr = 1.341 horsepower.

Landfill gas: A type of biogas that is generated by decomposition of organic material at landfill disposal sites. Landfill gas is approximately 50 percent methane.

Levelized life-cycle cost: The present value of the cost of a resource, including capital, financing and operating costs, expressed as a stream of equal annual payments. This stream of payments can be converted to a unit cost of energy by dividing the annual payment amount by the annual kilowatt-hours produced or saved. By levelizing costs, resources with different lifetimes and generating capabilities can be compared.

Lignin: Structural constituent of wood and (to a lesser extent) other plant tissues, which encrusts the cell walls and cements the cells together.

Megawatt: (MW) A measure of electrical power equal to one million watts (1,000 kW).

Mill residue: Wood and bark residues produced in processing logs into lumber, plywood, and paper.

Mill/kWh: A common method of pricing electricity in the U.S. Tenths of a U.S. cent per kilowatt hour.

MMBtu: One million British thermal units.

Moisture content, dry basis: Moisture content expressed as a percentage of the weight of oven-dry wood, i.e.:
[(weight of wet sample – weight of dry sample) / weight of dry sample] × 100

Moisture content, wet basis: Moisture content expressed as a percentage of the weight of wood as-received, i.e.:
[(weight of wet sample – weight of dry sample) / weight of wet sample] × 100

Moisture content: (MC) The weight of the water contained in wood, usually expressed as a percentage of weight, either oven-dry or as received.

Monoculture: The cultivation of a single species crop.

Net present value: The sum of the costs and benefits of a project or activity. Future benefits and costs are discounted to account for interest costs.

Nitrogen fixation: The transformation of atmospheric nitrogen into nitrogen compounds that can be used by growing plants.

Noncondensing, controlled extraction turbine: A turbine that bleeds part of the main steam flow at one (single extraction) or two (double extraction) points.

Old growth: Timber stands with the following characteristics: large mature and over-mature trees in the overstory, snags, dead and decaying logs on the ground, and a multi-layered canopy with trees of several age classes.

Organic compounds: Chemical compounds based on carbon chains or rings and also containing hydrogen, with or without oxygen, nitrogen, and other elements.

Particulate: A small, discrete mass of solid or liquid matter that remains individually dispersed in gas or liquid emissions. Particulates take the form of aerosol, dust, fume, mist, smoke, or spray. Each of these forms has different properties.

Photosynthesis: Process by which chlorophyll-containing cells in green plants concert incident light to chemical energy, capturing carbon dioxide in the form of carbohydrates.

Pilot scale: The size of a system between the small laboratory model size (bench scale) and a full-size system.

Present value: The worth of future receipts or costs expressed in current value. To obtain present value, an interest rate is used to discount future receipts or costs.

Process heat: Heat used in an industrial process rather than for space heating or other housekeeping purposes.

Producer gas: Fuel gas high in carbon monoxide (CO) and hydrogen (H_2), produced by burning a solid fuel with insufficient air or by passing a mixture of air and steam through a burning bed of solid fuel.

Public utility commissions: State agencies that regulate investor-owned utilities operating in the state.

Public Utility Regulatory Policies Act: (PURPA) A federal law requiring a utility to buy the power produced by a qualifying facility at a price equal to that which the utility would otherwise pay if it were to build its own power plant or buy power from another source.

Pyrolysis: The thermal decomposition of biomass at high temperatures (greater than 400° F, or 200° C) in the absence of air. The end product of pyrolysis is a mixture of solids (char), liquids (oxygenated oils), and gases (methane, carbon monoxide, and carbon dioxide) with proportions determined by operating temperature, pressure, oxygen content, and other conditions.

Quad: One quadrillion Btu (10^{15} Btu) = 1.055 exajoules (EJ), or approximately 172 million barrels of oil equivalent.

Recovery boiler: A pulp mill boiler in which lignin and spent cooking liquor (black liquor) is burned to generate steam.

Refractory Lining: A lining, usually of ceramic, capable of resisting and maintaining high temperatures.

Refuse-derived fuel: (RDF) Fuel prepared from municipal solid waste. Noncombustible materials such as rocks, glass, and metals are removed, and the remaining combustible portion of the solid waste is chopped or shredded. RDF facilities process typically between 100 and 3,000 tons of MSW per day.

Reserve Margin: The amount by which the utility's total electric power capacity exceeds maximum electric demand.

Return on investment: (ROI) The interest rate at which the net present value of a project is zero. Multiple values are possible.

Rotation: Period of years between establishment of a stand of timber and the time when it is considered ready for final harvest and regeneration.

Saturated steam: Steam at boiling temperature for a given pressure.

Shaft horsepower: A measure of the actual mechanical energy per unit time delivered to a turning shaft.

Silviculture: Theory and practice of controlling the establishment, composition, structure and growth of forests and woodlands.

SRIC: Short rotation intensive culture - the growing of tree crops for bioenergy or fiber, characterized by detailed site preparation, usually less than 10 years between harvests, usually fast-growing hybrid trees and intensive management (some fertilization, weed and pest control, and possibly irrigation).

Stand: (of trees) A tree community that possesses sufficient uniformity in composition, constitution, age, spatial arrangement, or condition to be distinguishable from adjacent communities.

Steam turbine: A device for converting energy of high-pressure steam (produced in a boiler) into mechanical power which can then be used to generate electricity.

Superheated steam: Steam which is hotter than boiling temperature for a given pressure.

Surplus electricity: Electricity produced by cogeneration equipment in excess of the needs of an associated factory or business.

Sustainable: An ecosystem condition in which biodiversity, renewability, and resource productivity are maintained over time.

Therm: A unit of energy equal to 100,000 Btus (= 105.5 MJ); used primarily for natural gas.

Thermochemical conversion: Use of heat to chemically change substances from one state to another, e.g. to make useful energy products.

Tipping fee: A fee for disposal of waste.

Ton, tonne: One U.S. ton (short ton) = 2,000 pounds. One Imperial ton (long ton or shipping ton) = 2,240 pounds. One metric tonne (tonne) = 1,000 kilograms (2,205 pounds). One oven-dry ton or tonne (ODT, sometimes termed bone-dry ton/tonne) is the amount of wood that weighs one ton/tonne at 0% moisture content. One green ton/tonne refers to the weight of undried (fresh) biomass material - moisture content must be specified if green weight is used as a fuel measure.

Topping and back pressure turbines: Turbines which operate at exhaust pressure considerably higher than atmospheric (noncondensing turbines). These turbines are often multistage types with relatively high efficiency.

Topping cycle: A cogeneration system in which electric power is produced first. The reject heat from power production is then used to produce useful process heat.

Transmission: The process of long-distance transport of electrical energy, generally accomplished by raising the electric current to high voltages.

Traveling grate: A type of furnace in which assembled links of grates are joined together in a perpetual belt arrangement. Fuel is fed in at one end and ash is discharged at the other.

Turbine: A machine for converting the heat energy in steam or high temperature gas into mechanical energy. In a turbine, a high velocity flow of steam or gas passes through successive rows of radial blades fastened to a central shaft.

Turn down ratio: The lowest load at which a boiler will operate efficiently as compared to the boiler's maximum design load.

Waste streams: Unused solid or liquid by-products of a process.

Water-cooled vibrating grate: A boiler grate made up of a tuyere grate surface mounted on a grid of water tubes interconnected with the boiler circulation system for positive cooling. The structure is supported by flexing plates allowing the grid and grate to move in a vibrating action. Ashes are automatically discharged.

Watershed: The drainage basin contributing water, organic matter, dissolved nutrients, and sediments to a stream or lake.

Watt: The common base unit of power in the metric system. One watt equals one joule per second, or the power developed in a circuit by a current of one ampere flowing through a potential difference of one volt. One Watt = 3.413 Btu/hr.

Wheeling: The process of transferring electrical energy between buyer and seller by way of an intermediate utility or utilities.

Whole-tree harvesting: A harvesting method in which the whole tree (above the stump) is removed.

Yarding: The initial movement of logs from the point of felling to a central loading area or landing.

Bibliography

A Bet on Ethanol, With a Convert at the Helm"; *New York Times*, October 8, 2006, p. 9.

A. Tizzani (2008-04-23). "Moto flex 'imita' carros e deve chegar até dezembro" (in Portuguese). UOL Carros. http://carros.uol.com.br/ultnot/2008/04/23/ult5498u65.jhtm.

Abuelsamid, S. (2008-12-02). "Volvo launches flex fuel S80 and C30 in Thailand to go with new E85 stations". AutoblogGreen. http://www.autobloggreen.com/2008/12/02/volvo-launches-flex-fuel-s80-and-c30-in-thailand-to-go-with-new/.

Aldrett-Lee, S. 2000. Catalytic hydrogenation of liquid ketones with emphasis on gas-liquid mass transfer. Ph.D. dissertation. Eggeman, T., Verser, D., and Weber, E. (2005), An Indirect Route for Ethanol Production US Department of Energy.

Allen, R.M. and Bennetto, H.P. 1993. Microbial fuel cells—Electricity production from carbohydrates. *Appl. Biochem. Biotechnol.*, 39/40, pp. 27-40.

Andrew Williamson. Cambodian Research Centre for Development (c2005). *Biofuel: A Sustainable Solution for Cambodia?*

Balabin, R. M. et al. (2007). "Molar enthalpy of vaporization of ethanol-gasoline mixtures and their colloid state". *Fuel* 86: 323.

Becker J, Boles E (2003) A modified Saccharomyces cerevisiae strain that consumes L-Arabinose and produces ethanol. *Appl Environ Microbiol.* 69(7):4144-50.

Bennetto, H. P., Stirling, J. L., Tanaka, K. and Vega C.A. (1983). Anodic Reaction in Microbial Fuel Cells Biotechnology and Bioengineering, 25, pp 559-568.

Bennetto, H.P. (1990). Electricity Generation by Micro-organisms Biotechnology Education, 1 (4), pp. 163-168.

Bialkowski, M.T.; Pekdemir, T.; Reuben, R.; Brautsch, M.; Towers, D. P.; Elsbett, G. (2005). "Preliminary Approach Towards a CDI System Modification Operating on Neat Rapeseed Oil" (PDF). *Journal of KONES* 12.

BIES L (2006) The Biofuels Explosion: Is Green Energy Good for Wildlife? *Wildlife Society Bulletin*: Vol. 34, No. 4 pp. 1203-1205

Biotechnology Industry Organization (2007). Industrial Biotechnology Is Revolutionizing the Production of Ethanol Transportation Fuel pp. 3-4.

Boone, D. and Mah, R. (2006) Transitional bacteria in anaerobic digestion of biomass, p. 35.

Bradley, M.W., Harris, N., Turner, K. 1982. Process for Hydrogenolysis of Carboxylic Acid Esters WO 82/03854, Nov. 11. *Preparation of esters by reaction of ammonium salts with alcohols.*

Brat D, Boles E and Wiedemann B (2009) Functional expression of a bacterila xylose isomerase in Saccharomyces cerevisie. *Appl. Environ. Microbiol.*

Briggs, Michael. "Widescale Biodiesel Production from Algae" UNH Biodiesel Group (2004).

Brinkman, N., Halsall, R., Jorgensen, S.W., and Kirwan, J.E., "The Development Of Improved Fuel Specifications for Methanol (M85) amd Ethanol (Ed85), SAE Technical Paper 940764.

Brown, L.R. (2006). Plan B 2.0 Rescuing a Planet Under Stress and a Civilization in Trouble W.W. Norton & Co, pp. 228-232.

Bullen, Ra; Arnot, Tc; Lakeman, Jb; Walsh, Fc (May 2006). "Biofuel cells and their development". *Biosensors and bioelectronics* 21 (11): 2015-45.

Burton, George; Holman, John; Lazonby, John (2000). *Salters Advanced Chemistry: Chemical Storylines (2nd ed.).* Heinemann.

Catherine Brahic. Hydrogen injection could boost biofuel production *New Scientist*, March 12, 2007.

Cedric Briens, Jan Piskorz and Franco Berruti, "Biomass Valorization for Fuel and Chemicals Production—A Review," 2008. International Journal of Chemical Reactor Engineering, 6, R2.

Chen, T., S.C. Barton, G. Binyamin, Z Gao, Y. Zhang, H.-H. Kim and A. Heller (2001). "A miniature biofuel cell". *J. Am. Chem. Soc.* 123 (35): 8630-8631.

Cheng, Ky; Ho, G; Cord-Ruwisch, R (May 2008). "Affinity of microbial fuel cell biofilm for the anodic potential.". *Environmental science and technology* 42 (10): 3828-34.

Choi Y., Jung S. and Kim S. (2000) Development of Microbial Fuel Cells Using Proteus Vulgaris Bulletin of the Korean Chemical Society, 21 (1), pp. 44-48

Cohen, B. (1931). The Bacterial Culture as an Electrical Half-Cell, Journal of Bacteriology, 21, pp. 18-19.

Conner AH, Lorenz LF. "Kinetic modeling of hardwood prehydrolysis. Part III. Water and dilute acetic acid Prehydrolysis of southern red oak, *Wood and Fiber Science*, 18(2): 248-263 (1986).

Crutzen, PJ, Mosier AR, Smith KA, Winiwarter W. "Nitrous oxide release from agro-biofuel production negates global warming reduction by replacing fossil fuels" *Atmospheric Chemistry and Physics*. Discuss., 7 11191-11205, (2007)

Cuong, A.P., Jung, S.J., Phung, N.T., Lee, J., Chang, I.S., Kim, B.H., Yi, H. and Chun, J. 2003. A novel electrochemically active and Fe(III)-reducing bacterium phylogenetically related to Aeromonas hydrophila, isolated from a microbial fuel cell. *FEMS Microbiol. Lett.*, Volume 223(1): 129-134.

David Pimintel, Cornell University professor of ecology and agricultural sciences, quoted in: *Consumer Reports*. Special Report: The Ethanol Myth, October 2006, p. 18.

De Oliveira, Marcelo E. Dias de, Burton E. Vaughan and Edward J. Rykiel Jr.; "Ethanol as Fuel: Energy, Carbon dioxide Balances, and Ecological Footprint." BioScience Vol. 55 No. 7, July 2005.

Decker, Jeff. Going Against the Grain: Ethanol from Lignocellulosics, *Renewable Energy World*, January 22, 2009.

Delaney, G.M., Bennetto, H.P., Mason, J.R., Roller, H.D., Stirling, J.L., and Thurston, C.F. 1984. Electron-transfer coupling in microbial fuel cells: 2. Performance of fuel cells containing selected micoorganism-mediator-substrate combinations. *J Chem. Tech. Biotechnol.*, 34B: 13-27.

DelDuca, M. G., Friscoe, J.M. and Zurilla, R.W. (1963). Developments in Industrial Microbiology. American Institute of Biological Sciences, 4, pp. 81-84.

Demain A, Newcomb M, Wu D (March 2005). "Cellulase, Clostridia, and Ethanol. Microbiology". *Molecular Biology Reviews* **69** (69): 124-154.

Diesendorf, Mark (2007). *Greenhouse Solutions with Sustainable Energy*, UNSW Press, p. 293.

Donald Sawyer (2008). "Climate change, biofuels and eco-social impacts in the Brazilian Amazon and Cerrado". *Philosophical Transactions of the Royal Society* 363: 1747.

DuPont, Genencor Form JV to Produce Cellulosic Ethanol. By: Sissell, Kara, Chemical Week, 0009272X, 5/12/2008, Vol. 170, Issue 15

Emma Marris (2006-12-07). "Drink the best and drive the rest". *Nature* 444: 670.

F.K. Agbogbo, M.T. Holtzapple (23 aug 2005). "Fixed-bed fermentation of rice straw and chicken manure using a mixed culture of marine mesophilic microorganisms.". *Bioresource Technology* 98 (8): 1586-1595.

Fairless D. (2007). "Biofuel: The little shrub that could - maybe". *Nature* 449: 652-655.

Fairley, Peter. Growing Biofuels - New production methods could transform the niche technology. *MIT Technology Review* November 23, 2005

Fargione *et al.* (2008). "Land Clearing and the Biofuel Carbon Debt". *Science* 319: 1235.

Farrell AE, Plevin RJ, Turner BT, Jones AD, O'Hare M, Kammen DM (2006-01-27). "Ethanol can contribute to energy and environmental goals". *Science* 311: 506-508.

Fergusen, T. and Mah, R. (2006) Methanogenic bacteria in Anaerobic digestion of biomass, p. 49.

Foody, B.E., Foody, K.J., 1991. Development of an integrated system for producing ethanol from biomass. In: Klass, D.L. (Ed.), Energy from Biomass and Waste. Institute of Gas Technology, Chicago, pp. 1225-1243

Ford, David N. (1994). Neil Schlager. ed. When Technology Fails: significant technological disasters, accidents, and failures of the twentieth century. Gale Research. pp. 267-270.

Frenz, J. et al. (1989). "Hydrocarbon Recovery and Biocompatibility of Solvents for Extraction from Cultures of Botryococcus braunii". *Biotechnology and Bioengineering* 34: 755-762.

Gardner, T. (2007-12-24). "US Ethanol Producers Covet Existing Oil Pipelines". Reuters. http://www.planetark.com/ dailynewsstory.cfm/newsid/46173/story.htm.

Gerpen, Van, John (2004 - 07). "Business Management for Biodiesel Producers, August 2002 - January 2004" (PDF). National

Renewable Energy Laboratory. http://www.nrel.gov/docs/
fy04osti/36242.pdf.

Gil, G.C., Chang, I.S., Kim, B.H., Kim, M., Jang, J.K., Park, H.S.,
Kim, H.J., 2003. Operational parameters affecting the
performance of a mediator-less microbial fuel. *Biosen.
Bioelectron.* 18, 327-334.

Goettemoeller, J., A. Goettemoeller (2007). Sustainable Ethanol:
Biofuels, Biorefineries, Cellulosic Biomass, Flex-Fuel Vehicles,
and Sustainable Farming for Energy Independence. Praire Oak
Publishing, Maryville, Missouri. p. 42.

Goettemoeller, Jeffrey; Adrian Goettemoeller (2007), Sustainable
Ethanol: Biofuels, Biorefineries, Cellulosic Biomass, Flex-Fuel
Vehicles, and Sustainable Farming for Energy Independence,
Prairie Oak Publishing, Maryville, Missouri, pp. 56-61.

Goldemberg, José (2008-05-01). "The Brazilian biofuels industry".
Biotechnology for Biofuels 1 (6): 4096.

Gordon Quaiattini. Biofuels are part of the solution *Canada.com*,
April 25, 2008.

Green Dreams J.K. Bourne JR, R. Clark National Geographic
Magazine October 2007 p. 41.

Growing Sustainable Biofuels: Common Sense on Biofuels, part 2
World Changing, March 12, 2008.

Guiry, M.D. and Blunden, G. (1991). "Seaweed Resources in Europe:
Uses and Potential.". *John Wiley and Sons Ltd.*

H.M. Treasury (2006). Stern Review on the Economics of Climate
Change p. 355.

Habermann, W. and E.-H. Pommer (1991). "Biological fuel cells
with sulphide storage capacity". *Appl. Microbiol. Biotechnol.*
35: 128-133.

Harris EE, Beglinger E, Hajny GJ, and Sherrard EC. "Hydrolysis of
Wood: Treatment with Sulfuric Acid in a stationary digester",
Industrial and Engineering Chemistry, 37(1): 12-23 (1945)

Hawkins, C.S.; Fuls, J.; and F.J.C. Hugo. "Engine Durability Tests
with Sunflower Oil in an Indirect Injection Diesel Engine." SAE
Paper 831357.

Heinberg, Richard (2003). *The Party's Over: Oil, War, and the Fate
of Industrial Societies*. Gabriola, BC: New Society Publishers.

Henderson, Bruce (June 11, 2007). "Driver ticketed for using biofuel:
Vegetable oil sticks him with $1,000 fine". The Charlotte

Observer. http://www.newsobserver.com/news/story/599471.html

Henry Fountain (2008-12-15). "Diesel made Simply From Coffee Grounds". New York Times. http://www.nytimes.com/2008/12/16/science/16objava.html.

Homer-Dixon, Thomas (2007) "The Upside of Down; Catastrophe, Creativity and the Renewal of Civilisation" (Island Press)

Huang, C. (2006). "Expression of mercuric reductase from Bacillus megaterium MB1 in eukaryotic microalga Chlorella sp. DT: an approach for mercury phytoremediation.". *Appl Microbiol Biotechnol.* 72 (1): 197-205.

Humanik, F. et al. (2007) Anaerobic digestion of animal manure, www.epa.gov,

Inderwildi, O.R. and D.A. King (2009). "Quo Vadis Biofuels". *Energy and Environmental Science* 2: 343.

Ingram, D. 2002. Ketonization of acetic acid. B.S. student report.

Inslee, J., H. Bracken (2007). "6. Homegrown Energy". *Apollo's Fire.* Island Press, Washington, D.C. pp. 153-155, 160-161.

Inslee, Jay, Bracken Hendricks (2007), *Apollo's Fire*, Island Press, Washington, D.C., pp. 153-155, 160-161.

Jalonick, M.C. (2008-05-22). "Congress enacts $290B farm bill over Bush veto". The Seattle Times. http://seattletimes.nwsource.com/html/politics/2004431073_apcongressfarmbill.html.

Jeffries, T.W., Jin YS (2004) Metabolic engineering for improved fermentation of pentoses by yeasts. *Appl Microbiol Biotechnol* 63: 495-509.

Jerger, D. and Tsao, G. (2006) Feed composition in Anaerobic digestion of biomass, p. 65. Richards, B. (1991). "High rate low solids methane fermentation of sorghum, corn and cellulose". *Biomass and Bioenergy* 1: 249-260.

Jewell, W. (1993). "Methane fermentation of energy crops: Maximum conversion kinetics and in situ biogas purification". *Biomass and Bioenergy* 5: 261-278.

Karhumaa, K., et al. (2006) Co-utilization of L-arabinose and D-xylose by laboratory and industrial Saccharomyces cerevisiae strains. Microb Cell Fact. 10; 5:18.

Karube, I., T. Matasunga, S. Suzuki and S. Tsuru (1976). "Continuous hydrogen production by immobilized whole cells of *Clostridium butyricum*". *Biocheimica et Biophysica Acta* 24 (2): 338-343.

Karube, I., T. Matasunga, S. Suzuki and S. Tsuru (1977). "Biochemical cells utilizing immobilized cells of *Clostridium butyricum*". *Biotechnology and Bioengineering* 19: 1727-1733.

Katzen, R. and Schell, D.J., "Lignocellulosic feedstock Biorefinery: History and Plant Development for Biomass Hydrolysis", pp 129-138 in Biorefineries - Industrial processes and Products, Volume 1, Kamm, B., Gruber, P.R., and Kamm, M., eds. Wiley-VCH, Weinheim, 2006.

Kelly, K.J., Bailey, B.K., Coburn, T.C., Clark, W., Lissiuk, P. "Federal Test Procedure Emissions Test Results from Ethanol Variable-Fuel Vehicle Chevrolet Luminas", SAE Technical Paper 961092.

Kemp, William. Biodiesel: Basics and Beyond. Canada: Aztext Press, 2006.

Kim, B.H., et al. (1999a) Direct electrode reaction of Fe (III) reducing bacterium, *Shewanella putrefacience. J Microbiol. Biotechnol.* 9:127-131.

Kim, B.H., et al. (2003) "Novel BOD (biological oxygen demand) sensor using mediator-less microbial fuel cell". *Biotechnology Letters* 25: 541-545.

Kim, H. (2002). "A mediator-less microbial fuel cell using a metal reducing bacterium, Shewanella putrefaciens". *Enzyme and Microbial Technology* 30: 145.

Kim, H.J., et al. (1999b) A microbial fuel cell type lactate biosensor using a metal-reducing bacterium, *Shewanella putrefaciens. J Microbiol. Biotechnol.* 9:365-367.

Klinke, H.B., Thomsen, A.B., Ahring, B.K. (2004) Inhibition of ethanol-producing yeast and bacteria by degradation products produced during pre-treatment of biomass. Appl Microbiol Biotechnol 66:10-26.

Knothe, G. "Historical Perspectives on Vegetable Oil-Based Diesel Fuels" (PDF). INFORM, Vol. 12(11), p. 1103-1107 (2001). http://www.biodiesel.org/resources/reportsdatabase/reports/gen/20011101_gen-346.pdf.

Kononova, M.M. Soil Organic Matter, Its Nature, Its role in Soil Formation and in Soil Fertility, 1961

L.W. Hillen et al. (1982). "Hydrocracking of the Oils of Botryococcus braunii to Transport Fuels". *Biotechnology and Bioengineering* 24: 193-205.

Lemmer, A. and Oeschsner, H. Co-fermentation of grass and forage maize, Energy, Landtechnik, 5/11, p. 56.

Lemos, William (2007-11-12). "Brazil's flex-fuel car production rises, boosting ethanol consumption to record highs". ICIS chemical business. http://www.icis.com/Articles/2007/11/12/9077311/brazils-flex-fuel-car-production-rises-boosting-ethanol-consumption-to-record-highs.html.

Leonard, Christopher (2007-01-03). "Not a Tiger, but Maybe a Chicken in Your Tank". *Washington Post* (Associated Press): p. D03. http://www.washingtonpost.com/wp-dyn/content/article/2007/01/02/AR2007010201057.html.

Lithgow, A.M., Romero, L., Sanchez, I.C., Souto, F.A., and Vega, C.A. 1986. Interception of electron-transport chain in bacteria with hydrophilic redox mediators. *J. Chem. Research*, (S):178-179.

Liu, H, Cheng S and Logan BE (2005). "Production of electricity from acetate or butyrate using a single-chamber microbial fuel cell". *Environ Sci Technol* 32 (2): 658-62.

Lynd, L.R. (1996). Overview and evaluation of fuel ethanol from cellulosic biomass: technology, economics, the environment, and policy. Annu Rev Energy Environ 21:403-465.

Makower, J. Pernick, R. Wilder, C. (2008) *Clean Energy Trends 2008* (Clean Edge, USA).

Márcia Azanha Ferraz Dias de Moraes (2007). "O mercado de trabalho da agroindústria canavieira: desafios e oportunidades" (in Portuguese). *Economia Aplicada* 11 (4).

Marshall, B. (October 2007). "Gas from the grass". *Field and Stream*: 40-42.

Martin, A.D. (2007) Understanding Anaerobic Digestion, Presentation to the Environmental Services Association, 16.10.07, www.esauk.org

McKenna, Phil (7 October 2006). "From smokestack to gas tank". *New Scientist* (Reed Business Information) 192 (2572): 28-29.

Melis, A. and Happe, T. (2001). "Hydrogen Production: Green Algae as a Source of Energy". *Plant Physiol.* 127: 740-748.

Metzger, P. C. Largeau (2005). "Botryococcus braunii: a rich source for hydrocarbons and related ether lipids". *Applied Microbiology and Biotechnology* 6 (25): 486-496.

Michael, Briggs (August 2004). "Widescale Biodiesel Production from Algae". UNH Biodiesel Group (University of New Hampshire). http://www.unh.edu/p2/biodiesel/article_alge.html.

Min, B., Cheng, S. and Logan B.E. (2005). Electricity generation using membrane and salt bridge microbial fuel cells, Water Research, 39 (9), pp. 1675-86.

Mohan, Venkata, S., Mohanakrishna, G., Srikanth, S., Sarma, P.N., 2008a. Harnessing of bioelectricity in microbial fuel cell (MFC) employing aerated cathode through anaerobic treatment of chemical wastewater using selectively enriched hydrogen producing mixed consortia. "Fuel". 87, 2667-2676.

——, 2008b. Bioelectricity generation from chemical wastewater treatment in mediatorless (anode) microbial fuel cell (MFC) using selectively enriched hydrogen producing mixed culture under acidophilic microenvironment. "Biochem. Engng. J." 39, 121-130.

——, 2008c. Bioelectricity production from wastewater treatment in dual chambered microbial fuel cell (MFC) using selectively enriched mixed microflora: Effect of catholyte. "Biores. Technol." 99(3), 596-603.

——, 2008d. Biochemical evaluation of bioelectricity production process from anaerobic wastewater treatment in a single chambered microbial fuel cell (MFC) employing glass wool membrane. "Biosen. Bioelectron." 23, 1326-1332.

——, 2008e. Influence of anodic biofilm growth on bioelectricity production in single chambered mediatorless microbial fuel cell using mixed anaerobic consortia. "Biosen. Bioelectron." 24(1), 41-47.

——, 2007. Bioelectricity production by meditorless microbial fuel cell (MFC) under acidophilic condition using wastewater as substrate: influence of substrate loading rate. "Current Sci." 92(12), 1720-1726.

Moore, Tony; Lakha, Raj (2006-11-20). *Tolley's Handbook of Disaster and Emergency Management, Third Edition: Principles and Practice* (3rd ed. (Hardcover) ed.). Butterworth-Heinemann. p. 71.

Mosier, N., Wyman C, Dale BE, Elander R, Lee YY, Holtzapple M, Ladisch M (2005) Features of promising technologies for pretreatment of lignocellulosic biomass. *Bioresour Technol* 96:673-686.

Mumford, T.F. and Miura, A (1988). "Porphyra as food: cultivation and economics". *In Lembi, C.A. And Waaland, J.R. (Ed.) Algae and Human Affairs*: 87-117.

National Ethanol Vehicle Coalition (2009-03-09). "New E85 Stations". NEVC FYI Newsletter (Vol. 15 Issue 5).

National Renewable Energy Laboratory (2006). *Nontechnical Barriers to Solar Energy Use: Review of Recent Literature*, Technical Report, NREL/TP-520-40116, September, 30 pages.

Ohgren, K., Bengtsson O, Gorwa-Grauslund MF, Galbe M, Hahn-Hagerdal B and Zacchi G (2006) Simultaneous saccharification and co-fermentation of glucose and xylose in steam-pretreated corn stover at high fiber content with Saccharomyces cerevisiae TMB3400. *J Biotechnol.* 126(4):488-98.

Oliver, R. Inderwildi, David A. King (2009). "Quo Vadis Biofuels". *Energy and Environmental Science* 2: 343.

Oliver, R. Inderwildi, Stephen J. Jenkins, David A. King (2008). "Mechanistic Studies of Hydrocarbon Combustion and Synthesis on Noble Metals". *Angewandte Chemie International Edition* 47: 5253.

Olsson, L., Hahn-Hägerdal B (1996) Fermentation of lignocellulosic hydrolysates for ethanol fermentation. Enzyme Microb Technol 18:312-331.

Owen, K., Coley., C.S. Weaver, "Automotive Fuels Reference Book", SAE International.

Palmqvist, E., Hahn-Hägerdal B (2000) Fermentation of lignocellulosic hydrolysates. I. Inhibition and deoxification. Bioresour Technol 74:17-24

Patrick, Barta. As Biofuels Catch On, Next Task Is to Deal With Environmental, Economic Impact *Wall Street Journal*, March 24, 2008.

Patzek, Tad (2006-07-22). "Thermodynamics of the Corn-Ethanol Biofuel Cycle (section 3.11 Solar Energy Input into Corn Production)" (PDF). Berkeley; Critical Reviews in Plant Sciences, 23(6):519-567 (2004). http://petroleum.berkeley.edu/papers/patzek/CRPS416-Patzek-Web.pdf.

Perlack, et al. 2005. Biomass as feedstock for a bioenergy and bioproducts Industry: the technical feasibility of a billion-ton annual supply. Oar Ridge National Laboratory Report ORNL/TM-2005/66, US Dept. of Energy, Oak Ridge, TN.

Pevsner, Nikolaus; John Grundy, Ian Richmond, Grace McCombie, Humphrey Welfare, Peter Ryder, Stafford Linsley (1992). *Northumberland* (2 ed.). Yale University Press. p. 103.

Policy Model. The authors consider that ethanol production in Brazil is a unique situation and it is not replicable, they think there is no other country where it makes sense to convert sugar or starch crops to ethanol, particularly the US.". *Two billion cars: driving toward sustainability*. Oxford University Press, New York. pp. 95-96.

Potter, M.C. (1911). Electrical effects accompanying the decomposition of organic compounds. Royal Society (Formerly Proceedings of the Royal Society) B, 84, p. 260-276.

Rabaey, K. and W. Verstraete (2005). "Microbial fuel cells: novel biotechnology for energy generations". *Trends Biotechnol* 23: 291-298.

Ratliff, E. (2007). "One molecule could cure our addiction to oil". *Wired Magazine* 15 (10).

Richards, B. (1991). "Methods for kinetic analysis of methane fermentation in high solids biomass digesters". *Biomass and Bioenergy* 1: 65-26.

Richards, B. (1994). "In situ methane enrichment in methanogenic energy crop digesters". *Biomass and Bioenergy* 6: 275-274.

Rocky Mountain Institute (2005). *Winning the Oil Endgame* p. 107.

Rome, Alfred Ward. *The Bulldozer in the Countryside*. (Cambridge, Cambridge University Press: 2001). pp. 45-55.

Ron Oxburgh. "Fuelling hope for the future." *Courier Mail*, August 15, 2007.

Ron Oxburgh. Through biofuels we can reap the fruits of our labours *The Guardian*, February 28, 2008.

Rouke Bosma, Prof. dr.ir J.Tramper, Dr. ir. R.H. Wijffels (PDF) (2003). "ULTRASOUND A new technique to harvest microalgae?". *Journal of Applied Phycology 15(2-3) 143-153*. Wageningen UR. http://www.springerlink.com/content/k03v3582334pt5j8/.

Ryan, Lisa; Turton, Hal (2007), *Sustainable Automobile Transport*, Edward Elgar Publishing Ltd, England, pp. 40-41.

Saeman, J.F. "Kinetics of wood saccharification: Hydrolysis of cellulose and decomposition of sugars in dilute acid at high temperature", *Industrial and Engineering Chemistry*, 37(1): 43-52 (1945).

Searchinger, T. et al. (2008). "Use of U.S. Croplands for Biofuels Increases Greenhouse Gases Through Emissions from Land-Use Change". *Science* 319: 1238.

Searchinger, T. et al. (2008). "Use of U.S. Croplands for Biofuels Increases Greenhouse Gases Through Emissions from Land-Use Change". *Science* 319: 1238.

Searchinger, Timothy et al. (2008). "Use of U.S. Croplands for Biofuels Increases Greenhouse Gases Through Emissions from Land-Use Change". *Science* 319: 1238.

Sergeeva, Y.E., Galanina LA, Andrianova DA, Feofilova EP. Lipids of filamentous fungi as a material for producing biodiesel fuel. *Applied Biochemistry and Microbiology* 2008: 44, 523-527

Sleat, R. and Mah, R. (2006) Hydrolytic Bacteria *in* Anaerobic digestion of biomass, p. 15.

Song, Y.C., Kwon, S.J., Woo, J.H. (2004) Mesophilic and thermophilic temperature co-phase anaerobic digestion compared with single-stage mesophilic- and thermophilic digestion of sewage sludge, Water Res. 2004 Apr.; 38(7):1653-62

Spath, P.L. and D.C. Dayton. "Preliminary Screening—Technical and Economic Assessment of Synthesis Gas to Fuels and Chemicals with Emphasis on the Potential for Biomass-Derived Syngas", NREL/TP510-34929, December, 2003, pp. 95.

Stephen, D. Kinrade, Jeffrey C.H. Donovan, Andrew S. Schach and Christopher T.G. Knight (2002), *Two substituted cubic octameric silicate cages in aqueous solution.* J. Chem. Soc., Dalton Trans., 1250-1252.

Stephen, Leahy. Can Sorghum Solve the Biofuels Dilemma? *IPS News*, May 13, 2008.

Strahan, David (13 August 2008). "Green Fuel for the Airline Industry". *New Scientist* (2669): 34-37. http://technology.newscientist.com/channel/tech/mg19926691.700-green-fuel-for-the-airline-industry.html.

Tachinardi, M.H. (2008-06-13). "Por que a cana é melhor que o milho" (in Portuguese). *Época* Magazine. http://revistaepoca.globo.com/Revista/Epoca/0,EMI5865-15273.html. Retrieved 2008-08-06. Print edition pp. 73.

Tainter, Joseph (1990) "The Collapse of Complex Societies" (Cambridge University Press)

The Royal Society (2008). p. 2 and 11.

Thomas, L. Friedman (2008). *Hot, Flat, and Crowded.* Farrar, Strauss and Giroux, New York, p. 190.

Tower, P., Wetzel, J., Lombard, X. (2006-03). "New Landfill Gas

Treatment Technology Dramatically Lowers Energy Production Costs". Applied Filter Technology. http://www. appliedfil tertechnology.com/Userfiles/Docs/AFT_SWANA_2006_ Paper_Rev1.pdf.

Turon, M. (1998-11-25). Ethanol as Fuel: An Evironmental and Economic Analysis. U.C. Berkeley, Chemical Engineering. http://www.turon.com/papers/ethanol.htm.

Vallios, I., and Tsoutsos, T, and Papadakis, G (2009). "Design of Biomass District Heating", *Biomass and Bioenergy*, 33(4) 659-678.

Volk, T.A., L.P. Abrahamson, E.H. White, E. Neuhauser, E. Gray, C. Demeter, C. Lindsey, J. Jarnefeld, D.J. Aneshansley, R. Pellerin and S. Edick (October 15-19, 2000). "Developing a Willow Biomass Crop Enterprise for Bioenergy and Bioproducts in the United States". *Proceedings of Bioenergy 2000*. Adam's Mark Hotel, Buffalo, New York, USA: North East Regional Biomass Program.

Wang, G.S., Pan XJ, Zhu JY, Gleisner R (2009). "Sulfite pretreatment to overcopme recalcitrabce of lignocellulose (SPORL) for robust Enzymatic Saccharification of hardwoods". *Biotechnology Progress* 25: 1086-1093.

Youngquist, W. Geodestinies, National Book company, Portland, OR, 499p.

Yue, P.L. and Lowther K. (1986). Enzymatic Oxidation of C1 compounds in a Biochemical Fuel Cell. The Chemical Engineering Journal, 33B, p 69-77

Zamorano, Marti (2006-12-22). "B-52 synthetic fuel testing: Center commander pilots first Air Force B-52 flight using solely synthetic fuel blend in all eight engines". *Aerotech News and Review*.

Zhu, J.Y., Pan XJ, Wang GS, Gleisner R (2009). "Sulfite pretreatment (SPORL) for Robust enzymatic saccharification of spruce and red pine". *Bioresource Technology* 100: 2411-2418.

Index